COASTAL UPWELLING

Its Sediment Record

Part A: Responses of the Sedimentary Regime to Present Coastal Upwelling

NATO CONFERENCE SERIES

I Ecology
II Systems Science
III Human Factors
IV Marine Sciences
V Air–Sea Interactions
VI Materials Science

IV MARINE SCIENCES

Volume 1 Marine Natural Products Chemistry
edited by D. J. Faulkner and W. H. Fenical

Volume 2 Marine Organisms: Genetics, Ecology, and Evolution
edited by Bruno Battaglia and John A. Beardmore

Volume 3 Spatial Pattern in Plankton Communities
edited by John H. Steele

Volume 4 Fjord Oceanography
edited by Howard J. Freeland, David M. Farmer, and Colin D. Levings

Volume 5 Bottom-interacting Ocean Acoustics
edited by William A. Kuperman and Finn B. Jensen

Volume 6 Marine Slides and Other Mass Movements
edited by Svend Saxov and J. K. Nieuwenhuis

Volume 7 The Role of Solar Ultraviolet Radiation in Marine Ecosystems
edited by John Calkins

Volume 8 Structure and Development of the Greenland–Scotland Ridge
edited by Martin H. P. Bott, Svend Saxov, Manik Talwani, and Jörn Thiede

Volume 9 Trace Metals in Sea Water
edited by C. S. Wong, Edward Boyle, Kenneth W. Bruland, J. D. Burton, and Edward D. Goldberg

Volume 10A Coastal Upwelling: Its Sediment Record
Responses of the Sedimentary Regime to Present Coastal Upwelling
edited by Erwin Suess and Jörn Thiede

Volume 10B Coastal Upwelling: Its Sediment Record
Sedimentary Records of Ancient Coastal Upwelling
edited by Jörn Thiede and Erwin Suess

COASTAL UPWELLING
Its Sedimentary Record

Part A: Responses of the Sedimentary Regime to Present Coastal Upwelling

Edited by
Erwin Suess
School of Oceanography
Oregon State University
Corvallis, Oregon

and
Jörn Thiede
Department of Geology
University of Kiel
Kiel, Federal Republic of Germany

Published in cooperation with NATO Scientific Affairs Division
PLENUM PRESS · NEW YORK AND LONDON

Library of Congress Cataloging in Publication Data

NATO Advanced Research Institute on Coastal Upwelling and Its Sediment Record (1981: Vila Moura, Portugal)

Responses of the sedimentary regime to present coastal upwelling.

(Coastal upwelling, its sediment record; pt. A) (NATO conference series. IV, Marine sciences; 10A)

"Published in cooperation with NATO Scientific Affairs Division."

Includes bibliographical references and index.

1. Marine sediments—Congresses. 2. Upwelling (Oceanography)—Congresses. I. Suess, Erwin. II. Thiede, Jörn, 1941– . III. North Atlantic Treaty Organization. Scientific Affairs Division. IV. Title. V. Series: NATO Advanced Research Institute on Coastal Upwelling and Its Sediment Record (1981: Vila Moura, Portugal). Coastal upwelling, its sediment record; pt. 1. VI. Series: NATO conference series. IV, Marine sciences; 10A.

QE47 C377] 551.3′6s [551.3′6] 83-8012

ISBN 978-1-4615-6653-3 ISBN 978-1-4615-6651-9 (eBook)

DOI 10.1007/978-1-4615-6651-9

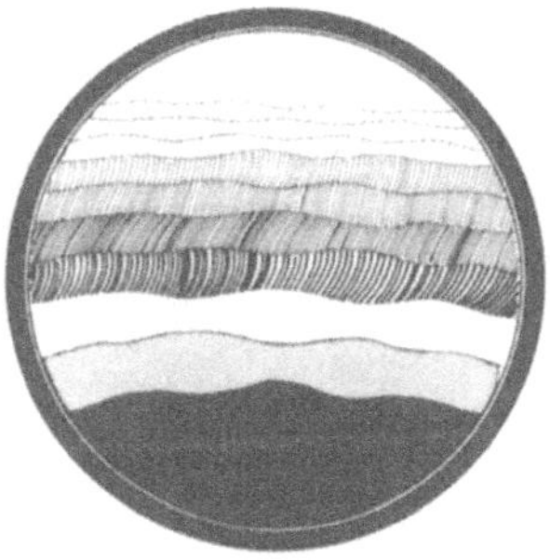

First half of the proceedings of a NATO Advanced Research Institute on Coastal Upwelling and Its Sediment Record, held September 1–4, 1981, at Vilamoura, Portugal

Softcover reprint of the hardcover 1st edition 1983

A Division of Plenum Publishing Corporation
233 Spring Street, New York, N.Y. 10013

PREFACE

NATO Advanced Research Institutes are designed to explore unresolved problems. By focusing complementary expertise from various disciplines onto one unifying theme, they approach old problems in new ways. In line with this goal of the NATO Science Committee, and with substantial support from the U.S. Office of Naval Research and the Seabed Assessment Program of the U.S. National Science Foundation, such a Research Institute on the theme of Coastal Upwelling and Its Sediment Record was held September 1-4, 1981, in Vilamoura, Portugal.

The theme implies a modification of uniformitarian thinking in earth science. Expectations were directed not so much towards finding the key to the past as towards exploring the limits of interpreting the past based on present upwelling oceanography. Coastal upwelling and its imprint on sediments are particularly well-suited for such a scientific inquiry. The oceanic processes and conditions characteristic of upwelling are well understood and are a well-packaged representation of ocean science that are familiar to geologists, just as the magnitude of bioproduction and sedimentation in upwelling regimes --among other biological and geological processes-- have made oceanographers realize that the bottom has a feedback role for their models.

The organization of these two volumes of proceedings reflects much of the initial intentions of this conference. This is the first time that the sedimentary response to coastal upwelling has been examined exclusively and in a joint effort by oceanographers and geologists. Experts in upwelling oceanography have kept sedimentologic, geochemical and paleoceanographic implications in mind and geologists relate findings and interpretations to their colleagues in terms of known oceanographic processes. The first volume examines those physical, chemical and biological phenomena which are unmistakably linked to the oceanography of upwelling and which have the potential of leaving an imprint in the sediment record. The goal was to identify and evaluate processes whose very existence, magnitude and products can be interpreted from the sediment record so that ultimately the evolution of upwelling regimes might be traced through time. The success of such attempts depends on finding appropriate answers to the following questions:

What controls the distribution and duration of upwelling centers?

What characteristic terrigenous, biogenous and authigenic constituents directly reflect temperature, salinity, nutrient make-up and current distributions of upwelling systems?

What is the relationship between bioproduction, vertical transfer, recycling and burial of biogenic components, particularly of organic matter?

What is the role of the well-developed oxygen minimum layer that impinges onto the sea floor in upwelling regimes and which imprints and modifies organic geochemical and inorganic geochemical signals?

The second volume explores the ancient sediment record for indications of upwelling as we know it today. Here the focus is on questions such as:

Are there sufficient independent micropaleontological, geochemical and sedimentological criteria for most of the large upwelling zones in today's ocean which could be identified and recognized in the fossil record?

How well do different time- and space-scales permit detailed interpretations of ancient upwelling records?

Is there solid evidence from the sedimentary regimes of fossil upwelling zones for which there is no present analogous oceanographic situation?

Each of the conference topics is introduced by a review or conceptual article designed to summarize up-to-date information within oceanography and geology. These are followed by articles which treat in detail circulation patterns of coastal upwelling; particulate organic matter production, transfer and preservation; patterns of dissolved nutrients; geochemistries of organic and inorganic geochemical constituents; and regional patterns of upwelling facies. The histories of Holocene upwelling regimes are examined as well as the upwelling records from the Pleistocene and Tertiary, the Mesozoic and Paleozoic. As with most conference proceedings, the theme is covered heterogenously. Some topics enjoy multiple contributions, others are less well covered, a few are not represented at all. The same is true for documentation; some contributions contain new and original data published here for the first time, others contain selected material previously published in different contexts. Still others summarize earlier and ongoing research under the conference's theme. Such variety, paired with the personal flavor of individual contributors, represent the true state-of-the-science in this field and the spirit of Vilamoura. This format should allow the reader rapid exposure to the theme of the conference as it has contributed to intensive interdisciplinary exchange among the conference participants. All participants and contributors gave generously and enthusiastically of their wisdom and knowledge and we extend to them, also in behalf of the conference sponsors, our gratitude.

Special thanks are due to our colleagues from the organizing committee, R.T. Barber (Beaufort), S.E. Calvert (Vancouver), J.H. Monteiro (Lisbon), E. Seibold (Bonn) and R.L. Smith (Corvallis), and the conference session chairmen without whose dedication the initial plan of the conference could not have been brought to a successful completion. Many thanks are also due numerous colleagues who diligently reviewed the individual contributions in preparation for this publication.

We gratefully acknowledge the superb clerical assistance of J.L. Dickson and M.J. Armbrust and the editorial advice and help of Z. Suess (all of Corvallis); their continued enthusiasm was invaluable in the task of editing the proceedings. We want to thank L. Schmidt of Plenum Publishing Corporation (New York) for continued and efficient work on the technical editing.

Finally, we acknowledge the financial contributions towards printing costs of these volumes by Exxon Research Corporation (Houston), British Petroleum Corporation (London), Deutsche Texaco A.G. (Wietze) and the OSU Foundation, Research Office and School of Oceanography (Oregon State University, Corvallis).

Erwin Suess Jörn Thiede*

Corvallis and Oslo
January, 1983

* Present address:
Geologisch-Paläontologisches Institut
der Universität
Kiel, Olshausenstrasse
Federal Republic of Germany

CONTENTS OF PART A

Geochemistry of Coastal Upwelling Systems

CONTENTS OF PART B

High Resolution Holocene Time Scales

Pleistocene Time Scales

Pre-Pleistocene Phanerozoic Time Scales

RESPONSES OF THE SEDIMENTARY REGIME
TO COASTAL UPWELLING

INTRODUCTION

Erwin Suess

School of Oceanography
Oregon State University
Corvallis, Oregon 97331, U.S.A.

Jörn Thiede

Department of Geology
University of Oslo
Blindern, Oslo 3, Norway

INTRODUCTION

Coastal upwelling is the process in which surface ocean waters are driven offshore by atmospheric and ocean forcing mechanisms and are replaced by nutrient-rich subsurface waters. Its consequences have profound biological effects with strong economic implications for regional fisheries (Smith, 1968; Ryther, 1969; Parsons, 1979; Hartline, 1981; Walsh, 1981). Coastal upwelling produces characteristic perturbations of the boundary current regimes along continental margins, whose long-term impact and changes also impart certain features to the underlying sediments.

The physical and biological aspects of coastal upwelling have been examined in several multi-institutional research projects discussed at recent international symposia and have since reached a new and more detailed level of understanding (Boje and Tomczak, 1978; Richards, 1981; Longhurst, 1981; Hempel, 1982). However, geologic aspects, in particular the responses of the sedimentary regime to coastal upwelling and the preservation of upwelling signals in the sedimentary record, have not previously been the subject of a comprehensive review and assessment. Therefore, an Advanced Research Institute was held in September 1981 in Vilamoura, Portugal to address such aspects of coastal upwelling. The objective was to gain insight into the processes which produce upwelling signals and which control their transfer to the geologic record. The ultimate goal would be to trace through time, by sedimentary signals, the evolution of such unique regimes in terms of known oceanographic processes. The processes which generate and transfer upwelling signals, operate on variable time and space scales and therefore have variable probabilities of producing a pronounced and lasting sedimentary imprint. It was the intent of the conference to identify those upwelling processes whose existence, magnitude, products, and regional extents may be read from the sediment record.

CIRCULATION

From recent work on circulation modelling of coastal upwelling systems, those attending the Advanced Research Institute learned that

Fig. 1. Eastern North Pacific, 6 September 1977, infrared image enhanced for water temperature; Columbia River mouth, Cape Blanco and Cape Mendocino are marked by arrows. Note upwelling centers off Oregon and northern California and entrainment of cold water plumes drifting westward; U.S. Navy Satellite San Diego; E.D. Traganza et al., A.F.G. Fiuza; K. Fisher et al.; all this volume.

such systems are extraordinarily responsive to atmospheric forcing both on a local and a more distant space scale. Thus, sediments may record detailed climatic signals of wind stress and prevailing wind regimes in their eolian dust components. Further, localized plumes of upwelled waters along a coast are knit together by a coherent poleward undercurrent system which interacts with the coastal and upper slope/shelf morphologies, and provides the source water during upwelling events. The resulting predictable upwelling pattern would encourage mapping of surface sediments in areas that are frequently traversed by fronts and cool water tongues. The routes of recurring fronts and cool water tongues were impressively shown in satellite images from the Oregon and northern California margin. The circulation here is characterized by entrainment of cold water masses into the California Current by synoptic scale eddies that could be traced as far as 300 km offshore (Fig. 1). Is it possible then that by this process the areal extent of upwelling signals in sediments off northern California is considerably larger than off Peru or other areas, where fronts and cool water tongues do not appear to be entrained in the offshore Peru Current system but instead are restricted to shorter distances in the offshore direction? Again, detailed mapping of sedimentary facies would seem worthwhile, although the question of areal resolution appeared a difficult one to answer. The poleward undercurrent was identified as the most important feature in the circulation system along eastern margins of ocean basins, not only as the source of upwelled waters but also in its role of allowing or preventing deposition of an upwelling sedimentary facies (Fig. 2). Off Peru and Oregon the depth of maximum onshore-offshore and alongshore current velocities is within the free water column and well off the bottom, whereas off northwestern Africa this depth coincides with the sea floor, resulting in strong bottom currents. One regime would therefore be favorable for the accumulation of an upwelling facies in the current shadow, the other would prevent continuous sediment accumulation and result in a reworked facies.

NUTRIENT MAKE-UP AND WATER PROPERTIES

On a global scale ocean-ocean fractionation explains different nutrient compositions of Atlantic and Pacific source waters. Similarly intra-ocean circulation, for example in the eastern Atlantic, explains the supply of North Atlantic Central Water (NACW) and South Atlantic Central Water (SACW) alternatingly feeding the northwest African upwelling system. Besides the different nutrient compositions and the biological consequences, the stable isotope ratios $^{18}O/^{16}O$ and $^{13}C/^{12}C$, as well as temperature and salinity of these source waters are contained and preserved in the benthic and planktonic signal carriers, the biogenic skeletons. Small changes in source-water properties of upwelling systems may be preserved in a much detailed fashion as was shown for planktonic foraminifers. Oxygen isotope ratios of *Globigerina bulloides* tests record the lowered

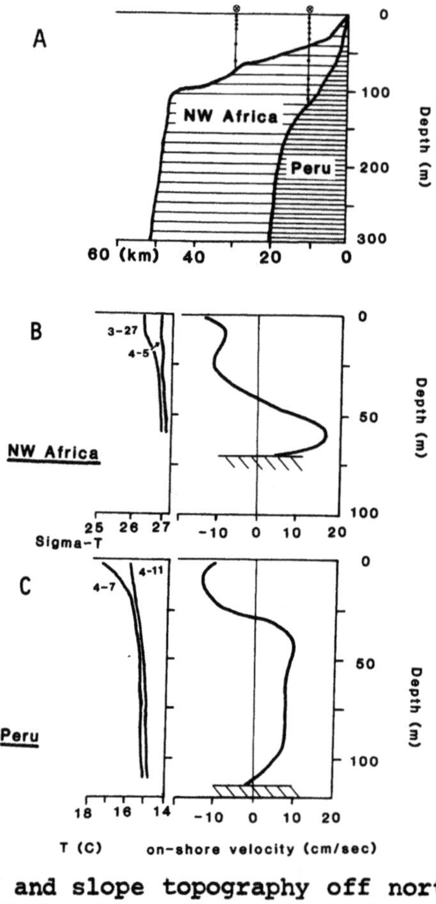

Fig. 2. Shelf and slope topography off northwest Africa (21°N) and Peru (15°S) with location of moored instrument arrays (A). Vertical profiles of sigma-t (density) and temperature (°C) before and during an upwelling event (B=northwest Africa: 27 March - 5 April 1974; C=Peru: 7 - 11 April 1977) and cross-shelf flows at maximum upwelling in both locations; strong onshore bottom currents off northwest Africa prevent sediment accumulation, whereas near-zero flow off Peru enhances sedimentation at the upper slope/ shelf-break. R.L. Smith, this volume.

winter sea surface temperatures off northwestern Africa and southern California when upwelling is in progress and *Globigerinoides ruber* does so for upwelling during the summer months (Fig. 3). Tentatively, the decreasing $\delta^{18}O$ of *Neogloboquadrina dutertrei* with increasing test size might record the offshore drift by adult forms during upwelling in passing from cooler to warmer surface waters. Benthic foraminiferal assemblages off northwestern Africa apparently record in their oxygen isotope signal the different temperature/salinity characteristics of NACW and SACW. First results were presented which indicate that the methodology has now been worked out to use $\delta^{18}O$-signals of diatomaceous SiO_2 for sea surface temperature recordings. Such information combined with floral compositions could provide sensitive upwelling indicators.

Another unique feature of upwelled waters is an internal feedback system whereby increased subsurface respiration, originating from high biological production, contributes to an increasing SiO_2 NO_3-ratio of the newly upwelled waters, as siliceous frustules sink to the bottom. Changes in this nutrient ratio may produce differently silicified diatom frustules or may affect "new" production, and even the entire succession of plankton in upwelling plumes. In addition, such an internal feedback transports more oxygen-deficient water onto the shelf and upper slope region, possibly enhancing sedimentary organic carbon preservation.

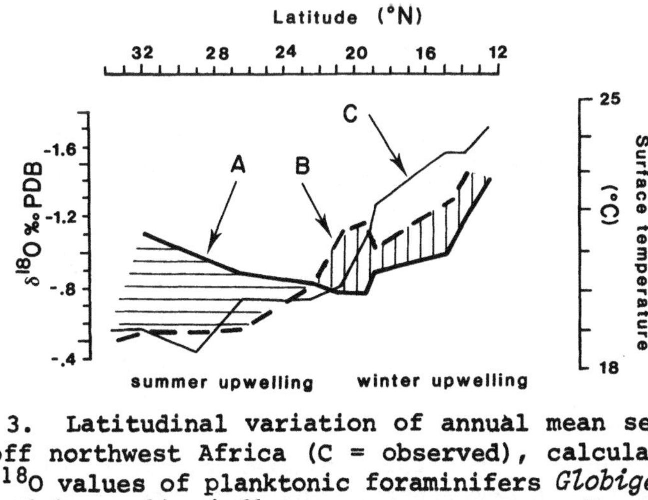

Fig. 3. Latitudinal variation of annual mean sea surface temperature off northwest Africa (C = observed), calculated (= B) from the mean $\delta^{18}O$ values of planktonic foraminifers *Globigerinoides ruber* white and *Globorotalia inflata* and recorded by 'upwelling species' *Globigerina bulloides* (= A). Note positive and negative $\delta^{18}O$-deviations for *G. bulloides* indicating the ranges of winter and summer upwelling, respectively. G. Ganssen and M. Sarnthein, this volume.

UNIQUE BIOLOGICAL FEATURES

Many of the signals of different source waters are recorded and preserved in the properties of organisms, i.e., $\delta^{18}O$ of species assemblages of skeletal plankton, $\delta^{13}C$ of skeletons and of organic matter. In addition, biological seeding of the source-waters determines trends in plankton succession and thus controls types of biogenic skeletons produced, and the degree of variability in organic matter to be expected in certain upwelling centers. A series of idealized zones have been identified which describe the phytoplankton-zooplankton-nutrient-temperature conditions as the upwelled waters flow offshore (Fig. 4). Geologists could hardly hope to recognize in the sedimentary record such lateral zonation related to different types of organic matter sources, but they could hope to find information related to dominantly SiO_2-rich or NO_3-rich source waters and to distal or proximal positions of upwelling plumes, by adaptations of shell- and skeleton-producing plankton communities.

Although these communities have yielded few upwelling-endemic species, their specialized assemblages allow to differentiate them clearly from the plankton over continental margins not affected by upwelling. Species typical of cool waters, --as inhabitants of the boundary currents--, coexisting with those typical of tropical-subtropical waters --as inhabitants of the undercurrent--, seem to be clearly related to coastal upwelling. The offshore drift of upwelled waters transports benthic organisms, either as meroplanktic larvae or as sessile organisms attached to floating materials from the neritic

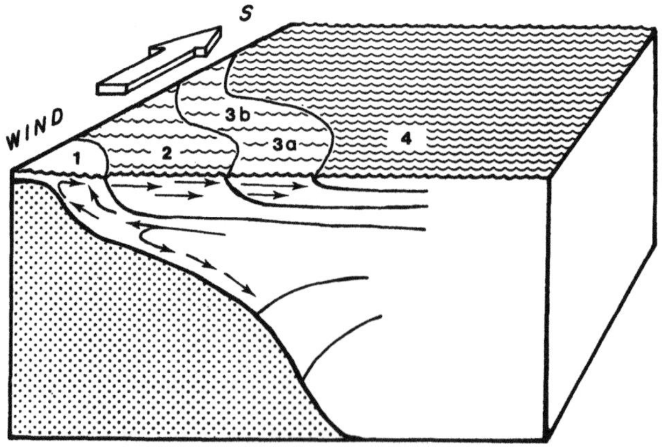

Fig. 4. Schematic for the idealized structure of an upwelling center in the northern hemisphere. The coastline is to the left, the arrows are hypothetical streamlines of the cross-shelf flow. The zones #1, #2, #3a, #3b, and #4 show changing nutrient-phytoplankton-zooplankton-temperature relationships during offshore drift of upwelled waters. B.H. Jones et al., this volume.

zone to the pelagic environment. Benthic foraminiferal faunas respond by changes in morphology, size and faunal composition to the intense midwater oxygen minimum at upwelling-affected continental margins.

Suspended matter fluxes in the marine environment appear to have a particularly significant impact on sedimentation processes in coastal upwelling regimes. First-order comparison between neritic and upwelling environments indicated that 20-60% of the organic matter produced at the sea surface reaches 100 m of water depth in the upwelling environments whereas only 15-35% of production reaches that same depth in neritic waters. Such relative differences in fluxes are of immense significance for organic matter sedimentation, and if they can be related to modes of particle transport, would contain much valuable ecological information. One of these modes which distinguishes the Peruvian upwelling system from other near-shore ecosystems, is by fast-sinking anchoveta fecal pellets. This transfer mechanism provides an important link between biological production in the euphotic zone and the sediments off coastal Peru.

IS THERE AN UPWELLING FACIES?

Holocene sediment studies were presented from the well-known regions of strong coastal upelling; i.e., northwestern Africa, southwestern Africa, off Peru, off Oregon/northern California, and off Somalia. Conference participants asked: Is there a general upwelling facies? Where, with regard to the shelf-slope morphology, is it being deposited? Why are facies patterns so different in these areas?

One clear step towards a better understanding of facies distribution patterns in all upwelling regimes was the realization that deposition occurs preferentially in the "shadow" of the undercurrent (Fig. 2) and that regions of "current shadows" and those of "high exposition" to the undercurrent are located at different sites along shelf-slope profiles in each particular example. This then results in different facies dominating certain locales of upwelling such as the upper slope mud lense off Peru at 11°S, inner shelf mud belt off Namibia, outer shelf sand and middle slope mud facies off northwestern Africa and off Peru at 8°S. A number of unique sedimentological features associated with these facies units bear directly on the original conference objective, i.e., extreme geotechnical properties, anomalously fine-grained upper slope sediments and development of laminae.

ORGANIC MATTER

Differentiating marine organic matter as a result of upwelling fertility from terrestrial organic matter carried by river runoff or as the result of fluvial nutrient input was an issue of major concern to some participants. Organic geochemists, though, assured that a distinction of such sources is, or will be possible in the future, through the use of terrestrial lipid marker compounds. In general, the use of organic geochemical markers (biomarkers) is a highly promising approach. Among the "marine" compounds there are those which can provisionally be interpreted as markers for particular organisms. Thus, dinosterol is presently recorded only for dinoflagellates, although specific carotenoid pigments and wax esters are also excellent markers for other organisms that might proliferate somewhere along the zonal structure of upwelling plumes. The poor chemical stability restricts biomarkers largely to young sediments. On the other hand, acyclic isoprenoid hydrocarbons, which are more resistant to biodegradation and physico-chemical diagenesis, appear to record long-term signals in sedimentary sequences.

Much of the success of using marker compounds depends on organic chemical surveys of organisms in today's upwelling regimes and their relative abundance and position in the food-web. Lists of marker compounds for appropriate plankton blooms or for zones of upwelled waters together with skewing functions for distributions subsequent to transformation in the water column and at the sediment surface, should be available sometime in the future. Kerogen, characterized mainly by its hydrogen and oxygen indices but also by its $\delta^{15}N$ signature, reveals much about the chemical environments prior to burial, such as oxygen minimum, microbial degradation and microbial biomass contributions. Factors controlling organic matter burial in upwelling regimes were considered in detail.

TRACE METALS

A long-standing interest in trace metal distributions of black shales and Holocene reducing sediments could not unambiguously be related to upwelling environments mainly because the ultimate metal sources, terrestrial or biogenic, remain unidentified. The Namibian shelf sediments, by being almost exclusively of biologic origin, still are the best-known modern environment accumulating black-shale type metal associations. A major step towards a better interpretation of these environments will be a better understanding of the sulfide equilibria in seawater. This involves modelling of polysulfides and the discovery of metal sulfide solubility minima as controlled by polysulfide speciation and organo-sulfur compounds.

An interesting phenomenon of "natural-metal-staining" of organic tissues in anoxic sediments was shown which might eventually prove useful in recognizing well-preserved organic tissue structures and their origin. Finally, evidence was given that sediments accumulating beneath upwelling regimes are sinks for particle reactive elements; the examples were based on significantly higher fluxes of

^{230}Th and ^{231}Pa to sediments off northwestern Africa than production of these nuclides in the oceanic water column. Moreover, on account of their anoxic chemistry, also dissolved uranium is preferentially incorporated into such sediments.

PHOSPHORITE GENESIS

High resolution uranium-series dating of phosphorites from the Peru upwelling region finally provides a tool for obtaining accurate growth rates and estimating phosphorus supply rates (Fig. 5). First results showed asymmetric growth, indicating supply of phosphorus from the sediment to the lower nodule surface. The apparently unique association of phosphorites and upwelling was complicated by the fact that in the recent and fossil sediment record there are phosphorite provinces which are not obviously associated with coastal upwelling. Some evidence was presented that the rare earth element signature here might provide a useful signal for differentiating between phosphorites of different origins.

EVIDENCE FROM THE GEOLOGIC RECORD

Limited space- and time-scale resolution in the geologic record is the main problem in relating the major results of recent upwelling

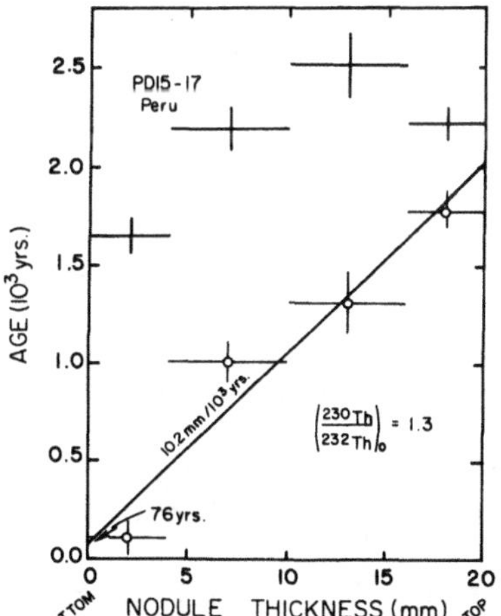

Fig. 5. Common-thorium corrected U-series ages versus thickness of Peru phosphorite nodule; the initital $^{230}Th/^{232}Th$ activity ratio (of 1.3) used for common-Th correction. Note asymmetric growth of nodule with phosphate supplied from the sediment column; the intercept = 76 years, the slope = 10.2 mm·ky^{-1}. W.C. Burnett et al., this volume.

research to ancient depositional environments. Holocene and Pleistocene phenomena can be resolved to within a few hundred or a few thousand meters and years in spatial and temporal dimensions, --as was shown for the monsoon-driven upwelling history of the Arabian Sea over the past 150,000 years--, but more ancient oceanographic conditions are usually not known to within a few tens of kilometers and millions of years. Marine black shales and fossils indicating cool water faunas in subtropical and tropical areas are usually the most readily identifiable indicators of upwelling along continental margins in the geologic past. Examples from the Paleozoic and Mesozoic eras seemed related to upwelling regimes generated in epi-continental seas and along ancient continental margins located within distinct climatic belts. All evidence points to settings analogous to recent coastal upwelling. The wide extent of Upper Mesozoic and Cenozoic, often laminated, diatomites bordering the North Pacific Ocean Basin, represents a sedimentary facies reflecting upwelling and high nutrient concentrations of coastal water masses over an unusually long period of time and huge areal extent. Long-term diagenetic effects obscure some chemical signals of upwelling. As knowledge of ancient geography and ancient circulation improves, areas of major coastal upwelling should become predictable as upwelling and its impact on sedimentation today is a well-constrained, spacially and geographically predictable, phenomenon.

REFERENCES

Boje, R. and Tomczak, M., 1978, "Upwelling Ecosystems," Springer-Verlag, Berlin-New York, 303 pp.

Hartline, B.K., 1981, Coastal upwelling: Physical factors feed fish, Science, 208:38-40.

Hempel, G., 1982, ed., "The Canary Current: Studies in an Upwelling System," Rapport et Process Verbeaux, Conseil International Exploration de Mer, 180:455 pp., Copenhagen.

Longhurst, A.R. (ed.), 1981, "Analysis of Marine Ecosystems," Academic Press, London, 741 pp.

Parsons, T.R., 1979, Some ecological, experimental and evolutionary aspects of the upwelling ecosystems, South African Journal of Science, 75:536-540.

Richards, F.A. (ed.), 1981, "Coastal Upwelling", Coastal and Estuarine Sciences 1, American Geophysical Union, Washington, 529 pp.

Ryther, J.H., 1969, Photosynthesis and fish production in the sea, Science, 166:72-76.

Smith, R.L., 1968, Upwelling, Oceanographic Marine Biology Annual Review, 6:11-46.

Walsh, J.J., 1981, A carbon budget for overfishing off Peru, Nature, 290:300-304.

CIRCULATION PATTERNS

CIRCULATION PATTERNS IN UPWELLING REGIMES

Robert L. Smith

School of Oceanography
Oregon State University
Corvallis, Oregon 97331, U.S.A.

ABSTRACT

Measurements of currents from over the continental margin off Oregon (45°N), northwest Africa (22°N) and Peru (15°S) are used to examine the circulation pattern in coastal upwelling regions. The three regions differ in latitude, in stratification, in the bathymetry of the shelf and slope, and in the strength and variability of the wind that drives the coastal upwelling. The mean wind during the observation periods (each about 50 days) was favorable for upwelling and the mean Ekman transport computed from the wind agrees, within a factor of two, with the mean offshore transport measured in the surface layer. However, the onshore flow in the lower layer does not balance the offshore flow in the surface layer on either the mean or "event" time scale. One concludes that the upwelling process is essentially three dimensional, and upwelling 'centers' may develop.

The mean alongshore currents on the shelf are stronger than the cross-shelf flow. The alongshore flow near the surface had a mean of about 20 cm s^{-1} equatorward, i.e., in the direction of the local wind during the observation periods. The deeper mean flow on the Oregon and Peru shelves was poleward, opposite to the mean wind, as observed in previous studies. This poleward flow leads to an Ekman layer (order of 10 m) with offshore flow adjacent to the bottom; thus off Oregon and Peru, the onshore flow that completes the upwelling circulation is not in a bottom Ekman layer but is observed to be maximum just below the surface Ekman layer. Off northwest Africa the strong equatorward flow throughout the water column leads to a relatively thick (order of 20 m) bottom Ekman layer that provides the compensatory onshore flow for the offshore flow in the surface Ekman layer. During periods of strong alongshore flow, the bottom stress dominates the Coriolis force on the inner (shallower) part of the northwest African shelf and the upwelling circulation pattern separates from the coast. Over the slope, a poleward undercurrent exists in all three regions. It is contiguous with the poleward flow on the shelf off Peru and Oregon, and in all three regions the core of the poleward undercurrent over the slope is near the depth of the shelf break. The poleward undercurrent thus provides a 'feed back' mechanism adjacent to the coastal upwelling zone. Interannual variability in the winds or currents, due to 'global phenomena', may significantly alter the 'mean' circulation patterns or their effects.

INTRODUCTION

As the editors point out in the introduction, physical and biological oceanographers jointly undertook several multi-national research projects during the 1970s. These studies were focused on relatively small subregions of the major coastal upwelling regions extending along the west coasts of the U.S.A., northwest Africa (Senegal, Mauritania, and Morocco), and Peru. Our understanding of the coastal upwelling process on time scales of days to weeks (the typical duration of an experiment or expedition) and on local scales (the order of 100 km by 100 km - the area within which a ship can make repeated daily measurements) is now much improved over a decade earlier. In particular, the wealth of data on the variability of the circulation due to wind variability (weather) absorbed much of the physical oceanographers' attention and the understanding of wind-driven circulation on continental shelves was greatly advanced [see review articles by Allen (1980) and Winant (1980)]. The variability of the coastal upwelling circulation on the weather event time scale is explicitly discussed in Halpern (1976); Huyer (1976); Halpern, Smith and Mittelstaedt (1977); Brink, Halpern and Smith (1980); and Smith (1981). In this paper I will focus on the 'mean' circulation; the 'means' are based on only a few months of observations, but they are validated by observations in other years. I begin with some fundamental concepts and then review results from the recent physical oceanographic studies which I believe may be relevant to interpreting the sediment record.

A SIMPLE CONCEPTUAL MODEL

The major coastal upwelling regions are located along the oceans' eastern boundaries, i.e., the west coasts of the continents, where the winds along the coast are predominantly equatorward as a result of the quasi-stationary mid-ocean atmospheric high pressure systems. Ekman (1905) showed that the wind-induced surface stress (τ) acting on the ocean can be balanced by the Coriolis force resulting from flow orthogonal (*cum sole*) to the applied wind stress. The magnitude of the transport ($MT^{-1}L^{-1}$, e.g., grams per second per cm alongwind) is τ/f where f is the Coriolis parameter. This transport occurs in a layer of thickness δ_E; δ_E is the order of 10s of meters and is a function of the stratification (and, hence, local heat exchange with the atmosphere), the stress in the fluid, and f. These results strictly apply only to water depths much greater than δ_E and away from coastal boundaries, but they lead to our present conceptual model for coastal upwelling. This model is clearly presented in Sverdrup, Johnson and Fleming (1942; pp. 500-503) and I paraphrase their description.

Consider a wind in the Northern Hemisphere which blows parallel to the coast, with the coast on the left, looking in the direction of the wind. At some distance from the coast, the surface water (layer of thickness δ_E) will be transported to the right of the wind (at rate τ/f) but at the coast all motion must be parallel to the coastline. The light surface water transported away from the coast must, owing to the continuity of the system, be replaced near the coast by heavier subsurface water. The upwelling thus leads to changes in the distribution of mass with denser water and lowered sea level along the coast. This creates a relative field of pressure (pressure gradient) across the shelf, "with which must (by geostrophy) be associated a current running parallel to the coast in the direction of the wind. Thus, the wind produces not only a pure wind current (Ekman transport) but also a relative (geostrophic) current that runs in the direction of the wind."

Charney (1955) showed that the response of a stratified fluid to forcing near a boundary, as in the case described above, would occur within a coastal zone with an offshore scale given by the Rossby baroclinic radius of deformation $R \sim HNf^{-1}$, where H is the water depth and $N^2 = (g/\rho)(\partial\rho/\partial z)$, the Brunt-Väisälä frequency (a measure of stratification). Theory (Yoshida, 1955) suggests that coastal upwelling, or the vertical replacement of the water moved offshore as Ekman transport, is confined to an offshore scale R. The other scale parameter, δ_E, the Ekman layer thickness, is usually assumed to have the form $U_*/(Nf)^{1/2}$, where U_* is the friction velocity $U_* = (\tau/\rho)^{1/2}$, for the values of N,f encountered in upwelling regions. This form holds for the surface Ekman layer caused by the wind stress (Pollard, Rhines and Thompson, 1973) and the bottom boundary Ekman layer caused by the bottom stress, i.e., frictional drag (Weatherly and Martin, 1978). Using typical values of $H(10^4 cm)$, $f(10^{-4}s^{-1})$, $\tau(1\ dyne\ cm^{-2})$ and $(1/\rho)(\partial\rho/\partial z) \sim (10^{-7}cm^{-1})$ for a mid-latitude continental shelf, we obtain $R \sim 10$ km and $\delta_E \sim 10$ m as scales for the width of the coastal upwelling zone and the thickness of the Ekman layer.

A simple circulation pattern, consistent with the above considerations, would have a fully developed offshore Ekman transport in a 10 m thick surface layer, 10 km from the coast. Inshore of 10 km active upwelling would be occurring, fed by onshore flow in a bottom Ekman layer which develops because of the alongshore geostrophic

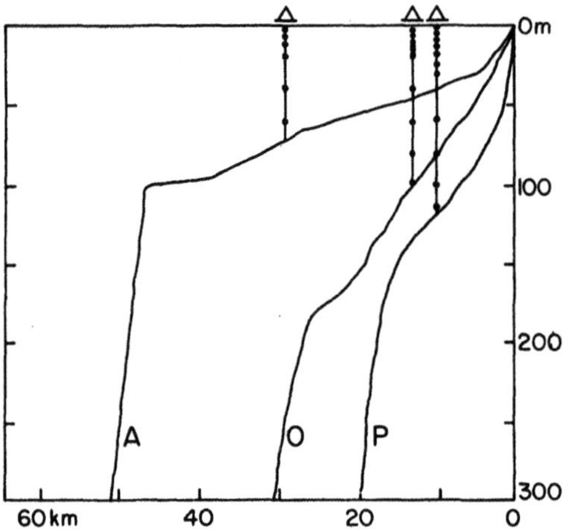

Fig. 1. Shelf and slope topography off northwest Africa (21° 40'N), Oregon (45°N) and Peru (15°S) with location of recording current meters at mid-shelf shown.

current in the direction of the wind. (The transport in a bottom Ekman layer is perpendicular, *contra solem*, to the alongshore current immediately above the layer.) If the wind forcing were steady over a long enough period (several days) the alongshore current would accelerate to the point where the bottom stress would balance the wind stress, and the surface Ekman layer transport out of the coastal region would be balanced by onshore transport within a bottom Ekman layer.

COMPARATIVE OBSERVATIONS

Does nature resemble this conceptual model? During the past decade extensive field experiments were made in the coastal upwelling regions off the west coast of North America (Oregon near 45°N), the northwest coast of Africa (Mauritania near 22°N) and the west coast of South America (Peru near 15°S) which enable us to test the conceptual model and further elucidate the circulation patterns. The topography of continental shelf and upper slope in the three regions is different (Fig. 1) as are the values of the basic parameters f, τ, and N^2 (Table 1). In each region, moored fixed level current meters recorded the horizontal flow throughout the water column at a site near the middle of the shelf (Fig. 1). Other moorings, with lesser vertical resolution in the surface layer, spanned the shelf and upper slope. The mean horizontal circulation at mid-shelf for each region, obtained from the arrays shown in Fig. 1, is shown in vector form in Fig. 2. The coordinate system used for drawing the vectors is such that upward/right in the plane of the figure is alongshore/onshore, i.e., the coastline runs nearly N-S for Oregon and Mauritania but NW-SE for Peru at 15°S. The presentation of the mean circulation as

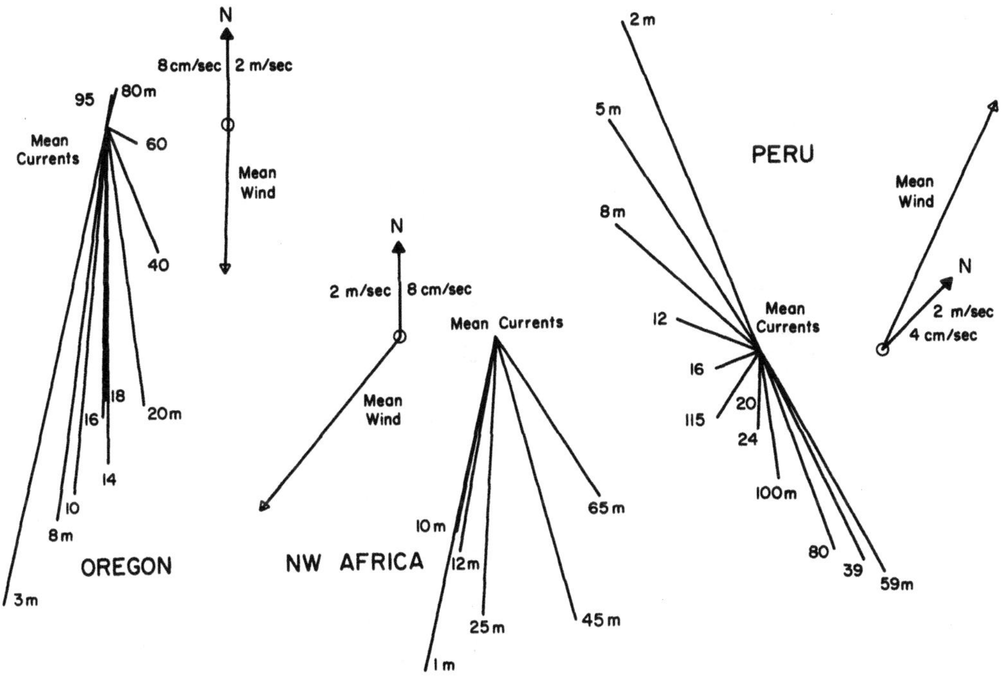

Fig. 2. Mean current and wind vectors from mid-shelf moorings (cf., Fig. 1) off Oregon (6 July to 27 August 1973), northwest Africa (10 March to 17 April 1974), and Peru (7 March to 13 May 1977).

vectors, in contrast to separate profiles of the onshore and alongshore component (Fig. 3), avoids risking the misconception that the upwelling process simply transfers water across the shelf and vertically. Indeed, an upwelling parcel of water moves great distances alongshore in the upwelling process.

The estimates of the baroclinic radius of deformation, R, and the hydrographic data obtained during the experiments indicate the moorings were in, or near the seaward edge of the upwelling zone (Fig. 4). The biological evidence (Huntsman and Barber, 1977; Peterson, Miller and Hutchinson, 1979; Brink et al., 1981) support this conclusion; the arrays were located where the ecosystem response to the upwelling process was clearly manifest. The circulation patterns observed at these sites determine the immediate source and fate (trajectory) of the water participating in the upwelling process and any biota advected with it. A comparison of the mean vector profiles in Fig. 2 is thus relevant.

The salient point to be gleaned from Figs. 2 and 3 are: 1) The mean offshore flow occurs in a surface layer of thickness about 20 m off Oregon, about 35 m off northwest Africa, and about 25 m off Peru. For Oregon and Peru this is less than the thickness of the euphotic

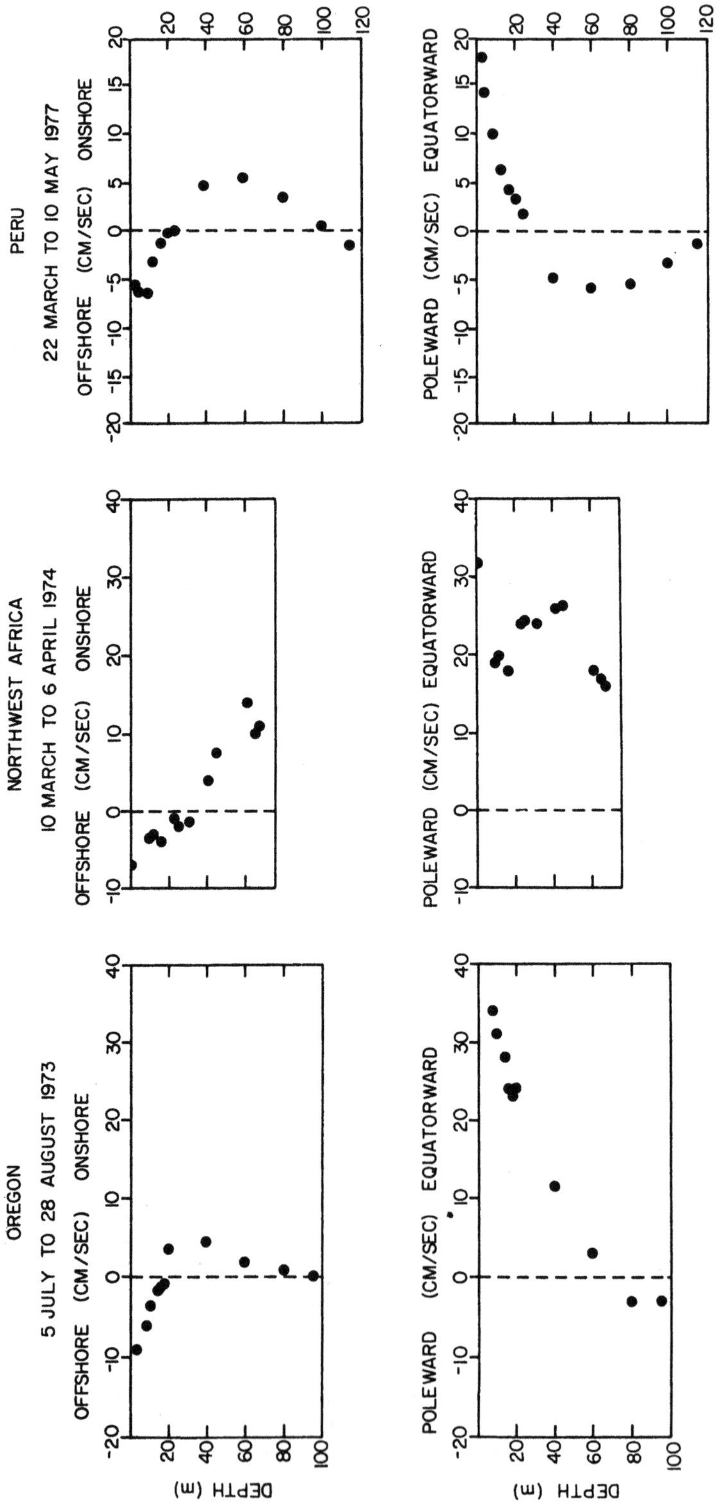

Fig. 3. Mean current profiles from the mid-shelf moorings (cf., Fig. 2) (from Barber and Smith, 1981).

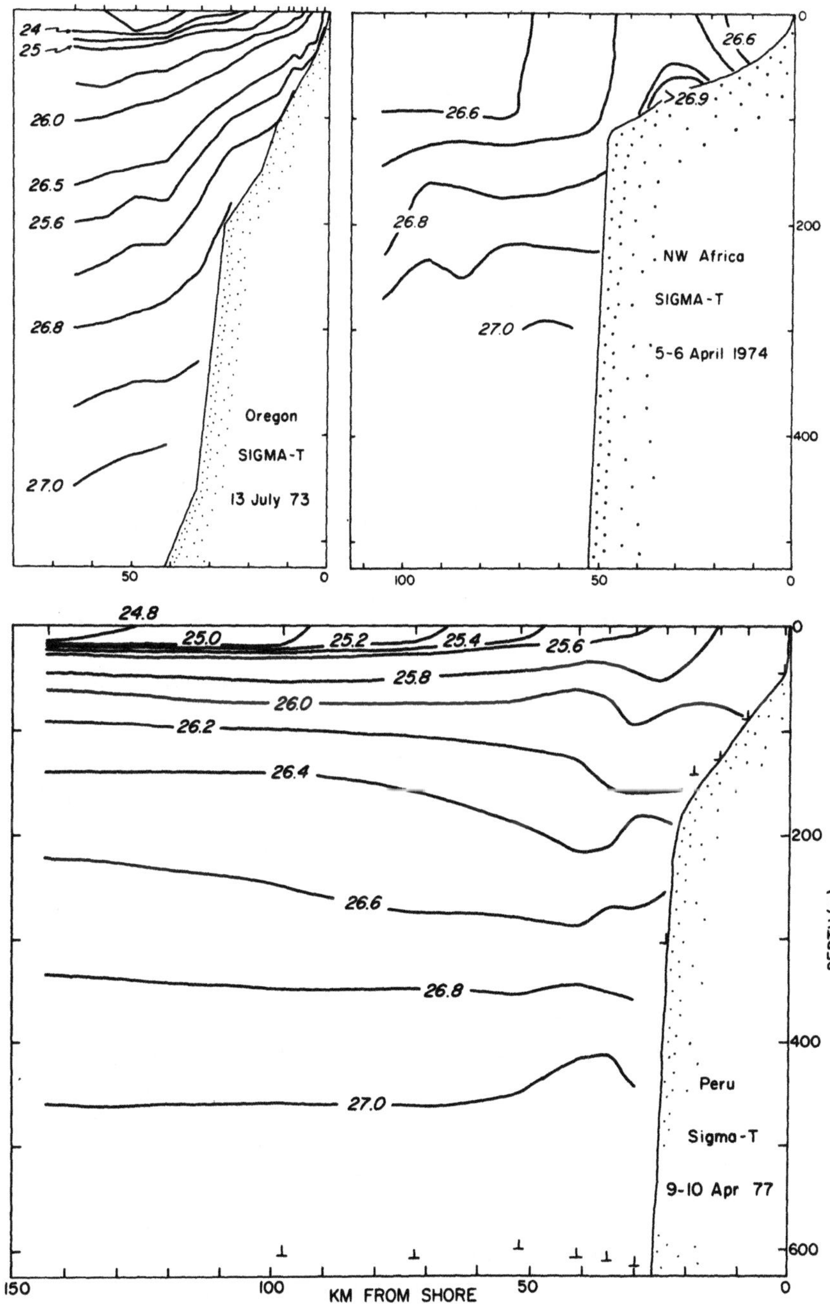

Fig. 4. Density sections across the shelf and slope during periods of strong upwelling favorable winds.

layer. These depths are in reasonable agreement with the estimated Ekman depth δ_E (Table 1). Although δ_E would vary with wind intensity and stratification, the cross shelf flow pattern during an intense upwelling event was very similar to the mean (Fig. 5). Smith (1981) computed the mean transport in the surface Ekman layer from the observations and compared it with the theoretical value (τ/f) using the alongshore wind stress estimated from anemometers at, or near, the moorings. There was agreement within a factor of two.

2) The compensatory onshore flow, feeding the upwelling inshore, is not occurring in a bottom Ekman layer off Oregon and Peru. Indeed, in those regions there is a mean poleward flow (counter to the wind) at depth; this drives an offshore Ekman transport near the bottom. The onshore compensatory flow in the Oregon and Peru upwelling regions is maximum near mid-depth in the water column (40 m in 100 m off Oregon; 59 m in 121 m off Peru) and the source water for the upwelling process is not in contact with the bottom boundary except very near the coast. In contrast, the flow pattern for northwest Africa looks much like our initial conceptual model with alongshore flow in the direction of the wind and onshore flow in the layer adjacent to the bottom with the thickness (35 m) expected for the bottom Ekman layer (cf., Table 1). The source water for the upwelling is carried in the bottom Ekman layer across the entire shelf (Mittelstaedt, Pillsbury and Smith, 1975; Fig. 6). The onshore transport below the surface layer balances, but only within a factor of two, the offshore surface layer transport off northwest Africa, but not off Oregon or Peru (Table 1 and Smith, 1981) where the mean onshore transport was much greater than the mean offshore transport in the surface layer. The latter misbalance suggests a strong three-dimensionality to the upwelling, i.e., alongshore variations in the alongshore flow may be important in the upwelling process. Upwelling 'centers' are further evidence of this.

POLEWARD UNDERCURRENTS

A poleward undercurrent, flowing counter to the wind and near-surface flow, is a ubiquitous feature of the major coastal upwelling regions. The undercurrent hugs the continental slope off the upwelling coasts of the Americas and Africa; sometimes, and in some regions, it extends onto the continental shelf. The mean vectors from mid-shelf off Oregon and Peru reveal its presence there (Fig. 2). Off northwest Africa at 22°N, it is confined to the slope (Fig. 6) and was never manifest at the mid-shelf site. Off Peru, however, it dominated the entire flow on the shelf beneath the thin surface Ekman layer (Fig. 7). The array of current meters did not extend sufficiently offshore and deep to provide an equivalent picture for the 1973 Oregon experiment, but an experiment in 1978 revealed broad poleward flow with a maximum near the inner slope (Fig. 8). Data from an array with better spatial resolution off Washington, 200 km

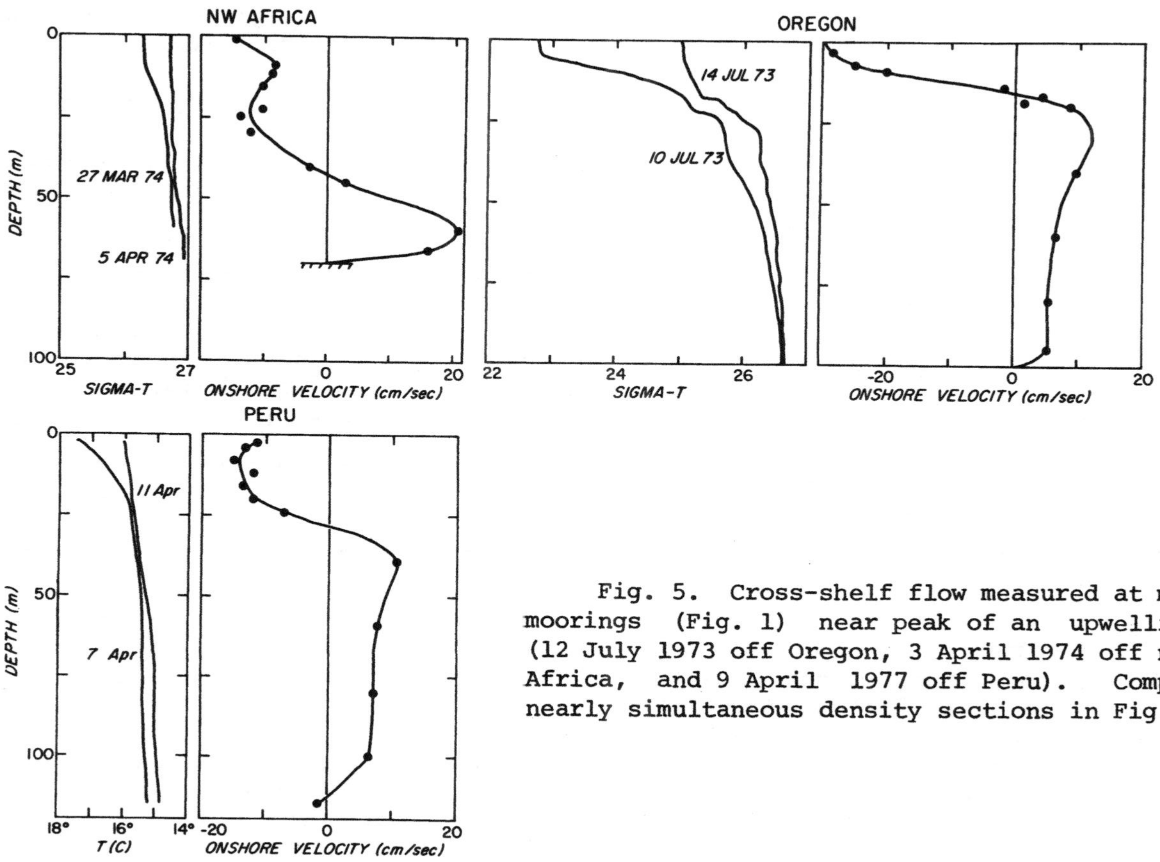

Fig. 5. Cross-shelf flow measured at mid-shelf moorings (Fig. 1) near peak of an upwelling event (12 July 1973 off Oregon, 3 April 1974 off northwest Africa, and 9 April 1977 off Peru). Compare with nearly simultaneous density sections in Fig. 4.

Table 1. Some parameters and the mean values of some variables on three coastal upwelling shelves, based on data from mid-shelf sites (Fig. 1) for the same period used for the vector mean currents and wind (Fig. 2). Because H and N^2 change appreciably with distance offshore, R, the Rossby radius, as computed here can only be considered a rough estimate of scale. N^2 was computed from the density difference between surface and bottom at mid-shelf (cf., Fig. 4). τ_w and τ_b are taken from Allen and Smith (1981). The Ekman layer thickness, δ_E, were estimated as 1.7 U_* $(Nf)^{-1/2}$, following Pollard et al. (1973); U_* was estimated from the appropriate τ [and, for computing δ_{Eb}, also from the speed above the boundary layer (Smith, 1981)].

	Oregon	Northwest Africa	Peru
Coriolis parameter, f (sec^{-1})	1.03×10^{-4}	0.54×10^{-4}	-0.38×10^{-4}
Mid-shelf depth, H (m)	100	73	121
Stratification, N^2 (sec^{-2})	1.5×10^{-4}	0.4×10^{-4}	0.4×10^{-4}
Rossby radius, R (km)	13	9	20
Alongshore wind stress, τ_w (dynes cm^{-2}, ±s.d.)	-0.51 ± 0.74	-1.50 ± 0.95	0.79 ± 0.40
Bottom stress, τ_b (dynes cm^{-2}, ±s.d.)	0.05 ± 0.18	0.53 ± 0.44	-0.05 ± 0.12
Surface Ekman layer, δ_{Es} (m)	10	36	30
Bottom Ekman layer, δ_{Eb} (m) [and from Smith, 1981]	4 [10]	21 [24]	8 [13]

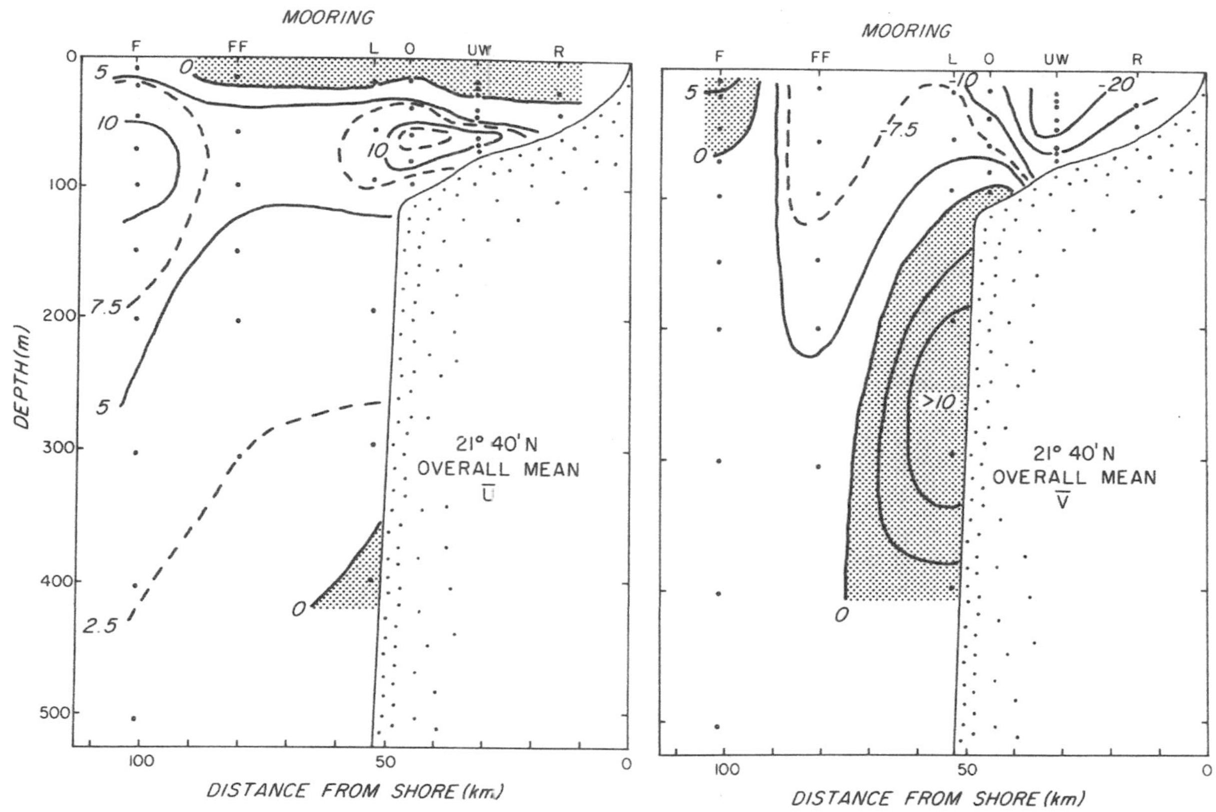

Fig. 6. Overall mean of cross-shelf (u) and alongshore (v) flow off northwest Africa. Shaded areas are offshore and poleward in the respective panels (from Mittelstaedt et al., 1975).

north of the Oregon site, show the undercurrent with jet-like core in which flow exceeds 10 cms^{-1} over the inner slope, just beyond the shelf edge, in July and August, 1972 (Hickey, 1979, see her Fig. 10).

There is not yet a completely satisfactory explanation for the existence and strength of the poleward undercurrent. Its importance to the ecology and the distribution of sediments is clear. Firstly, when the poleward undercurrent is in contact with the bottom (as on the shelf off Peru, the shelf and slope off Oregon, and the slope off northwest Africa) the bottom drag causes an offshore Ekman transport. Thus, there is a tendency for fine grain material to be moved offshore. Secondly, it can transport sedimentary material appreciable distances away from their source (and in a direction opposite to prevailing wind and surface current). This was shown by Karlin (1980) for the undercurrent off northern California, Oregon, and Washington. It is likely that the poleward undercurrent is at least quasi-continuous along the west coast of the U.S. Off Peru the poleward undercurrent is continuous from at least 5°S to at least 15°S (Brockmann et al., 1980) and is also the source for the upwelling water (see Fig. 7). Thirdly, the undercurrent provides a feedback loop for the biota. The high primary productivity in coastal upwelling regions occurs near the surface in waters which are moving downwind (equatorward) and offshore. The detritus from the high productivity, and from the higher trophic levels which it supports, sink out of the euphotic zone to deeper water where the inorganic nutrients are gradually regenerated by bacterial action. The poleward undercurrent flows beneath the region of highest productivity and transports the detrital products and nutrients back upstream relative to the flow in

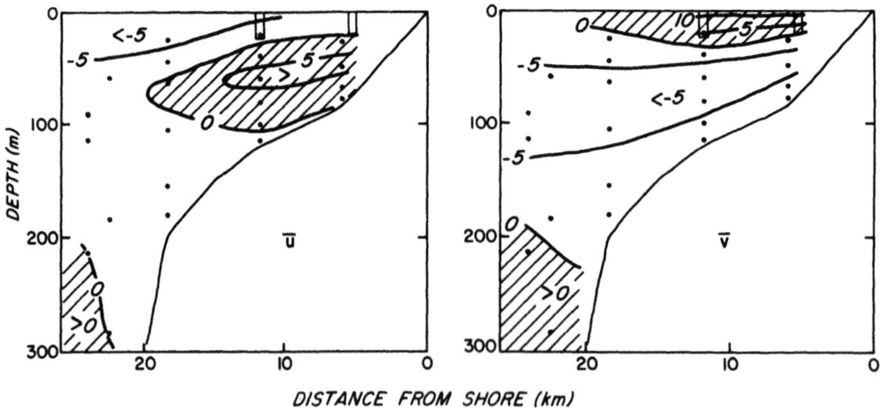

Fig. 7. Mean cross-shelf (u) and alongshore (v) flow off Peru for period 22 March to 10 May 1977. Shaded areas are onshore and equatorward in the respective panels. Units are cm s^{-1} (from Brink et al., 1980).

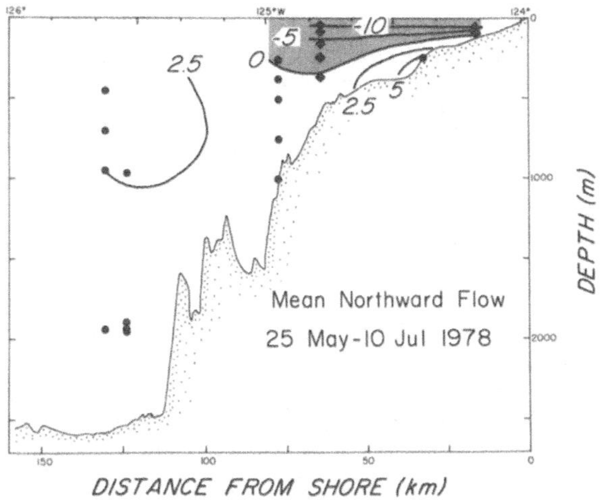

Fig. 8. Alongshore flow off Oregon at 45°20'N. Shaded areas are equatorward. Units are cm s^{-1}.

the euphotic layer. Thus, a natural feed-back process is established and the nutrients may eventually be returned to the euphotic zone by onshore flow (compensating for the offshore Ekman transport above) and be upwelled to again repeat the cycle. One can imagine that the biota may also take advantage of this conveyer belt, thus remaining in regions of high productivity, i.e., upwelling 'centers'.

SHALLOW SHELVES

Most of the continental shelves where coastal upwelling occurs at present are relatively steep, deep at the seaward edge, and narrow. This may not always have been so in the past and it is therefore instructive to consider the northwest Africa shelf to elucidate the effect of the shallowness on the upwelling process. The flow is strongly wind-driven, with bottom friction playing an important role (Allen and Smith, 1981). The entire water column on the northwest African shelf is frictionally dominated at times (see Badan-Dangon, 1981, for a thorough discussion of the dynamics), consisting of two nearly adjacent Ekman layers superimposed on a barotropic alongshore current. [J.D. Smith and Long (1976) show that a similar situation arises off Washington during winter when the predominant winds are from the south or southwest. There is no coastal upwelling then, but the basic dynamics driving the flow on the shelf is similar to that off northwest Africa; the Ekman layers have the opposite direction to those during upwelling.]

The Ekman layer thickness increases as stratification decreases and U_* increases. Wide shallow shelves, with low stratification and high alongshore winds (and hence strong near-bottom alongshore currents), should have circulation patterns similar to those off northwest Africa. On even shallower shelves, the upwelling circulation pattern may not even occur: The Ekman transport perpendicular to the wind would not develop since the wind-induced surface stress would be transferred directly to the bottom boundary layer and hence balanced by a bottom stress, without any rotational effects, i.e., the bottom stress would dominate the Coriolis term in the equation of motion when $\delta_E/H>>1$.

The criterion which determines the flow characteristics is the ratio δ_E/H or U_*/Hf. In the case of pure wind forcing (as in our initial conceptual model) where $\delta_E/H<<1$, surface and bottom Ekman layers develop, with offshore and onshore transport respectively, separated by a region in which the geostrophic alongshore current is dominant. As $\delta_E/H \rightarrow 1$, i.e., as the Ekman layers merge, the rotational effects are still strong, and an appreciable angle still exists between the near surface and near bottom currents (Smith and Long, 1976). For the northwest African shelf, and for the Oregon and Washington shelves during the winter (the non-upwelling season), $\delta_E/H \sim 1$. Smith and Long (1976) show that as U_*/Hf becomes >2 the rotational effects diminish; for $\delta_E/H>>1$ all the flow is parallel to the coast in the direction of the wind. During strong wind events, $\delta_E/H > 5$ for inner 25 km of the northwest African shelf. At these times the upwelling circulation pattern seems to separate from the coast. The density section from northwest Africa shown in Fig. 4 is during strong upwelling favorable winds: Note the lighter water inshore. Barton, Huyer and Smith (1977) show warming inshore and cooling at mid-shelf after a few days of strong upwelling favorable winds (see their Fig. 3). During these periods $U_*/Hf > 5$ inshore, and presumably the warming inshore is the result of the breakdown of the upwelling circulation pattern. There is mixing from top to bottom inshore, but the rotational effect, which leads to the cross-shelf circulation essential to "upwelling", is negligible. This must happen very near the coast in all regions--but on a broad, shallow shelf the upwelling would be displaced far from the coast. The upwelling, and the high productivity arising from the recirculation of nutrients, would presumably occur at or near the outer edge of the shelf where the depths would be sufficient for the Ekman layers to develop.

VARIABILITY: SPATIAL AND INTERANNUAL

The relatively localized (in both space and time) studies of the past decade were made within persistent large scale coastal upwelling systems. By the end of the decade, the spatial inhomogenity (upwelling 'centers') and the interannual variability within these systems had become apparent and were recognized as important aspects of the coastal upwelling process. (Upwelling 'centers' are, in a way, the spatial analogue to the temperal upwelling 'events' which were so

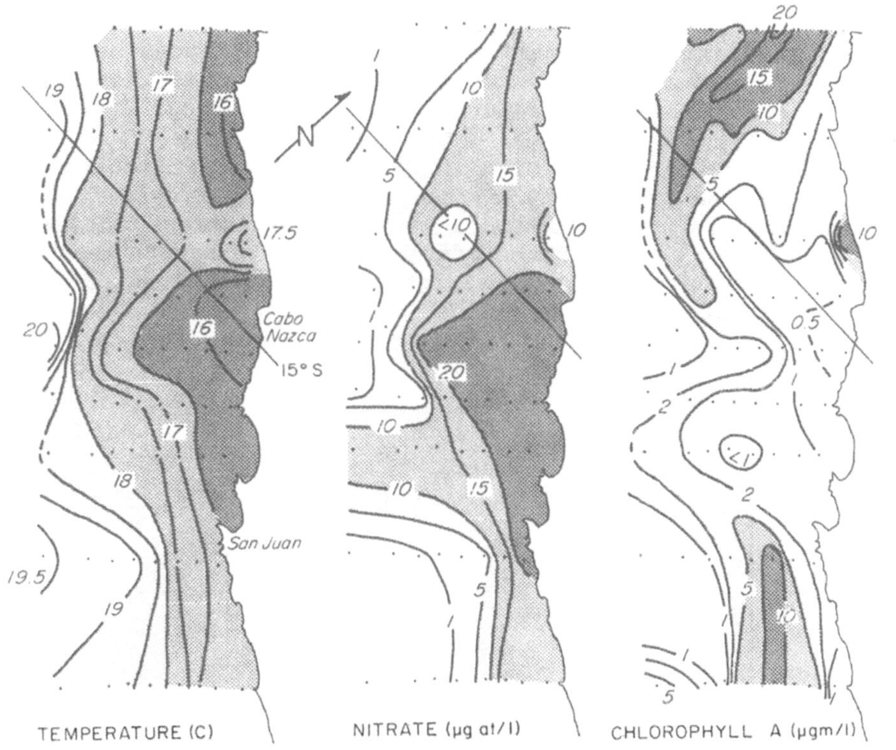

Fig. 9. Ship-based maps of surface values off Peru near the 15°S upwelling 'center'. Length of North arrow represents 25 km.

thoroughly studied in the 1970's.) These aspects will, one hopes, be more adequately addressed observationally in the 1980's. Some progress, via numerical modeling, has been made toward understanding the mechanisms, e.g., see Preller and O'Brien (1980) on the spatial variability of upwelling and O'Brien, Busalacchi and Kindle (1981) for a review of El Niño (interannual variability) models. I will only comment briefly on some observations, made in the regions discussed above, which are relevant to the topic of variability.

The apparent spatial inhomogeneity of coastal upwelling can be seen off Peru in Fig. 9. The persistence of the upwelling 'center' near Cabo Nazca is well documented: It appears in atlases of sea surface temperature (e.g., Zuta and Urquizo, 1972) and was clearly depicted by Gunther (1936). Both wind and topography may cause 'centers' to occur: The wind field is not uniform along upwelling coasts. Fig. 10 shows both the latitudinal and seasonal variability of wind stress off North America between 32° and 50°N; one should expect the Ekman transport and resulting coastal upwelling to vary accordingly. It is also well known, both from observations and theory, that upwelling is intensified equatorward of capes as a consequence of the enforced curvature of the horizontal flow (Arthur, 1965). The sea surface temperature (SST) patterns near Cape Blanco, Oregon (Fig. 11), and those near Cabo Nazca, Peru (Fig. 9), are likely the result of both local wind intensification near the capes and the effect of the topography on the circulation itself. However, variations in upwelling could occur without variations in the wind and the coastline: Even if the offshore Ekman transport in the surface layer remains unchanged, variations in the alongshore flow, caused by variations in the bottom topography, could induce variations in the vertical flow. An alongshore convergence (or diver-

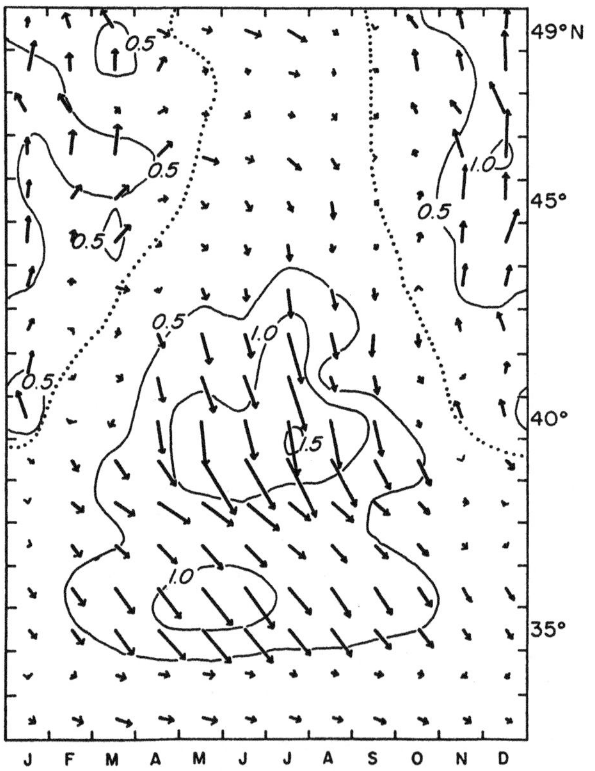

Fig. 10. Long-term mean monthly wind stress for one degree squares along the west coast of the United States, computed from ship reports from 1857-1972 by Nelson (1977). Vectors show the direction and magnitude of the monthly mean stress at each latitude. The dotted line shows the boundary between stress vectors with a southward component and those with a northward component. Magnitude of the stress is contoured at intervals of 0.5 dynes/cm^2.

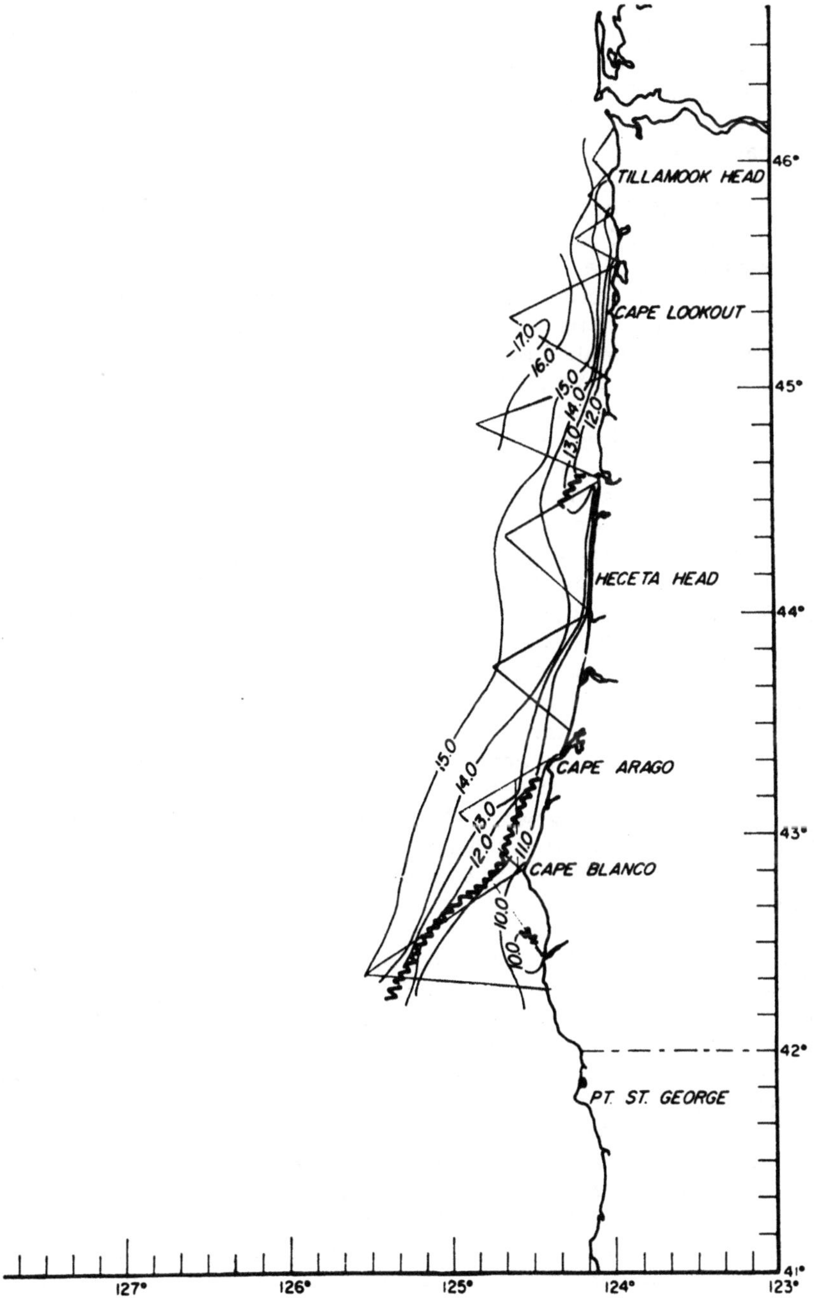

Fig. 11. Aircraft-based map of sea surface temperature along the Oregon coast, 12 August 1969. Heavy jagged lines indicate color discontinuity.

gence) could reduce (or increase) the amount of upwelling needed to compensate for the offshore Ekman transport. (This type of three-dimensionality could explain the misbalance, observed off Oregon and Peru, between the offshore surface layer transport and the deeper onshore transport.) Thus the bottom topography, acting on a quasi-steady alongshore flow, could induce geographically persistent upwelling 'centers'.

The wind and the weather may differ from year to year--and so would the amount of upwelling. Off Oregon (45°N) the winds have a strong seasonal cycle (see Fig. 10) and the upwelling circulation pattern occurs for only about half the year. A very late or very early onset of the upwelling season (the period of southward winds) in relation to the peak spring river runoff might affect the sediment record, especially if the shift persisted for years. In contrast to Oregon, the winds off central Peru (15°S) are always equatorward (upwelling favorable) with seldom a single day of poleward winds. The contrast is apparent in the annual statistics for the years during which the intensive experiments took place off Oregon (1973) and Peru (1976-77). Off Oregon the <u>annual mean</u> wind stress is actually unfavorable for upwelling, due to the predominant strength of the winter storms (poleward winds), and is 0.16 dynes cm^{-2} poleward with a standard deviation of 0.81 dynes cm^{-2}. Off Peru the annual mean is 0.88 dynes cm^{-2} equatorward (favorable for upwelling) with a standard deviation of only 0.44 dynes cm^{-2}. Nevertheless, one of the more dramatic examples of interannual variability in the ocean occurred in the Peru coastal upwelling region in 1976: El Niño. Fig. 12 shows a year long current and temperature series off Peru from the mid-shelf site shown in Fig. 1. The wind and sea level are from San Juan, a coastal station 50 km away. The anomalous warming during April to June (the austral autumn) 1976 and strong poleward flow are the physical signals of El Niño. The economic signal is a collapse of the anchoveta fishery.

What causes El Niño? It had once been thought that El Niño was due to weak local winds resulting in weak coastal upwelling. This is not the case and, indeed, the coastal winds at Callao (12°S) are actually more strongly favorable for upwelling during El Niño (Enfield, 1981). Wyrtki (1975) presented a conceptual model, the basis for most contemporary theories, that explains El Niño as the dynamic response of the equatorial part of the Pacific Ocean to atmospheric forcing. During a prolonged period of excessively strong southeast (the upwelling favorable direction) trade winds in the eastern tropical Pacific, warm water is "piled up" in the western Pacific. El Niño results from the reaction of the equatorial Pacific to the relaxation of the southeast trades back to normal, or below normal, strength. The water from the western Pacific "sloshes" back when the trades relax and this leads to a deepening of the surface layer (with its warm, nutrient-poor water) off Ecuador and Peru. Local coastal upwelling still continues--but instead of drawing cool, nutrient-rich

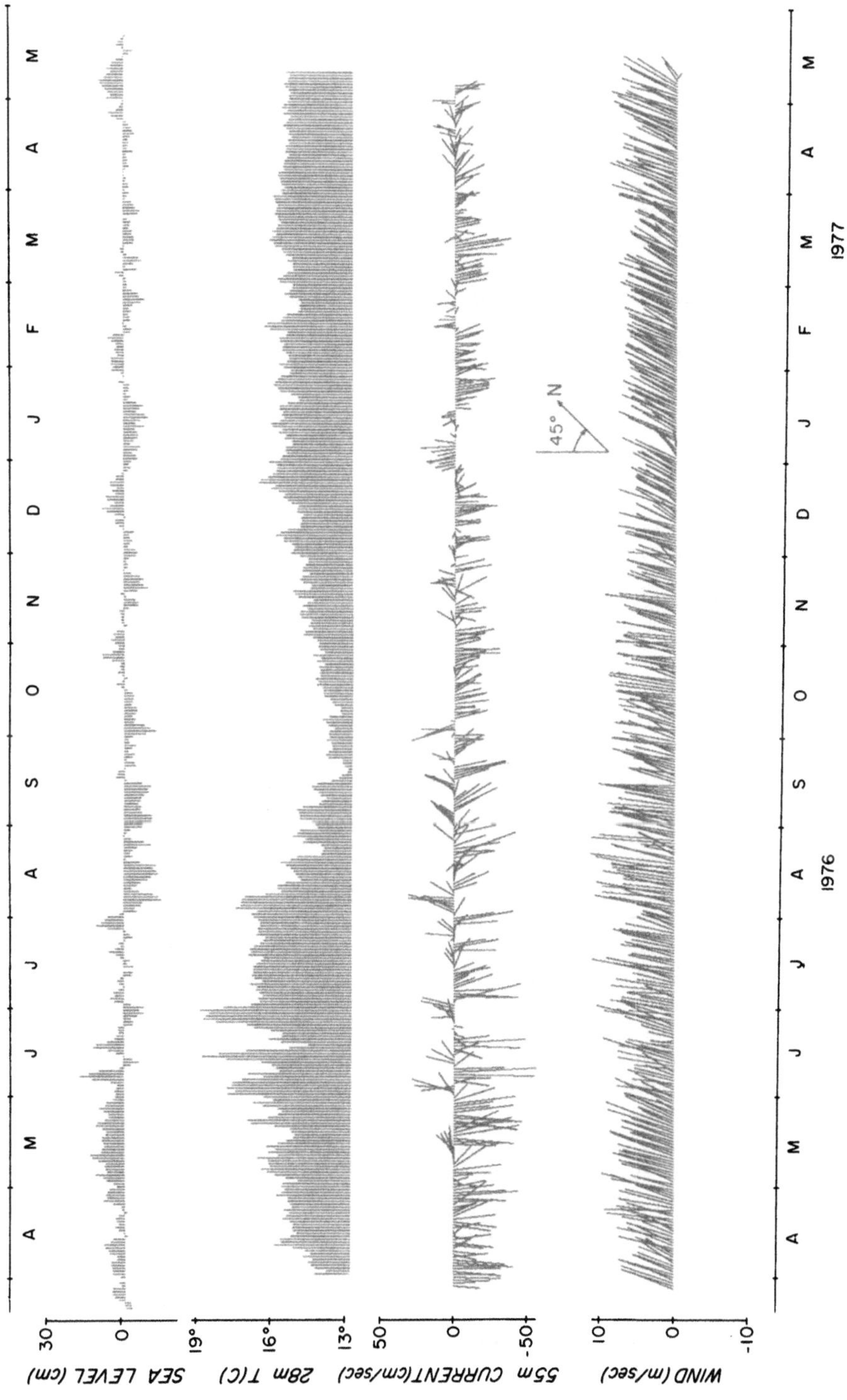

Fig. 12. Time series from March 1976-May 1977 from Peru coast (15°S). Adjusted sea level at San Juan, 28 m temperature and 55 m current over the shelf (mid-shelf mooring, 121 m bottom depth), and wind measured at San Juan. Daily values are from low-passed filtered hourly series. Vertical axis is the alongshore direction for current and wind vectors.

water to the surface, the upwelling process doesn't extend below the now deeper surface layer of warm, nutrient-poor water. The fisheries and the ecosystem are perturbed.

Recently, in 1965, in 1972-73, and in 1976-77, there have been severe El Niños. The one in 1972-73, coming after years of high fishing pressure on the anchoveta, virtually decimated the Peru anchoveta fishery, and the anchoveta fishery did not recover as after previous El Niños. Ominously, there was a reproductive failure in the anchoveta associated with the El Niño of 1976 and the fishery statistics for 1977 were even poorer than 1973. The heavy fishing coupled with environmental catastrophes of 1972-73 and 1976 has reduced the Peru anchoveta fishery from the world's largest to insignificance in a few years. The effects should appear in the sediment record.

SUMMARY

Measurements of the horizontal flow field in three major coastal upwelling regions (Oregon near 45°N, northwest Africa near 22°N, and Peru near 15°S) show circulation patterns which are similar in the surface layer, where the observed mean offshore transport agrees with the Ekman transport computed from the wind stress, but which differ in the subsurface alongshore flow and the compensatory onshore flow.

The circulation pattern on the relatively wide and shallow northwest Africa shelf resembles the simple conceptual model: Offshore flow in a surface Ekman layer, equatorward alongshore flow in the direction of the wind throughout the water column, and compensatory onshore flow in a bottom Ekman layer. The observed cross-shelf transports in the two Ekman layers balance within a factor of two. During periods of very strong equatorward winds, the inner (shallower) half of the shelf becomes frictionally dominated, i.e., the surface and bottom stresses nearly balance and no Ekman layers develop there, and the upwelling circulation pattern is displaced toward the shelf edge.

On the Oregon and Peru shelves, the alongshore component of the flow at depth is poleward and the bottom Ekman layer transport therefore, is directed offshore. The onshore flow, which supplies the upwelling near the coast, occurs just below the surface Ekman layer and is maximum at mid-depth. (No evidence for a mean multi-cell cross-shelf circulation pattern is observed, except the bottom Ekman layer off Oregon and Peru.) The observed surface layer offshore and the subsurface onshore transports do not balance, although both are significantly correlated with the alongshore component of the wind stress (Smith, 1981). Presumably, the mass balance is completed by convergences and divergences in the alongshore flow, and lesser or greater upwelling (vertical flow) occurs. Thus, upwelling 'centers' may develop; alongshore (spatial) variations within the major upwelling regions are readily apparent in maps of mean sea surface temperature.

The mean alongshore currents on the shelf are stronger than the onshore-offshore currents and, hence, an upwelling parcel of water may move great distances alongshore in the upwelling process. The alongshore flow near the surface has a mean of about 20 cm s^{-1} equatorward and is in the direction of the local wind. The deeper mean flow on the Oregon and Peru shelves is poleward, opposite to the mean wind. Over the slope, a poleward undercurrent exists in all three regions. It is contiguous with the poleward flow on the shelf off Peru and Oregon, and in all three regions the core of the poleward undercurrent over the slope is near the depth of the shelf break. The biota may use the poleward undercurrent as a 'feedback' mechanism adjacent to the coastal upwelling zone. Sediment originating on the shelves of coastal upwelling regions may be carried 'upstream', relative to the equatorward wind and surface currents, by the poleward undercurrent.

ACKNOWLEDGMENTS

Many of the observations discussed in this paper were obtained in the CUEA (Coastal Upwelling Ecosystem Analysis) experiments, which were supported by the U.S. National Science Foundation. I am indebted to my CUEA colleagues, whose data and ideas I have used. Synthesis and rethinking has been supported by National Science Foundation grants OCE-7925019 and OCE-8024116.

REFERENCES

Allen, J.S., 1980, Models of wind-driven currents on the continental shelf, Annual Review of Fluid Mechanics, 12:389-433.

Allen, J.S. and Smith, R.L., 1981, On the dynamics of wind-driven shelf currents, Philosophical Transactions of the Royal Society of London, A, 302:617-634.

Arthur, R.S., 1965, On the calculation of vertical motion in eastern boundary currents from determinations of horizontal motion, Journal of Geophysical Research, 70:2799-2803.

Badan-Dangon, A.R.F., 1981, "On the Dynamics of Subinertial Currents off Northwest Africa," Ph.D. Dissertation, Oregon State University, Corvallis, 167 pp.

Barber, R.T. and Smith, R.L., 1981, Coastal upwelling ecosystems, in: "Analysis of Marine Ecosystems", A.R. Longhurst, ed., Academic Press, London, 31-68.

Barton, E.D., Huyer, A. and Smith, R.L., 1977, Temporal variation observed in the hydrographic regime near Cabo Corveiro in the northwest African upwelling region, February to April 1974, Deep-Sea Research, 24:7-23.

Brink, K.H., Halpern, D. and Smith, R.L., 1980, Circulation in the Peruvian upwelling system near 15°S, Journal of Geophysical Research, 85:4036-4048.

Brink, K.H., Jones, B.H., Van Leer, J.C., Mooers, C.N.K., Stuart, D.W., Stevenson, M.R., Dugdale, R.C. and Heburn, G.W., 1981, Physical and biological structure and variability in an upwelling center off Peru near 15°S during March, 1977, in: "Coastal Upwelling," F.A. Richards, ed., Coastal and Estuarine Sciences 1, American Geophysical Union, Washington, 473-495.

Brockmann, C. Fahrbach, E., Huyer, A. and Smith, R.L., 1980, The poleward undercurrent along the Peru coast: 5 to 15°S, Deep-Sea Research, 27:847-856.

Charney, J.G., 1955, The generation of oceanic currents by wind, Journal of Marine Research, 14:477-498.

Ekman, V.W., 1905, On the influence of the earth's rotation on ocean currents, Arkiv för Matematik, Astronomi, och Fysik, 2:1-53.

Enfield, D.B., 1981, Thermally driven wind variability in the planetary boundary layer above Lima, Peru, Journal of Geophysical Research, 86:2005-2016.

Gunther, E.R., 1936, A report on oceanographical investigations in the Peru coastal current, Discovery Reports, 13:107-276.

Halpern, D., 1976, Structure of a coastal upwelling event observed off Oregon during July 1973, Deep-Sea Research, 23:495-508.

Halpern, D., Smith, R.L. and Mittelstaedt, E., 1977, Cross-shelf circulation on the continental shelf off northwest Africa during upwelling, Journal of Marine Research, 35:787-796.

Hickey, B., 1979, The California Current system--hypotheses and facts, Progress in Oceanography, 8:191-279.

Huntsman, S.A. and Barber, R.T., 1977, Primary production off northwest Africa: the relationship to wind and nutrient conditions, Deep-Sea Research, 24:25-33.

Huyer, A., 1976, A comparison of upwelling events in two locations: Oregon and northwest Africa, Journal of Marine Research, 34: 531-546.

Karlin, R., 1980, Sediment sources and clay mineral distributions off the Oregon coast, Journal of Sedimentary Petrology, 50:543-560.

Mittelstaedt, E., Pillsbury, R.D. and Smith, R.L., 1975, Flow patterns in the northwest African upwelling area, Deutschen Hydrographischen Zeitschrift, 28:145-167.

Nelson, C.S., 1977, Wind stress and wind stress curl over the California Current, NOAA Technical Report NMFS SSRF-714, U.S. Department of Commerce, Washington, D.C., 87 pp.

O'Brien, J.J., Busalacchi, A. and Kindle, J., 1981, Ocean Models of El Niño, in: "Resource Management and Environmental Uncertainty: Lessons from Coastal Upwelling Fisheries", M.H. Glantz and J.D. Thompson, eds., Wiley-Interscience, New York, 159-212.

Peterson, W.T., Miller, C.B. and Hutchinson, A., 1979, Zonation and maintenance of copepod populations in the Oregon upwelling zone, Deep-Sea Research, 26:467-494.

Pollard, R.T., Rhines, P.B. and Thompson, R.O.R.Y., 1973, The deepening of the wind-mixed layer, Geophysical Fluid Dynamics, 3:381-404.

Preller, R. and O'Brien, J.J., 1980, The influence of bottom topography on upwelling off Peru. Journal of Physical Oceanography, 10:1377-1398.

Smith, R.L., 1981, A comparison of the structure and variability of the flow field in three coastal upwelling regions: Oregon, northwest Africa, and Peru, in: "Coastal Upwelling," F.A. Richards, ed., Coastal and Estuarine Sciences 1, American Geophysical Union, Washington, 107-118.

Smith, J.D. and Long, C.E., 1976, The effect of turning in the bottom boundary layer on continental shelf sediment transport, Memoires Societe Royale des Sciences de Liege, 6e serie, 10:369-396.

Sverdrup, H.U., Johnson, M.W. and Fleming, R.H., 1942, "The Oceans", Prentice-Hall, Englewood Cliffs, 1087 pp.

Weatherly, G.L. and Martin, P.J., 1978, On the structure and dynamics of the oceanic bottom boundary layer, Journal of Physical Oceanography, 8:557-570.

Winant, C.D., 1980, Coastal circulation and wind-induced currents, Annual Review of Fluid Mechanics, 12:271-301.

Wyrtki, K., 1975, El Niño--the dynamic response of the Pacific Ocean to atmospheric forcing, Journal of Physical Oceanography, 5:572-584.

Yoshida, K., 1955, Coastal upwelling off the California coast, Records of Oceanographic Works in Japan, 2:8-20.

Zuta, S. and Urquizo, W., 1972, Temperatura promedio de la superficie del mar frente a la costa Peruana, periodo 1928-1969, Boletin Instituto del Mar del Peru-Callao, 2:459-520.

OBSERVATIONS OF A PERSISTENT UPWELLING CENTER OFF POINT CONCEPTION, CALIFORNIA

Burton H. Jones

Department of Biological Sciences
University of Southern California
Los Angeles, California 90089, U.S.A.

Kenneth H. Brink, Woods Hole Oceanographic Institution
Woods Hole, Massachusetts, U.S.A.
Richard C. Dugdale, University of Southern California
Los Angeles, California 90089, U.S.A.
David W. Stuart, The Florida State University
Tallahassee, Florida, U.S.A.
John C. Van Leer, University of Miami
Miami, Florida, U.S.A.
Dolors Blasco, Bigelow Laboratory for Ocean Science
West Boothbay Harbor, Maine, U.S.A.
James C. Kelley, San Francisco State University
San Francisco, California, U.S.A.

ABSTRACT

Centers of intensified upwelling occur throughout the world and are often associated with topographic irregularities along the coasts and continental shelves. Preliminary observations from a study near Point Conception, California, as well as previous observations from Peru, near 15°S, provide us with some generalizations about coastal upwelling centers. The mean structures of these centers appear to be well-defined regions of low temperature, high nutrient concentrations and low phytoplankton abundance, with characteristic gradients of each of these variables away from the center. The biological productivity of these centers results in the generation of organic particles. The impact of this near-surface productivity on the sediments will depend on the physical structure of the upwelling centers and the variability of this structure.

INTRODUCTION

Centers of intensified upwelling have been observed in many of the world's upwelling regions. Such features are characterized by an area near the coast which is colder and generally more nutrient rich than the adjacent waters. Examples of such structures are found along the coasts of Peru (Gunther, 1936; Ryther et al., 1966; Walsh, et al., 1971; Brink et al., 1981); Somalia (Smith and Codispoti, 1980); southwest Africa (Hart and Currie, 1960; Andrews and Hutchings, 1980; Hutchings, 1981); Baja California (Walsh et al., 1974; 1977); northwest Africa (Cruzado and Salat, 1981); and California (Sverdrup and Allan, 1939; Breaker and Gilliland, 1981). Our knowledge of coastal upwelling centers and hence our basis for forming questions about the structure of these centers, relies largely on the observations of the center at 15°S off Peru. The motivation for the study of the upwelling structure off Point Conception, California came directly from questions left unanswered by the study of the Peruvian upwelling center. For this reason, we present here first a synopsis of the results from the study of the upwelling center at 15°S and then attempt to make comparisons between the Peruvian upwelling center and a persistent upwelling center located at 34°30'N off Point Conception, California.

Cabo Nazca Upwelling Center (15°S)

The surface temperature map of the mean center at 15°S (Fig. 1) shows an area of persistently colder water with an alongshore scale of roughly 25 km, which is comparable to its extent offshore. Of the 13 aircraft maps from March-May, 1977, that were averaged to compile Fig. 1, only one failed to show cold water near the coast in the general region of the mean center. This is perhaps not surprising since the winds were persistently upwelling-favorable (equatorward) and the mean near-surface currents had an offshore component consistent with mean upwelling. The surface thermal structure was reflected in the density and primary nutrient distributions. For example, Fig. 2 demonstrates the close relation between surface nutrients and temperature. Also, the distribution of phytoplankton biomass, as indicated by chlorophyll *a*, tends to complement the temperature structure in that the coldest central water tends to have low phytoplankton biomass and the warmer outer waters tend to have a larger standing stock (see also Boyd and Smith, in press). However, the phytoplankton biomass-temperature relation appears to be only partly illuminated in Fig. 2. Much farther offshore, where the water is yet warmer, the chlorophyll *a* content again decreased apparently in response to grazing pressure and to nutrient depletion.

The fluctuations in the intensity (horizontal temperature contrast) and size of the upwelling center were largely controlled by variations in the local alongshore wind stress. Also, the structure tended to be spatially smooth and well organized during strong winds,

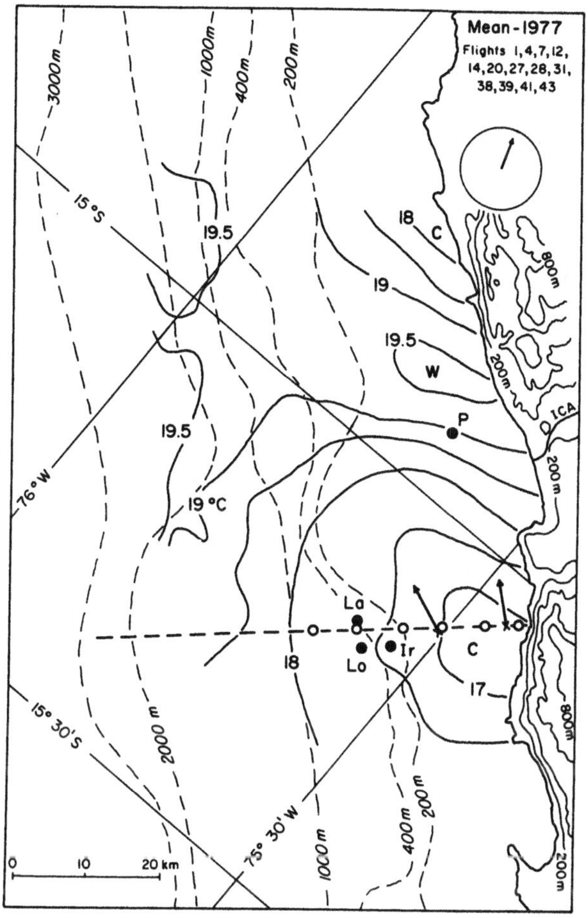

Fig. 1. Mean sea surface temperature at 15°S off Peru from 13 aircraft maps between 21 March and 6 May 1977. The circle in the upper right-hand corner contains the mean wind stress vector. The radius of the circle is 1 dyne cm^{-2}. The two vectors on the map represent the mean current velocities at 4.5 m of depth at these positions. The length of the vectors is the distance that the flow would traverse in 12 hours. The dashed line perpendicular to the coast is referred to as the C-line; Brink et al. (1981).

and relatively chaotic during weaker winds. Presumably, the surface chemical and biological structures underwent fluctuations at least as dramatic as those of sea surface temperature.

The upwelling center structure, as defined by alongshore temperature gradients, was apparently confined to the near-surface region. Current meter and CTD observations indicate that the surface structure is undetectable at depths much below 30 m. For comparison, the surface mixed layer during the two-week March 1977 study period was typically about 10 m deep, and only occasionally did it penetrate to about 25 m depth.

The mean currents at midshelf in the upwelling center are shown in Fig. 3. The onshore-offshore component u is qualitatively what might be expected in a region of persistent upwelling (see also Smith, this volume): offshore flow in a surface Ekman layer, and onshore flow of the upwelling source water between about 25 and 100 m depth. The alongshore flow structure, v, differs from those of Oregon (Kundu and Allen, 1976) and of northwest Africa (Barton, Huyer, and Smith, 1977) in that a poleward mean undercurrent exists at depths as shallow as 15 m. Apparently, mean, locally wind-driven alongshore flow exists only in the surface Ekman layer. This mean flow could provide a way for drifting particles, such as diatoms, to recycle through the upwelling center. For example, a particle initially upwelled near the coast would be advected offshore and equa-

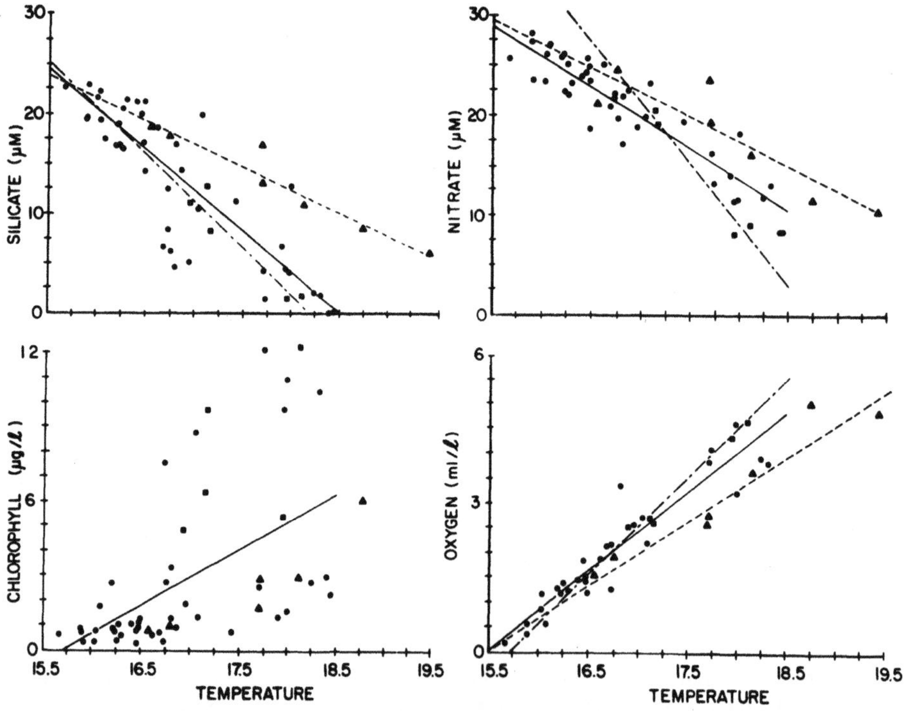

Fig. 2. Chemical variables related to temperature in the upper 25 m of the water column off Peru near 15°S. Data is from the period of 12-30 March 1977. The circles represent data from routine hydrographics, and the solid lines are the linear regressions for the two variables. The remaining data points and lines are from two drogue-following studies in the area; Brink et al. (1981).

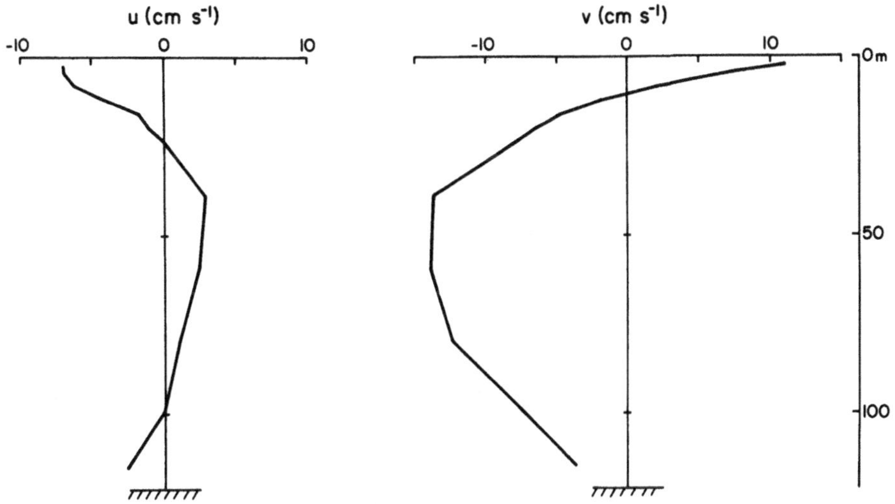

Fig. 3. Mean velocity profiles at mid-shelf (12 km from the coast) along the C-line. The mean is for the period 12-30 March 1977. The cross-shelf component, u, is positive for onshore flow. The alongshore flow, v, is positive for equatorward flow; Brink et al. (1981).

torward until it sank below about 20 m, where it would be advected back in the general direction of the upwelling center. However, the fluctuations in both velocity components, especially u, were substantial relative to the mean, and this unsteadiness would tend to scatter the phytoplankton considerably relative to their mean trajectory.

Mean hydrographic sections through the upwelling center (Fig. 4) show a different aspect of the structure. The distributions of temperature (to which most of the density structure was attributed), nitrate and silicate in the upper 100 m are about what might be expected in a region of persistent upwelling. Isopleths tilted upwards toward the coast, as well as surface values tended to decrease monotonically offshore. Enhanced near-bottom concentrations of silicate over the shelf suggest either resuspension or regeneration in the bottom boundary layer. Pronounced surface fronts are not expected to appear in mean sections because averaging of such transient features tends to smear them out. The mean chlorophyll a section (Fig. 4e) shows high values only near the surface, with a maximum near the shelf break. The tendency for lower values offshore supports the above assertion that chlorophyll a content decreased at sufficiently high surface temperatures. The maximum mean chlorophyll a concentration, about 3 μg/ℓ, appears to be anomalously low for this region relative to other years of observation (Ryther et al., 1966; Walsh et al., 1971; Dugdale et al., 1977; Jones, 1977). The location of the chlorophyll a maximum appears to be more related to frontal location and to the interplay of growth and advection rates than to a direct topographic control due the shelf-slope relief.

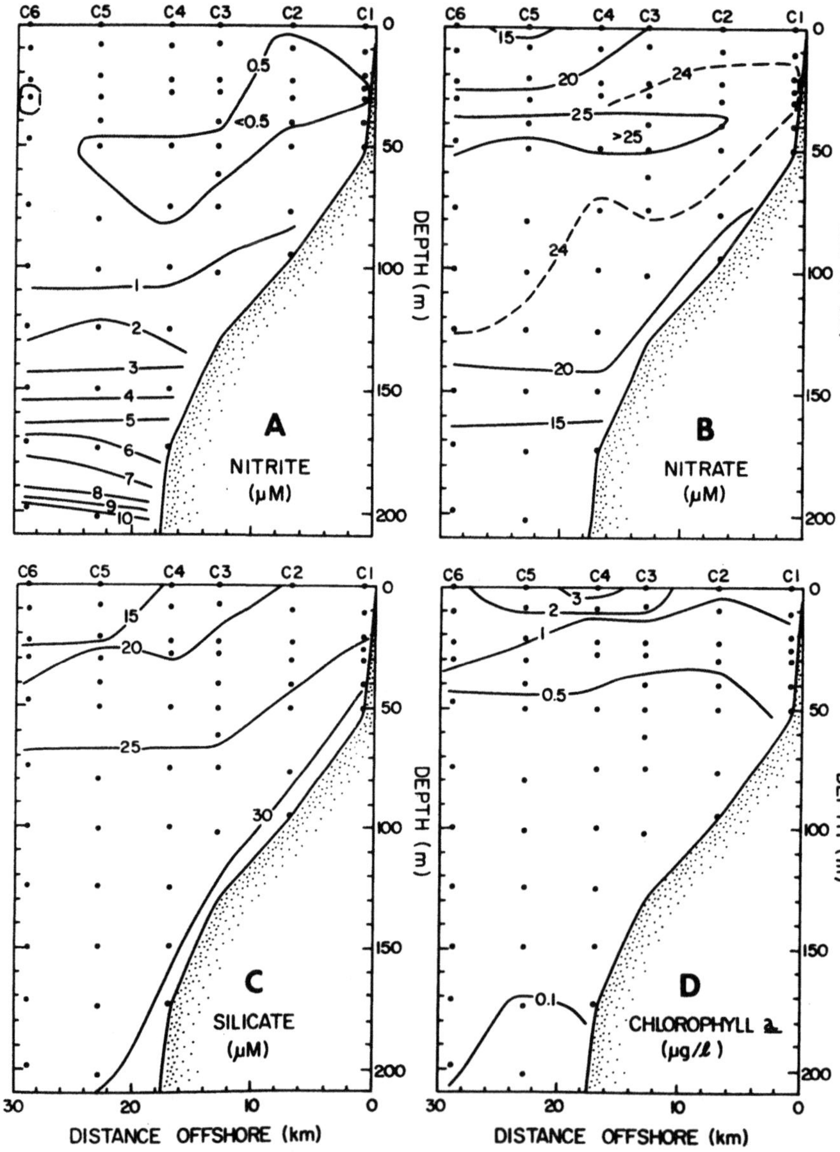

Fig. 4. Mean hydrographic sections from the C-line off Peru for the period 7 March-16 May 1977; Brink et al. (1981).

In some ways, the 15°S Peru upwelling center appears to represent a classic example of locally intensified upwelling. It is quite persistent, both during March-May, 1977, and over interannual periods. Observations of this center show it to be a coupled biological, chemical and physical phenomenon.

The observations presented by Brink et al. (1981) and other authors provide a basis for our understanding of upwelling centers. However, the detailed structure and the processes associated with this feature have been only partially understood. The Organization of Persistent Upwelling Structures (OPUS) program was initiated with the goal of resolving the dynamics and structure of such features and determining their generality or uniqueness. The OPUS participants developed a set of hypotheses regarding the structure of an upwelling center, based on the CUEA observations of the feature at 15°S off Peru.

These hypotheses are best described with the help of the diagram in Fig. 5. A thin surface layer of offshore and equatorward flow is expected near the surface with the flow diverging away from the cold core, but skewed equatorward by the alongshore component of surface flow. A series of idealized zones can be identified which describe the conditions we expect to observe as the water flows offshore. Zone I is the zone of intense upwelling. The water is generally cold, rich in nutrients and low in phytoplankton biomass. The phytoplankton are likely to be growing at less than optimal rates and zooplankton are unlikely to be abundant. As the water progresses outward (Zone II) it begins to warm, and the phytoplankton begin to adapt to both high nutrient concentrations and near-surface light intensities, increasing their rates of photosynthesis and growth. As

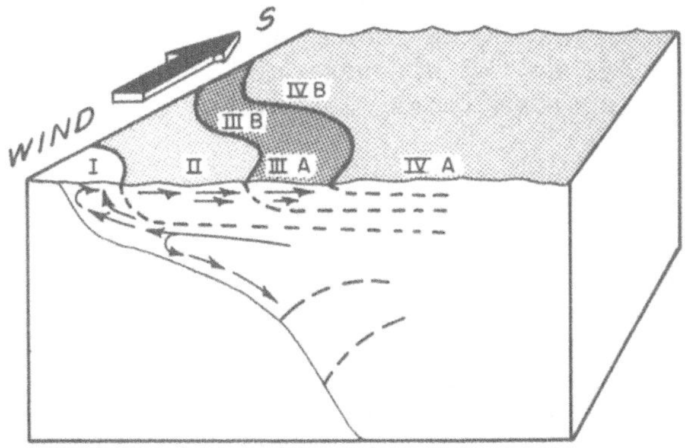

Fig. 5. A schematic for the idealized structure of an upwelling center in the northern hemisphere. The coastline is on the left. The arrows are hypothetical streamlines of the cross-shelf flow. The details of the labelled zones are in the text.

a result, the nutrient concentrations are declining and phytoplankton biomass is likely to be increasing. As the water moves into Zone III, nutrients are beginning to become depleted and phytoplankton biomass is likely to attain its maximum concentrations. Zooplankton biomass may be relatively high in this region, taking advantage of the availability of abundant prey. Physically, this zone may be a convergent frontal region acting to increase the local concentration of phytoplankton. Diatoms are likely to be most abundant here. However, dinoflagellates and possibly *Mesodinium rubrum* may also occur in great abundance near the surface, particularly if this is a frontal zone or is well stratified. In regions of narrow continental shelves, Zone III is likely to lie near or beyond the shelf break. Beyond this region, in Zone IV, at least one of the major plant nutrients is likely to become depleted, phytoplankton biomass declines relative to the preceeding zone and zooplankton abundance is also likely to decline. The relative abundance of diatoms in the phytoplankton community diminishes due to nutrient limitation, sinking, and grazing. Gradually the transition into more oceanic conditions will occur. Decreases in the rates of growth and photosynthesis of the phytoplankton will result from depletion of nutrients and probably to the changing composition of the community. Nutrient dependency in this region will be primarily on regenerated rather than freshly upwelled nutrients.

To begin to test the hypotheses and to evaluate the suitability of the Point Conception, California area as a field site for further refinement of upwelling centers, the OPUS-1 field program took place in the spring of 1981 (Fig. 6). The program was carried out aboard the RV "Velero IV" in three 2-week legs beginning March 2 and continuing through April 15, 1981. Aircraft support, including remote sensing of sea surface temperature, was provided by the National Center for Atmospheric Research Queen Air for the period of March 18-31, 1981. The sea surface temperature and wind records of the NODB buoy located about 20 km of Punta Santa Maria and 30 km north of Point Arguello are included in the discussion. The results that follow are from these various observational platforms.

METHODS

The methods used during the March-April, 1981 OPUS cruise were similar to those used in other expeditions to study such features. However, the major effort was concentrated on the upwelling center rather than on the whole upwelling area and on determining the interactions of the center with adjacent regions. Hydrographic lines were established to sample down the estimated axis of the upwelling structure and to provide a three-dimensional description of the features. Temperature, salinity, and density were profiled with a Plessy 9040 CTD. These casts were made to 350 meters when the water depth allowed. Water sampling to depths of 200 meters was performed with

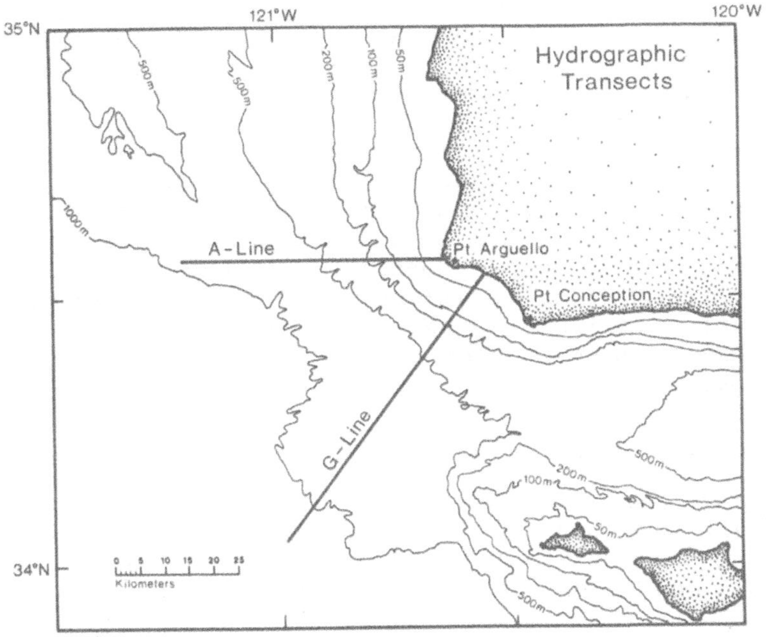

Fig. 6. The OPUS study area in the region of Point Conception and Point Arguello, California.

7-liter Niskin bottles. Nutrient chemistry was performed on these samples with an Autoanalyzer as described by Codispoti et al. (1976). Salinity was measured using a Guildline Autosal salinometer. Dissolved oxygen was measured using the Winkler method (Carpenter, 1964). Chlorophyll a and phaeopigments were measured by the fluorometric technique described by Holm-Hansen et al. (1967).

Underway mapping was usually performed at night. The ship traversed a zig-zag path through the area crossing the major gradients of the upwelling structure. Water was continuously sampled from approximately 3 meters depth; chlorophyll fluorescence was monitored with a Turner 111 fluorometer and nutrients were sampled using an Autoanalyzer as described above. Batch samples were taken from the underway system every 15 to 30 minutes for measurement of the extracted chlorophyll.

OBSERVATIONS FROM THE POINT CONCEPTION-POINT ARGUELLO FEATURE

Wind Conditions

Normally, March and April represent the time of year when the monthly mean large scale winds near Point Conception change from downwelling favorable (poleward) to upwelling favorable (equatorward). This transition corresponds to the seasonal strengthening of the high pressure over the north Pacific. Our observation period of March 2-April 15, 1981, appears to have come very near this transition (Fig. 7a). The wind stress was generally equatorward until March 18, poleward from March 18-22, and strongly equatorward until the end of the observations. Throughout the entire period, fluctuations associated with atmospheric frontal passages were observed. Comparison of the low pass filtered NOAA buoy winds with various short coastal wind records show that low frequency wind fluctuations were extremely coherent and of nearly the same magnitude in the entire region north of Point Conception. South of this promontory, the north-south wind component was severly distorted by the local topography, and the alongshore wind stress record in Fig. 7a is not representative.

Observed Structure of the Upwelling Features

The average sea surface temperature, derived from underway maps, for each of the three legs of the cruise (Fig. 8) demonstrate that we did, in fact, capture the development of a well-defined upwelling feature. The most obvious signal is the decline in the surface tem-

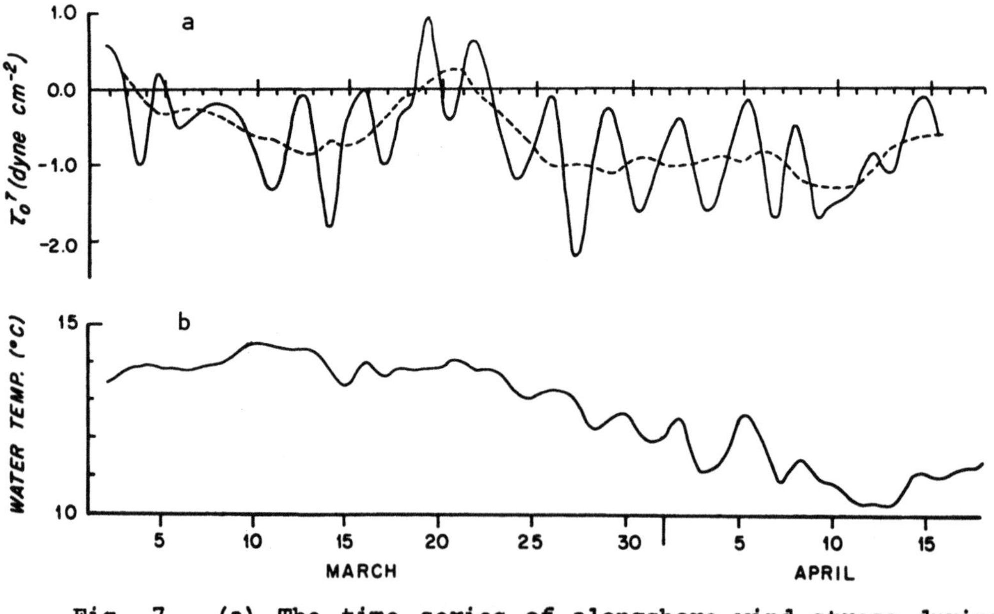

Fig. 7. (a) The time series of alongshore wind stress during the OPUS cruise. The data was taken from NOAA buoy located off Purissima Point, just north of the study area. Positive wind stress is northward.

(b) The time series of sea surface temperature recorded at the NOAA buoy for the period of the OPUS field program.

perature over the six-week period. The minimum in mean temperature for Leg 1 was 12.9°C, and by Leg 3 the minimum in mean temperature in the area was 9.4°C. This is consistent with the sea surface temperature record at the buoy off Punta Santa Maria (Fig. 7b).

A weakly defined nearshore temperature minimum was observed between Points Arguello and Conception during Leg 1, however, the overall structure during this period seems to have been related to advection from north of Point Arguello. Again on Leg 2 the overall input of colder water seems to have come from north of Point Arguello. The average temperature is only slightly less than during Leg 1. In contrast, the mean structure during Leg 3 contains a well-defined cold center just south of Point Arguello. The structure appears to continue to show some influence of advection from the north; however, it extends basically to the south with a slight skew to the east. It can also be seen that there are relatively strong temperature gradients to either side of the apparent upwelling center.

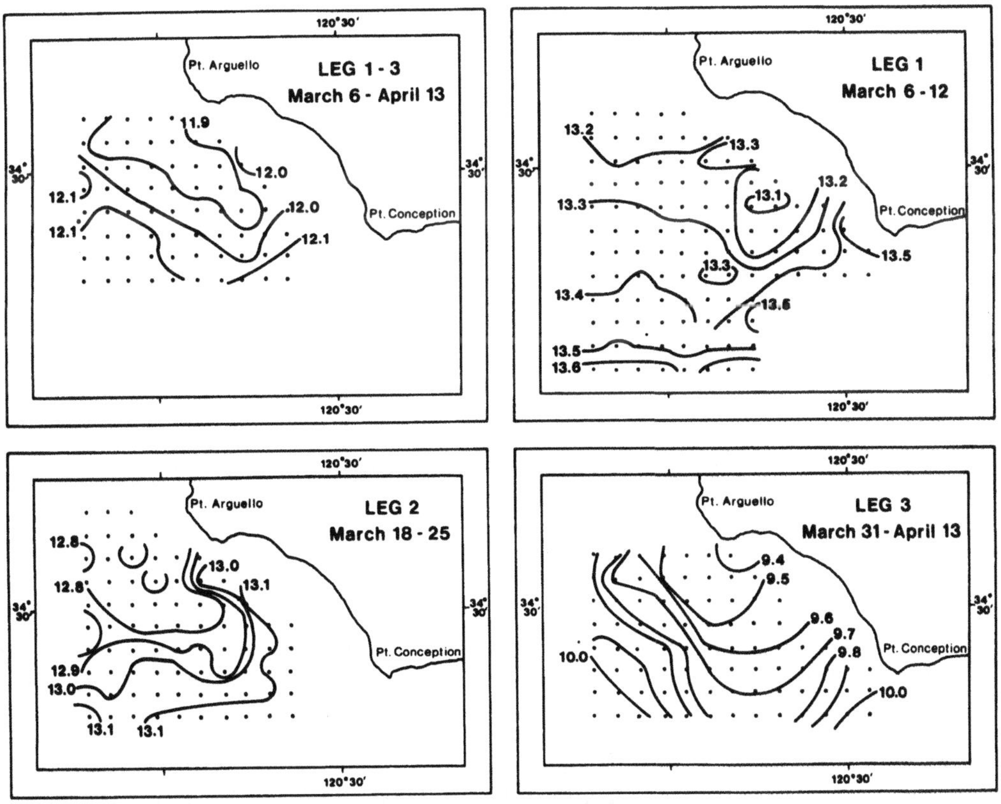

Fig. 8. Mean sea surface temperature maps from the Point Conception region during the OPUS expedition. The map in the upper left represents the mean for the whole cruise. The other three maps are for each of three 2-week legs; 1981.

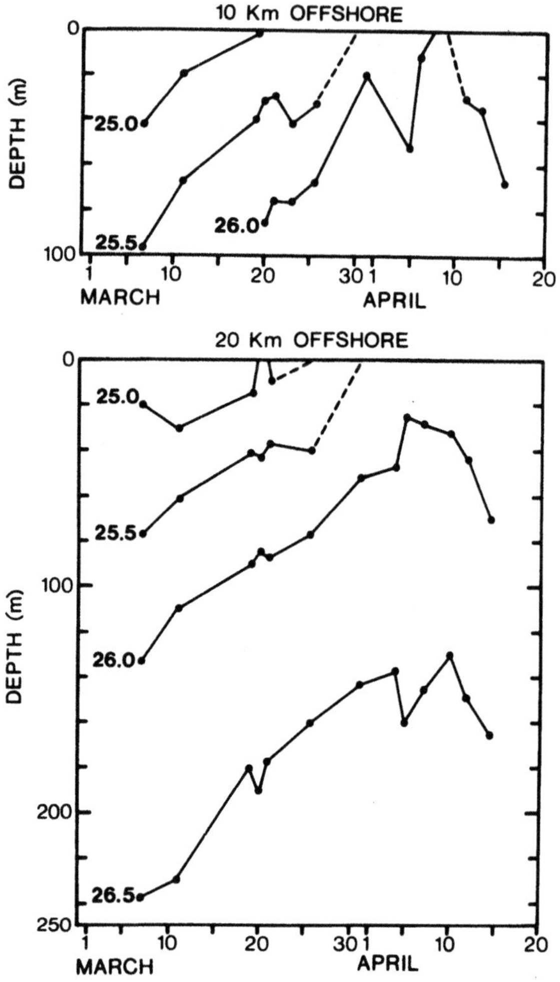

Fig. 9. The isopycnal depth time series at 10 and 20 km offshore along the G-line, which is perpendicular to the coast and midway between Points Conception and Arguello; 1981.

Additional support for the presence of local upwelling southeast of Point Arguello, and not just the advection of colder water from the north, is given by the time series of isopycnal depths from the G-line at 10 and 20 kilometers off the coast (Fig. 9). It is clear from both of these time series that the isopycnals became monotonically more shallow over the period of March 7 to April 12, 1981. The wind stress for this period was to the south with the exception of March 19-22. The period of northward wind stress was reflected in the near-surface isopycnal records, as well as in the near surface horizontal structure.

The Seasonal Spin-up of the Point Arguello Upwelling Center

Early, weak upwelling. In early March, near the beginning of the OPUS observational period, weak upwelling was apparent in the vicinity of Point Arguello. On March 11, 1981, both a section and an underway map were made. Alongshore wind stress (Fig. 7a) showed an equatorward maximum on March 10 which then declined somewhat on March 11. A hydrographic section was made along the C-line (northwest of and parallel to the G-line) on March 11. The isopycnals were generally flat below 60 m, but showed some elevation near the coast above 50 m depth. The isopleths of nutrients and chlorophyll generally show the same subsurface pattern, except that chlorophyll has a near-surface maximum at about 35-40 km from the coast.

Period of reversed wind direction. During 18-22 March 1981 the alongshore wind direction reversed and stresses in the northward direction were up to 1 dyne/cm^2 (Fig. 7a). Upwelling was not apparent against the coast in the underway map from the night of March 20-21 (Fig. 10), but a cold spot did exist offshore south of Point Arguello. The temperature-nutrient signature of this feature appears to be characteristic of recently upwelled water, i.e., nutrients were negatively correlated with temperature. The chlorophyll distribution, on the other hand, was opposite to that expected for an upwelling feature. That is, the chlorophyll concentration was higher in the colder region as were nutrients, and temperature was relatively low. A set of CTD sections from March 21 (Fig. 11) shows that this feature was in fact a dome which was resolvable to about 50 m depth. From the underway map and set of sections, it is not clear what the life time or origin of this particular feature is, but the presence of the higher chlorophyll in the center suggests that the feature originated at least 2 or 3 days earlier. Underway maps on the following two nights also showed a cold, higher nutrient center offshore from Point Arguello.

Period of sustained equatorward wind direction. Winds had been blowing steadily equatorward since 22 March and were as strong as 1.6 dynes/cm^2 on 30 March. Both a CTD section and an underway map were made on 31 March. The section along the G-line showed very strong upwelling as indicated by the upward tilt of the isopleths of all variables near the coast (Figs. 12 and 13). The density section shows that the thermocline (lying between the 25.5 and 26.0 sigma-T isopycnals) had reached the surface and moved a few kilometers offshore. Nutrient distributions showed similar patterns with maximum nitrate concentrations of more than 25 μg-at/ℓ nearshore. Silicate was in excess of 30 μg-at/ℓ near the coast. Chlorophyll near the surface was positively correlated with temperature, increasing from less than 0.25 μg/ℓ nearshore to about 2 μg/ℓ at 25 km offshore.

The map from that night (Fig. 14) was confined to within 15 km of the coast. The map showed a temperature minimum, 10.5°C with cor-

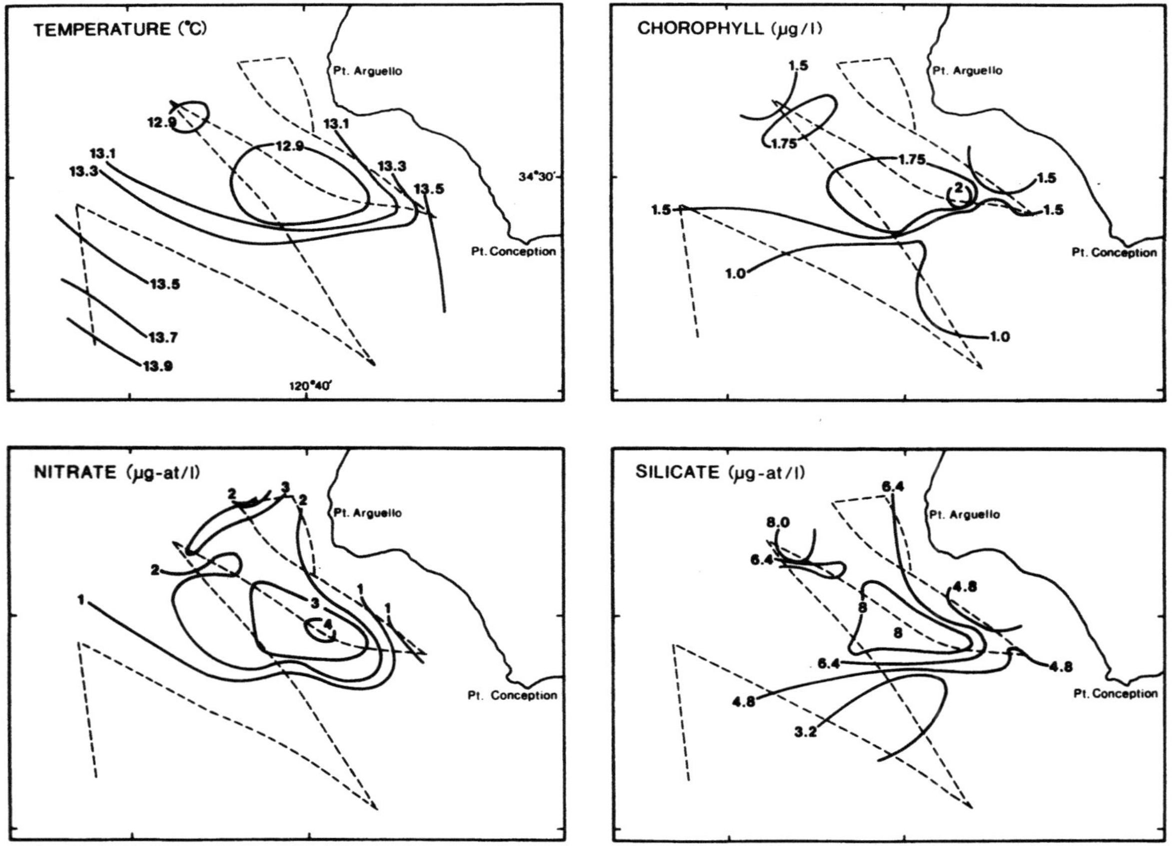

Fig. 10. The surface map results for the night of 20-21 March 1981. The solid lines are the contours for the specified variable and the dashed lines show the ship track.

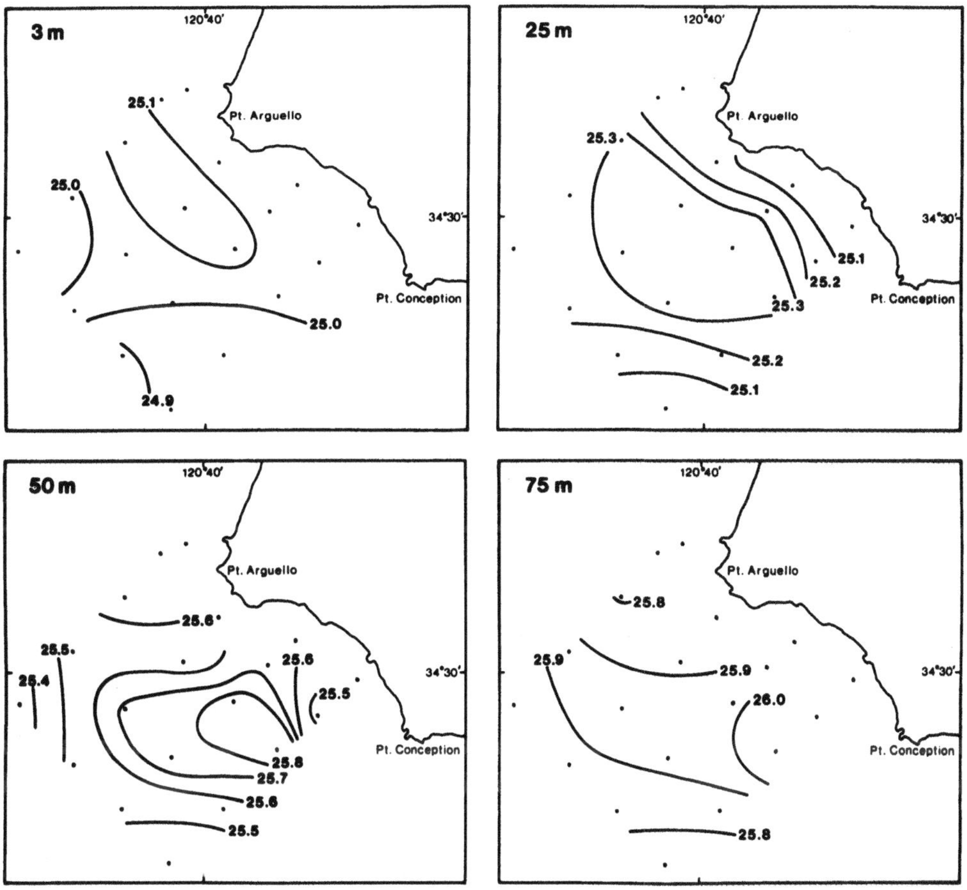

Fig. 11. Maps of isopycnals at 4 depths from 21-22 March 1981. The maps are derived from 5 hydrographic sections between Point Arguello and Point Conception. Station positions are indicated by dots.

responding high nutrient concentrations (nitrate was 25 µg-at/ℓ and silicate was 32 µg-at/ℓ) occurring between Points Arguello and Conception. To the southeast of this temperature minimum, a very sharp temperature front was observed off Point Conception. The maximum temperature gradient across this front was 0.65°C per kilometer.

Local wind continued to be equatorward into April, although the fluctuations associated with frontal passages also continued (Fig. 6). The map on April 4-5 (Fig. 15) showed again the presence of a cold center below Point Arguello with sea surface temperatures below 9.5°C and nitrate concentrations greater than 24 µg-at/ℓ at the center. Chlorophyll concentration was negatively correlated with nutrient concentration, but the ratio of chlorophyll increase to nitrate

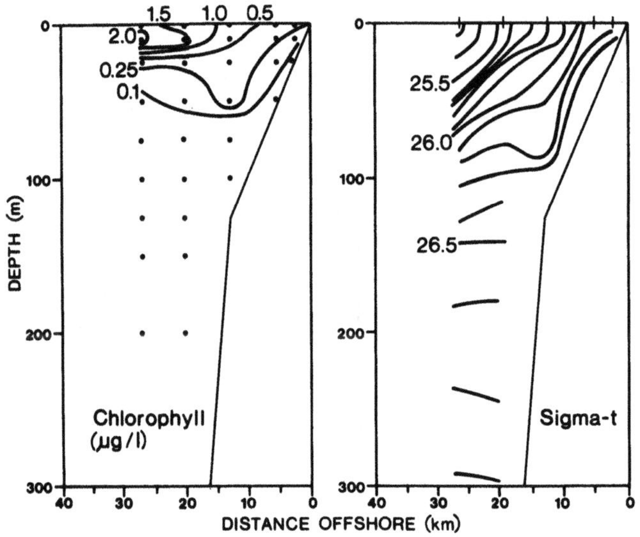

Fig. 12. Density and chlorophyll profiles from the G-line on March 31, 1981.

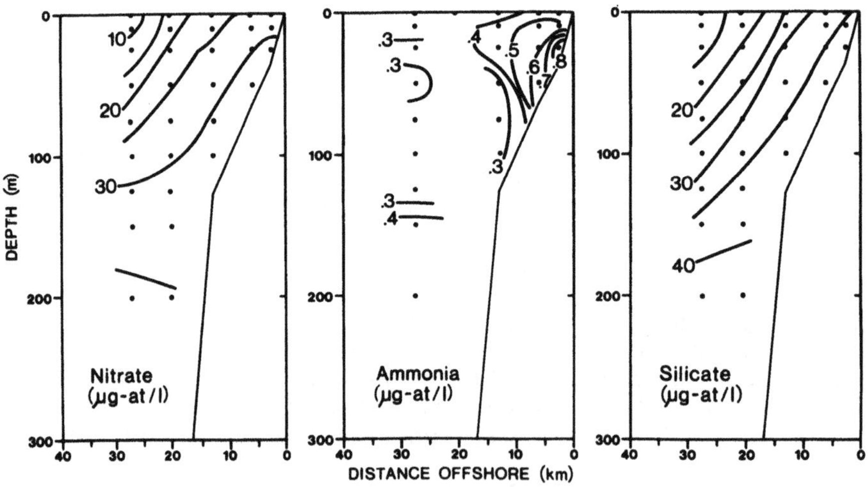

Fig. 13. Nutrient profiles from the G-line on March 31, 1982.

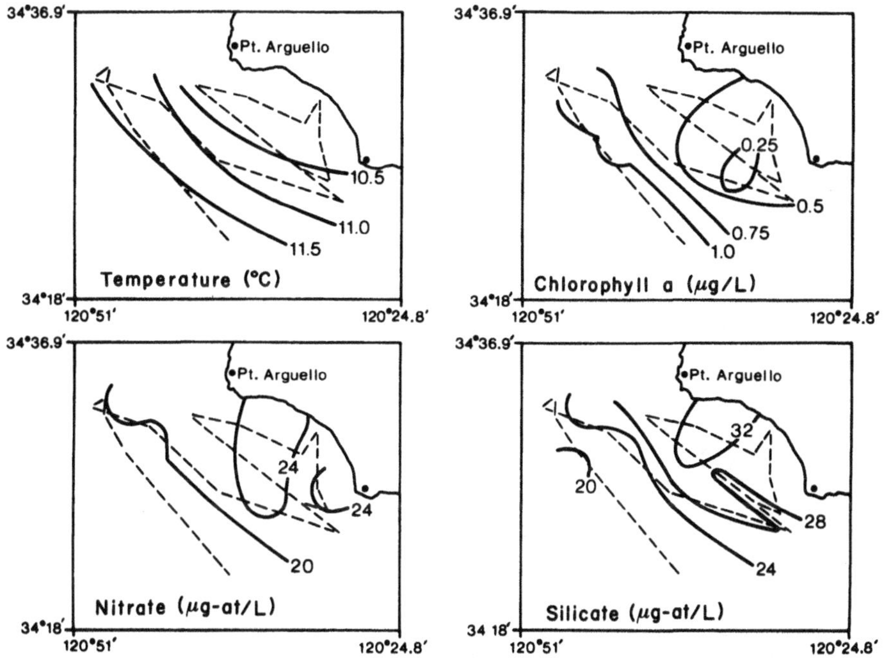

Fig. 14. Results of underway mapping on the night of 31 March 1981. Details are the same as in Fig. 10.

decrease was much less than the "optimal" ratio of about 1 μg chl./ℓ μg-at-N.

The first indication of a significant phytoplankton bloom in response to the observed high nutrient concentrations in the upwelling was on 8 April in the Santa Barbara Channel north of Santa Rosa Island. Later efforts to resolve the extent of the high chlorophyll region showed the high chlorophyll water extending west toward Point Arguello. By 12-13 April not only were there significant temperature and nutrient signatures associated with the upwelling center, but also a discernible chlorophyll signature. As can be seen in Fig. 16, there was a core of cold water extending to the southeast from Point Arguello. High chlorophyll values occur to the southeast of Point Arguello and just below Point Conception. By looking at the nitrate-temperature and the chlorophyll-temperature regressions (Fig. 17) one can see that the temperature range of the highest chlorophyll concentrations occurred between 10.0 to 10.8°C, where the points of the nitrate-temperature regression were less tightly correlated than for both colder and warmer temperatures, suggesting the influence of patchy phytoplankton growth on this water. This is also true for two high chlorophyll points (Fig. 17) at and above 11.0°C. Both of these points co-occurred with nitrate values which fell below the other observations in this temperature range.

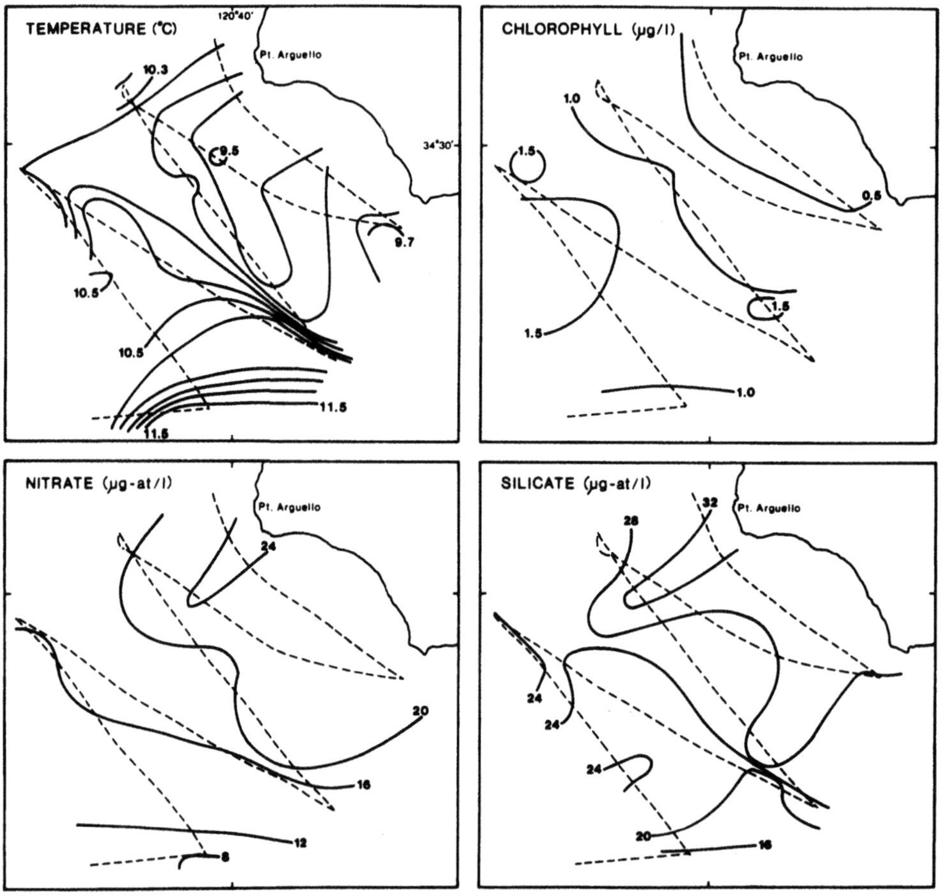

Fig. 15. Results of underway mapping on the night of 4 April 1981.

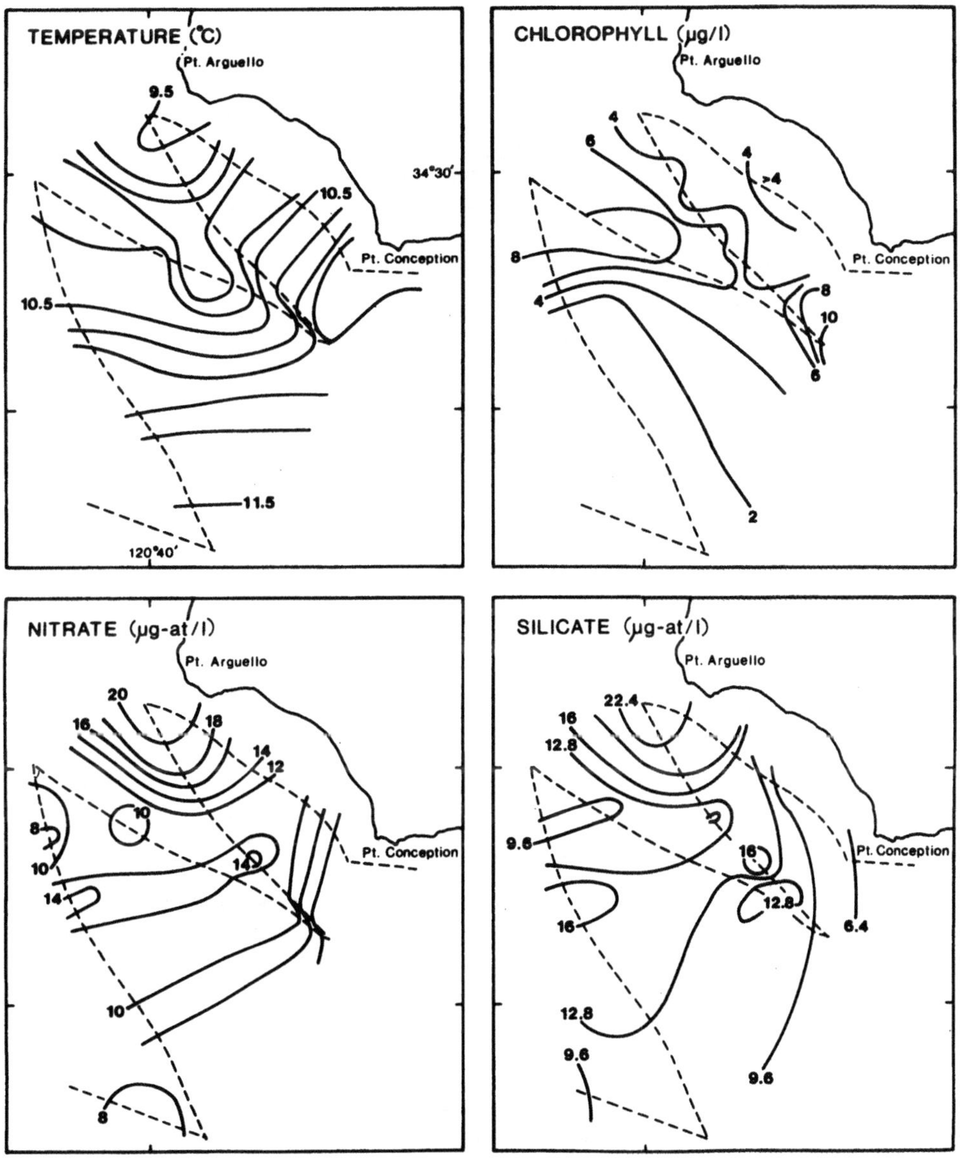

Fig. 16. Results of underway mapping on the night of 12 April 1981.

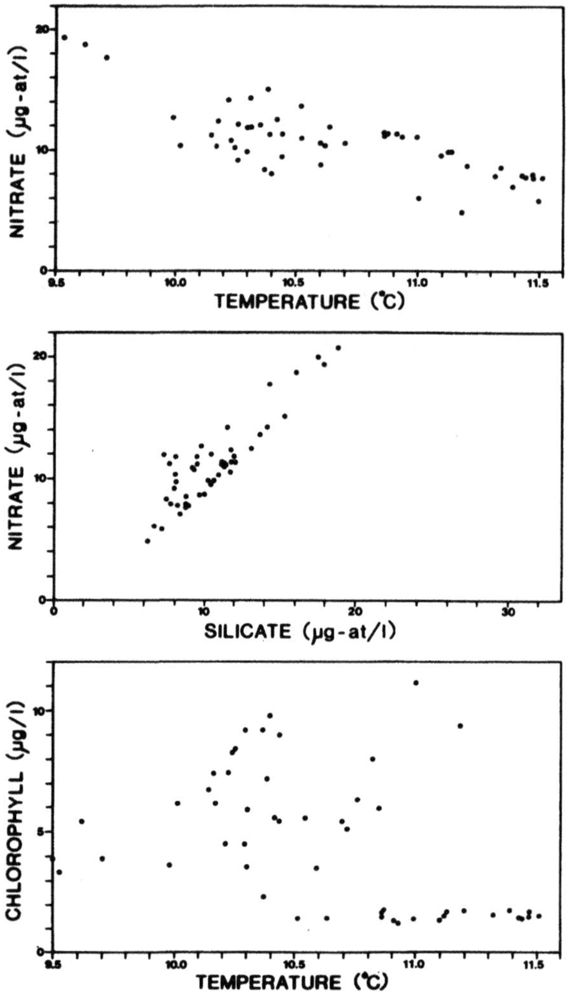

Fig. 17. Nitrate plotted with respect to temperature and silicate for the map on 12 April 1981. Chlorophyll is also plotted with respect to temperature.

DISCUSSION

During the OPUS-1 cruise a strong, very low frequency upwelling event was observed. The rising of isotherms and formation of the surface upwelling front took place during this period. A cold, nutrient-rich upwelling center developed south of Point Arguello. Prior to April 8, despite the high nutrient concentrations in the upwelled water, the maximum chlorophyll concentrations in the area were only about 3 μg/ℓ. Late in the observation period, April 8 and after, chlorophyll concentrations increased to values as high as 10-20 μg/ℓ.

The results from this pilot study do not allow an exhaustive comparison with the upwelling center off Peru at 15°S, however some basic comparisons can be made. The surface isotherms surrounding the cold center at 15°S generally closed against the coast, i.e., the system was "closed" at the surface. However, off Point Arguello the isotherms often did not intersect the coast to the west and north, suggesting continuity of the structure at the surface at Point Arguello with the upwelling occurring to the north.

In the Peru system, near-surface silicate concentrations have been observed to decrese to undetectable levels while measurable concentrations of nitrate are still present (Dugdale, 1972; Brink et al., 1981). This suggests the potential limitation of diatom growth by low silicate concentrations. Further support for silicate limitation in the Peru system has been provided by physiological measurements. Dugdale et al. (1981) have shown that nitrate uptake appeared to be regulated by ambient silicate concentrations off Peru in March-May, 1977. However, off Point Conception nitrate concentrations fell below measureable concentrations while significant concentrations of silicate (6-8 μg-at/ℓ) remained. These observations suggest that nitrogen limitation of phytoplankton growth is likely to be the major form of nutrient limitation in the Point Conception-Point Arguello area.

The local wind direction along the Peru coast near 15°S is favorable for upwelling, i.e., equatorward, throughout the year. Near Point Conception, the monthly mean wind direction is also favorable for upwelling throughout the year. However, during the winter months, frequent reversals in the wind direction occur resulting in episodic upwelling. Beginning in March, the short time-scale and wind stress variations result in fewer direction reversals and upwelling becomes a steadier, more seasonably persistent process.

CONCLUSIONS

It is clear from the early results obtained off Point Conception, that there exists a persistent upwelling center having the following characteristics:

a. sea surface temperature lower than the surrounding waters;
b. surface nutrient concentrations higher than in the surrounding waters;
c. chlorophyll concentrations lower than in the surrounding waters. These characteristics are consistent with those observed in the Peru center at 15°S.

The strength of the upwelling and the size of the upwelling center appear to be related to the local winds. This is definitely true for the Peru observations. It also appears to be true for the Point

Conception-Point Arguello area, but it has not been statistically resolved at this point.

The structure of the upwelling centers is complex. The Peru system is apparently strongly influenced by the Peru undercurrent and the layered structure of the cross-shelf flow adds complexity to the cross-shelf dynamics. Analogous concepts are also likely to be true for the Point Conception area. In the Point Conception area, added complexity may be added by the apparent openness of the upwelling feature to the north.

In regions of narrow shelves, such as both Point Conception and the Peru center at 15°S, the maximum phytoplankton biomass and primary productivity often appear to be located at or beyond the shelf break.

While we do not have current meter observations of the Point Conception area, the observations from Peru and the likely vertical complexity of the currents off Point Conception suggest that the advection of particles is not straightforward.

a. A portion of the phytoplankton are possibly carried shoreward by the upwelling water as they sink.
b. The alongshore distribution of the phytoplankton below the surface layer is likely to be smeared by the vertical current structure. That is, the coastal upwelling center may have an enhanced nutrient supply and associated increased productivity resulting in a discernible instantaneous surface signature, but may not leave as distinct a sedimentary "footprint".

ACKNOWLEDGEMENTS

We gratefully acknowledge the support of the U.S. National Science Foundation through grants OCE 80-21029 (Stuart), OCE 80-21028 (Brink) and OCE 80-22631 (Jones, Dugdale and Blasco). Observations from the NOAA buoy were kindly provided to D. Stuart by the NOAA Data Buoy Office, Bay St. Louis. C. Platt assisted with the manuscript preparation and D. Hadley drafted the figures.

REFERENCES

Andrews, W.R.H. and Hutchings, L., 1980, Upwelling in the Southern Benguela Current, Progress in Oceanography, 9:1-81.

Barton, E.D., Huyer, A. and Smith, R.L., 1977, Temporal variation observed in the hydrographic region near Cabo Corbeiro in the northwest Africa upwelling regime, February-April, 1974, Deep-Sea Research, 24:7-24.

Boyd, C.M. and Smith, S.L., in press, Interactions of plankton populations in the Peruvian upwelling region, Deep-Sea Research.

Breaker, L.C. and Gilliland, R.P., 1981, A satellite sequence on upwelling along the California Coast, in: "Coastal Upwelling," F.A. Richards, ed., Coastal and Esutarine Sciences 1, American Geophysical Union, Washington, 87-94.

Brink, K.H., Jones, B.H., Van Leer, J.C., Mooers, C.N.K., Stuart, D.W., Stevenson, M.R., Dugdale, R.C., and Heburn, G.W., 1981, Physical and biological structure and variability in an upwelling center off Peru near 15 degrees S. during March, 1977, in: "Coastal Upwelling," F.A. Richards, ed., Coastal and Estuarine Sciences 1, American Geophysical Union, Washington, 473-495.

Carpenter, J.H., 1964, The Chesapeake Bay Institute technique for the Winkler dissolved oxygen method, Limnology and Oceanography, 10:141-143.

Codispoti, L.A., Bishop, D.D., Friebertshauser, M.A., and Friederich, G.E., 1976, JOINT-II RV "Thomas G. Thompson" Cruise 108, Bottle Data, April-June 1976, Coastal Upwelling Ecosystems Analysis Data Report 35, 370 pp.

Cruzado, A. and Salat, J., 1981, Interaction between the Canary Current and the bottom topography, in: "Coastal Upwelling," F.A. Richards, ed., Coastal and Estuarine Sciences 1, American Geophysical Union, Washington, 167-175.

Dugdale, R.C., 1972, Chemical oceanography and primary productivity in upwelling regions, Geoforum, 2:47-61.

Dugdale, R.C., Goering, J.J, Barber, R.T., Smith, R.L., and Packard, T.T., 1977, Denitrification and hydrogen sulfide in the Peru upwelling system during April, 1977, Deep-Sea Research, 24:601-608.

Dugdale, R.C., Jones, B.H., MacIsaac, J.J., and Coering, J.J., 1981, Adaptation of nutrient assimilation, in: "Physiological Basis of Phytoplankton Ecology," T. Platt, ed., Canadian Bulletin of Fisheries and Aquatic Sciences, 210:234-250.

Gunther, E.R., 1936, A report on oceanographical observations in the Peru coastal current, Discovery Reports, 13:107-276.

Hart, T.J. and Currie, R.I., 1960, The Benguela Current, Discovery Reports, 31:123-289.

Holm-Hansen, O., Lorenzen, C.J., Holmes, R.W., and Strickland, J.D.H., 1967, Fluorometric determinaiton of chlorophyll, Journal Conseil Permanente International pour l'Exploration de la Mer, 30:3-15.

Hutchings, L., 1981, The formation of plankton patches in the Southern Benguela Current, in: "Coastal Upwelling," F.A. Richards, ed., Coastal and Estuarine Science 1, American Geophysical Union, Washington, 496-506.

Jones, B.H., 1977, "A Spatial Analysis of the Autotrophic Response to Abiotic Forcing in Three Upwelling Ecosystems: Oregon, Northwest Africa, and Peru," Ph.D. Dissertation, Duke University, Durham, North Carolina, 262 pp.

Kundu, P.K. and Allen, J.S., 1976, Some three-dimensional characteristics of low frequency current fluctuations near the Oregon coast, Journal of Physical Oceanography, 6:181-199.

Ryther, J.H., Menzel, D.W., Hulburt, E.M., Lorenzen, C.J., and Corwin, N., 1966, Scientific results of the Southeast Pacific Expedition, "Anton Bruun" Report No. 4, Texas A&M Press, 12 pp.

Smith, S.L. and Codispotti, L.A., 1980, Southwest monsoon of 1979; chemical and biological response of Somali coastal waters, Science, 209:597-600.

Sverdrup, H.U. and Allen, W., 1939, Distribution of diatoms in relation to the character of water masses and currents off Southern California in 1938, Journal of Marine Research, 2:131-144.

Walsh, J.J., Kelley, J.C., Dugdale, R.C., and Frost, B.W., 1971, Gross features of the Peruvian upwelling system with special reference to possible diel variation, Investigacion Pesqueras, 35:25-42.

Walsh, J.J., Kelley, J.C., Whitledge, T.E., MacIsaac, J.J., and Huntsman, S.A., 1974, Spin-up of the Baja California upwelling system, Limnology and Oceanography, 19:553-572.

Walsh, J.J., Whitledge, T.E., Kelley, J.C., Huntsman, S.A., and Pillsbury, R.D., 1977, Further transition states of the Baja California upwelling ecosystem, Limnology and Oceanography, 22: 264-280.

Zuta, S., Rivera, T. and Bustamante, A., 1978, Hydrologic aspects of the main upwelling areas off Peru, in: "Upwelling Ecosystems," R. Boje and M. Tomczak, eds, Springer-Verlag, Berlin, 235-257.

NUTRIENT MAPPING AND RECURRENCE OF COASTAL UPWELLING CENTERS BY SATELLITE REMOTE SENSING: ITS IMPLICATION TO PRIMARY PRODUCTION AND THE SEDIMENT RECORD

Eugene D. Traganza, Vitor M. Silva, Dana M. Austin,
Walter L. Hanson, and Sherman H. Bronsink

Department of Oceanography
Naval Postgraduate School
Monterey, California 93940, U.S.A.

ABSTRACT

Satellite infrared (IR) images of the eastern North Pacific show surface thermal features which appear to be mesoscale eddies and fronts within the California Current and coastal upwelling zone. Coordinated satellite and shipboard measurements show that some of these are produced by interaction between the California Current and wind-induced upwelling near points and capes along the coastal boundary. The process of coastal upwelling as viewed from space is unlike the simple structure expected from classical concepts and may develop a cyclonic circulation, an anticyclonic circulation, or surface flows which extend far offshore as giant cold water plumes embedded in the California Current. Observations of a cyclonic upwelling system off Point Sur, California, show that these recurrent injections of nutrient-rich, cold waters produce the same distinctive structure in the nutrient field. When present, microplankton concentrate in the sharp thermal and nutrient fronts that define these systems. The combined effect of the distinctive advective process and redistribution of nutrients is a significant factor in determining the distribution of primary production and biological patchiness along the coast and in the California Current system. Since ships are too slow to synoptically study these large and rapidly changing upwelling systems, phytoplankton distribution and the nutrient structure of the sea surface is inferred from satellites. By calibrating satellite infrared data against *in situ* nutrient measurements, sea surface nutrient maps can be produced so that estimates of the true nutrient distribution and potential sites of biological activity can be made. Satellite ocean color imagery shows this relationship between chlorophyll-biomass blooms and sharp nutrient gradients. This upwelling center appears to operate like a "natural chemostat." Such recurrent upwelling at favored locations along the coastal boundary may be reflected in the sediment record.

INTRODUCTION

Satellite images show recurring upwelling fronts and eddies associated with topographic features along the California coast. The high biological production resulting from these recurring systems may be reflected in the sediment record. Here we report the use of infrared satellite data to map their sea surface nutrient content, and color imagery to show that phytoplankton blooms occur near the sharp nutrient gradients which delineate their boundaries. The coldest and most nutrient-rich upwellings frequently appear as cyclonic systems off Point Sur and Cape Mendocino. As viewed from space, the Point Sur upwelling system appears to form by the intrusion of relatively warm California Current waters (possibly eddies) into the coastal upwelling zone where they interact with cold upwelling water to form a cyclonic circulation characterized by strong thermal (and nutrient) gradients or fronts. Both cyclonic and anticyclonic systems recur equatorward and poleward of points and capes. Occasionally surface flows extend hundreds of kilometers seaward from Point Sur and Cape Mendocino with the appearance of "giant plumes" (Traganza, Conrad and Breaker, 1981) embedded in the mean flow of the slow ($\sim$ 25 cm s^{-1}) equatorward flowing California Current. These different frontal systems produce a large scale redistribution of nutrients and are therefore important in determining the production and distribution of biomass along the coastal boundary and in the California Current. It is possible that these patterns of distribution are reflected in the sediment record. Off Point Sur, recurrent injections of cold, nutrient-rich waters upwell from depths of tens to hundreds of meters, frequently become cyclonic and produce the same cyclonic structure in the surface thermal and nutrient fields. These systems can form in a few days and become tightly curled cyclonic features in less than 48 hours as shown in Fig. 1 from 9 and 11 June, 1980.

PROCEDURE

Satellite infrared imagery of the sea surface temperature (SST) field, ocean color imagery of "chlorophyll" biomass and underway shipboard measurements of temperature, fluorescence and nutrients were obtained during this upwelling event off Point Sur, California, viz., 9-11 June, 1980. Estimates of microplanktonic (0.5 to 200 μm) phytoplankton based on chlorophyll and total microplankton (bacteria, phytoplankton and microzooplankton) based on adenosine triphosphate (ATP) were made along with the physical and chemical data (Bronsink, 1980; Hanson, 1980). Surface nitrate and phosphate concentrations were determined every 2 minutes by continually sampling seawater pumped from a keel intake (at 2.5 m depth) to the shipboard laboratory. Temperature of the intake water was sensed by a thermistor and recorded continuously on a strip chart recorder. The equipment was calibrated against thermometer readings of the sea surface. Vertical temperature profiles were obtained by expendable bathythermograph (XBT) drops at intervals of 2.5 to 5 km along a cross shelf cruise track. Surface concentrations of phytoplankton biomass were estimated from continuous measurement of the fluorescence of seawater pumped to the shipboard laboratory. Discrete samples (250 to 275 ml) were taken every 0.5 hr to calibrate the fluorescence record in terms of plant pigment, principally chlorophyll a. This combination of satellite imagery, frequent *in situ* sampling, and vertical temperature profiles from XBT drops made it possible to locate this coastal upwelling feature and transect it from different directions.

Satellite derived sea surface temperatures were corrected by applying radiative transfer theory and linear correlations with *in situ* temperatures. The corrected SST field was then used to infer the sea surface nutrient field from strong linear correlation between *in situ* nutrient concentrations and temperatures (Silva, 1982). Following earlier investigators (Maul et al., 1978), atmospheric water vapor corrections were calculated by comparing SST gradients observed by the satellite with *in situ* SST gradients measured aboard the RV "Acania". The corrected satellite SST's agreed with *in situ* values within 0.5°C (SD) when tested against temperature data from other portions of the ship's track. The satellite thermal images (Fig. 2) were converted to thermal and nutrient maps of the sea surface (Figs. 3, 4 and 5) within the upwelling feature off Point Sur. When the inferred nutrient concentrations were tested against nutrient concentrations as sampled by the ship in other representative regions of the upwelling feature, standard deviations of 6% and 4% were computed for nitrate and phosphate, respectively.

Infrared Radiometry

The usefulness of satellite IR observations to infer the sea surface temperature depends on the ability of the radiometer to view the sea with little error introduced by the atmosphere. This error, which can be substantial even in a relatively clear air, is mainly induced by atmospheric attenuation of the infrared radiance due to absorption by water both as vapor and as clouds.

The cloud-free satellite observations, present during this study, minimized the atmospheric errors. Errors can originate at the sea surface due to the very strong absorption of infrared photons. As a consequence, the IR radiation measured by the radiometer origi-

Table 1. Calibration table for satellite images from 9 and 11 June 1980†

Color Grade		AVHRR Count		Satellite Temp (°C)		Corrected Temp (°C)		Nitrate (μM)		Phosphate (μM)	
K	L	103	105	12.90	12.03	14.92	15.17	*	*	0.69	0.48
J	K	104	106	12.47	11.59	14.31	14.48	*	*	0.85	0.68
I	J	105	107	12.03	11.15	13.68	13.80	0.73	*	1.02	0.87
						$\overline{13}.37$		$\overline{1}.83$		$\overline{1}.11$	
H	I	106	108	11.59	10.70	13.06	13.10	2.93	1.82	1.19	1.07
						$\overline{12}.75$	$\overline{12}.76$	$\overline{4}.05$	$\overline{3}.27$	$\overline{1}.28$	$\overline{1}.17$
G	H	107	109	11.15	10.26	12.43	12.41	5.16	4.76	1.36	1.27
						$\overline{12}.11$	$\overline{12}.06$	$\overline{6}.30$	$\overline{6}.26$	$\overline{1}.45$	$\overline{1}.37$
F	G	108	110	10.70	9.81	11.79	11.70	7.44	7.79	1.54	1.48
						$\overline{11}.48$	$\overline{11}.35$	$\overline{8}.56$	$\overline{9}.28$	$\overline{1}.63$	$\overline{1}.58$
E	F	109	111	10.26	9.36	11.16	11.00	9.67	10.78	1.71	1.68
						$\overline{10}.84$	$\overline{10}.65$	$\overline{10}.82$	$\overline{12}.27$	$\overline{1}.80$	$\overline{1}.78$
D	E	110	112	9.81	8.91	10.51	10.29	11.96	13.80	1.88	1.88
						$\overline{10}.19$	$\overline{9}.94$	$\overline{13}.11$	$\overline{15}.29$	$\overline{1}.97$	$\overline{1}.98$
C	D	111	113	9.36	8.46	9.87	9.58	14.25	16.83	2.05	2.09
						$\overline{9}.55$	$\overline{9}.22$	$\overline{15}.40$	$\overline{18}.36$	$\overline{2}.14$	$\overline{2}.19$
B	C	112	114	8.91	8.00	9.22	8.85	16.55	19.94	2.23	2.30
						$\overline{8}.90$	$\overline{8}.49$	$\overline{17}.70$	$\overline{21}.48$	$\overline{2}.32$	$\overline{2}.40$
A	B	113	115	8.46	7.54	8.57	8.12	18.85	23.05	2.40	2.51
							$\overline{7}.76$		$\overline{24}.59$		$\overline{2}.61$
	A		116		7.00		7.39		26.17		2.72

†First and second columns under each heading are for June 9 and 11 respectively.

*Undetectable

$^{-}$Mean temperature or nutrient concentration of adjacent color grades which are assigned to isopleths (Figs. 3, 4, and 5).

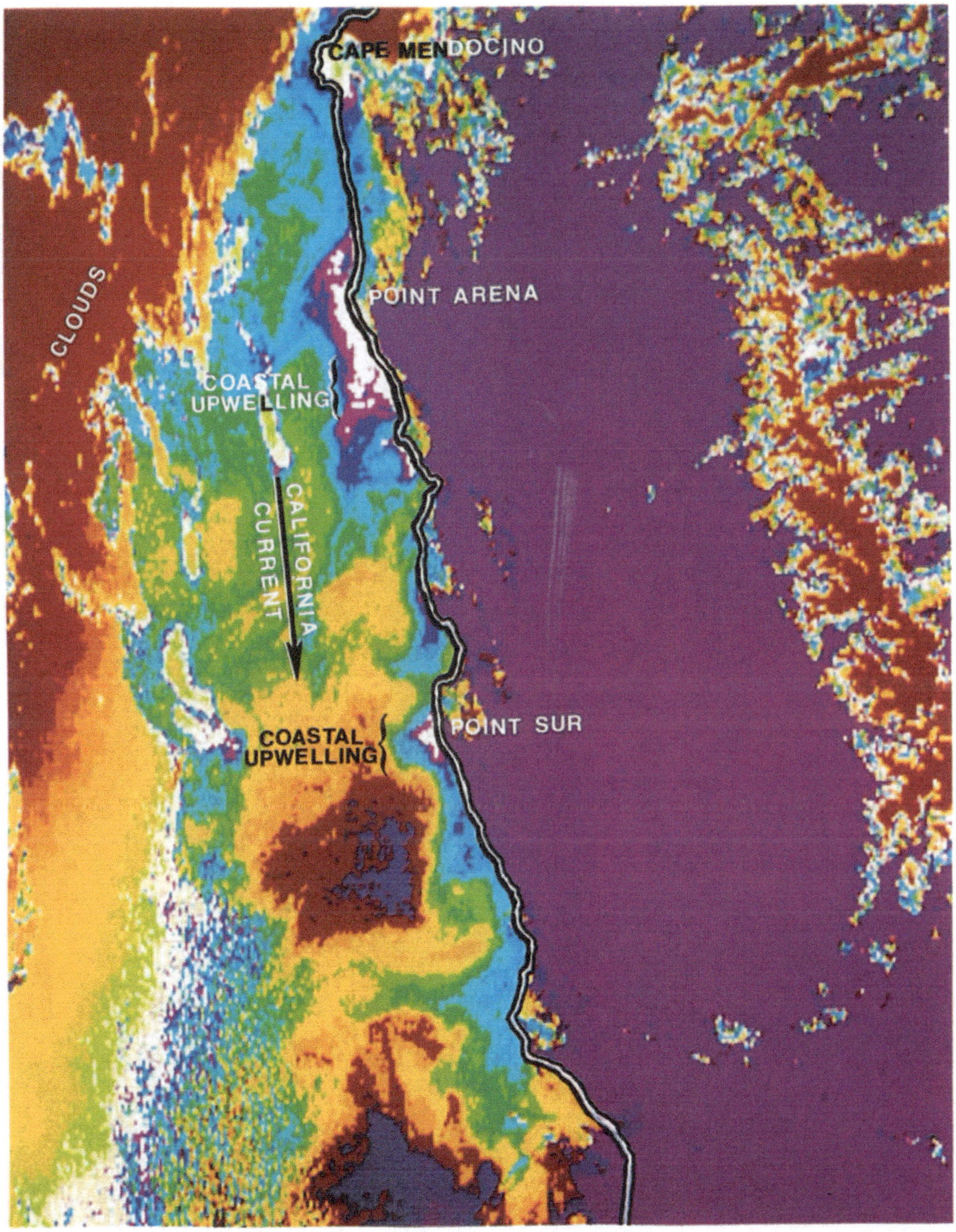

Fig. 1A. Upwelling thermal patterns in the eastern north Pacific in false color as derived from NOAA-6 AVHRR (Advanced Very High Resolution Radiometer) imagery for June 9, 1980 (color grades represent 1 radiometric count or *ca.* 0.7°C; spatial resolution is *ca.* 1 km). The scenes are cloud-free except for scattered thin stratus over land (varigated) and in the northwest oceanic region (medium red).

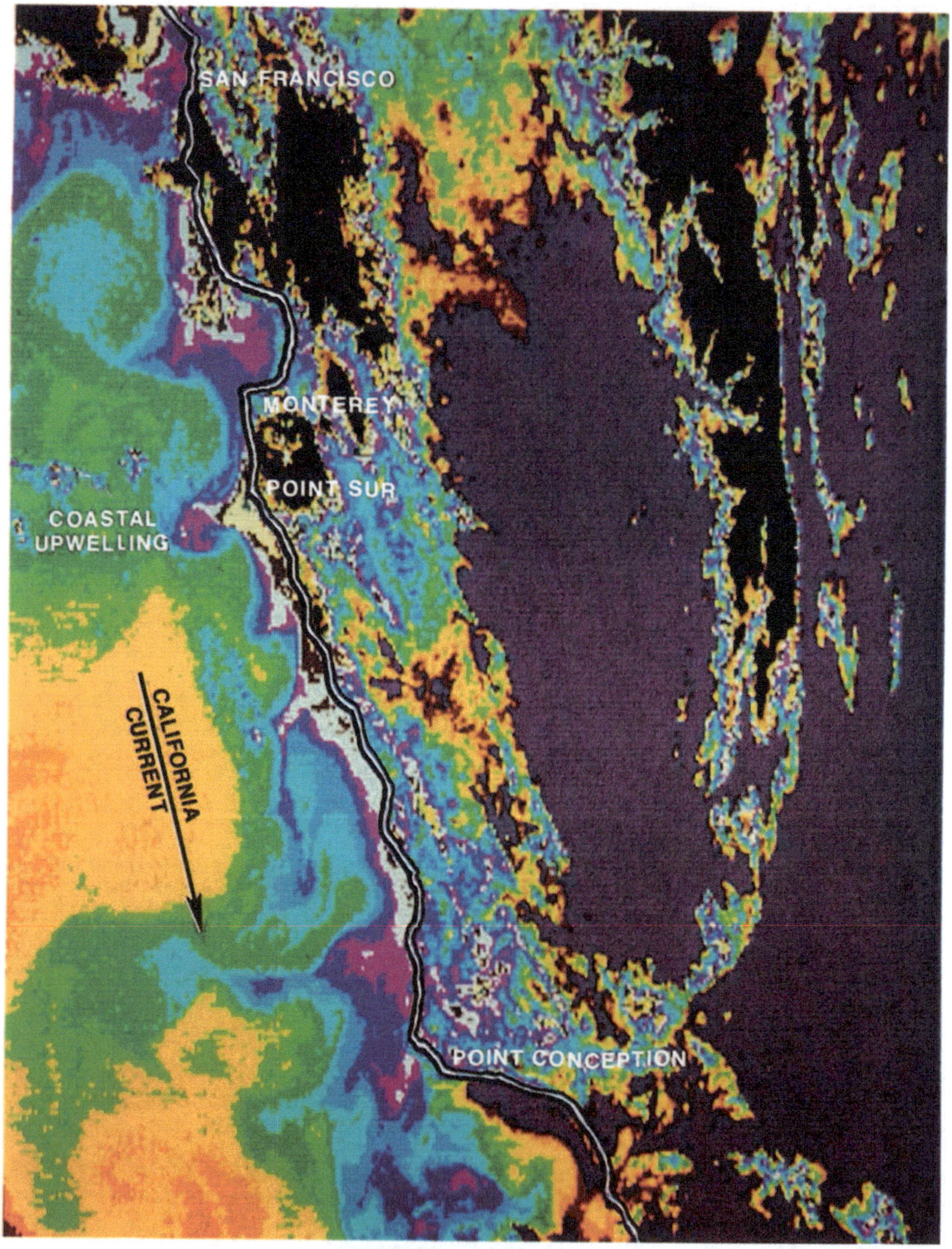

Fig. 1B. Upwelling thermal patterns in the eastern north Pacific in false color as derived from NOAA-6 AVHRR (Advanced Very High Resolution Radiometer) imagery for June 11, 1980 (color grades represent 1 radiometric count or *ca.* 0.7°C; spatial resolution is *ca.* 1 km). The scenes are cloud-free except for scattered thin stratus over land (varigated) and in the northwest oceanic region.

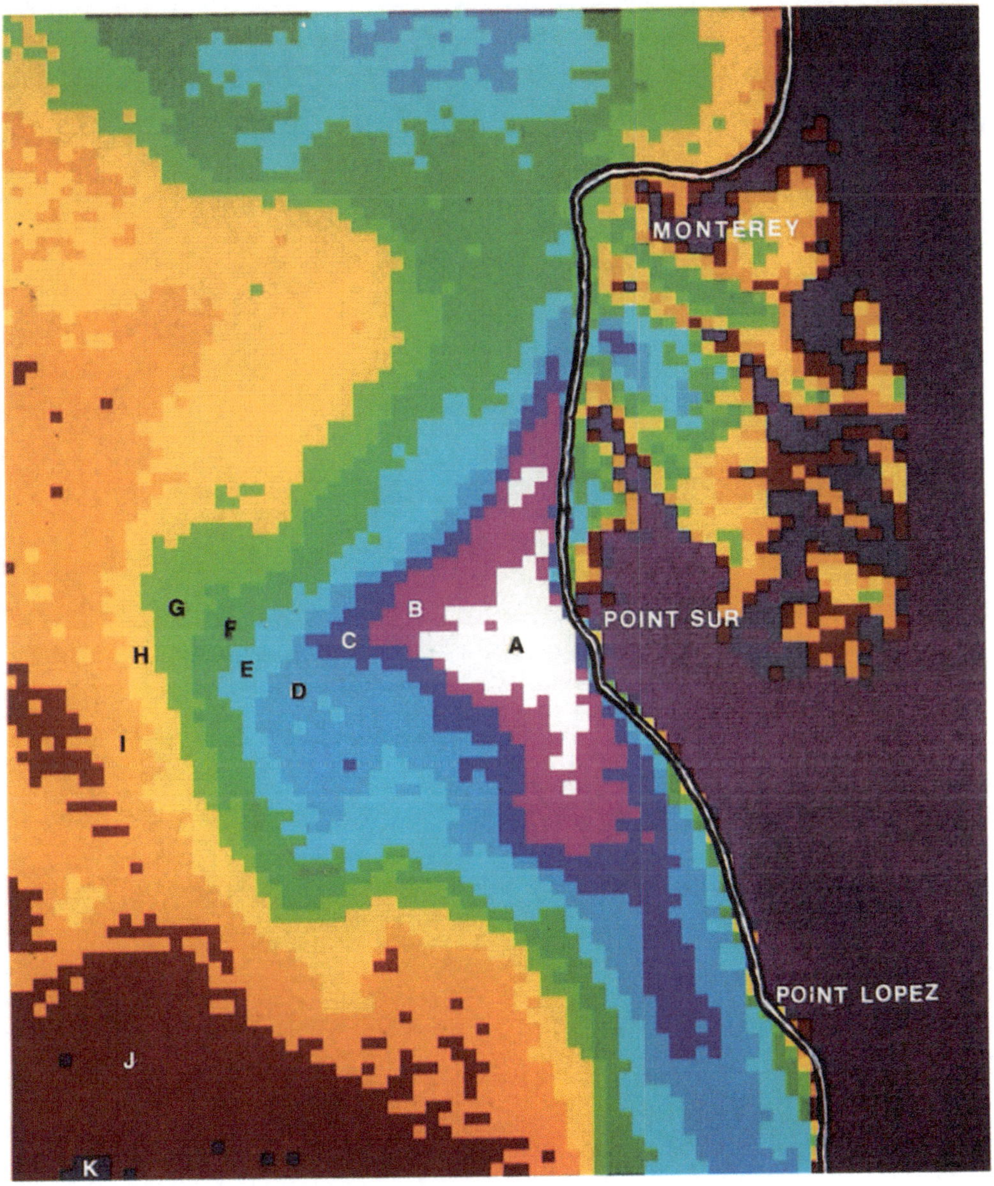

Fig. 2. IR image sequence in false color showing the formation of a cyclonic upwelling feature off Point Sur, California between June 9 (A) and 11 (B, p. 68), 1980. Letters associated with color grades represent average pixel temperatures measured by the radiometer (see Table 1 for AVHRR count, temperature, and associated nitrate and phosphate values).

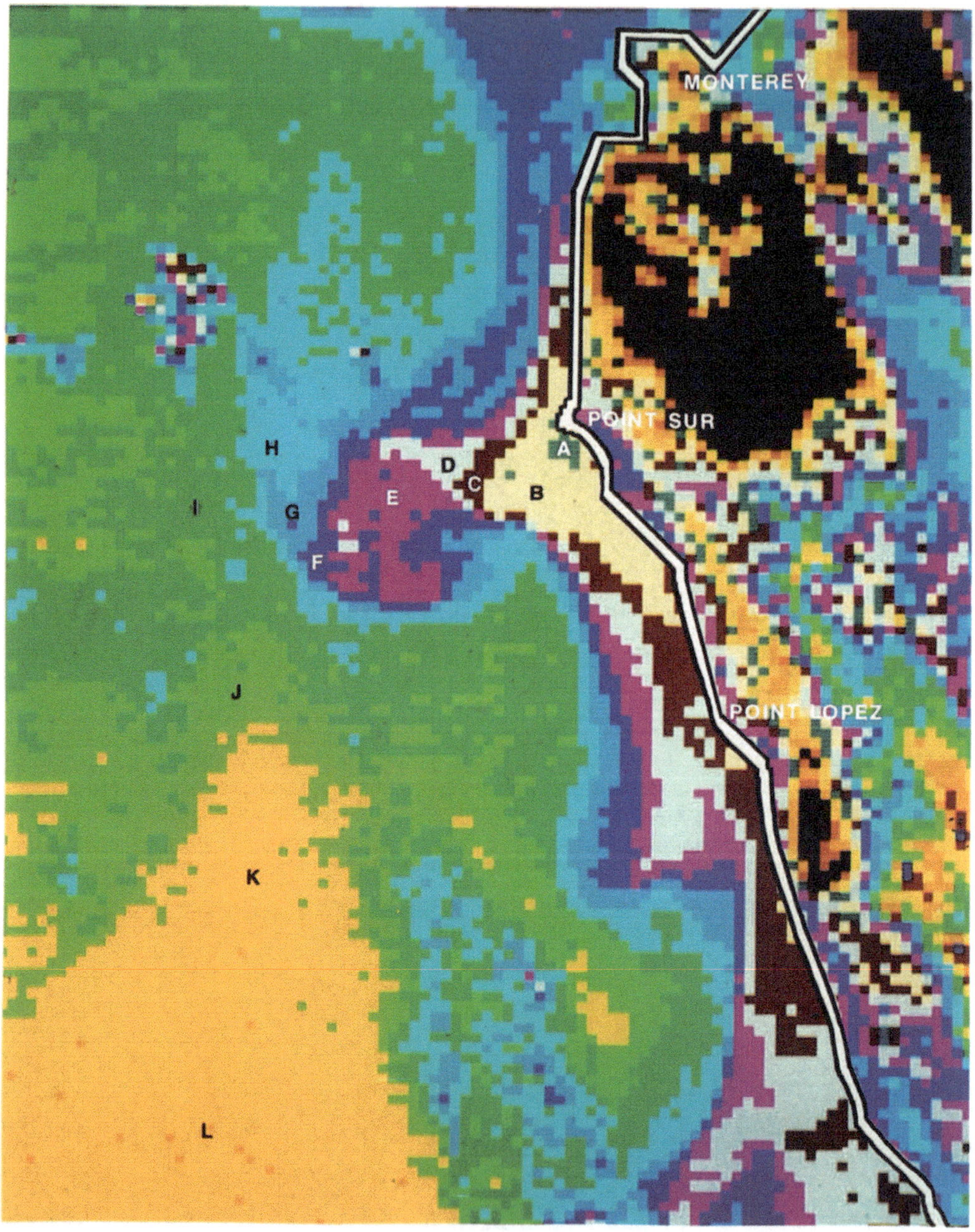

Fig. 2(B)

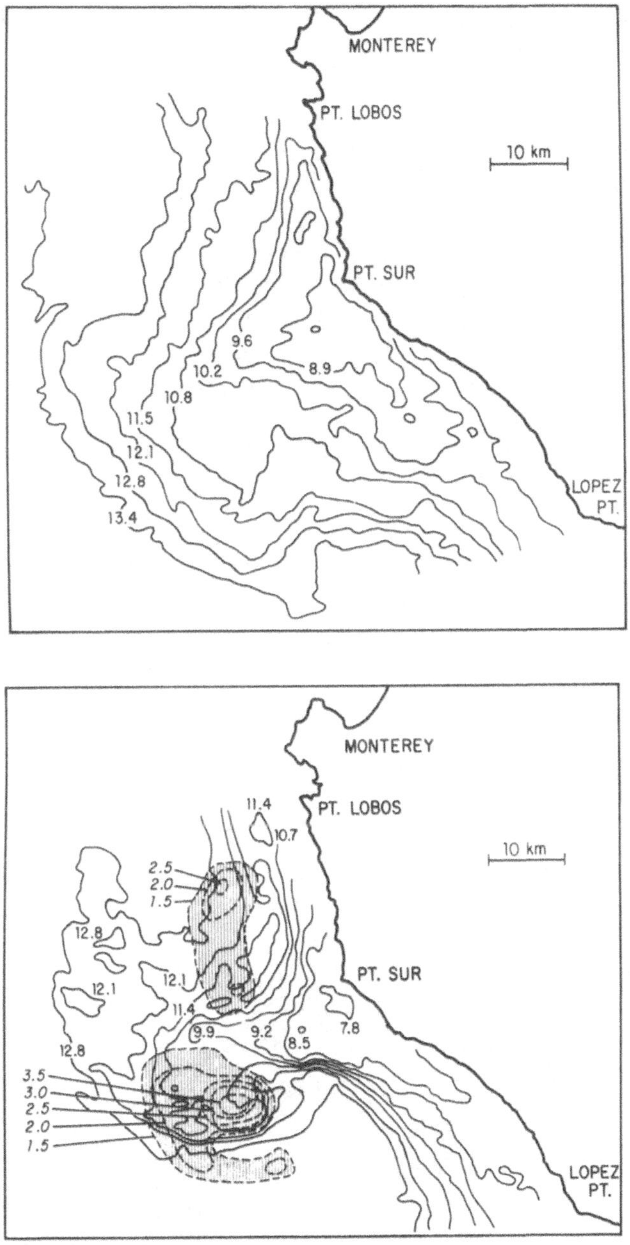

Fig. 3. Sea surface temperature maps in °C, June 9 (A, upper) and 11 (B, lower), 1980, inferred from satellite IR imagery. Contour interval is one radiometric unit of measurement (1 count ≈ 0.7°C). Phytoplankton pigments, primarily chlorophyll a (hatched area), are shown in lower figure with a contour interval of 0.5 mg m^{-3}.

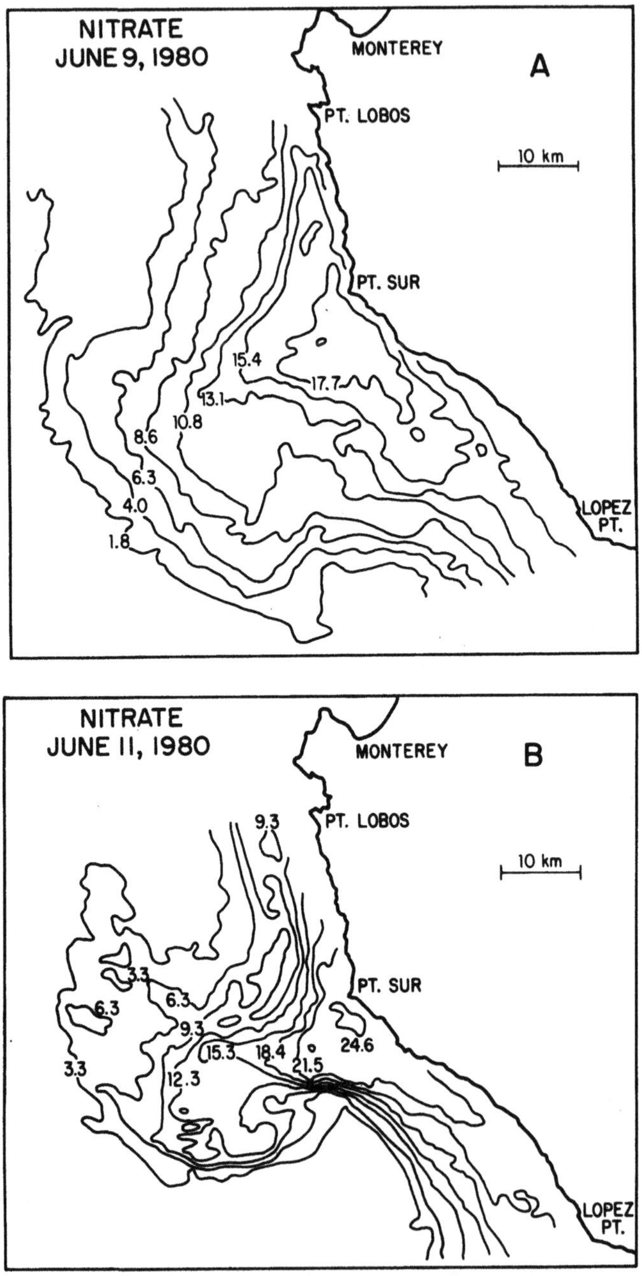

Fig. 4. Sea surface nitrate maps, June 9 (A, upper) and 11 (B, lower), 1980, generated by correlation with sea surface temperature distribution given by IR imagery. (Contour interval is one radiometric unit, ≃ 0.7°C; isopleths are in μM.)

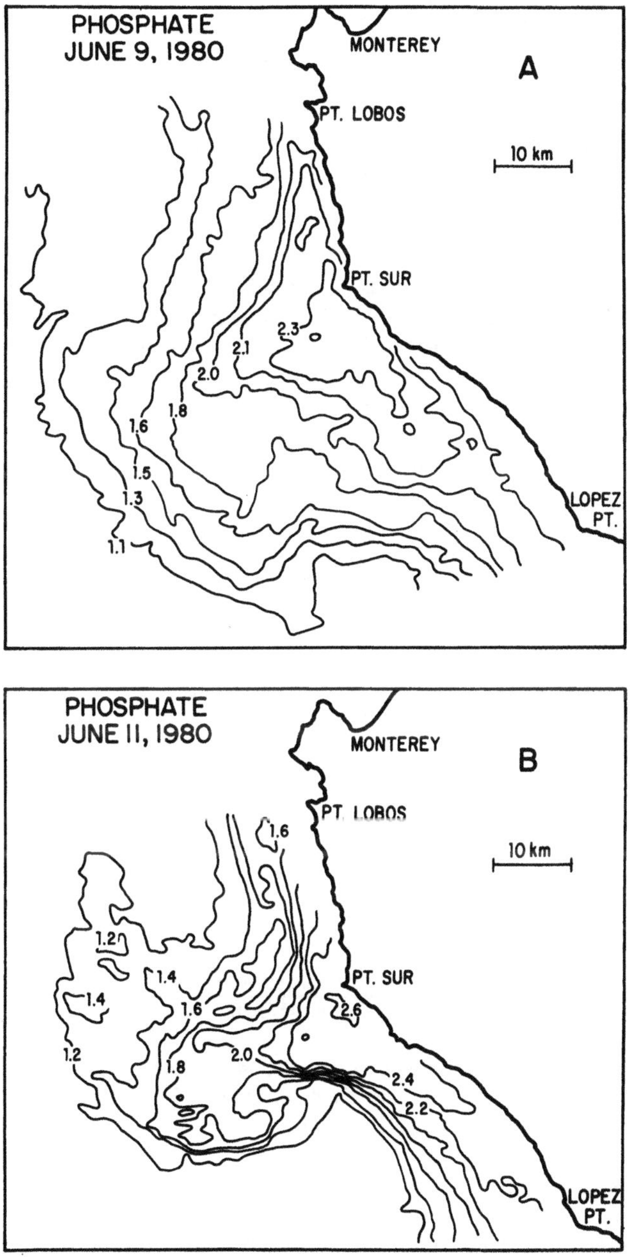

Fig. 5. Sea surface phosphate maps, June 9 (A, upper) and 11 (B, lower), 1980, generated by correlation with sea surface temperature distribution given by IR imagery. (Contour interval is one radiometric unit, ≈ 0.7°C; isopleths are in µM.)

nates in the upper 0.006 mm of the sea surface, and may not represent deeper temperatures obtained by the thermistor at 2.5 m. The errors originating in the atmosphere and at the ocean surface were corrected indirectly in this study by correlation between radiances derived from temperatures measured by the satellite radiometer and radiances derived from measurements of temperature *in situ*. Both the water vapor gradient and the gradient of difference between satellite derived and *in situ* temperatures were assumed to be negligible.

Planck Function

The sea surface emits thermal infrared spectral radiance (L) as a function of the wavelength (λ), the sea surface temperature (T) and zenith incidence angle (θ) according to the equation:

$$L(\lambda,\theta,T) = \varepsilon(\lambda,\theta) \cdot \beta(\lambda,T)$$

where $\beta(\lambda,T)$ is the Planck Function,

$$\beta(\lambda,T) = \frac{2hc^2}{\lambda^5 \cdot [\exp(hc/kT^{\lambda})-1]} \text{ Watts} \cdot \text{steradian}^{-1} \cdot \text{m}^{-3}$$

where

h = Planck's constant = 6.626196×10^{-34} joule · sec
λ = wavelength of emited radiation, meters
c = speed of light = 2.997925×10^{8} m · sec^{-1}
k = Boltzmann's constant = 1.380622×10^{-23} joule · K^{-1}
T = temperature, K, °Kelvin

The Planck Function gives the spectral radiance emitted by an ideal blackbody. However, the sea surface is not an ideal blackbody and emits less radiation. The factor $\varepsilon(\lambda,\theta)$ accounts for this fact, and is called Emissivity,

$$\varepsilon(\lambda,\theta) = \frac{\beta(\text{sea surface, K})}{\beta(\text{blackbody, K})}$$

Because its averaged value over the IR spectral window (10.5 µm to 12.5 µm) ε remains near *ca.* 0.98 for θ *ca.* 0° to 40°; this value was used in the computations of radiances.

Radiative Transfer Theory

Absolute measurements of sea surface temperature can, in theory, be produced from solutions of the radiative transfer equation,

$$L(T_\lambda) = L_0(T_s) \cdot \tau(p_0) + L_*$$

where

$L(T_\lambda)$ = radiance measured by the satellite radiometer in watts • steradian^{-1} • m^{-3}
T_λ = temperature obtained by inverting the Planck's function for the measured value of radiance in K
$L_O(T_s)$ = total radiance emitted from the surface at temperature T_s in watts • sterandian^{-1} • m^{-3}
$\tau(p_O)$ = atmospheric transmittance between the pressure levels *ca.* 0 (top of the atmosphere) and p_O (sea level)
L_* = path radiance due to the isotropic thermal emission of photons by the atmosphere along the propagation path from the sea surface to the radiometer in watts • sterandian^{-1} • m^{-3}

To derive the parameters τ, the atmospheric transmittance, and L_*, the path radiance, linear regression analyses were performed on both sets of temperature-derived radiances versus elapsed distance along the cruise track. Sharper frontal features were selected along the track where thermal differences were large enough to minimize the relative errors due to satellite measured thermal noise (e.g., caused by variable moisture over the area and due to a coarse AVHRR [Advanced Very High Resolution Radiometer] temperature resolution of 0.5°C). The ratio of the slopes of the regression lines gives the transmittance, τ, assuming the sea surface temperature gradient is much larger than the moisture gradient, i.e., $|\nabla L_O| >> |\nabla L_*|$, then by the radiative transfer equation:

$$\tau = \nabla L/\nabla L_O$$

The path radiance, L_*, was estimated from the intercepts with the Y-axis of the satellite and *in situ* radiances regression lines, respectively, L' and L_O', knowing that

$$L = m_1 x + L'$$

and

$$L_O = m_2 x + L_O'$$

where m_1 and m_2 are the slopes of the satellite and *in situ* radiance regression lines. Radiances were obtained from *in situ* and satellite temperatures by the Planck function. Regression lines were generated for the radiances versus elapsed distance, x, along the portions of the cruise track with strong temperature gradients and nearest in time to the satellite overpass. The slopes and y-intercepts of regression lines were computed from the equations:

$$m = \frac{\Sigma x_i y_i - N\Sigma x_i y_i}{(\Sigma x_i)^2 - N\Sigma x_i^2}$$

$$b = \bar{y} - m\bar{x}$$

Taking into account the relationship between the slopes of both lines, $\tau = m_1/m_2$, and substituting in the radiative transfer equation, the equation for the path radiance is obtained:

$$L_* = L' - \tau \cdot L_O'$$

Finally, knowing both transmittance, τ, and path radiance, L_*, sea surface temperatures as measured by the satellite radiometer were corrected using the radiative transfer equation.

Least squares linear regressions, designed to minimize the sum of the squares of the deviations of the data points from the straight line of best fit, were performed for nitrate versus phosphate, nitrate versus temperature and phosphate versus temperature.

Infrared (10.5 to 12.5 µM) data recorded during the satellite orbits were processed at the NASA-Ames Research Center, Moffet Field, California, in a computer equipped with the Interactive Digital Image Manipulation System (IDIMS). This is a comprehensive software package developed by Electromagnetic Systems Laboratories, Inc. With IDIMS the scene was selected, magnified and converted to false color from the digital radiometer data. Some 8" x 10" color Polaroid points, 35-mm slides and a computer printout ("picprint"), with the recorded "count" values (radiometric units of measurement) were obtained (Silva, 1982). A computer program developed by Lundell (1980) was used to navigate on the digital printout which had no geographical coordinates. The purpose of this program is to determine the line and pixel number of a geographical location given the location's latitude and longitude, a landmark location (line, pixel, latitude and longitude) and the period and inclination of the satellite orbit.

With the image navigational problem solved, it was possible to select portions of the RV "Acania" cruise track on the picprint in order to compare the digitized satellite temperatures with the *in situ* values. The sea surface temperature, nitrate and phosphate maps were generated by using a zoom transfer optical scope to magnify and linearly stretch the thermal pattern in the IR satellite image so that spatial features may be matched and traced on a navigational chart. Each isopleth (Figs. 3, 4, and 5) is the average of the temperature or nutrient concentration of pixel values on each side (Table 1). The pixel or picture element represents a resolution of *ca.* 1.1 km for the NOAA satellite.

RESULTS

On June 9 the cold thermal surface feature off Point Sur was similar in shape to the local bottom topography as delineated by the

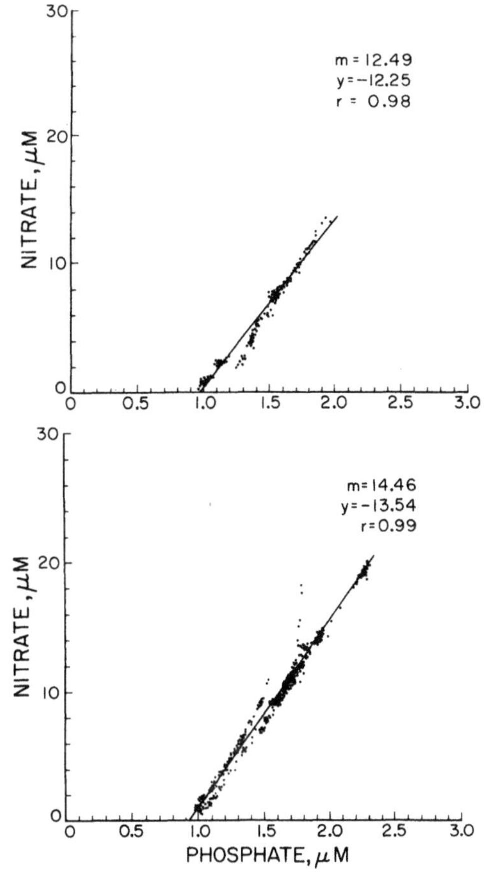

Fig. 6. Nitrate versus phosphate with regression lines for June 9 (A, upper) and 11 (B, lower), 1980.

break in slope at *ca.* 150 m. By June 11 the feature developed the cyclonic appearance with its equatorward edge over the axis of the Sur Canyon. The nearshore 8.57°C water (shown in white, code A, in Fig. 2A) moved seaward (while warming) toward the center of the feature (code D in Fig. 2B) and was replaced with colder 7.39°C water at its original location. Inferred nitrate and phosphate concentrations (Table 1) increased from 18.85 and 2.40 μM to 26.17 and 2.72 μM, respectively. Compared to eight other events, nutrient concentrations were in a middle-to-high range and were highly correlated with respect to each other (Fig. 6). Both nutrients were negatively correlated to temperature with regression slopes in a mid range (Figs. 7 and 8). The large range of nutrient concentrations and the near theoretical (Redfield, 1958; Goldman, McCarthy and Peavy, 1979) ΔNO_3 to ΔPO_4 ratio indicate the biochemical newness of the water in this feature.

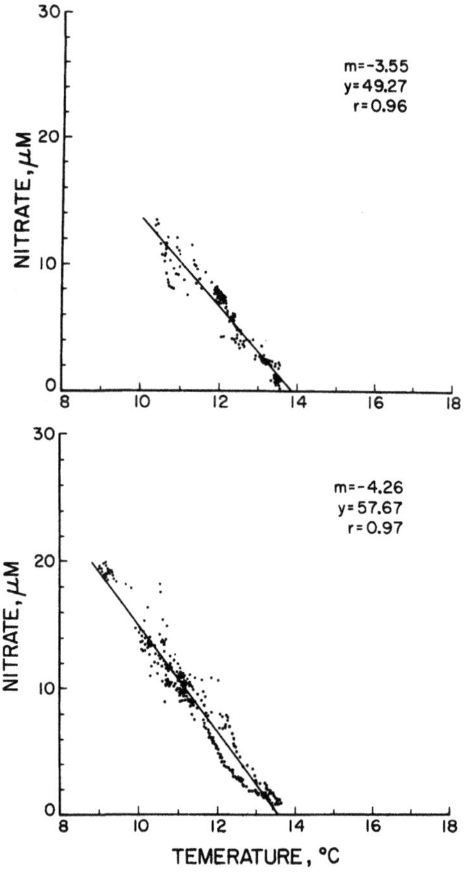

Fig. 7. Nitrate versus temperature with regression lines for June 9 (A, upper) and 11 (B, lower), 1980.

Biomass concentrations were highest (6.25 mg Chl m^{-3}) near the thermochemical fronts on the equatorward side (Fig. 3B) but fell to low values (<0.5 mg Chl m^{-3}) in low temperature water inside the feature and in the warm (low nutrient) water outside its frontal boundary. Because this upwelling feature developed and moved so rapidly, chlorophyll contours (Fig. 3B) can only represent the area (cross-hatched) in which phytoplankton were highly concentrated. Resampling of the area during the latter phase of the survey verified that maximal concentrations of biomass were in this area of sharp thermal and nutrient gradients. However, convincing information showing the "bloom" location was provided after this predictive figure was constructed, when the color image (Fig. 9) from June 12, 1980, was processed using the algorithm outlined in Gordon et al. (1980). The time-integrated distribution pattern illustrated in Fig. 3B shows very good agreement with the pattern inferred from the processed color image.

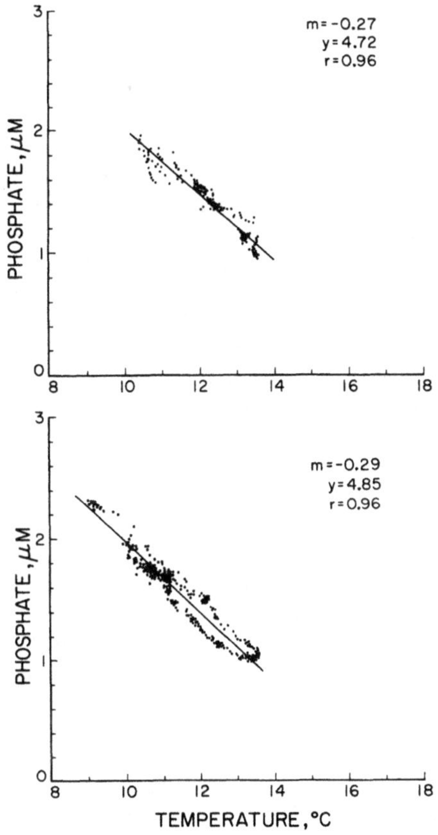

Fig. 8. Phosphate versus temperature with regression lines for June 9 (A, upper) and 11 (B, lower), 1980.

DISCUSSION

The satellite inferred view of coastal upwelling shows that there are distinctively structured systems which redistribute nutrients and phytoplankton and must to a large extent determine productivity along the continental boundary. When inferring nutrient structure of coastal upwelling systems from satellite infrared data, it is important to understand that the covariance between nutrients and temperature, even when well correlated, may differ in slope in different features or in the same feature at different times. Even when water upwells from the same depth, correlations may differ in similar features because of differences in the initial conditions of the source water mass, viz., temperature, nutrient concentration, layer depth and upwelling velocity. Or, differences may result when water upwells from deeper layers as a result of a longer duration and/or stronger wind stress (or other forcing function). A seasonal cycle may be expected (Bolin and Abbott, 1963; Bolin, 1964) as the result of a seasonal variation of forces which drive upwelling (Bakun, 1973) and a seasonal influx of Transitional Water formed by mixing Subarctic Pacific Water and Pacific Equatorial Water (Barham, 1957).

Even with such variability, if the correlation of nutrients and temperatures are monitored in a given feature, and if the temperature field is obtained by satellite infrared sensors, the nutrient field can be inferred from satellites. Judging from the high correlations observed in recently formed and persistent upwelling systems, deviations from the covariance of nutrients and temperature (which must occur as a result of mixing, heat and salt transfer, and biological activity) were not large over the time of each study (2-3 days). In the coastal upwelling zone, vertical velocities are much larger than open ocean values (Horne, Bowman and Okubo, 1978). This intensified motion and mixing may explain the high correlations observed between nutrients and temperature even in the presence of high concentrations of phytoplankton.

Previously, we reported (Traganza et al., 1981) the tendency for phytoplankton to concentrate in the thermal gradient on the equatorward edge of cyclonic features. This may be the result of a combination of physical entrainment and increased growth rates. Because these upwelling features persist and phytoplankton blooms are located in their sharp nutrient gradients, they may act as "natural chemostats" in which the specific growth rate of phytoplankton is sustained in proportion to the nutrient flux or dilution rate and the total biomass is proportional to the concentration of the nutrient supply. This intriguing possibility underscores the biogeochemical importance of these systems and their role in determining primary production and biomass distribution. In June 1980, it appears that such a feature had a poleward and equatorward concentration of phytoplankton. However, the equatorward side was the more concentrated region. Studies are needed to investigate the spatial and temporal variability of frontal circulation and the interacting water masses.

Horne et al. (1978) suggest several dynamic processes in frontal zones which may affect cross-frontal mixing while yet maintaining sharp frontal gradients. These authors point to the common misunderstanding that frontal zones must represent a barrier to mixing between two juxtaposed water masses. Any process which could account for cross-frontal mixing could account for "biological patchiness" along the frontal boundary. Whatever the physical mechanisms which distribute or concentrate phytoplankton along a thermochemical front, they appear to be placed in an optimum location. The resulting concentration of biomass at favored locations (equatorward of capes and points) may be reflected in the sediment record along the continental boundary.

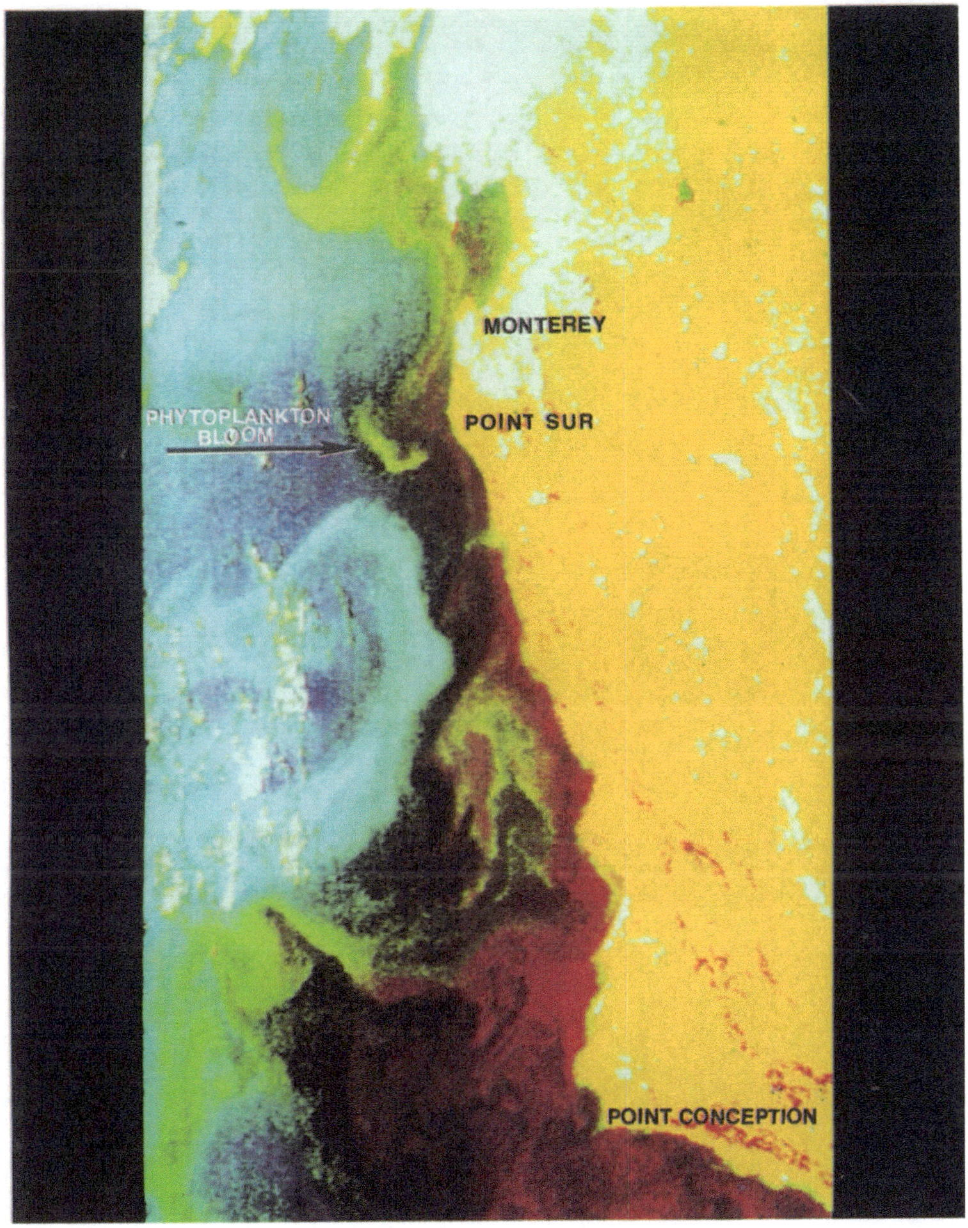

Fig. 9A. Phytoplankton pigments distribution in the eastern north Pacific, June 12, 1981, as derived from Nimbus-7 Coastal Zone Color Scanner (CZCS) imagery (A, raw color; B, calibrated by algorithm, p. 78).

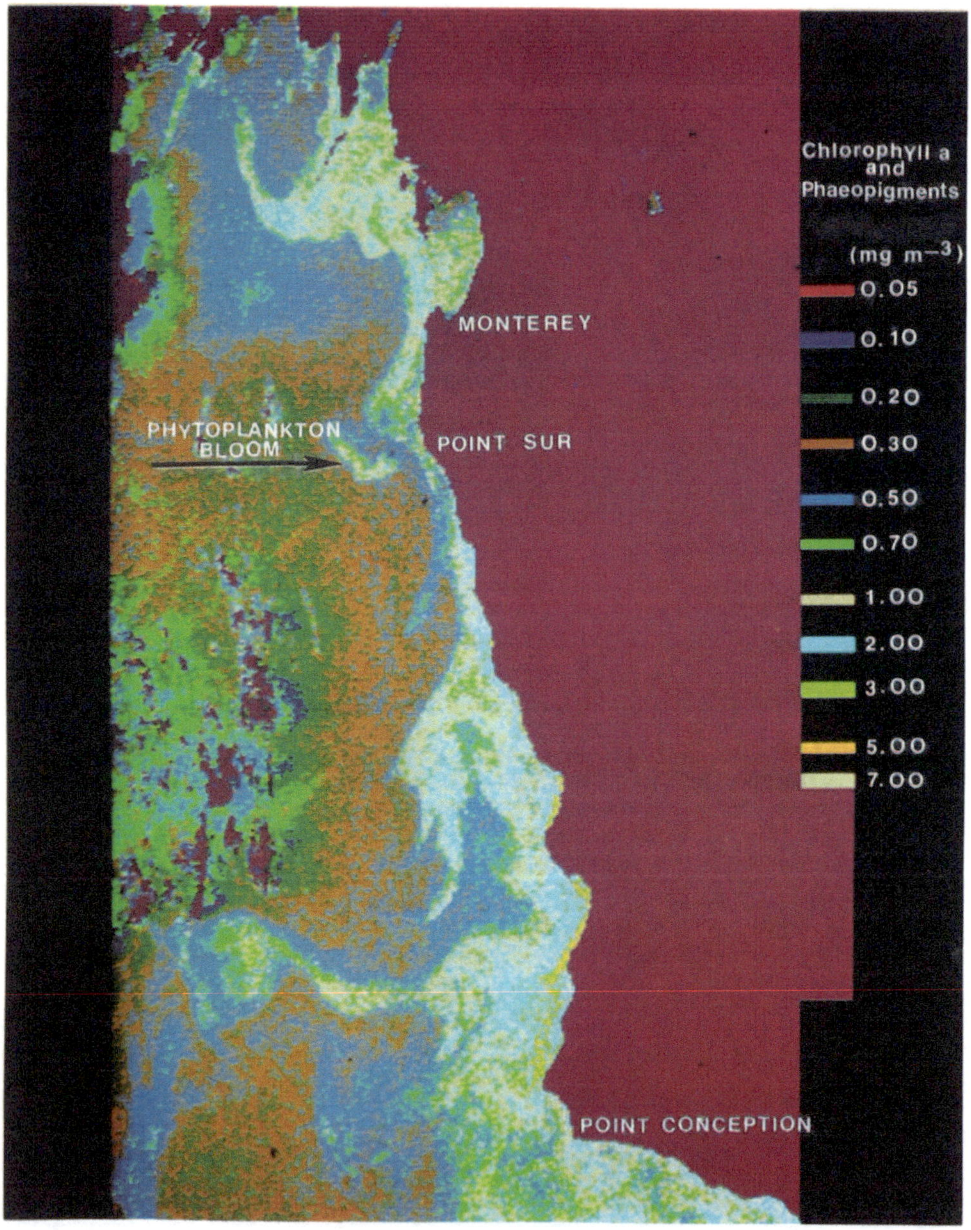

Fig. 9B. Color-coded phytoplankton pigment distribution; compare this chlorophyll pattern in the upwelling feature off Point Sur, California with phytoplankton pigment distribution measured by *in situ* fluorescence measurements made during the two days just prior to this orbit (Fig. 3B). (This CZCS image has been processed using simplified methods which do not remove either spatial distortions or geometrically induced variations in Rayleigh radiance. The present comparisons should, therefore, be regarded as qualitative.)

CONCLUSIONS

It is feasible to infer sea surface nutrient concentrations from satellite derived thermal patterns in regions of strong upwelling where the nutrient to temperature correlations are very high. However, some errors are to be expected. The computer program developed by Lundell (1981) to navigate on the satellite IR digital printout of radiometric units or "picprint" can introduce an error in the position of the cruise track on the order of two lines and two pixels, a distance of ∿2 km. In our study, with the identification of three landmarks on the picprint (Point Sur, Pfeiffer Point and Point Lopez), there were R.M.S. errors of *ca.* 0.9 lines and 1.3 pixels for the 9 June 1980 cruise track, and *ca.* 1.0 lines and 2.1 pixels for the 11 June 1980 cruise track. These errors are smaller than expected due to the proximity of the landmarks, and navigation did not introduce significant errors in the thermal calibration. However, the process of averaging the transmittance and path radiance values obtained for each sharp temperature gradient along portions of the cruise track did add errors. Taking this into account, standard deviations were on the order of ±0.27°C and ±0.13°C for the calibrated temperatures of June 9 and 11, 1980 (Figs. 3A and 3B). With these errors for the thermal maps, when the nutrient:temperature regression equations (Table 2) are used to infer nutrient concentrations, one can expect errors of up to 6% and 4% for the nitrate and phosphate maps respectively on 9 June 1980 and 3% and 2% for the nitrate and phosphate maps respectively on 11 June 1980. Additional errors due to imperfect temperature vs. nutrient correlations are negligible when applied to the strong upwelling system. However, the quality of

Table 2. Linear regression analyses of nutrients vs. temperature

Day	Linear Regression Equations	Correlation Coefficient
9 June	N = -3.55T + 49.27	-0.96
	P = -0.27T + 4.72	-0.96
11 June	N = -4.26T + 57.67	-0.97
	P = -0.29T + 4.85	-0.96
29 October*	N = -1.27T + 25.26	-0.71
	P = -0.09T + 2.26	-0.70

N=nitrate, μM; P=phosphate, μM; T=temperature, °C.
* Data from Silva (1982).

the result for any given upwelling system will depend upon the age and uniformity of the upwelled surface waters (compare correlation values in June and October, in Table 2).

It seems reasonable to assume that in an upwelling zone, the major patterns of nutrient concentrations can be inferred from satellite IR imagery and limited *in situ* thermal and nutrient data. With the development of a suitable biochemical model, this near real-time nutrient analysis of surface waters could provide a basis for predicting phytoplankton blooms within ecosystems in the California Current and coastal upwelling zone.

ACKNOWLEDGEMENTS

I thank E.J. Green and ONR Code 422 NSTL Station, Bay St. Louis, Mississippi for support of this research; J.L. Mueller, Naval Postgraduate School, Monterey, California, who provided the CZCS images and helpful critique, A. Bakun, NMFS, Monterey, California for "upwelling index" data; L. Breaker (NESS) and R. Wriggley (NASA) for the Tiros-N infrared images.

REFERENCES

Bakun, A., 1973, Coastal upwelling indices of the west coast of North America, National Oceanic and Atmospheric Administration Technical Report NMFS SSRF-671.

Barham, E.G., 1957, The ecology of sonic scattering layers in the Monterey Bay area, Hopkins Marine Station Pacific Grove, California, Technical Report No. 1.

Bolin, R.L., 1964, Hydrographic data from the area of the Monterey submarine canyon, 1951-1955, Final Report, NGO ONR-25127, NSFG 911, NSF61780, Hopkins Marine Station, Pacific Grove, California.

Bolin, R.L. and Abbott, D.P., 1963, Studies on the marine climate and phytoplankton of the central coastal area of California, 1954-1960, CalCOFI Report 9:23.

Bronsink, S.H., 1980, "Microplankton ATP-biomass and GTP-productivity Associated with Upwelling off Point Sur, California," M.S. Thesis, Naval Postgraduate School, Monterey, 70 pp.

Goldman, J.C., McCarthy, J.J. and Peavy, D.G., 1979, Growth rate influence on the chemical composition of phytoplankton in oceanic waters, Nature, 279:210-215.

Gordon, H.R., Clark, D.K., Mueller, J.L., and Hovis, W.A., 1980, Phytoplankton pigments from the Nimbus-7 coastal zone color scanner: comparisons with surface measurements, Science, 210:60-63.

Hanson, W.L., 1980, "Nutrient Study of Mesoscale Thermal Features of Point Sur, California," M.S. Thesis, Naval Postgraduate School, Monterey, 182 pp.

Horne, E.P., Bowman, M.J. and Okubo, A., 1978, Crossfrontal mixing and cabbeling, in: "Oceanic Fronts in Coastal Processes," M.J. Bowman and W.E. Esaias, eds., Springer-Verlag, Berlin, 105-113.

Lundell, G., 1981, "Rapid Oceanographic Data Gathering: Some Problems in Using Remote Sensing to Determine the Horizontal and Vertical Thermal Distributions in the Northeast Pacific Ocean," M.S. Thesis, Naval Postgraduate School, Monterey, 188 pp.

Maul, G.A., DeWitt, P., Webb, W.C., Yanaway, A., and Baig, S.R., 1978, Geostationary satellite observations of Gulf Stream meanders: infrared measurements and time series analysis, Journal of Geophysical Research, 83(c12):6123.

Redfield, A.C., 1958, The biological control of chemical factors in the environment, American Scientist, 46(3):205-227.

Silva, V.M., 1982, "Thermal Calibration of Satellite Infrared Images and Correlation with Sea Surface Nutrient Distribution," M.S. Thesis, Naval Postgraduate School, Monterey, 65 pp.

Traganza, E.D., Conrad, J.C. and Breaker, L.C., 1981, Satellite observations of a cyclonic upwelling system and giant plume in the California Current, in "Coastal Upwelling," F.A. Richards, ed., Coastal and Estuarine Science 1, American Geophysical Union, Washington, 228-241.

UPWELLING PATTERNS OFF PORTUGAL

Armando F.G. Fiúza

Oceanography Group, Department of Physics/Geophysical Center, University of Lisbon
Rua da Escola Politécnica 58
1200 Lisboa, Portugal

ABSTRACT

Remotely sensed sea surface temperatures are used to characterize the Portuguese coastal upwelling from the distribution and evolution of its surface thermal signature. During the summer, upwelled waters occupy the surface layers over the whole western shelf and part of the upper slope of Portugal, and their areal extent pulsates onshore-offshore in response to cycles of northerly winds, reaching 30-50 km from the coast under calm conditions and extending to 100-200 km during and shortly after strong north winds.

Three regions are examined in detail, each showing a different upwelling pattern related to characteristic topographical constraints. On the wide, flat meridional shelf north of the Nazaré Canyon, upwelling is fairly homogeneous alongshore. Off the southern half of the Portuguese west coast, three-dimensionality is induced on the upwelling distribution by a mesoscale protrusion of the coastline which, associated with the nearshore deep features of the Lisboa and Setúbal canyons, apparently causes an offshore separation of the waters upwelled north of the Bay of Setúbal. South of Cape Sines, isotherms again tend to follow the depth contours, but thermal gradients are closer to the shore in relation to the steep shelf of this region. The zonal south coast of Portugal, the Algarve, is affected directly by upwelling only under locally favorable, westerly winds which blow only occasionally. However, during cycles of moderate to strong north winds over the west coast, even with calm wind conditions in the Algarve, the cold upwelled waters of the west coast apparently turn around Cape São Vicente and then seem to flow eastward along the shelf break, although the same pattern could be generated by shelf edge upwelling. During a particularly strong northwest wind event, upwelled waters were observed to cover the whole Algarve shelf and then proceed offshore from near the mouth of the Guadiana River at the Spanish-Portuguese border, again following the shelf edge of the Gulf of Cadiz. During the upwelling spin-down, a coastal countercurrent seemed to carry warm surface waters to the west, eventually reaching São Vicente, and even proceeding northward along the west coast when winds reversed to southerly. The decoupling between the shelf and the upper slope waters of Algarve is attributed to a strong (≈500 m high) bathymetric "step" extending along the whole Algarve shelf edge.

INTRODUCTION

Upwelling takes place along the west coast of Portugal during the summer under the fairly strong and steady northerly "Portuguese trades" (Fiúza, Macedo and Guerreiro, 1982). The Oceanography Group of the University of Lisbon has been applying remote sensing techniques to the study of this upwelling system. With the support of the Portuguese Air Force, systematic airborne surveys using an infrared radiometer were conducted over two areas of the continental margin off the Portuguese west coast, one between 39°10'N and 40°15'N (in 1976-1977), and another extending from 37°43'N to 38°40'N (in 1979-1980). Satellite thermal infrared pictures have been obtained for the corresponding periods from the University of Dundee ground receiving station. A preliminary analysis of some of these observations is presented here with the objective of describing the upwelling patterns over the whole continental shelf and upper slope region off Portugal through the configuration, persistence, and evolution of their thermal surface signatures in three "typical" areas, two off the west coast and the other off the southernmost Portuguese province, Algarve.

GENERAL TOPOGRAPHY OF THE STUDY AREA

The most remarkable features of the bathymetry of the Portuguese continental margin are easily identifiable in Fig. 1. They include the typical physiographic aspects of the continental shelf, the slope, and the rise. The margin is cut in some places by important submarine canyons (Nazaré, Lisboa, Setúbal, São Vicente, Portimão) which generally define boundaries between regions with relatively similar topographical conditions. North of the Nazaré Canyon the shelf is wide and very flat and the shoreline extends almost meridionally up to the Spanish border (≈42°N). Between Nazaré and the system of the Lisboa and Setúbal canyons, the shelf is more irregular and the coast presents strong indentations associated with prominent capes and large embayments.

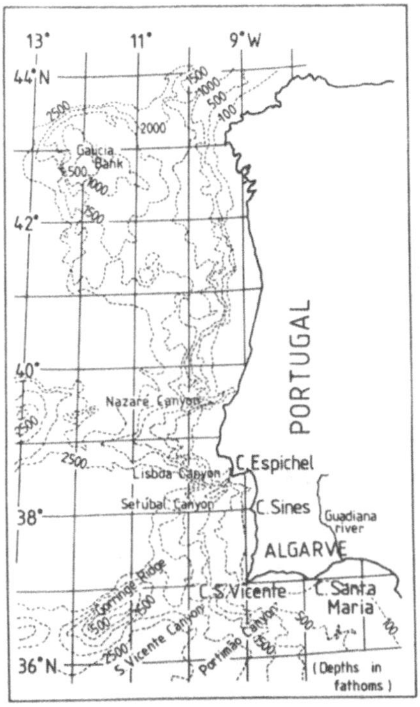

Fig. 1. General bathymetric chart of the Portuguese continental margin; depth contours are in fathoms (1 fathom = 1.83 m).

South of the Setúbal Canyon, the shelf is fairly flat up to Cape Sines but then becomes very steep up to the S. Vicente Canyon, practically without a shelf break. Between these canyons the coast is again oriented very nearly along the local meridian. Off the whole of the southern Portuguese coast (which is zonal except for the protrusion of the Cape Santa Maria area) the shelf extends with a small inclination down to an extremely sharp edge at about 100-130 m depth, defined by a sudden step down to the 700 m contour. This pronounced feature, which is well shown in very detailed charts, extends around the southwest tip of Portugal, reaching about 10 km north of Cape S. Vicente.

GENERAL CONFIGURATION OF THE SST FIELD DURING THE UPWELLING SEASON

Almost daily TIROS-N infrared pictures of the Portugal area were obtained for the period 30 July-11 September 1979, corresponding to the peak of the upwelling season. Each of these clearly shows a more or less extended band of cold upwelled waters covering the western shelf and upper slope of the Iberian Peninsula (see examples in Fig. 2). The outer limit of the upwelled waters (as defined by the sepa-

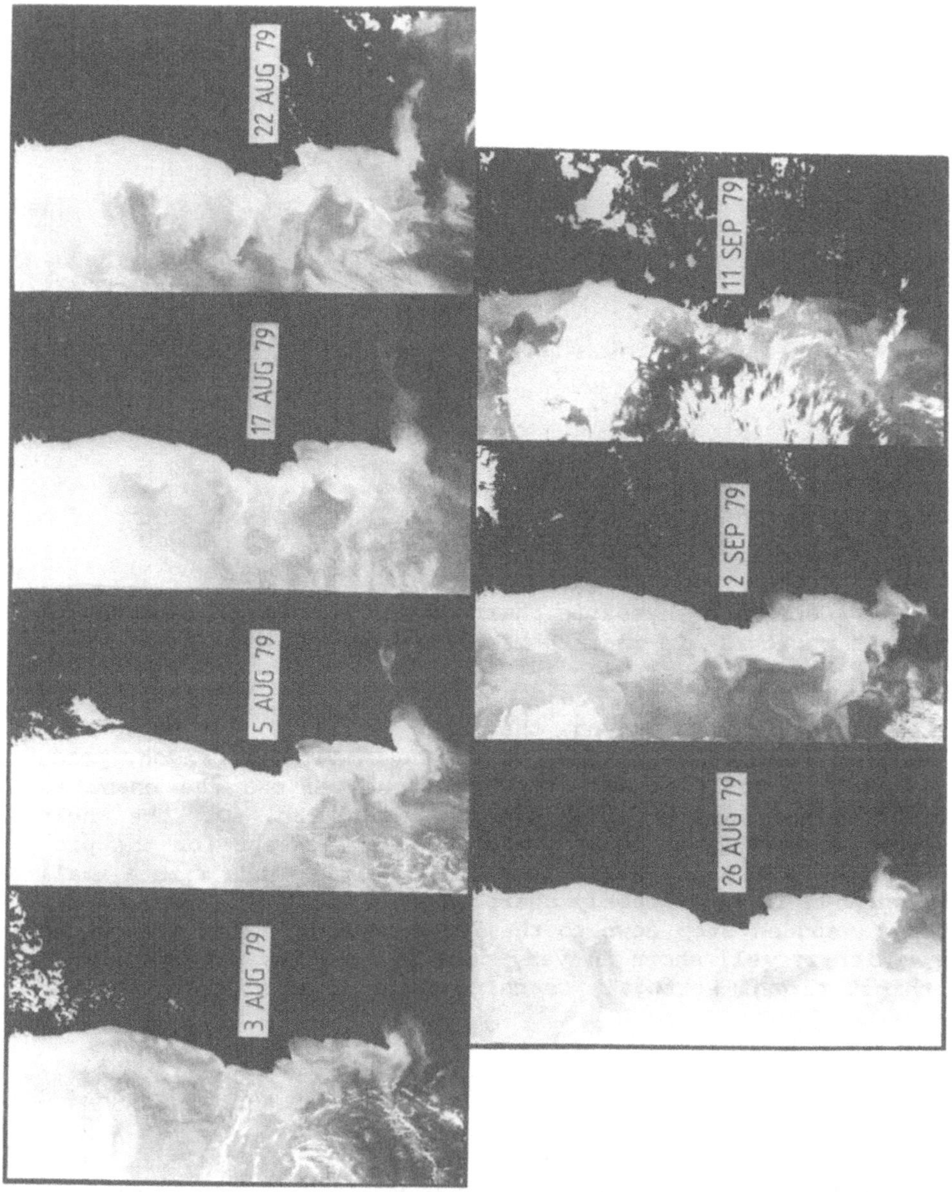

Fig. 2. TIROS-N thermal infrared images obtained during the summer of 1979: a) 3 August; b) 5 August; c) 17 August; d) 22 August; e) 26 August; f) 2 September; g) 11 September. Clear tones represent colder waters.

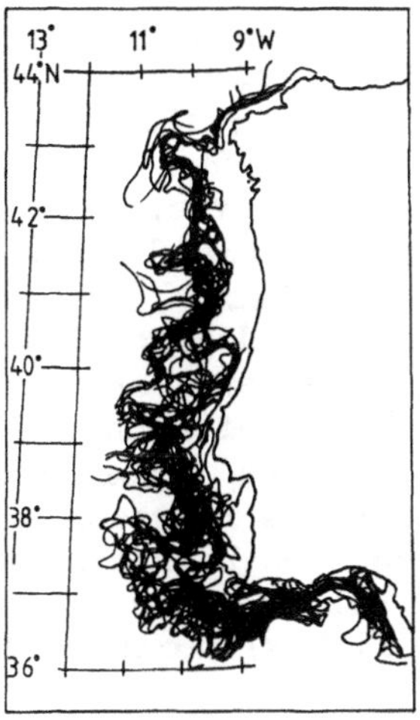

Fig. 3. Positions of the outer limit of the upwelled waters as determined from almost daily TIROS-N infrared images relative to the period 30 July-3 September 1979; with geographic coordinates.

ration between their lighter tones and the dark signature of the warm oceanic waters) was contoured in each of these satellite images using a transparent overlay with a compatible geographic coordinates grid (the same as in Fig. 1) and the results are illustrated in Fig. 3 as a joint plot of these individual "external fronts".

During periods of weak northerly winds, the upwelled waters reached up to only 30-50 km from the west coast, roughly reaching the 1000 m (≈500 fathoms) depth contour. These cold waters extended further seaward, up to 100 to 200 km from the shore in response to stronger pulses of the upwelling favorable north wind. Their outline still tended to keep some "memory" of the contour of the upper slope region although it was often deformed by eddies and filaments. During such periods the upwelled waters turned counterclockwise round Cape S. Vicente and migrated eastwards to about 6°30'W, receding back to their former position under weak winds as will be described below in more detail.

DETAILED SURFACE STRUCTURE OF THE UPWELLING IN SELECTED AREAS

The satellite infrared images show considerable thermal structure within the waters upwelled off the Portuguese coast (Fig. 2). Three areas will now be analyzed in more detail: the Nazaré area, the Sines area and the Algarve region. The first two were surveyed radiometrically with an aircraft during the summers of 1977 and 1980, respectively, whereas for the southern region only the 1979 satellite imagery was available.

The airborne observations were conducted with a Barnes PRT-5 instrument operating in the narrow range 9.5-11.5 micrometer at a flight level of only 150 m, generally under clear sky conditions. The radiometer was calibrated before and after each flight and its readings were corrected for the effect of ambient temperature change as recommended by Kraan (1977). Field tests during the observational program have shown that atmospheric corrections are negligible at such low flight levels. Wind conditions during the upwelling seasons of these years were very similar and consisted of strengthening-weakening cycles of the northerly wind with a duration of 6-7 days each (see example in Fig. 6a). The sea temperature samples obtained with the aircraft are considered to be representative of upwelling related patterns as they corresponded, both in the cases of Sines and Nazaré, to conditions of calms (1 flight), and of weak (2 flights), moderate (2 flights) and strong (2 flights) north wind stresses.

Nazaré Area

Sea surface temperature (SST) data from seven flights conducted between 13 July and 6 September 1977 were used to obtain averages and

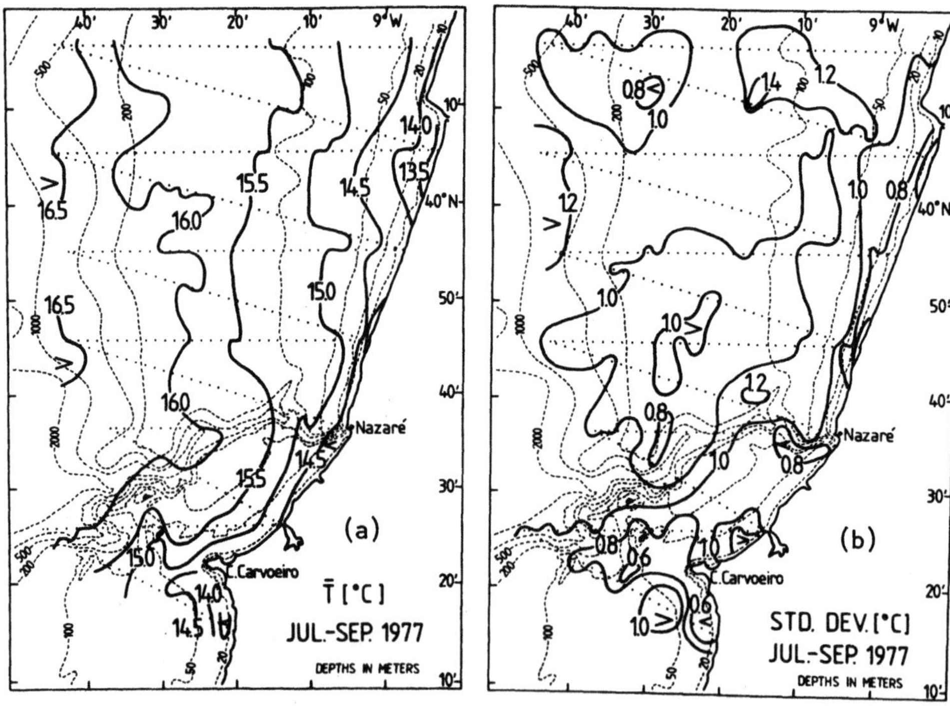

Fig. 4. a) Mean sea surface temperature (SST) field in the Nazaré area obtained from airborne surveys conducted in the summer of 1977.

b) Standard deviations for the same set as in (a).

standard deviations which were plotted at every nautical mile (1 n.m. = 1853 m) along the flight tracks. The mean thermal field is displayed in Fig. 4a and shows isotherms essentially aligned with the bottom contours, with temperature values increasing monotonically offshore. Standard deviations (Fig. 4b), giving an estimate of local variability, were smaller at the coast, where upwelling was more intense and steady, and tended to be somewhat larger at the mid-shelf, probably in association with cross-shelf excursions of a meridional band of stronger thermal gradients (upwelling front?) which was observed in the SST charts for some of the individual flights (not shown here). These also showed a guide effect of the Nazaré Canyon for oceanic warm waters during the slackening of the north wind.

Sines Area

From 30 July to 22 August 1980, a sequence of seven airborne SST surveys was conducted in the Sines area. Mean surface temperatures

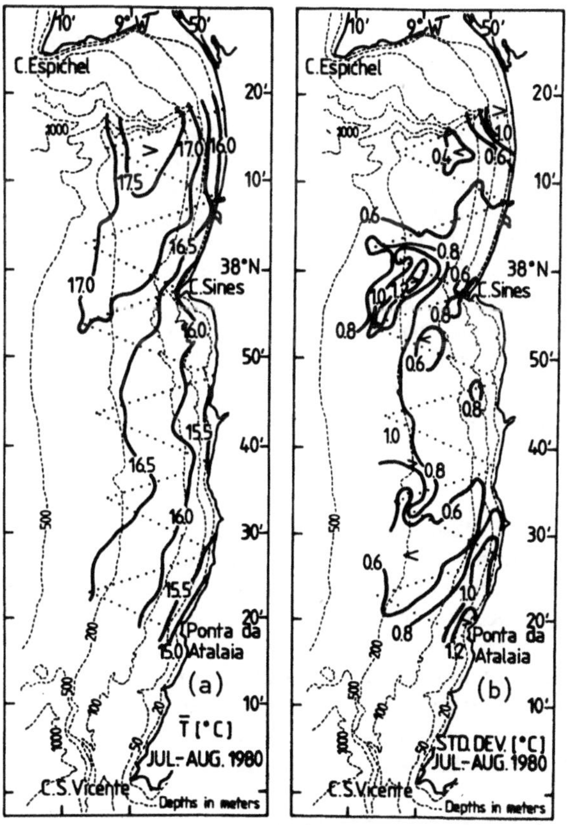

Fig. 5. a) Mean sea surface temperature (SST) field in the Sines area obtained from airborne surveys in the summer of 1980.
b) Standard deviations relative to the same set as in (a).

were computed from these observations and the corresponding chart is shown in Fig. 5a. The mean thermal field shows essentially the same patterns as those observed in the results of the individual flights (not shown here). The coastal morphology north from Cape Espichel (38°24'N) and the local bathymetry result in strong three dimensionality on the structure of the upwelling in the Bay of Setúbal (Cape Espichel - Cape Sines). In fact, the dominant feature of the SST in this area is a warm intrusion between cool coastal and offshore waters, which may correspond either to an anticyclonic eddy or more probably to the result of southward advection of waters upwelled further north (see pictures of 3 and 5 August 1979 in Fig. 2). The influence of the coastal configuration may be important, as the zonal extension of the topographic perturbation constituted by the Setúbal Peninsula is about twice as large as the internal Rossby radius of deformation (≃16 km in this area during the upwelling season), which provides the characteristic offshore scale for coastal upwelling dynamics (e.g., Mooers, Collins and Smith, 1976). The bottom topography around Cape Espichel is dominated by the deep canyons of Lisboa and Setúbal which almost reach the coastline, thus creating locally a strong bathymetric discontinuity. The association of both topographical perturbations in this area apparently offsets the coastal upwelling occurring north and south of Cape Espichel.

Standard deviations were also computed from the SST data of these flights and are shown in Fig. 5b. Maximum variability, perhaps associated with migrations of the inner upwelling front, is observed near the coast in the areas where upwelling was strongest, particularly near Ponta da Atalaia. The localized offshore variability maximum in front of Sines is apparently related to oscillations of the southern limit of the warm feature observed in the Bay of Setúbal. The closeness to the shore of the "inner" front in this region, as opposed to the mid-shelf position observed in the Nazaré area, is perhaps related to the narrower and steeper shelf off the Sines - Cape S. Vicente coast.

Algarve Area

The selection of TIROS-N infrared pictures relative to the summer of 1979, presented in Fig. 2, illustrates typical sequences of SST distributions observed off Portugal during a complete upwelling cycle, and particularly the evolution of the upwelling along the Algarve coast. In Fig. 6a low-passed (24 hours running mean) north-south wind stresses computed from hourly winds measured in Lisboa are shown for the period covered by the satellite pictures of Fig. 2. These "stresses" were computed by multiplying the magnitude of the wind vector by its meridional component.

These Lisbon winds are considered to represent reasonably well the general wind regime along the Portuguese west coast. Along the Algarve coast, however, the atmospheric circulation is fairly dif-

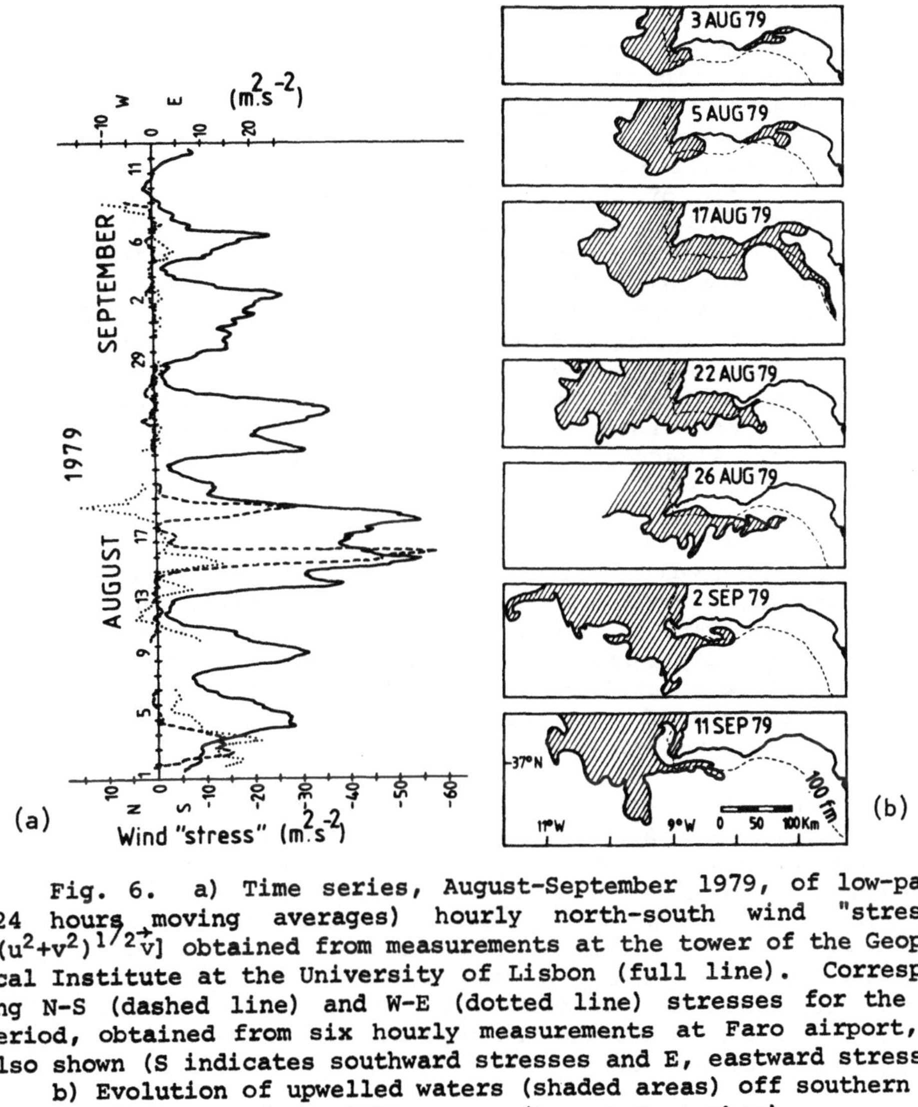

Fig. 6. a) Time series, August-September 1979, of low-passed (24 hours moving averages) hourly north-south wind "stresses" $[(u^2+v^2)^{1/2}\vec{v}]$ obtained from measurements at the tower of the Geophysical Institute at the University of Lisbon (full line). Corresponding N-S (dashed line) and W-E (dotted line) stresses for the same period, obtained from six hourly measurements at Faro airport, are also shown (S indicates southward stresses and E, eastward stresses).

b) Evolution of upwelled waters (shaded areas) off southern Portugal during part of the 1979 summer (August-September).

ferent from that on the west coast due in part to its more southern latitude and also to the coastal mountain range that extends zonally between about 8°30'W and the mouth of the Guadiana River (7°25'W) at the Portugal-Spain border. Therefore, wind stresses representative for Algarve were similarly calculated by applying a 4-point running mean on the six hourly wind measurements available from Faro airport, a few kilometers west of Cape Santa Maria. Both east-west and north-south stresses were computed in this case and are shown in Fig. 6a.

During the first days of August 1979, winds locally favorable to upwelling were observed to increase along both the west coast (northerly winds) and the south coast (westerly winds). They were related to a southern position of the Azores anticyclonic cell and a well-developed thermal low centered on the Iberian Peninsula. The corresponding satellite picture taken on 3 August (Fig. 2) is typical of the early stage of the coastal upwelling under a strengthening favorable wind. Colder waters appear along the west coast, more intensively south of all capes from where they extend southwards as cold plumes. These are probably entrained by an equatorward coastal current; such jets are characteristic features of the dynamics of coastal upwelling regimes and have been systematically found in other upwelling regions (e.g., Mooers et al., 1976). Cold waters turn anticlockwise around Cape S. Vicente (itself a strong upwelling center under local north winds) and upwelling starts to occur along the shore of eastern Algarve. Fig. 6b shows in detail the evolution of the events in this area.

The picture of 5 August shows further intensification and extension of the cold upwelling plumes along the west coast under the continuing, moderate, north winds. Off Algarve (detail in Fig. 6b), there was an eastward progression of the S. Vicente plume of freshly upwelled waters, apparently concentrating over the mid-shelf/upper slope area, whereas upwelling became more intense along its eastern coast, with a plume of cold water starting to detach seawards near the mouth of the Guadiana River.

The winds then weakened for a couple of days, but increased again between 7 and 9 August in relation to an intensification of the Azores anti-cyclone. There is a gap then in the Faro winds but the corresponding satellite infrared pictures (not shown) indicate a continuation of the eastward flow of cold waters off western Algarve until they reach Cape Santa Maria and merge into the waters upwelled along the eastern Algarve. A conspicuous long, cold plume spreads offshore from near the Guadiana River towards the southeast up to about 36°N, closely following the edge of the continental shelf (the 100 fathoms depth contour).

This situation off Algarve is analogous to that shown in the infrared picture of 17 August (Fig. 2, detail in Fig. 6b) which illustrates the situation at the highest peak of the 1979 upwelling season. The corresponding strong northerly winds (Fig. 6a), blowing at that time over the whole Portugal coast, were associated with a high pressure ridge extending northeastward from the Azores area into northern France and showing strong gradients due to interaction with a low pressure trough over the North Atlantic at the latitudes of the British Isles. On that image, and very similar to the pictures obtained on 15, 16, 18 and 19 August, the shelf and the upper slope areas off western Algarve and the shelf off eastern Algarve were covered with upwelled water which extended offshore from the Guadiana as a plume, or better, as a front showing several small-scale eddies, with a wave length of about 25 km. The circular, clear-cut boundary between the warm, oceanic and the cold, upwelled waters off Cape Santa Maria and eastern Algarve suggests the existence of a warm core eddy of about 70 km diameter in this area. The eastern boundary of this anticyclonic eddy is apparently defined by the Guadiana front whose instabilities are perhaps a manifestation of strong shears between the eddy to the west and a warm current flowing to the northwest along the Spanish coast.

Following the peak situation, the atmospheric circulation then consisted of a succession of weakening-strengthening cycles of north winds along the west coast of Portugal while very weak winds prevailed along the Algarve coast up to the end of the study period (Fig. 6a). These conditions were controlled by a generally northward position of the Azores anticyclone which was periodically broken by interaction with the polar front, originating pulses of northeasterly flows over the western Iberian Peninsula and very weak easterly winds over the Algarve area.

The last two satellite infrared pictures in Fig. 2 illustrate the response of the Portuguese coastal ocean to such wind cycles. After the peak upwelling situation of 15-19 August, the wind decayed rapidly, reaching a minimum on 22 August. Meanwhile, the area off Algarve occupied by the upwelled waters shrank progressively to the west and the Guadiana front broke down and disappeared. The satellite picture obtained on 22 August (Figs. 2 and 6b) illustrates this upwelling "spin-down" pattern.

At the west coast, a 5-day north wind cycle followed until 28 August, while light winds continued to prevail in Algarve (Fig. 6a). The daily satellite images show a zonal tongue of cold waters extending eastwards along the Algarve shelf break, interleaving with warmer waters both inshore and offshore, through small eddies with scales of 20-25 km. This feature is shown for 26 August in Figs. 2 and 6b. It is apparently due to the eastward continuation of the west coast jet along the steep topography defining the shelf edge around Cape S. Vicente and off the southern Portugal coast, indicated by the 100 fathoms contour in Fig. 6b. It is to be noted that upwelling is now absent from the Algarve coastline as a consequence of the lack of locally favorable winds.

The cold water intrusion then reduced its extension while the warm waters close to the Algarve shore progressed to the west as the northerly winds fell off. With some oscillations due to the secondary pulses of the north wind on the west coast (30 August-3 September and 5-7 September) this pattern proceeded, associated with the general upwelling spin-down situation, and is illustrated for 2 and 11 September in Figs. 2 and 6b. The coastal warm countercurrent reached Cape S. Vicente (2 September) and under weak winds even turned clockwise around it and then flowed northward along the southwest coast of Portugal (11 September). The general pre-upwelling event pattern of SSTs is then reached. The permanence of some small, cold eddies and filaments off southwest Algarve over the shelf break area indicates a bathymetric control that was also evident during the upwelling spin-up phases, even with local upwelling taking place along the Algarve coast (7-10 August, 13-19 August).

CONCLUSIONS

Local winds, the continental-shelf/upper-slope bathymetry and the coastal morphology largely determine the upwelling patterns off Portugal. On the wide and flat northern shelf, upwelling is fairly two-dimensional, isotherms tend to parallel the bottom contours, coldest water is generally observed at the coast, and variability is slightly stronger near the mid shelf. These features are generally similar to those observed in intensively surveyed upwelling regions with comparable morphological features (e.g., northwest Africa, Oregon).

Off the southern part of the west coast of Portugal three-dimensionality results in the Bay of Setúbal area from the association of large coastal protrusions with the pronounced submarine topography of the Lisboa-Setúbal Canyons; as a consequence, there is an offset of the cold waters upwelled further north relative to those upwelled at the coast in that area. To the south of Cape Sines, however, the pattern of the upwelled waters is more regular, due to an even topography, though the shelf is much steeper than in the northern region and there is virtually no shelf break. This morphology is reflected on the compression of the thermal gradients towards the shore, relative to the distributions observed on the Portugal northern shelf.

In contrast to the general upwelling conditions that occur along the (meridional) west coast, practically during the whole summer under the predominant northerly winds, at the (zonal) southern coast upwelling takes place only occasionally when favorable westerly winds blow locally. However, during north wind cycles, which may not even reach this southern area, upwelled waters are carried over the Algarve shelf break probably by an easterly extension of an apparent equatorward coastal upwelling current flowing along the west coast. During wind events comprising a moderate to strong northerly component at the west coast and a westerly component at the south coast, this cold flow merged into waters upwelled locally along the Algarve coastline and reached up to the Spanish-Portuguese border, the Guadiana River, from where it left the coast and extended as a cold front to the southeast over the shelf break for more than 100 km.

During the upwelling spin-down periods, the Guadiana front was dislocated towards the Spanish coast and disappeared quickly while a coastal westward countercurrent carrying oceanic warm waters seemed to develop over the Algarve shelf. This reached Cape S. Vicente and even turned clockwise around it and then flowed northward along part of the west coast when winds calmed down completely or reversed to southerly. A tongue of upwelled waters tended to remain offshore over the shelf break off southwest Algarve.

It is probable that the mesoscale upwelling dynamics off Algarve is dominated by the strong bathymetric step (≈500 m high) defining the shelf edge in this region. The cold water, which is apparently transported from the west coast upwelling, seems to "feel" this step as a coast and "ignores" the real Algarve shore where upwelling only takes place when induced locally by westerly winds. Without such winds, the Algarve coastal circulation seems to be predominantly westwards. Off Algarve the surface thermal features suggest some interaction between cold upwelled waters and warm-core mesoscale eddies.

Obviously, the Portuguese upwelling system deserves more *in situ* research which, coupled with remote sensing techniques like those exclusively employed in this paper, might provide adequate insight on the dynamics of the features described here.

ACKNOWLEDGEMENTS

This investigation was supported by the Junta Nacional de Investigação Científica e Tecnológica (JNICT), Lisboa, through research contract no. 47.78.47 between JNICT and the Faculty of Sciences of Lisbon University, and by the Instituto Nacional de Investigação Científica, Lisboa (Geophysical Center of Lisbon University).

The continued support provided by the Portuguese Air Force, with equipment, aircraft, officers and crews, to conduct the airborne remote sensing surveys, is gratefully acknowledged.

The high quality infrared satellite pictures used here were provided by P. Baylis of the satellite ground receiving station of the Department of Electrical Engineering and Electronics, University of Dundee, United Kingdom.

REFERENCES

Fiúza, A.F.G., Macedo, M.E. and Guerreiro, M.R., 1982, Climatological space and time variation of the Portuguese coastal upwelling, Oceanologica Acta, 5(1):31-40.

Kraan, C., 1977, Bepaling van de temperatuur van het zeeoppervlak mit infrarode straling. Een kritische evaluatie van de methode.

(Determination of sea surface temperature from infrared radiation. A critical evaluation of the method.), Wetenschappelijk Rapport, 77-5, KNMI, De Bilt, 99 pp.

Mooers, C.N.K., Collins, C.A. and Smith, R.L., 1976, The dynamic structure of the frontal zone in the coastal upwelling region off Oregon, Journal of Physical Oceanography, 6(1):3-21.

STABLE-ISOTOPE COMPOSITION OF FORAMINIFERS: THE SURFACE AND BOTTOM WATER RECORD OF COASTAL UPWELLING

Gerald Ganssen and Michael Sarnthein

Geologisch-Paläontologisches Institut der Universität
Olshausenstrasse 40-60
D-2300 Kiel, Federal Republic of Germany

ABSTRACT

Sea-surface temperature (SST), salinity gradients, and enhanced productivity of coastal upwelling regions leave a marked imprint on the oxygen and carbon stable isotopic composition of foraminiferal shells. For a better understanding of these isotopic signals, we systematically analyzed planktonic and benthic species from surface sediment samples and plankton tows from the eastern Atlantic continental margin (12-35°N). Due to seasonal variations of the Trade Wind regime, centers of coastal upwelling where SST is much below normal migrate markedly in latitude, from ~12°N during late winter to ~30°N during late summer, and continue to be active throughout the year at 20-23°N. As a result, nearshore seasonal temperature variations are low north of 20°N, but unusually high near 15°N. The $\delta^{18}O$ values of shells of *Globigerinoides sacculifer* and *Globigerinoides ruber* (pink and white) delineate the pattern of SST during summer and $\delta^{18}O$ values for *Globorotalia inflata* reflect the SST during winter. Oxygen isotopes of *Globigerina bulloides* document the temperature ranges of the upwelling seasons. The combined isotopic record of *G. ruber* white and *G. inflata* clearly reveals the latitudinal variations of seasonality and the annual mean SST, thus documenting the long-term position of upwelling cells during the different seasons in the sediments.

Tests of *G. ruber* pink and *G. inflata* show a depletion of ^{13}C in regions of summer and winter upwelling, respectively. The latitudinal variation of $\delta^{13}C$ values for *G. ruber* white, *G. sacculifer*, and *G. bulloides* is not related to upwelling intensity. A decrease in $\delta^{13}C$ also occurs in the benthic species *Uvigerina finisterrensis* and *U. peregrina* in water depths near 500 m where supply of organic carbon is strongest.

The nutrient-rich, less saline South Atlantic Central Water which feeds the southern and central upwelling zones off northwest Africa can be distinguished from more nutrient-depleted North Atlantic Central Water by low $\delta^{18}O$ values in living (stained) shells of *Uvigerina sp.* and *Hoeglundina elegans.*

INTRODUCTION

The stable isotopic composition of foraminiferal shells should reflect the distributional pattern of coastal upwelling in the sediments. For example, upwelling-induced seasonal lowerings of sea-surface temperatures may be recognized by an increase of the $\delta^{18}O$ values (approximately 0.25°/₀₀/°C) in shells of planktonic species with seasonally restricted occurrence. In addition, the enhanced productivity of upwelling cells (Smith, 1968) may promote abundances of certain planktonic species. The oxygen isotopic composition of these shells may record the temperature range of the upwelling seasons only. The upwelling-induced high productivity should influence particularly the $^{13}C/^{12}C$ ratio in both planktonic and benthic shells (Berger, Diester-Haass and Killingley, 1978). Furthermore, the stable isotope signature of foraminiferal shells can also reflect, to some degree, the inherited isotopic characteristics of the different upwelled water masses. Finally, Wefer, Dunbar and Suess (this volume) show that contrary to the general distribution (Berger, Killingley and Vincent, 1978), $\delta^{18}O$ values are high in smaller shells of species from upwelling regions because of the high rates of reproduction characteristic of cold fertile upwelling waters.

Some factors may limit the stable-isotope record in sediments from the continental margin. For example, major parts of the continental shelf and upper slope are exposed to erosion and non-deposition by the poleward undercurrent (Ganssen, 1982). Accordingly, the areas which lie directly underneath the upwelling plumes are in many cases bare of any isotopic record in the sediments (compare Fütterer, this volume). Benthic foraminiferal shells from the deeper slope environment, beyond 500 m, might still reflect the upwelling-generated flux of organic matter in their carbon-isotope compositions. The oxygen isotopes of benthic shells at these depths are no longer influenced by upwelling but by the local deep-water character and by effects of (upwelling-related) locally increased calcium carbonate dissolution. Foraminiferal ooze at the deeper slope probably contains a mixture of planktonic shells from both inside and outside the upwelling centers because of lateral shell transport. Accordingly, the upwelling signal of plankton-shells is possibly dampened and in addition, may depend on the width of the continental shelf, i.e., the distance planktonic shells are transported from their origin to their place of deposition.

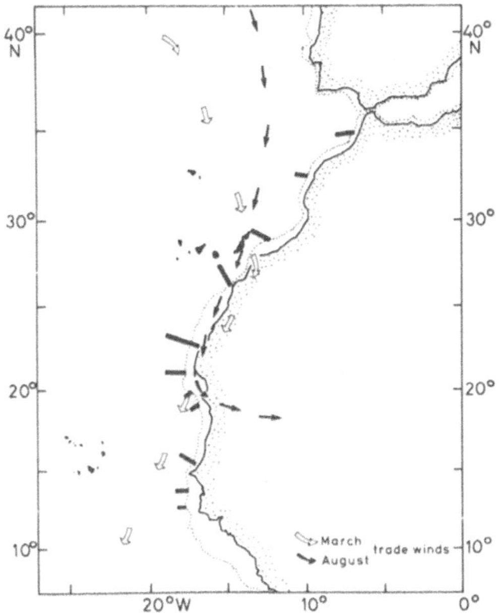

Fig. 1. Location of transects (black bars) with sediment samples for carbon and oxygen isotopic determinations. Position of trade winds during winter and summer is shown by arrows (after R. Bryson, Madison, Wisconsin, pers. comm.).

In this paper, we examine the coastal upwelling system off northwest Africa, an upwelling which mainly occurs over the shelf and upper slope. We have mapped in detail the stable isotopic composition of modern foraminifers in the surface sediments from the continental slope and in plankton hauls from sea-surface water. In the following, we aim to determine the response of stable isotopes to the different structures of oceanic circulation, particularly to the influences of winds, latitudes, and seasons on the coastal upwelling regime.

ENVIRONMENTAL SETTING

The position and seasonal variation of coastal upwelling off northwest Africa is controlled by the trade winds that shift seasonally in latitute (Fig. 1) and by coastal topography. Upwelling cells form anomalies of low sea-surface temperatures (SST) near 12°N during late winter (February-April), near 30°-32°N during late summer (July-October), and between 20° and 23°N almost perennially (Schemainda, Nehring and Schulz, 1975; Wooster, Bakun and McLain, 1976; Speth,

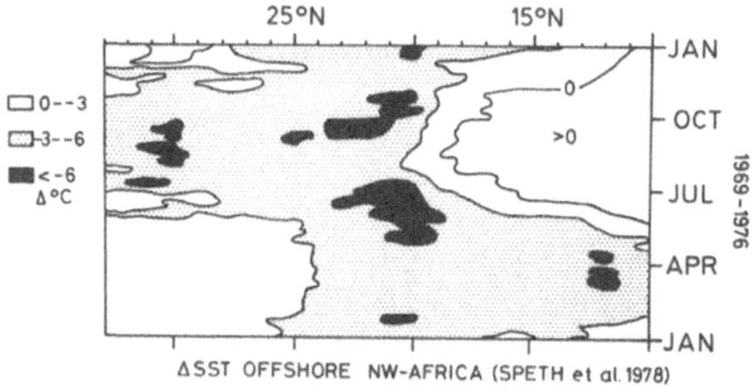

Fig. 2. Mean sea-surface temperature differences between northwest African coastal areas and mid-Atlantic for the period 1969 until 1976. Negative values indicate coastal temperatures colder than mid-ocean. Differences reflect the appproximate intensity of coastal upwelling (modified after Speth et al., 1978).

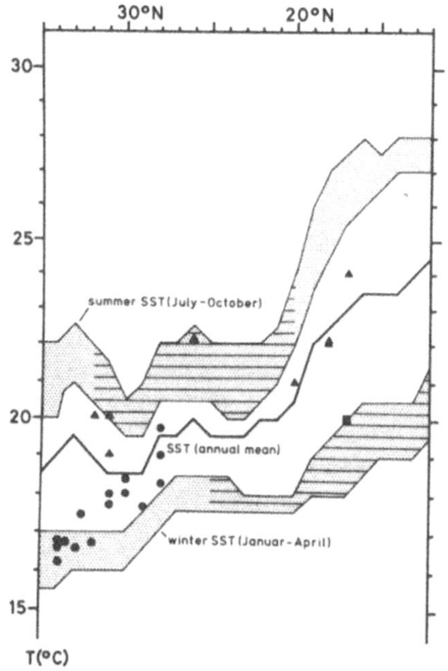

Fig. 3. Average range of sea-surface temperatures offshore northwest Africa up to an offshore distance of approximately 200 km. Data from Robinson, Bauer and Schroeder (in press). Upwelling regions indicated by horizontal hatching (compare Fig. 2). Measured SST during plankton hauls are shown by dots (March), triangles (October), and squares (January).

Detlefsen and Sierts, 1978; Fig. 2). As a result, the seasonal fluctuations of nearshore SST are smoothed north of 20°N and exaggerated to exceed 8°C near 15°N where winter upwelling increases the 'normal' seasonal SST anomalies (Fig. 3).

MATERIALS AND METHODS

The surface sediment samples are from 44 Reineck Kastengreifers and 67 van Veen grabs collected between 100 and 4300 m water depth on the northwest African continental margin (Fig. 1) during a number of cruises of RV "Meteor" and RV "Valdivia" 1971-1980 (Seibold 1972; Seibold and Hinz, 1976; Ganssen, 1982). Samples comprise a 0.5-1.0 cm thick layer of the actual sediment surface which was carefully scraped off from the rest of the sample immediately after recovery (Lutze, 1980). The samples were stored in ethanol stained with Rose Bengal, in order to separate "living" (stained) from "dead" (unstained) benthic specimens (Walton, 1952; Lutze, 1964; Lees, Buller and Scott, 1969). In addition, samples of living planktonic foraminifers were collected during "Meteor" cruises 51, 53, and "Valdivia" cruise Oct. 1979 by two different methods:

1) Seawater was pumped through a 100 µm plankton net by a pump in the hold of RV "Meteor" and RV "Valdivia". From 5 to 40 cubic meters of seawater from -4m water depth were filtered for each sample (details of method in Thiede, 1975 and Ganssen, 1982)

2) Vertical plankton hauls from 10 cubic meters of water each were taken with a Hydrobios multi-net (100 µm mesh size) at 5 depth intervals down to 200 m (Weikert and John, 1981). Several (5-7) hauls were taken at each station in order to collect larger samples. All plankton samples were fixed in ethanol-seawater solution avoiding an alteration of the isotopic composition of the shells (Ganssen, 1981).

Both sediment and plankton samples were washed in 63 µm and 100 µm sieves, respectively, and dried at 40°C. All samples were sieved to narrow sand-size intervals (planktonic species: 315-400 µm) to minimize isotopic variation resulting from changing test size (Berger et al., 1978b; Duplessy, 1978). The planktonic foraminiferal species were selected for the isotopic analyses on the basis of the regional distribution patterns presented by Pflaumann in Sarnthein et al. (1982). *Globigerina bulloides* was selected as main representative of cool, upwelling-influenced water, and *Globigerinoides sacculifer* (with a sack-like end chamber) and *G. ruber* (pink and white variation) as representatives of tropical warm water. *Globorotalia inflata* was chosen to reflect conditions of upwelled subsurface waters. Living (stained) specimens of *Uvigerina peregrina*, *U. finisterrensis*, *Hoeglundina elegans* , and *Trifarina elongatastriata* (315-400 µm and, if necessary, up to 630 µm) were selected as benthic species for isotope analysis because they are spread over wide depth ranges along the continental slope (Lutze, 1980). Dead specimens of these species were analyzed for comparison. Three to fifty individuals of each species were picked with special attention to a good state of preservation in order to avoid contamination by admixed fossil specimens. The samples were cleaned in an ultrasonic bath to remove adhering fine-grained particles, and reacted with orthophosphoric acid at 50°C under vacuum. The extracted CO_2 gas was separated by three freezing steps and analyzed in an on-line VG Micromass 602 D mass spectrometer (list of 600 isotopic analyses in Ganssen, 1982).

All oxygen and carbon isotopic data are referred to PDB by the standard notation (Craig, 1957) and are calibrated to the NBS standard through an intermediate laboratory standard (Solnhofen Limestone). The analytical precision from this working carbonate standard run before each analytical session is ± 0.1°/₀₀ (1σ) for oxygen and ±0.05°/₀₀ (1σ) for carbon (Ganssen, 1982).

No direct isotopic determinations of the carbon and oxygen isotopic composition of seawater are available from the eastern Atlantic continental margin. Such values would enable us to relate the $\delta^{18}O$ of foraminiferal shells formed in or near equilibrium with ambient seawater directly to the actual temperature and salinity or to species-specific metabolism. Because of the lack of such data on seawater, we computed the isotopic equilibrium of shell formation by applying the paleotemperature equation of Epstein et al. (1953):

$$T = 16.5 - 4.3\,(\delta^{18}O_c - \delta^{18}O_w) + 0.14\,(\delta^{18}O_c - \delta^{18}O_w)^2 \qquad (1)$$

where $\delta^{18}O_c$ equals the $\delta^{18}O$ of carbonate and $\delta^{18}O_w$ of modern seawater (both values relative to the PDB-CO_2 standard). Other paleotemperature equations (Craig, 1965; Shackleton, 1973; Duplessy, 1978) do not differ significantly in their results from those calculated by equation (1).

The $\delta^{18}O_w$ values used in equation (1) were deduced from the correlation lines of Craig and Gordon (1965) relating the $\delta^{18}O$ composition of seawater (SMOW) to the salinity in the present ocean by the equations

$\delta^{18}O_{wt} = 0.59S - 20.68$ (2) for the whole of Atlantic seawater

and $\delta^{18}O_{ws} = 0.11S - 3.15$ (3) for Atlantic surface water only,

where S = salinity (°/₀₀). Salinity and temperature data employed for this study are based on unpublished data supplied by courtesy of the Deutsches Ozeanographisches Datenzentrum (DOD), Hamburg, and on Robinson et al. (in press). The $\delta^{18}O_w$ values of equations (2) and (3) were corrected by -0.22°/₀₀ in order to take into account the isotopic difference between SMOW and PDB standard scales.

$\delta^{13}C$ values correlate with *in situ* water temperatures during calcite precipitation (Emrich, Ehhalt and Vogel, 1970), i.e., they change by +0.35°/₀₀ when the SST increases by 10°C. We computed the temperature related $\delta^{13}C$ correction for the five species examined (Fig. 5) according to the specific range of seasonal temperatures as recorded by $\delta^{18}O$.

STABLE ISOTOPIC COMPOSITION OF PLANKTONIC SPECIES: THE SURFACE-WATER RECORD OF UPWELLING

Data

The oxygen isotopic data of five dominant planktonic species (Sarnthein et al., 1982) from both surface-sediment samples and plankton tows are compared in Fig. 4A-E with the range of SST observed offshore northwest Africa during late winter and late summer (see Fig. 3). All species plot as distinct data groups on the latitude-temperature diagram. The $\delta^{18}O$ values for *G. sacculifer* and white *G. ruber* are largely consistent with the range of observed summer temperatures. $\delta^{18}O$ values for pink *G. ruber* are generally lighter (by 0.5°/₀₀) than the converted summer temperatures.

The $\delta^{18}O$ values for *G. inflata* correspond to temperatures lying slightly below the range of the observed winter temperatures (up to 0.6°/₀₀). This holds true for most profiles, independent of latitudes. *G. bulloides* has $\delta^{18}O$ values which agree approximately with the observed winter temperatures near 12°-19°N and with the summer temperatures near 29°-32°N, and which average the annual temperatures near 20°-27°N and at 34°N. This apparently complicated distribution pattern closely corresponds to the seasonally shifting latitudinal position of upwelling activity as shown in Fig. 2. Hence, *G. bulloides* --as the only species analyzed-- may only form the bulk of its shell carbonate inside the upwelling centers south of 32°N (confirming Prell and Curry, 1981). North of 33°N, *G. bulloides* reflects the annual average of the temperature record possibly because it may be transported to this location mainly by the North Atlantic Transitional Water, i.e., from perennial habitats outside upwelling cells (Bé and Tolderlund, 1971).

$\delta^{18}O$ values for samples from plankton tows (Fig. 4A-E) are generally consistent with the data range observed from the surface sediment samples. This coincidence holds true only for tows which originate during the main season that is indicated for a species by the sediment data of Fig. 4A-E. The matching data range tends to exclude admixtures from sub-fossil shell sediments and from far-reaching lateral shell transport.

Distinct latitudinal variations of the $\delta^{13}C$ values are common to all planktonic species examined (Fig. 5). *G. sacculifer* and pink and (less distinctly) white *G. ruber* show the highest $\delta^{13}C$ values of

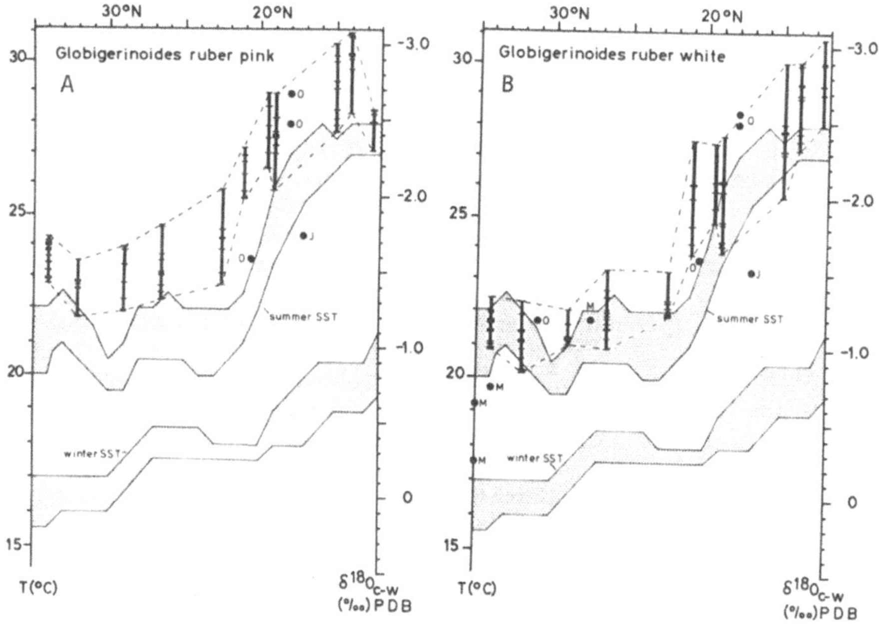

Fig. 4. Latitudinal oxygen isotopic variations of planktonic foraminifers from sediment samples (vertical bars with dashes for each individual analysis) and plankton tows (solid dots) collected during January (J), March (M), and October (O). $\delta^{18}O_{c-w}$ is the difference of $\delta^{18}O$ for foraminiferal carbonate minus $\delta^{18}O$ for the local surface-water composition (equation 3), calculated to eliminate the influence of local salinity variations. Dotted fields show range of observed winter and summer SST.

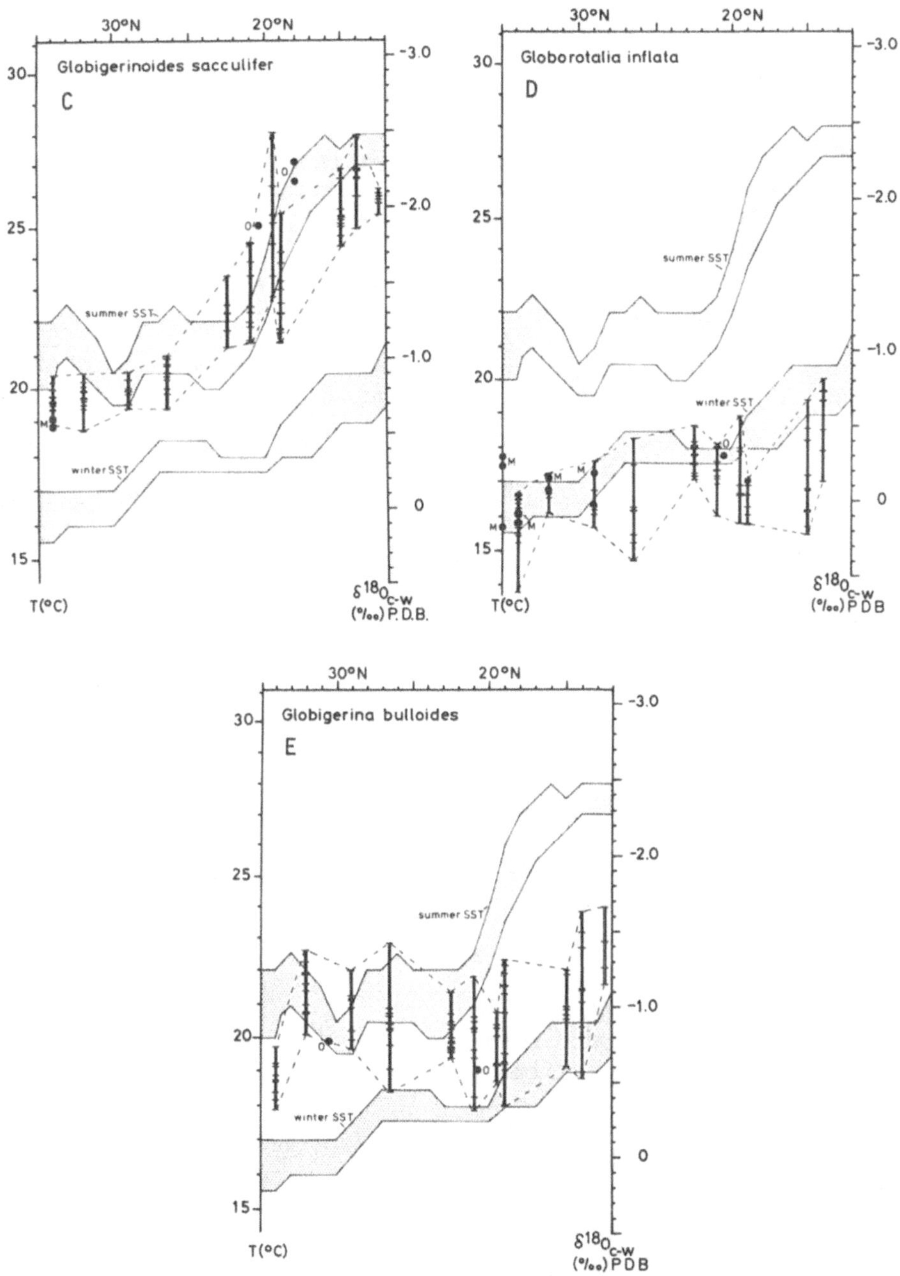
30°N
20°N
Globigerinoides sacculifer
C
summer SST
winter SST
T(°C)
δ18O c-w (‰) P.D.B.
Globorotalia inflata
D
summer SST
winter SST
T(°C)
δ18O c-w (‰) PDB
Globigerina bulloides
E
summer SST
winter SST
T(°C)
δ18O c-w (‰) PDB

all species and a relative $\delta^{13}C$ minimum in the perennial and northern (summer) upwelling regions (as may be expected from the interpretations of Berger et al, 1978a). *G. inflata*, generally less enriched in ^{13}C, is relatively depleted in ^{13}C in the perennial and southern (winter) upwelling regions. *G. bulloides*, the species reflecting most distinctly the temperature regime of upwelling in its $\delta^{18}O$ values, is drastically depleted in its average $\delta^{13}C$ values, by 2.5°/₀₀ relative to *G. sacculifer*. However, *G. bulloides* is relatively enriched in ^{13}C only in the regions with perennial and winter upwelling where the upwelled water is richest in nutrients according to Fraga (1974) and Codispoti and Friederich (1979).

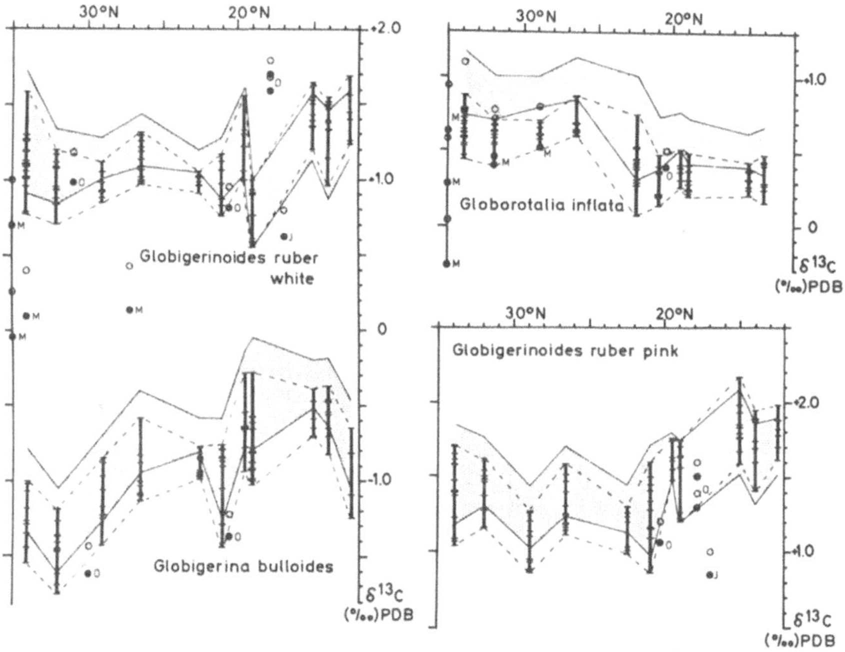

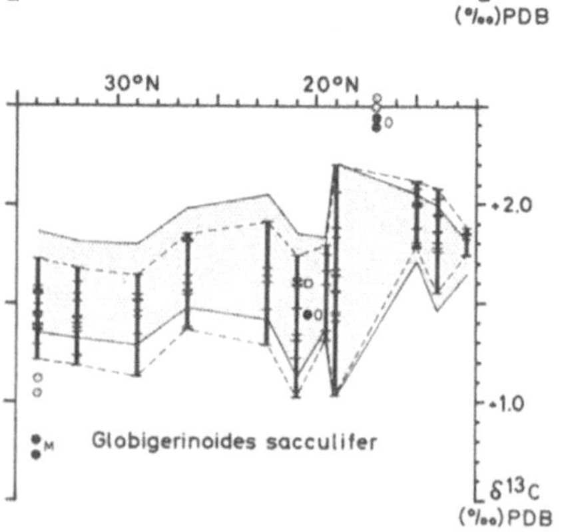

Fig. 5. Latitudinal $\delta^{13}C$ variations in planktonic foraminiferal shells from sediment samples (vertical bars with dashes for each individual analysis) and plankton tows (solid dots) collected offshore northwest Africa during January (J), March (M), and October (O). Dotted fields and circles show range of $\delta^{13}C$ values after correction for local SST (after Emrich et al., 1970).

The temperature corrected $\delta^{13}C$ range of all species from sediment samples largely coincides with the (corrected) $\delta^{13}C$ values from those plankton tow samples originating from seasons which are characteristic for the major occurrence of each species (Fig. 5). We did not observe a systematic enrichment of ^{13}C for samples from surface sediments relative to those from plankton tows as suggested by Williams, Bé and Fairbanks (1981): out of 28 plankton samples, 4 are enriched, 11 depleted, and 13 equal to the sediment samples.

Discussion

The fidelity with which the stable isotopic record mirrors coastal upwelling is tied to two basic questions. One concerns the degree to which planktonic foraminifers precipitate calcite in isotopic equilibrium with seawater or whether metabolism affects fractionation. The other problem arises from the different seasonal habitats of each species, particularly the water depths determining the temperature-salinity conditions of calcification. Systematic examination of shell collections from stratified plankton tows and traps has recently provided the following crucial information to be compared with our data (Berger, Bé and Vincent, 1981).

Williams et al. (1981, values corrected for conversion factor SMOW-PDB: 0.22°/₀₀) and Deuser et al. (1981) reported that the composite oxygen isotopic composition of *G. ruber* white was slightly depleted (~0.2°/₀₀) compared to the predicted seasonal $\delta^{18}O$ of seawater. In addition, *G. ruber* pink was found to be lighter by 0.66°/₀₀. Both species variations were shown by these authors to produce the calcite shells mainly during the summer and close to the sea surface, with grain-size means similar to the size range (315-400 μm) used for our isotopic analyses.

G. inflata was observed dwelling in the upper 200 m subsurface water near Bermuda and calcifying mainly above the thermocline (<75 m). It showed a composite $\delta^{18}O$ record being in equilibrium with the temperature level of the late winter season (Fairbanks, Wiebe and Bé, 1980; Deuser et al., 1981).

G. bulloides was found living (in cycles?) both close to the sea surface and in deeper water (Fairbanks et al., 1980). Its composite $\delta^{18}O$ value lies very near to the annual mean of the equilibrium line of surface water, despite an obvious disequilibrium ('vital effect') of samples from single seasons amounting to -0.5 to -0.7°/₀₀ (Deuser et al., 1981; Kahn and Williams, 1981).

G. sacculifer has rarely been collected in plankton tows so far. Fairbanks et al. (1980) observed its habitat as being centered in the 100 m thick surface layer. According to Duplessy, Blanc and Bé (1981) about 80% of the shell of *G. sacculifer* is precipitated in shallow water and only the rest is in greater water depths, which results in slightly higher $\delta^{18}O$ values compared to the surface-water equilibrium.

The results of Fig. 4A-E demonstrate that the cited (seasonal) average $\delta^{18}O$ values for plankton samples from the central North Atlantic match with surprising detail the $\delta^{18}O$ data for our surface sediment samples collected from the continental margin over a lati-

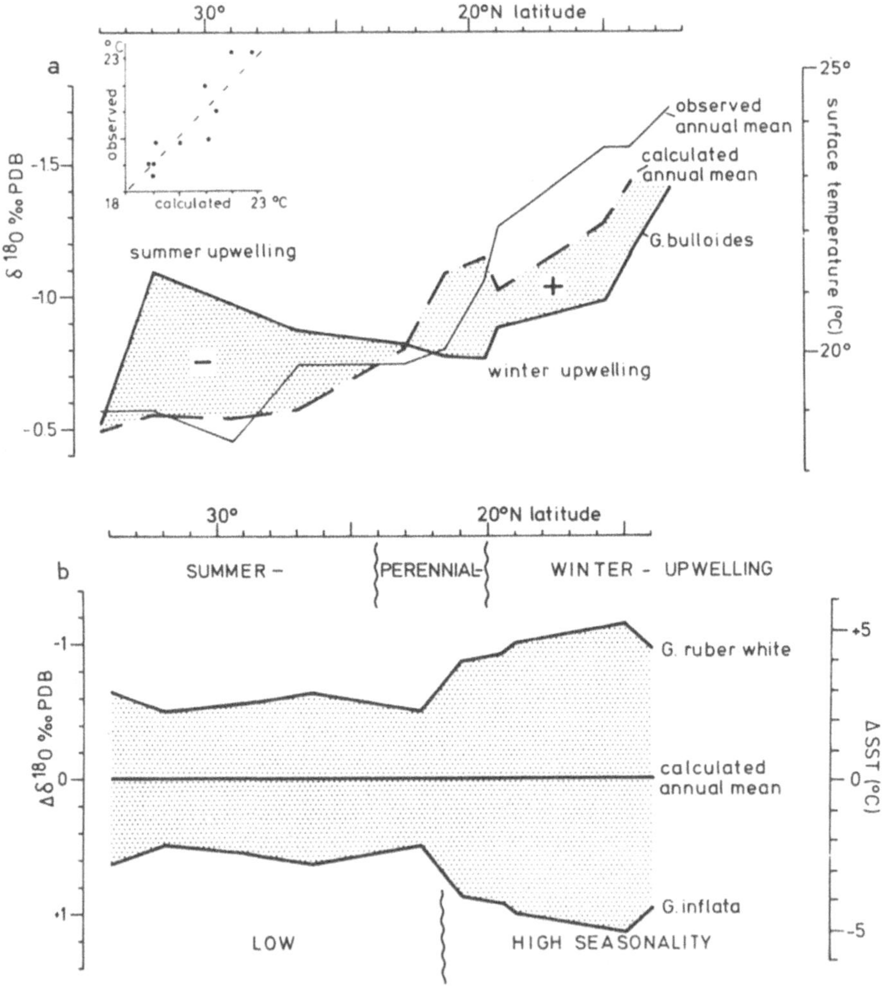

Fig. 6. (A) Latitudinal variation of annual mean SST offshore northwest Africa as calculated from the mean of the combined $\delta^{18}O_{c-w}$ values for *G. ruber* white and *G. inflata* shells (Fig. 4B, D). Observed annual mean SST from Robinson et al. (in press). Note positive and negative $\delta^{18}O$ deviations of the means of *G. bulloides* (Fig. 4E) indicating the ranges of winter and summer upwelling.

(B) Latitudinal variation of $\delta^{18}O$ deviations of *G. ruber* white and *G. inflata* from the calculated mean values (Fig. 6A) reflecting the amplitude of seasonal SST contrast.

tudinal range of almost 3000 km. This holds true for each species examined. Systematic discrepancies between values for plankton and sediment samples as reported by Williams et al. (1981) are not observed in our data. The composite $\delta^{18}O$ values of *G. bulloides*, which have a latitudinally different response to model seasonal equilibria south of 34°N, apparently document and average the $\delta^{18}O$ equilibria of the main high fertility seasons.

In summary, we observe a double response of $\delta^{18}O$ values for plankton to coastal upwelling (Fig. 6A and 6B). The calculated mean for $\delta^{18}O$ values of *G. ruber* white plus *G. inflata* represents the observed annual mean SST and its gradient very well. The actual deviations from the mean of $\delta^{18}O$ for these two species document the range of seasonal SST fluctuations. The positions of summer and winter upwelling plumes are recorded because upwelled waters are cooler than surrounding waters and, accordingly, diminish the seasonal SST signal during summer or enlarge it during winter. In addition, winter and summer upwelling centers are clearly documented by enriched or depleted $\delta^{18}O$ values, respectively, for *G. bulloides* relative to the calculated means for the $\delta^{18}O$ values of *G. ruber* plus *G. inflata*.

It is far more difficult to assess our $\delta^{13}C$ data from plankton shells in terms of upwelling signals than to interpret the $\delta^{18}O$ values. The complex relationships of the carbon cycle and ^{13}C depletion and enrichment in upwelling systems have been considered in detail by a number of authors (e.g., Deuser and Hunt, 1969; Kroopnick, 1974; Williams, Sommer and Bender, 1977; Berger et al., 1978a; Fontugne and Duplessy, 1978; Prell and Curry, 1981). The data of Kroopnick (1980, Geosecs station 113, 3-695 m water depth) show that the dissolved oxygen (ml/l) and the $\delta^{13}C$ of total dissolved inorganic carbon (ΣCO_2) can be correlated by the equation

$$\delta^{13}C = 0.6\ O_{2\ \mathrm{diss.}} - 0.75 \quad (4)$$

On the basis of these considerations and using hydrographic data (oxygen values, $\delta^{13}C$ data of ΣCO_2) from various sources (Deutsches Ozeanographisches Datenzentrum, unpublished; Kroopnick, 1980), we calculated that the $\delta^{13}C$ of ΣCO_2 amounts to 0.0 to 0.5°/₀₀ near 12°-22°N and 0.5 to 1.5°/₀₀ near 22°-34°N in the source regions of upwelled water, which lie between 100 and 200 m depth. As is to be expected, the $\delta^{13}C$ of CO_2 in this source water is up to 2.0°/₀₀ lighter than that of average North Atlantic surface waters (about +2°/₀₀ $\delta^{13}C$; Kroopnick, 1980; Williams et al., 1981).

The tests of planktonic foraminifers dwelling in the upwelling water should generally record the calculated ^{13}C depletion in their carbon isotope composition. The latitudinal variations of the temperature-corrected $\delta^{13}C$ values for the "summer" species, *G. ruber* white and *G. sacculifer*, lie below the general noise background (except for a "light" profile of *G. ruber* white at 19°N). Latitudinal

Table 1. Comparison of $\delta^{13}C$ range for planktonic foraminifers from sediments of the northwest African continental margin and from plankton tows and traps in the central Atlantic. ($\delta^{13}C$ values, in ‰, are corrected for a uniform temperature range of 25°C)

	NW Africa Sediments 315-400 mm	Average Difference (NW Africa vs. Atlantic and Bermuda)	Bermuda 125-500 µm (Deuser et al., 1981)	Bermuda 125-500 µm (Williams et al., 1981)	Central Atlantic >250 µm (Erez and Honjo 1981)
G. ruber pink	+1.0 - +2.1	+0.85	-0.2 - +1.3	-0.2* - +1.2**	+0.4 - +1.3
white	+0.3 - +1.4	+0.7		-0.3 - +0.6***	
G. sacculifer	+1.0 - +2.2	+0.2			+1.3 - +1.5
G. inflata: outside	+0.3 - +1.1	+0.25	+0.3 - +0.6		
inside upwelling	+0.4 - +0.7	+0.1			
G. bulloides	-1.6 - 0.0	+1.2	-2.1 - -1.9		

* 210-270 µm ** 360-500 µm *** 240-360 µm

$\delta^{13}C$ variations are distinct for *G. bulloides*; but $\delta^{13}C$ increases only in the southern (winter) upwelling regions, particularly after the temperature related $\delta^{13}C$ correction. Accordingly, these three species do not document the postulated $\delta^{13}C$ gradient associated with latitudinally changing supply of upwelled water which is considered equal to the changing coastal upwelling intensity. However, the $\delta^{13}C$ records of the "winter" species *G. inflata* and of the "summer" species *G. ruber* pink do record the respective regions of winter and summer upwelling by $\delta^{13}C$ signals of about -0.5°/₀₀.

Our conclusions corroborate, in part, the positive correlation of $\delta^{13}C$ in *G. ruber* with upwelling reported by Berger et al. (1978a), but also the findings of Prell and Curry (1981) who demonstrated that the $\delta^{13}C$ values of various other species do not correlate with the upwelling signals.

It is interesting to note that most $\delta^{13}C$ values for our samples (both plankton and surface sediment) are enriched by up to 1.2°/₀₀ relative to isotopic data from plankton collected in the central North Atlantic, i.e., outside this upwelling region (Table 1). An equal $\delta^{13}C$ range occurs only for the carbon isotopic composition of *G. inflata* from the southern (winter) upwelling region. We cannot directly explain these general discrepancies, which possibly follow a regular trend but definitely do not correlate with the upwelling temperature gradients. Rapid gas fractionation during the warming of upwelled water masses and exchange with atmospheric CO_2 may be important factors dampening the potentially depleted ^{13}C signal as suggested by Prell and Curry (1981). An extensive consumption of CO_2 by plankton blooms near the sea surface was shown by Codispoti et al. (1982) and may be the major factor for the general ^{13}C enrichment which we observe for most planktonic foraminiferal species nearshore and which is opposed to the initially postulated ^{13}C depletion within the zones of upwelling water.

STABLE ISOTOPIC COMPOSITION OF BENTHIC SPECIES: THE BOTTOM-WATER RECORD OF COASTAL UPWELLING

Data

The bathymetric distribution of stable isotopic values for living (stained) benthic species is shown on N-S transects along the northwest African upper continental slope (Figs. 7 and 8). The observed $\delta^{18}O$ values of *Uvigerina peregrina*, *U. finisterrensis*, and *Hoeglundina elegans* were corrected for the local bottom-water temperatures (Lutze, 1980) by calculating an estimated $\delta^{18}O_w$ with equation 1. Both genera were mapped on one plot in order to improve the sample coverage which suffered from the fact that many samples did not contain living (stained) specimens. The merging of the data is legitimate because the difference in oxygen isotope values between the

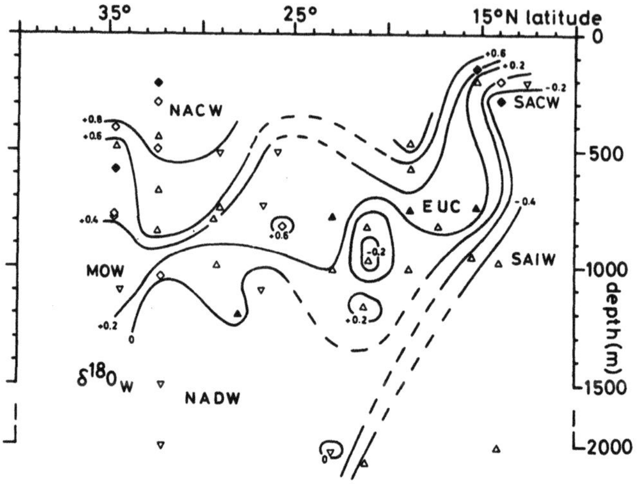

Fig. 7. Latitudinal distribution pattern of estimated $\delta^{18}O_w$ values for living (stained) *Uvigerina sp.* (Δ) and *Hoeglundina elegans* (∇) in °/₀₀ PDB. (◇) = mean value of *Uvigerina sp.* plus *H. elegans*. Solid symbols represent mean of multiple analyses. The isotopic offset of $\delta^{18}O$ values for *H. elegans* was corrected by -0.5°/₀₀ relative to *Uvigerina sp.* (see Fig. 9) which is assumed to calcify in isotopic equilibrium with ambient seawater (Shackleton, 1973). EUC = Equatorward Undercurrent; MOW = Mediterranean Outflow Water; NACW = North Atlantic Central Water; NADW = North Atlantic Deep Water; SACW = South Atlantic Central Water; SAIW = Subantarctic Intermediate Water.

two genera is rather constant and is almost independent of temperature variation (Fig. 9), with *H. elegans* being offset from the values of *Uvigerina sp.* by about +0.5°/₀₀ (the actual range lies between 0.62°/₀₀ at 3°C and 0.41°/₀₀ at 13°C, i.e., within the scatter of data). Accordingly, the systematic variation of the isotopic disequilibrium of calcite (*Uvigerina sp.*) and aragonite (*Hoeglundina elegans*) as suggested by Sommer and Rye (1978) is possibly only half the value they have in our data and very close to the range of scatter.

The distribution of estimated $\delta^{18}O_w$ values in Fig. 7 enables us to separate essentially three domains along the northwest African upper continental slope. $\delta^{18}O$ is heavy in the area above 700-900 m between 15° and 32°N, and generally light below 1000 m (by about 1°/₀₀). ^{18}O depletion is also marked (by about 1.2°/₀₀) south of 15°N below 100-200 m water depth.

The $\delta^{13}C$ values of (living/stained) *U. finisterrensis* and *U. peregrina/pygmaea* (species identification according to Lutze, 1980) were plotted together in Fig. 8 because the offset of the two species of about 0.3°/₀₀ appeared sufficiently constant (Ganssen, 1982).

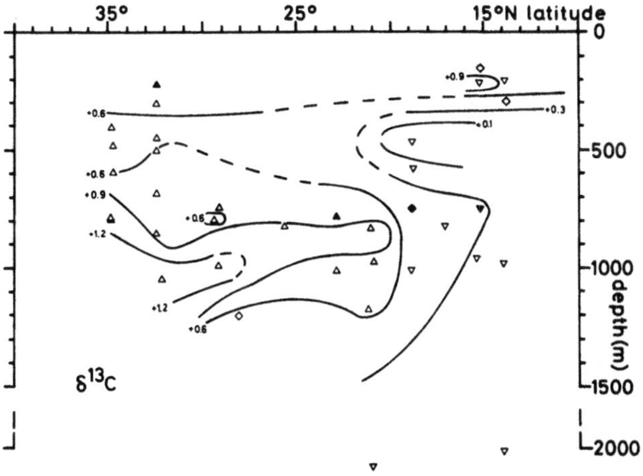

Fig. 8. Latitudinal distribution pattern of bottom-water temperature corrected $\delta^{13}C$ values for living (stained) *U. peregrina/pygmaea* (∇) and *U. finisterrensis* (Δ). (◇) = mean value of *U. peregrina* plus *finisterrensis*. Solid symbols represent mean of multiple analysis. The carbon isotopic values of *U. peregrina/pygmaea* were spliced with those of *U. finisterrensis* by adding 0.3°/₀₀ (Fig. 9; Ganssen, 1982).

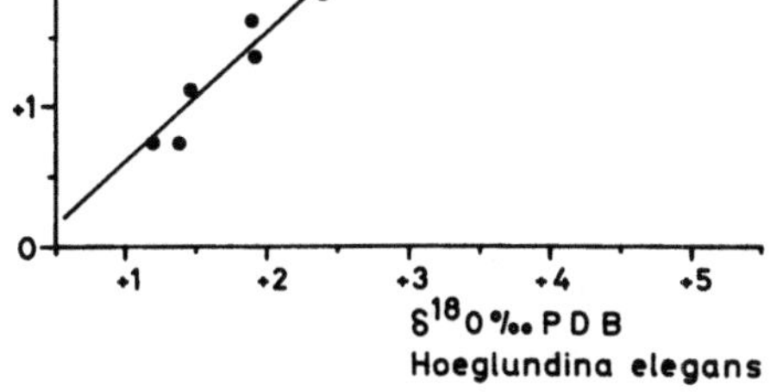

Fig. 9. Correlation of $\delta^{18}O$ values for living (stained) shells of *Uvigerina sp.* vs. *H. elegans*. $r^2 = 0.96$.

However, *H. elegans* had to be disregarded because its strong disequilibrium fractionation relative to *Uvigerina sp.* is irregular. All $\delta^{13}C$ values were corrected for the effects of local bottom-water temperatures (Lutze, 1980) by 0.35°/₀₀ per 10°C (after Emrich et al., 1970). North of 15°N $\delta^{13}C$ values are lowest near 400-500 m water depth and south of 15°N, they are low generally below 200 m. $\delta^{13}C$ is high (up to 1°/₀₀) near the surface above 250-300 m water depth and along an intermediate zone being centered between 750 and 1100 m and gradually attenuating from north to south.

Discussion

As expected, the pattern of estimated $\delta^{18}O_w$ of benthic foraminifers enables us to distinguish different water masses feeding the coastal upwelling system. One can readily separate the more saline (35.2->36.0°/₀₀) and somewhat nutrient-depleted North Atlantic Central Water (NACW) in the north (>+0.8°/₀₀ $\delta^{18}O_w$) from the less saline (<35.4°/₀₀) but nutrient enriched South Atlantic Central Water (SACW) in the south (<+0.2°/₀₀ $\delta^{18}O_w$) in Fig. 7. (A compilation of the bottom-water stratification is given in Sarnthein et al., 1982.) We do not yet understand why this $\delta^{18}O$ signal is apparently lost by the time it reaches the surface and should be recorded in the tests of planktonic foraminifers (Fig. 6). The $\delta^{13}C$ in the gas content of these undercurrents (and the foraminifer shells) is related linearly to the apparent oxygen utilization (AOU: Redfield, 1942) (equation 4). Therefore it depends on both the local, upwelling-controlled release of carbon and nutrients by oxidation and the inherited oxygen deficit of the water masses. The latter obviously determines the maximum intensity of the oxygen minimum layer south of 17°N. Accordingly, the $\delta^{13}C$ decrease (Fig. 8) is more distinct below the southern upwelling region than below the northern. However, this inherited signal of upwelled water is only weakly observed by our sample coverage.

The other feature reflected by the stable isotopes of living benthic foraminifers is a broad, undulating shearing zone of the various intermediate waters with opposite current directions near 800-1000 m water depth. There the Subantarctic Intermediate Water (SAIW; <35°/₀₀ salinity) interferes with the equatorward flowing Equatorial Undercurrent (EUC) and overrides the Mediterranean Outflow Water (MOW; 1000-1500 m depth). At this depth, horizontal undulations in the benthic estimated $\delta^{18}O_w$ values are possibly a result of the rough, canyon-controlled bottom morphology. However, at this depth range we note that the $\delta^{13}C$ values show a distinct maximum just near the surface of the MOW and the EUC, but no signal related to the nearby coastal upwelling is evident. Thus we believe that most of the benthic carbon isotopic signals related to present upwelling are confined to the sediments above the critical zone of transition and can only rarely be detected in sediments of, for example, the continental rise.

In addition, we know from a comparison of oxygen isotopes from living (stained specimens of *Trifarina elongatastriata*) and dead benthic foraminifers (Ganssen, 1982) that in the critical depths of the undercurrents (200-600 m) erosion prevails over large distances. Thus a suitable benthic isotopic documentation of coastal upwelling is strongly impeded. However, in case the sea level drops to the shelf edge during glacial stages, the whole upwelling system moves seaward and a much stronger $\delta^{13}C$ signal on the deeper slope and the rise can be expected.

CONCLUSIONS

The following conclusions are drawn from a detailed stable isotopic study of some modern planktonic and benthic foraminiferal species from the northwest African upwelling region and from comparison with hydrographic data.

(1) Oxygen isotopes in planktonic specimens from carefully sampled surface sediments do not significantly differ from those of living species collected in plankton tows during their dominant growing season. Carbon isotopes vary a little more, but not in a systematic way.

(2) The $\delta^{18}O$ values of planktonic foraminifers in surface sediments reflect the major growing seasons of the species. Oxygen isotopes of *G. ruber* white and pink and *G. sacculifer* display the range of SST during late summer, those of *G. inflata* the SST during winter, and those of *G. bulloides* the temperature ranges of the different upwelling seasons; i.e., G. bulloides indicates winter SST in the south, summer SST in the north, and annual mean SST near 21°-25°N, as well as north of the major coastal upwelling zone at 34°N.

(3) Coastal upwelling and its seasonal variability with latitude are recorded in the sediment by combining the $\delta^{18}O$ values of different species. The means of the combined $\delta^{18}O$ values of *G. ruber* white and *G. inflata* correlate very well with the annual mean SST; the $\delta^{18}O$ difference between the two species reflects the maximum seasonal temperature contrast, i.e., the seasonality. Both high seasonality and enriched $\delta^{18}O$ values for *G. bulloides* relative to the annual mean temperature values document upwelling during winter, whereas low seasonality and a relative ^{18}O depletion indicate upwelling during summer.

(4) The "winter" species *G. inflata* and the "summer" species *G. ruber* pink are lower in $\delta^{13}C$ by 0.5°/₀₀ within the regions of winter and summer upwelling, respectively. The $\delta^{13}C$ compositions of *G. sacculifer*, *G. ruber* white, and *G. bulloides*, however, are not correlated with upwelling.

(5) The aragonitic shells of *Hoeglundina elegans* are precipitated in a constant $\delta^{18}O$ disequilibrium of +0.5°/₀₀ relative to the calcitic shells of *Uvigerina sp.*

(6) The northwest African coastal upwelling zone is fed by North Atlantic Central Water north of approximately 22°N and by South Atlantic Central Water in the south. The two water masses can be distinguished; enriched $\delta^{18}O$ and $\delta^{13}C$ values in living (stained) benthic foraminiferal shells characterize the range of NACW and depleted values characterize the range of SACW at 200-600 m water depth.

ACKNOWLEDGEMENTS

We thank G.F. Lutze and U. Pflaumann for their advice and help with the foraminiferal analyses, and H. Erlenkeuser and H. Cordt who kindly supported our isotopic analyses with assistance and discussions. The authors gratefully acknowledge the reviews of W.H. Berger, C. Hemleben, W. Prell, and H. Thierstein. Sediment surface samples from RV "Meteor" and RV "Valdivia" were provided by G.F. Lutze and F.W. Haake. We are indebted to I. Bornhöft and D. Müller for technical assistance. Facilities of a VG Micromass 602 D were provided by the ^{14}C-Laboratory of the Institut für Kernphysik, Universität Kiel (H. Willkomm). This study received generous support by the Deutsche Forschungsgemeinschaft.

REFERENCES

Bé, A.W.H. and Tolderlund, D.S., 1971, Distribution and ecology of living planktonic foraminifera in surface waters of the Atlantic and Indian Oceans, in: "Micropaleontology of Oceans," B.M. Funnell and W.R. Riedel, eds., Cambridge University Press, 105-149.

Berger, W.H., Diester-Haass, L. and Killingley, J.S., 1978a, Upwelling off Northwest-Africa; the Holocene decrease as seen in carbon isotopes and sedimentological indicators, Oceanologica Acta, 1:3-7.

Berger, W.H., Killingley, J.S. and Vincent, E., 1978b, Stable isotopes in deep-sea carbonates: Box Core ERDC-92, West Equatorial Pacific, Oceanologica Acta, 1(2):203-216.

Berger, W.H., Bé, A.W.H. and Vincent, E., eds., 1981, Oxygen and carbon isotopes in foraminifera, Palaeogeography, Palaeoclimatology, Palaeoecology, Special Issue 33, 277 pp.

Codispoti, L.A. and Friederich, G.E., 1979, Local and mesoscale influences on nutrient variability in the northwest African upwelling region near Cabo Corbeiro, Deep-Sea Research, 25:751-770.

Codispoti, L.A., Friederich, G.E., Iverson, R.L., and Hood, R.W., 1982, Temporal changes in the inorganic carbon system of the south-eastern Bering Sea during spring 1980, Nature, 296:242-245.

Craig, H., 1957, Isotopic standards for carbon and oxygen and correction factors for mass spectrometric analysis of CO_2, Geochimica et Cosmochimica Acta, 12:133-149.

Craig, H., 1965, The measurement of oxygen isotope paleotemperatures, in: "Stable Isotopes in Oceanographic Studies and Paleotemperatures," Spoleto, 1965, E. Tongiorgi, ed., Consiglio Nazionale delle Richerche, Laboratorio di Geologia Nucleare, Pisa, 3-24.

Craig, H. and Gordon, L.I., 1965, Deuterium and oxygen 18 variations in the ocean and the marine atmosphere, in: "Stable Isotopes in Oceanographic Studies and Paleotemperatures," Spoleto, 1965, E. Tongiorgi, ed., Consiglio Nazionale delle Richerche, Laboratorio di Geologia Nucleare, Pisa, 9-130.

Deuser, W.G. and Hunt, J.M., 1969, Stable isotope ratios of dissolved inorganic carbon in the Atlantic, Deep-Sea Research, 16:221-225.

Deuser, W.G., Ross, E.H., Hemleben, C., and Spindler, M., 1981, Seasonal changes in species composition, numbers, mass, size, and isotopic composition of planktonic foraminifera settling into the deep Sargasso Sea, Palaeogeography, Palaeoclimatology, Palaeoecology, 33:103-127.

Duplessy, J.-C., 1978, Isotope Studies, in: "Climatic Change," J. Gribbin, ed., Cambridge University Press, 46-67.

Duplessy, J.-C, Blanc, P.L. and Bé, A.W.H., 1981, Oxygen-18 enrichment of planktonic foraminifera due to gametogenic calcification below the euphotic zone, Science, 213:1247-1250.

Emrich, K, Ehhalt, D.H. and Vogel, J.C., 1970, Carbon isotope fractionation during the precipitation of calcium carbonate, Earth and Planetary Science Letters, 8:363-371.

Epstein, S., Buchsbaum, R., Lowenstam, H., and Urey, H.C., 1953, Revised carbonate-water isotopic temperature scale, Geological Society of America, Bulletin, 64:1315-1326.

Erez, J. and Honjo, S., 1981, Comparison of isotopic composition of planktonic foraminifera in plankton tows, sediment traps and sediments, Palaeogeography, Palaeoclimatology, Palaeoecology, 33:129-156.

Fairbanks, R.G., Wiebe, P.H. and Bé, A.W.H., 1980, Vertical distribution and isotopic composition of living planktonic foraminifera in the western North Atlantic, Science, 207:61-63.

Fontugne, M. and Duplessy, J.C., 1978, Carbon isotope ratio of marine plankton related to surface water masses, Earth and Planetary Science Letters, 41(3):365-371.

Fraga, F., 1974, Distribution des masses d'eau dans l'upwelling de Mauritanie, Tethys, 6:5-10.

Ganssen, G., 1981, Isotopic analysis of foraminiferal shells: interference from chemical treatment, Symposium Oxygen and Carbon Isotopes in Foraminifera, W.H. Berger et al., eds., Palaeogeography, Palaeoclimatology, Palaeoecology, 33:271-276.

Ganssen, G., 1982, "Dokumentation von Küstenauftrieb anhand stabiler Isotopen in Schalen rezenter Foraminiferen vor West-Afrika," Ph.D. Thesis, University of Kiel.

Kahn, M.I. and Williams, D.F., 1981, Oxygen and carbon isotopic composition of living planktonic foraminifera from the northeast Pacific Ocean, Palaeogeography, Palaeoclimatology, Palaeoecology, 33:47-69.

Kroopnick, P., 1974, Correlations between C^{13} and ΣCO_2 in surface waters and atmospheric CO_2, Earth and Planetary Science Letters, 22(4):397-403.

Kroopnick, P., 1980, The distribution of ^{13}C in the Atlantic Ocean, Earth and Planetary Science Letters, 49:469-484.

Lees, A., Buller, A.T. and Scott, J., 1969, Marine carbonate sedimentation processes, Connemara, Ireland, Reading University Geological Report, 2:1-64.

Lutze, G.F., 1964, Zum Färben rezenter Foraminiferen, Meyniana, 14: 43-47.

Lutze, G.F., 1980, Depth distribution of benthonic foraminifera on the continental margin off NW Africa, "Meteor"Forschungs-Ergebnisse, C33:31-80.

Prell, W.L., and Curry, W.B., 1981, Faunal and isotopic indices of monsoonal upwelling: Western Arabian Sea, Oceanological Acta, 4(1):91-98.

Redfield, A.C., 1942, The processes determining the concentration of oxygen, phosphate, and other organic derivatives within the depths of the Atlantic Ocean, Papers on Physical Oceanography and Meteorology, 9(2):1-22.

Robinson, M.K., Bauer, R.A. and Schroeder, E.H., in press, Atlas of North-Atlantic-Indian Ocean monthly mean temperature and mean salinity of the surface layer, Naval Oceanographic Office, Washington.

Sarnthein, M., Thiede, J., Pflaumann, U., Erlenkeuser, H., Fütterer, D., Koopmann, B., Lange, H., and Seibold, E., 1982, Atmospheric and oceanic circulation patterns off northwest Africa during the past 25 million years, in: "Geology of the Northwest African Continental Margin," U. von Rad et al., eds., Springer Verlag, Heidelberg, 545-604.

Schemainda, R., Nehring, D. and Schulz, S., 1975, Ozeanologische Untersuchungen zum Produktionspotential der nordwestafrikanischen Wasserauftriebsregion 1970-1973. Geodätische Geophysikalische Veröffentlichungen, 4(16):1-88.

Seibold, E., 1972, Cruise 25/1971 of RV "Meteor": Continental margin of West Africa. General report and preliminary results, "Meteor"Forschungs-Ergebnisse, C10:17-38.

Seibold, E. and Hinz, K., 1976, German cruises to the continental margin of North West Africa in 1975: General reports and preliminary results from "Valdivia" 10 and "Meteor" 39, "Meteor" Forschungs-Ergebnisse, C25:47-80.

Shackleton, N.J., 1973, Attainment of isotopic equilibrium between ocean water and the benthonic foraminifera genus *Uvigerina*: Isotopic changes in the ocean during the last glacial, Colloques Internationaux du Centre National de la Recherche Scientifique, 219:203-209.

Smith, R.L., 1968, Upwelling, in: "Oceanography and Marine Biology Annual Review," H. Barnes, ed., G. Allen & Unwin, Ltd., London, 6:11-46.

Sommer, M.A., II, and Rye, D.M., 1978, Oxygen and carbon isotope internal thermometry using benthic calcite and aragonite foraminifera pairs, in: "Short Papers of the Fourth International Conference, Geochronology, Cosmochronology, Isotope Geology," U.S. Geological Survey Open File Report No. 78-701:408-410.

Speth, P., H. Detlesen and Sierts, H.-W., 1978, Meteorological influences on upwelling off northwest Africa, Deutsche Hydrographische Zeitschrift, 31(3):95-104.

Thiede, J., 1975, Shell- and skeleton producing plankton and nekton in the Eastern North Atlantic Ocean, "Meteor"Forschungs-Ergebnisse, C20:33-79.

Walton, W.R., 1952, Techniques for the recognition of living foraminifera, Contribution from the Cushman Foundation for Foraminiferal Research, 3:56-60.

Weikert, H. and John, H.C., 1981, Experience with a modified Bé multiple opening closing plankton net, Journal of Plankton Research, 3(2):167-177.

Williams, D.F., Sommer, M.A. and Bender, M.L., 1977, Carbon isotopic compositions of Recent planktonic foraminifera of the Indian Ocean, Earth and Planetary Science Letters, 36(3):391-403.

Williams, D.F., Bé, A.W.H. and Fairbanks, R.G., 1981, Seasonal stable isotopic variations in living planktonic foraminifera from Bermuda plankton tows, Palaeogeography, Palaeoclimatology, Palaeoecology, 33:71-102.

Wooster, W.S., Bakun, A. and McLain, D.R., 1976, The seasonal upwelling cycle along the eastern boundary of the North Atlantic, Journal of Marine Research, 34,2:131-141.

PARTICULATE AND DISSOLVED CONSTITUENTS IN THE WATER COLUMN

ON NUTRIENT VARIABILITY AND SEDIMENTS IN UPWELLING REGIONS

Louis A. Codispoti

Bigelow Laboratory for Ocean Sciences
McKown Point, West Boothbay Harbor, Maine 04575, U.S.A.

ABSTRACT

The relationships between upwelling intensity, euphotic zone nutrient concentrations, and biological productivity are non-linear, and these non-linearities must be taken into account in interpreting the sedimentary record. For example, the strong nutrient gradients within and between basins in the present day ocean would produce considerable differences in the nutrient concentrations of waters rising to the surface even if upwelling dynamics were otherwise the same. In the recent geologic past, these gradients may have varied significantly.

The sediments in many upwelling regions are in contact with oxygen deficient waters. The resulting lack of bioturbation may allow these sediments to preserve signals with frequencies as high as those caused by seasonal variations in the nutrient supply, and it is likely that lower frequency variations, such as the recent change in the nitrite distribution off Peru, may be correlated with changes in the sediment make-up. Shorter term variability *per se* may not leave a recognizable sedimentary imprint, but regions with well developed short-term variability may in general export more biogenic material to the sediments than less variable systems.

INTRODUCTION

The presence of greater or lesser amounts of biogenic material (e.g., biogenic silica) in past and present upwelling regions is often explained by invoking changes in upwelling intensity. While this hypothesis may sometimes be correct, upwelling *per se* is a physical process. Many of the relationships between upwelling intensity, biological production and the export of biogenic material to the sediments are non-linear (e.g., Wroblewski, 1977; Walsh, 1981; Codispoti, in press). For example, in upwelling source waters found at similar depths the quantity of plant nutrients can vary greatly over space and time which suggests that upwelling events which are physically identical could differ greatly in their biological effects (Codispoti and Friederich, 1978; Codispoti, Dugdale and Minas, 1982; Minas, Codispoti and Dugdale, 1982).

The confusion that can arise from invoking a direct relationship between upwelling intensity and biological production is exemplified by considering the El Niño phenomenon. In the past, it was thought that the reduced productivity off Peru during El Niños arose from a drastic reduction in upwelling intensity, but it is now known that upwelling intensity in the upper ~100 m is not greatly reduced and may even increase along some sections of the coast during these periods (Enfield, 1981). The reductions in biological productivity arise largely from water mass changes (including lowered nutrient concentrations) associated with the wave dynamical process that is thought to cause El Niño (Enfield, 1981; Guillén and Calienes, 1981). It has also been shown that increased upwelling intensity may sometimes be associated with decreases in photic zone nutrient concentrations and/or productivity (Huntsman and Barber, 1977; Wroblewski, 1977; Codispoti and Friederich, 1978; Codispoti, in press). These studies suggest that there is an optimal upwelling intensity beyond which factors such as increased depths of the mixed surface layer or decreases in nutrient concentration, associated with changes in water mass structure, may act to reduce biological productivity.

The nutrients (ammonia, nitrate, nitrite, dissolved silicon, and reactive phosphorus) which will be discussed in this paper are frequently reduced to very low concentrations by phytoplankton growth. The availability of the nitrogenous forms may be one of the main controls on organic production in the sea (Ryther and Dunstan, 1971), while silicon limitation may be important because it affects diatoms, a group that has high growth rates (Dugdale, 1972). Without fast-growing species such as diatoms, the growth potential of an upwelling system could be seriously reduced because such systems are sometimes bordered by fronts where sinking occurs. Thus, the phytoplankton may have only a few days to thrive on the nutrients in an upwelled water parcel (e.g., Wroblewski, 1977). Although the availability of these nutrient chemicals and of sufficient light for photosynthesis are not the only controls on biological production in upwelling systems, quite often nutrient concentrations are positively correlated with the growth rates of phytoplankton (Jones, 1978), even at high concentrations (Huntsman and Barber, 1977). Nutrient limitation of the specific growth rates of indigenous phytoplankton or of productivity supported by locally recycled nutrients may be rare (Goldman, McCarthy and Peavey, 1979), but limitation of total growth rates by changes in phytoplankton populations and of "new" (exportable) production is common (Dugdale and Goering, 1967).

On the assumption that only lower frequency types of variability will be reflected in the sedimentary record, this paper will discuss seasonal or longer period nutrient variability and differences in the patterns of short term variability. The wealth of recent information on the day-to-weeks variability scale (mesoscale), that has accumulated in recent years (e.g., Jones, 1972; Codispoti and Friederich, 1978; Codispoti, 1981; Codispoti et al., 1982), will not be emphasized.

BACKGROUND

Since we shall suggest that it might be interesting to consider the variability in the large scale nutrient distributions when interpreting the sedimentary record, it may be helpful to discuss first how the vertical gradients that contribute to the general character of the marine nutrient distribution are established. Despite compli-

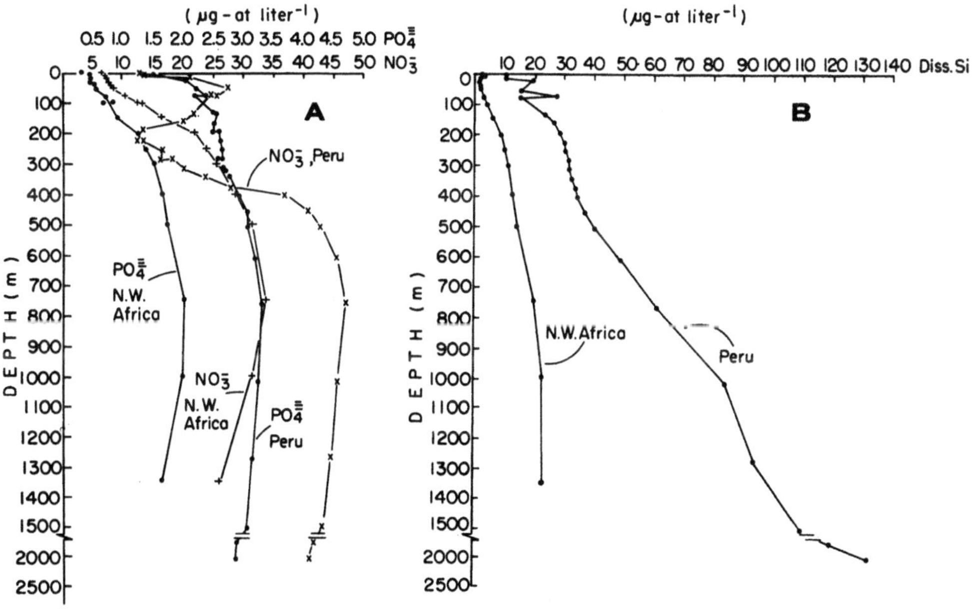

Fig. 1. (A) Nitrate and reactive phosphorus concentrations versus depth, at a station in the northwest African upwelling region (JOINT-I station 52, 21°03'N, 17°52'W, 21 March 1974) and at a station in the Peruvian upwelling region (JOINT-II "Melville" station 406, 15°10'S, 76°06'W, 11 May 1977). The nitrate minimum at about 200 m off Peru is a consequence of denitrification.

(B) Dissolved silicon versus depth profiles for the same stations. The shallow maxima and minima off Peru were common features arising from the interleaving of different water masses and from nutrient regeneration within the upwelling system.

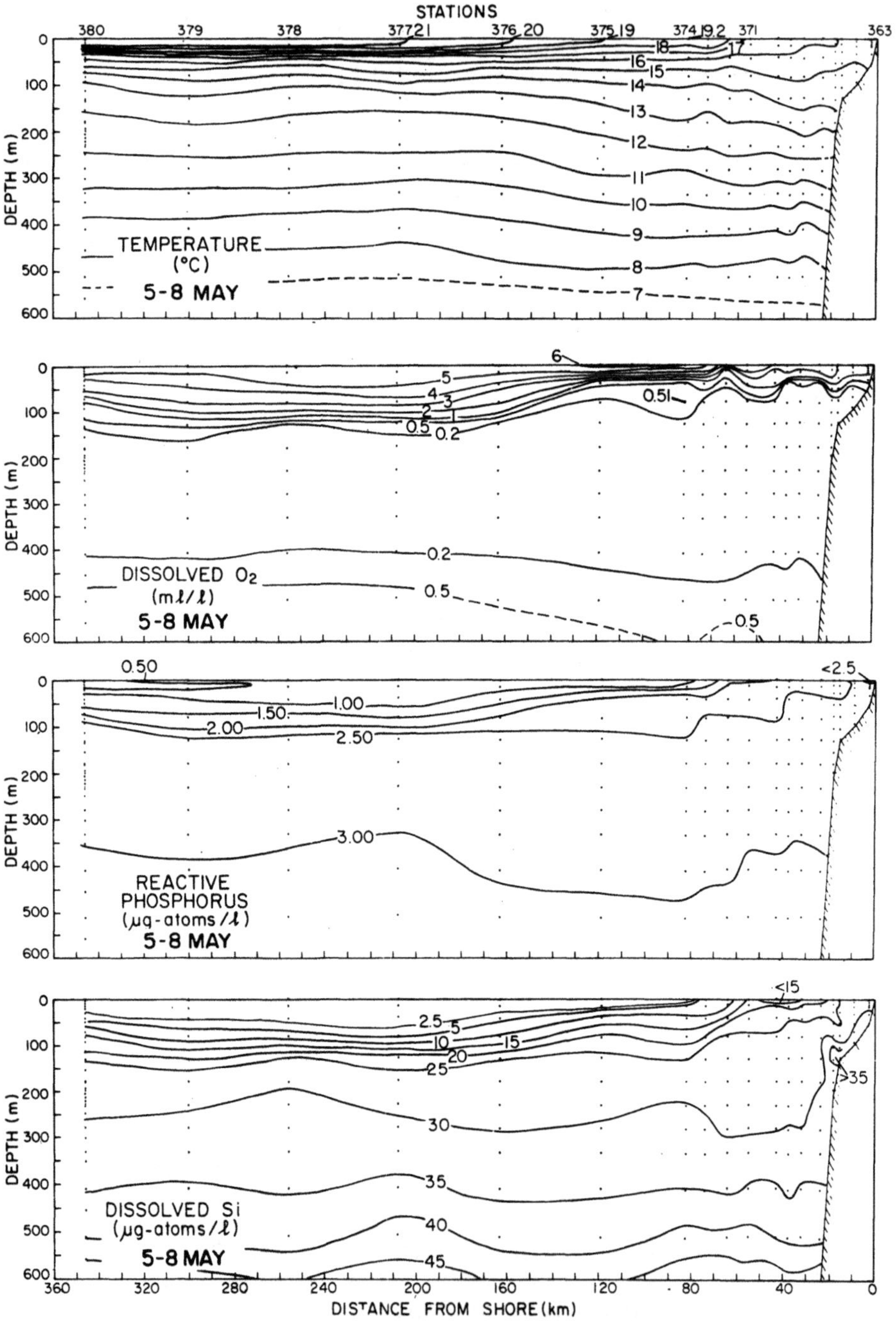

Fig. 2. Temperature, dissolved oxygen, reactive phosphorus, and dissolved silicon distributions in a section normal to the coastal isobaths in the portion of the Peruvian upwelling system near 15°S, JOINT-II, 1977.

cating local factors such as the interleaving of water masses with different histories (Fig. 1), the occurence of denitrification (Figs. 1-3) and the "piling up" of nutrient-poor waters on the western boundaries of the Atlantic and Pacific (Fahrbach and Meincke, 1979; Voituriez, 1981; Codispoti, in press), the general shapes of the vertical nutrient profiles can be explained by vertical changes in nutrient uptake and regeneration, and by differences in water mass residence times beneath the sea surface.

It is well known that the regeneration of nitrate and reactive phosphorus tend to be in the constant proportion of $\Delta N/\Delta P \sim 16/1$ (by atoms), and that the regeneration of these species is intimately associated with the metabolism and oxygen consumption of organisms (Redfield, Ketchum and Richards, 1963). As a consequence, the vertical distributions of nitrate and reactive phosphorus tend to parallel each other (Figs. 1 and 4), although this relationship can be complicated by the occurrence of denitrification (Figs. 1 and 3), and by the accumulation of more reduced species such as urea (Whitledge, 1981), ammonia (Fig. 3) (Dugdale and Goering, 1967), and nitrite (Vaccaro, 1965) during the initial stages of nitrate regeneration (Fig. 3). Since nitrate and reactive phosphorus regeneration occur in association with respiration, their maxima tend to be found close to the oxygen minimum zone. The nutrient maxima will usually be found a little deeper in the water column (Fig. 4), and they should only be expected to coincide with the oxygen minimum when the effects of "preformed nutrients" (Redfield et al., 1963) and the increased solubility of oxygen in high latitude water are taken into account. The term, "preformed nutrient," is a measure of the nutrient content of a water parcel when it sinks beneath the surface layer. Differences in preformed nutrient content should not cause large differences in the relationship between the vertical nitrate and reactive phosphorus curves because "average" seawater contains these chemicals in almost the same ratio as that required by marine biota (15:1 vs 16:1 by atoms; Redfield et al., 1963).

The principles used to explain the formation of the oxygen minimum zone should apply, at least in a general way, to the explanation of the nitrate and reactive phosphorus maxima. Basically, these explanations invoke the interaction of a rapid decrease in respiration rates with depth (and by implication, rapidly decreasing rates of nutrient regeneration) in the upper $\sim$1 km of the ocean, and a rapid increase in the residence times of water over this depth interval (Wyrtki, 1962; Packard and Garfield, in press). Wyrtki (1962) suggests that "the oxygen minimum itself lies in the upper part of the layer of smallest advection because the consumption of oxygen decreases almost exponentially with depth." Some recent studies show that the shapes of the respiration vs. depth curves can differ from the exponential decrease described by Wyrtki (Codispoti and Packard, 1980; Suess, 1980; Packard and Garfield, in press). Nevertheless, the general tendency does appear to be a more or less exponential

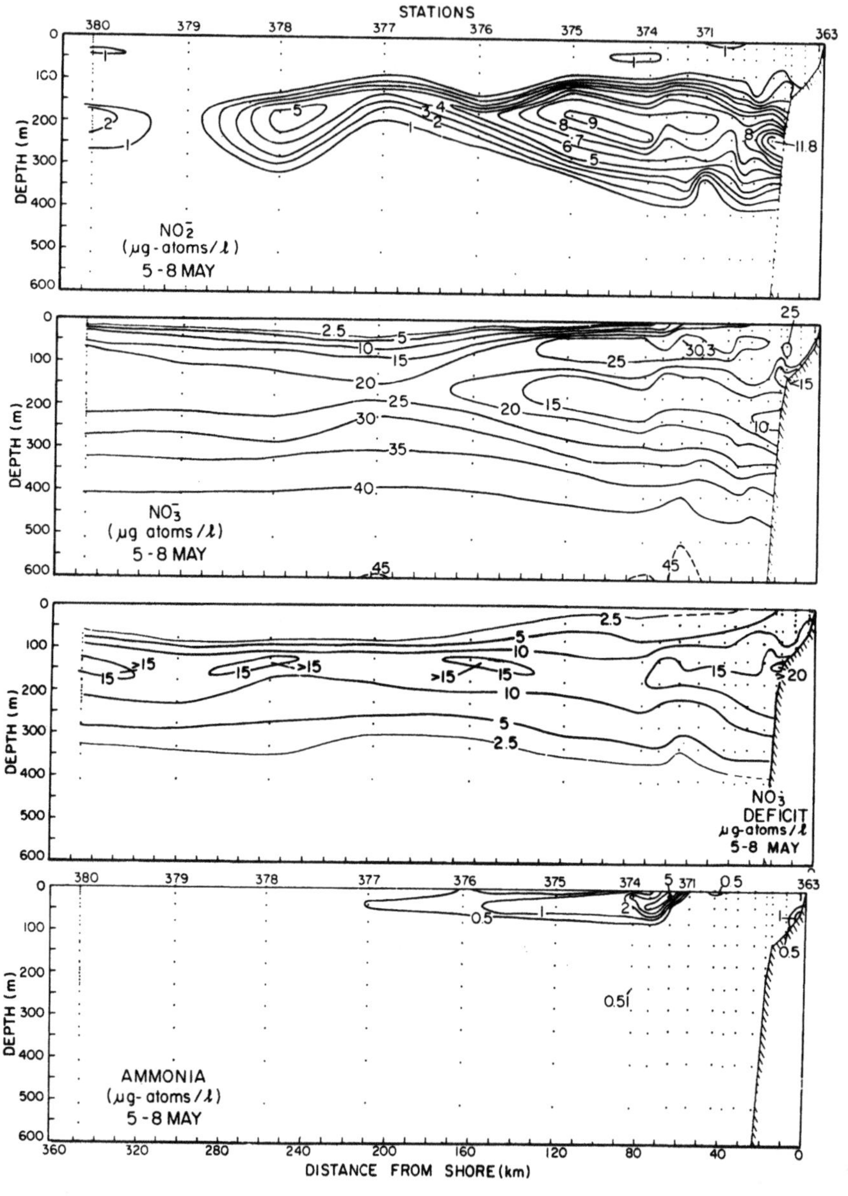

Fig. 3. Nitrite, nitrate, nitrate deficit and ammonia distributions in the same section shown in Fig. 2. The shallow nitrite maxima at stations 374 and 380 are a consequence of nitrification or the excretion of nitrite by phytoplankton (Vaccaro, 1965), but the deeper, or "secondary maximum" that is associated with a nitrate minimum, is a consequence of nitrate reduction and denitrification in the oxygen deficient waters that exist between about 100-400 m (Fig. 2). Nitrate deficits are a measure of the amount of nitrate or nitrite reduced to N_2. Note the association between the subsurface ammonia maximum and the convergence of isolines at the surface in Fig. 2. A time series diagram for nitrate deficits at a station on this section is given by Packard et al. (this volume).

decrease in respiration rates with depth with the curve being shifted towards higher values in regions of high surface productivity. Wyrtki's explanation also correctly emphasizes the fact that oxygen mimima (and nitrate and reactive phosphorus maxima) arise from the interaction of currents and mixing and the food supply which fuels respiration and nutrient regeneration.

As Fig. 1 shows, dissolved silicon vs. depth profiles differ significantly from the reactive phosphorus and nitrate profiles. In many regions, a tendency to form a maximum in association with the oxygen minimum is replaced by a tendency for a monotonic increase with depth. This situation arises at least partially because the pathway of silicon through the biota differs fundamentally from that of nitrogen and phosphorus. Large amounts of silicon are required only by certain specific groups of organisms (e.g., diatoms, radiolarians, silicoflagellates) whereas nitrogen and phosphorus are universal requirements. Because of this and because the organisms which consume silicon-rich organisms convert little of the biogenic silica into soluble forms, relative to the amounts of soluble nitrogen (e.g., NH_3 and urea) and phosphate that they excrete (see Whitledge and Packard, 1971), silicon is enriched in rapidly sinking fecal material and is recycled less effectively than nitrogen and phosphorus in the surface layer (Broecker, 1974). Silicon regeneration is fundamentally a solution process and only a portion of the dissolution

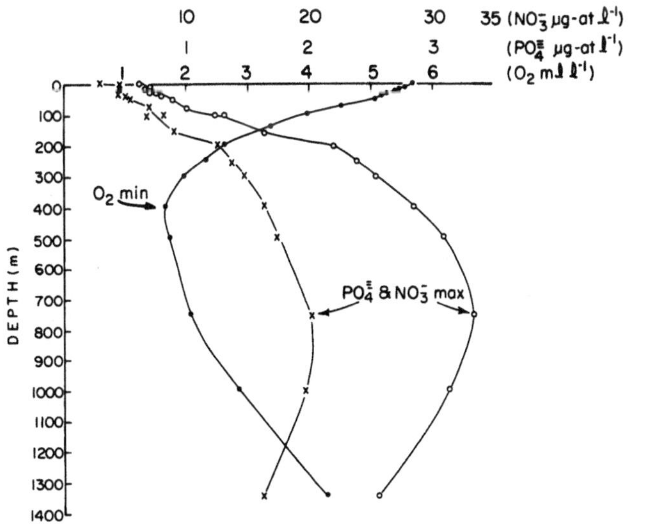

Fig. 4. Nitrate, reactive phosphorus, and dissolved oxygen concentrations versus depth at a station in the northwest African upwelling region (JOINT-I station 52). The vertical displacement between the nutrient maxima and oxygen minimum discussed in the text is clearly indicated in this figure.

of biogenic silica can be correlated with respiratory processes (Berger, 1970).

Hurd (1972; 1973) has discussed the factors that control the dissolution of silica, and he has shown that the solution potential decreases markedly with decreasing temperature over the 25-3°C temperature range. It might be wise to consider this and the fact that silica dissolution is not as intimately associated with the biota as are reactive phosphorus and nitrate regeneration, when interpreting the sedimentary record. The biota can compensate for changes in temperature (at least partially) by changes in population density and species composition, or by changes in the biochemistry of individual species. It is possible that a decrease in temperature would decrease phosphorus and nitrate regeneration rates less effectively than silica dissolution rates.

DISCUSSION

The zones of maximum and minimum nutrient concentrations in the ocean are not well correlated with the sites of nutrient sources and sinks (Codispoti, 1979), and it is generally agreed that the inter-ocean nutrient fractionation suggested in Table 1 and Fig. 1 arises largely from the nature of the vertical nutrient gradients and the ocean's circulation patterns. "Estuarine" basins with a surface outflow and a deep inflow tend to be enriched in nutrients at the expense of "lagoonal" basins which supply deep waters and receive surface waters in return (Redfield et al., 1963; Berger, 1970; Broecker, 1974). Thus, the North Atlantic Ocean which is the site of substantial deep water formation is rich in oxygen and poor in nutrients. The Mediterranean Sea is even poorer in nutrients (Table 1) because it is a lagoonal basin attached to the lagoonal North Atlantic. Its waters may be thought of as having been subjected to a two-stage fractionation process. Conversely, the North Pacific Ocean can be thought of as lying at the head of a global-scale estuary, and it has the highest nutrient concentrations found in any of the major ocean basins. The Bering Sea is an "estuary" attached to the estuarine North Pacific. This factor and the topographic isolation of its deepest waters (Sayles, Aagaard and Coachman, 1979) lead to an increase in reactive phosphorus and dissolved silicon concentrations in relation to the deep waters of the adjacent Pacific with silicon concentrations achieving very high values (Table 1). Nitrate concentrations are somewhat lower than in the adjacent ocean possibly because of denitrification on and in the sediments (Tsunogai et al., 1979).

Perhaps the most remarkable example of the effect of circulation and stratification on nutrient concentrations is provided by comparing the "estuarine" Black Sea with the adjacent Mediterranean (Redfield et al., 1963). The shallow sill depth of the Black Sea when combined with the fresh water supply leads to extreme stratifi-

Table 1. High nutrient concentrations observed during upwelling off the Aleutians and northwest Africa, and maximum nutrient concentrations observed in portions of the Mediterranean Sea and Bering Sea regardless of depth.

Region	Nitrate µg-at ℓ^{-1}	Phosphate µg-at ℓ^{-1}	Silicate µg-at ℓ^{-1}	Reference
Surface Values:				
Aleutian Islands ∿ 53°N; 170°W	30	2.5	70	a
NW Africa (Sta. 14)	12	1.0	7	b
∿ 22°N (Sta. 161)	11	1.0	8	b
Maximum Values Regardless of Depth:				
Bering Sea	45	3.5	240	c
Mediterranean Sea Crete/Libya	7	0.5	11	d

[a] Hood and Kelley (1976)
[b] Friebertshauser et al. (1975)
[c] Tsunogai et al. (1979)
[d] Miller et al. (1970)

cation and the development of sulphate reduction in its depths (Caspers, 1957). Dissolved silicon and reactive phosphorus concentrations in this basin may exceed 300 and 10 µg-atoms/ℓ, respectively. Because of the anoxic conditions, inorganic nitrogen is present primarily as ammonia (Richards, 1965) and ammonia concentrations may attain values on the order of 100 µg-atoms/ℓ (RV "Thompson" cruise 46; Lowman and Codispoti, 1973).

The greater fractionation of dissolved silicon suggested by the data in Table 1 is only to be expected because of the differences between silicon's vertical distribution and the vertical distributions of nitrate and reactive phosphorus. The exceptionally high Si/NO_3^- and Si/PO_4^{-3} ratios found in waters that upwell near the Aleutian Islands (Table 1; and Hood and Kelley, 1976) might arise from the combination of relatively long residence times for the subsurface waters in this region, and the tendency of the deep and intermediate water of this region to have relatively high upward transports (Reid, 1965; Reid and Lynn, 1971). This is what one would ex-

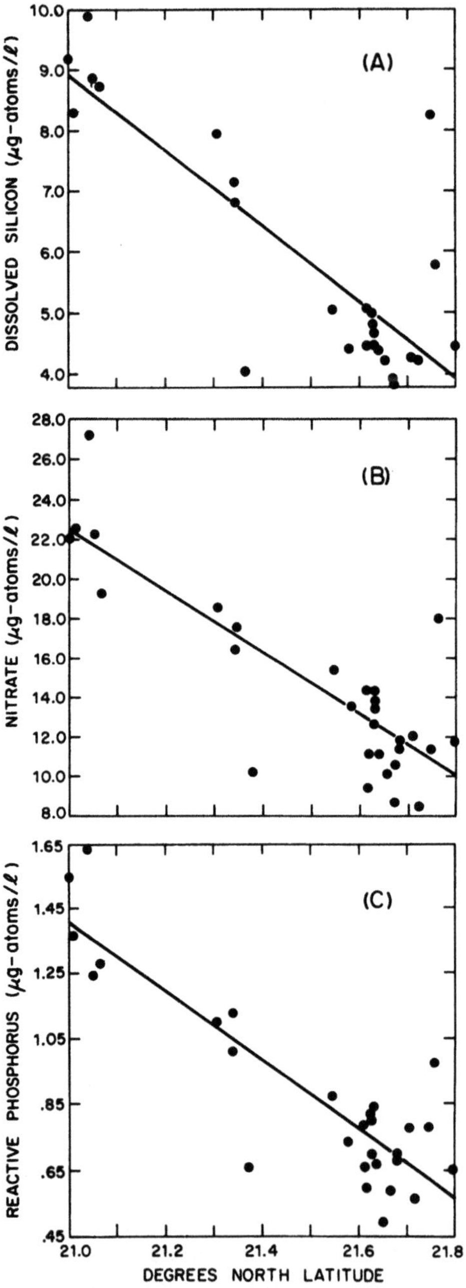

Fig. 5. Dissolved silicon (A), nitrate (B), and reactive phosphorus (C) versus latitude near the shelf-break off northwest Africa at depths greater than 150 m and at sigma-t values between 26.6 and 26.8 (inclusive). This figure is taken from Codispoti and Friederich (1978).

pect at the head of the oceanic scale "estuary" represented by the North Pacific.

Such inter-ocean nutrient distribution patterns suggest that fractionation can be changed by any process that alters vertical nutrient gradients or major circulation features in the world's oceans. Consequently, the prospect of large differences in the inter-basin nutrient fractionation is almost a certainty if one goes back far enough in the geologic record and considers paleoceans that existed during periods with climates and ocean basin morphologies that differ from those in the present ocean. Perhaps the more interesting question is: What is the shortest time interval over which significant changes in the global nutrient distribution might occur? Certainly climatic changes that have occurred within the past 25,000 years or so have the potential to change the boundaries between water masses with markedly different nutrient contents (e.g., Fig. 5), and some of the contributions draw attention to possible effects of a more northward penetration of nutrient rich South Atlantic Central water into the northwest African upwelling region (Ganssen and Sarnthein; Labracherie et al., both this volume). The question really boils down to the sensitivity of the fractionation process. Stated another way, do changes in the vertical particle flux, such as those suggested by some recent studies, and factors such as the closing of the Bering Strait during the last ice age affect nutrient distributions in a way that could produce noticeable changes in the sedimentary record of upwelling regions over the last 100,000 years?

An argument for the "steady-state" hypothesis would note that the inter-ocean gradients that exist presently can be established over periods of no more than a few thousand years and diffusion along these gradients might balance the losses and gains, thereby producing a steady-state global nutrient distribution (Broecker, 1974). This camp would also point out that the fractionation process is largely driven by the formation of North Atlantic Deep Water, and some evidence suggests that the North Atlantic has been "lagoonal" and the Pacific "estuarine" since the late Eocene (Ramsay, 1977; Schopf, 1980).

Nevertheless, it does not seem prudent at the present state of knowledge to dismiss the possibility that significant variations have been played on the major steady-state theme. For example, Codispoti (1979) pointed out that the total volume of water exchanged between the Atlantic and Indo-Pacific basins is only a small fraction of the water circulating around Antarctica, and that the flow from the Pacific into the Atlantic via the Bering Strait might represent a significant amount of the exchange between the oceans. He also speculated on the possible consequences of the opening and closing of the Bering Strait and showed that closing during the last ice age might have had a significant effect on the global dissolved silicon distribution. Changes in the morphology of other crucial straits might also have significant effects. For example, the upper North Atlantic Deep Water is formed in the Mediterranean Sea (Defant, 1961), so changes in the sill depth of the Straits of Gibralter might also have an effect on the nutrient fractionation process. In addition, vertical nutrient gradients are a function of biological production and community structure (Margalef and Estrada, 1981), and a significant change in nutrient distributions has been observed off Peru in recent years that appears to be related to changes in ecosystem structure (Codispoti and Packard, 1980).

Before turning to types of nutrient variability which have relatively high frequency signals, compared to the global nutrient fractionation process, it may be useful to point out that these shorter period signals are most likely to be preserved in sediments underlying waters which are oxygen deficient ($O_2 \leq \sim 0.1$ mℓ/ℓ) or anoxic. Bioturbation in the sediments underlying oxygenated waters is likely to obscure signals that have frequencies of less than a few thousand years. For reasons that have been discussed by Codispoti (in press), the Peruvian, eastern tropical North Pacific, and Arabian Sea upwelling systems contain large volumes of oxygen deficient water (Codispoti and Richards, 1976; Deuser, Ross and Mlodzinska, 1978; Codispoti and Packard, 1980), and smaller oxygen deficient regions exist off southwest Africa (Calvert and Price, 1971) and California (Emery, 1954).

At least three types of nutrient variability with frequencies of less than one cycle per year occur off Peru. The El Niño phenomenon is the best known of these types. The decreases in biological production associated with this perturbation do not arise from a reduction in upwelling as previously thought. Upwelling continues but the nutrient supply is reduced mainly because of a reduction in the nutrient content of the upwelling source waters (Enfield, 1981; Guillén and Calienes, 1981). Although opinions may differ on whether some years should be classified as El Niño periods or merely as being anomalously warm, most investigators would agree that 1891, 1925-26, 1940-41, 1957-58, 1965, 1972-73, and 1976 were El Niño periods. Years with significant cold anomalies also occur off Peru, and there is some evidence that primary production may be reduced during these periods as well. Reduced nutrient supplies are an unlikely cause for these reductions, and Guillén and Calienes (1981) suggest reduced stability, rapid sinking of organisms, or more intense grazing pressure as possible causes for these reductions.

In 1917 and 1976 massive dinoflagellate blooms occurred off Peru (Dugdale et al., 1977; Cushing, 1981), and it is reasonable to suggest that these occurrences would be correlated with changes in the nutrient regime and would leave signals in the sedimentary record. Silica deposition may have decreased, since dinoflagellates do not require silicon, and perturbations of an ecosystem may lead to increased rates of organic carbon burial (Walsh, 1981).

So-called secondary nitrite maxima (Fig. 3) occur in the oxygen deficient waters that are found in some upwelling regions, and an increase in the coastal portion of the secondary nitrite maximum off Peru in recent years has been described by Codispoti and Packard (1980). This increase (Fig. 6) appears to have arisen after the 1972 El Niño, an event that marks the massive decline in the population of Peruvian anchoveta (Cushing, 1981; Glantz and Thompson, 1981), and it may be a signal of increased denitrification rates arising from changes in the distribution of organic matter within the water column (Codispoti and Packard, 1980). Certainly, removal of the bulk of the anchoveta population has the potential to change the ecosystem structure sufficiently to affect the nutrient distribution. The first nitrite observations reported from Peru were made in the late 1950's (Wooster, Chow and Barrett, 1965), and the changes noted after 1972 have persisted, so this change may represent a fairly low frequency type of variability. Of particular interest to this discussion is Walsh's (1981) suggestion that organic carbon sedimentation may have increased by almost an order of magnitude as a consequence of the

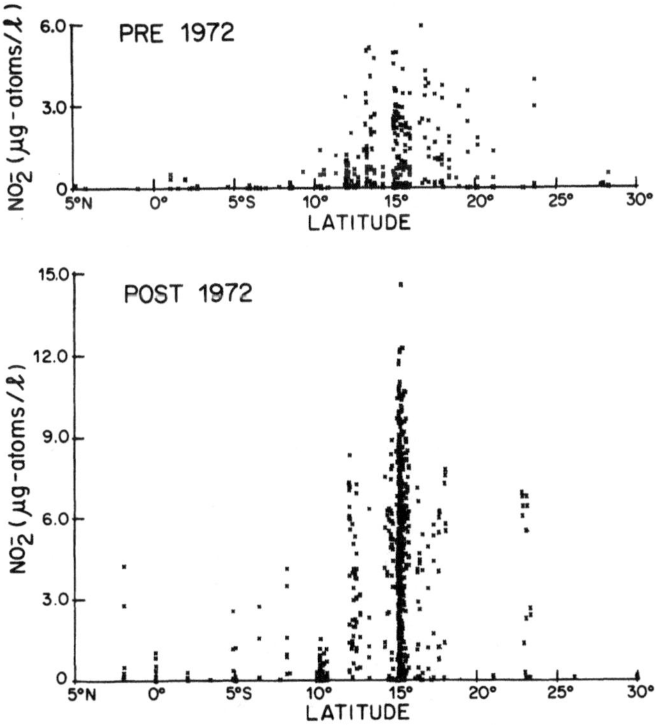

Fig. 6. Nitrite versus latitude within 200 km of the coast off western South America before and after 1972. Values were used only from water parcels with dissolved oxygen concentrations of less than 0.25 mℓ ℓ^{-1}, and sigma-t values between 26.2 and 26.8. Data are from the two 1° squares closest to the coast, so the offshore boundary varies somewhat with latitude (from Codispoti and Packard, 1980).

anchoveta decline. Since 1972, three documented occurrences of hydrogen sulphide in the open waters of coastal Peru have also been reported, and secondary nitrite maxima have been found closer to the equator than in earlier times (Codispoti and Packard, 1980). Whether or not these occurrences represent a type of long term change or are the result of intensified observations is not yet clear.

One possible consequence of increased denitrification off western South America is increased reactive phosphorus/nitrate ratios in the rising waters. As Fig. 3 suggests, the rising waters contain appreciable nitrate deficits, and an inspection of all available data from the region suggests that maximum nitrate deficits have risen somewhat after 1972. The occurrence of nitrate deficits in waters rising to the surface has been suggested as a mechanism for limiting marine production (Piper and Codispoti, 1975), and possibly reveals an important non-linearity in the nutrient-regeneration/primary-production relationship. If the inter-ocean fractionation process "goes too far" or if an upwelling system becomes "too efficient" as a nutrient trap, reactive phosphorus and dissolved silicon values may continue to increase, but nitrate concentrations may decrease as a consequence of denitrification.

Some cores from sediments underlying oxygen deficient and anoxic waters retain seasonal signals (Gross et al., 1963; Schrader et al., 1980), so it is possible that the sediments underlying the oxygen deficient waters in upwelling regions may contain memories of seasonal variations in the nutrient supply. Most upwelling regions have some seasonal variability in their nutrient regimes, but the Somalia and Arabian upwelling zones in the northern Indian Ocean probably exhibit some of the strongest seasonal variability in upwelling that can be found anywhere (Smith and Bottero, 1977; Smith and Codispoti, 1980). This variability is mainly due to the monsoonal character of the winds and currents in the northern Indian Ocean.

During the Northeast Monsoon, conditions off Somalia are oligotrophic, but during the Southwest Monsoon vigorous upwelling results in high surface nutrient concentrations near 10°N and sometimes near 5°N as well. Although the finding of two upwelling sites during some years and not during others may be a reflection of interannual differences, some recent studies suggest that this situation might be largely due to the extremely rapid changes that occur during the Southwest Monsoon. Strong upwelling occurs on the northern edges of zones where the "Somali Current" turns offshore, and Evans and Brown (1981), in a recent re-examination of sea surface temperature data from this region, conclude that whether a single or dual gyre nature is observed may be largely a function of when and how observations are made. Their study demonstrates that the gyres that are associated with upwelling can move rapidly along the coast, and in particular that there is a tendency for the southern gyre and associated upwelling to migrate rapidly northward during a portion of the Southwest Monsoon period and to coalesce with the northern gyre in some years. In addition, their study suggests significant year to year variability in the motion of the two gyres.

Perhaps as a consequence of the rapid changes that can occur off Somalia during the Southwest Monsoon, fish kills have been found in this region (Foxton, 1965), and it may be that seasonal and interannual differences are recorded in some of the sediments in this region, even though none of the overlying waters are truly oxygen deficient (minimum concentrations are about 0.5 mℓ/ℓ). For example, the rapid changes may inhibit the development of large benthic fauna, and may encourage massive transports of biogenic matter to the sediments by causing mass mortalities.

Off Arabia, there is also a strong seasonal upwelling signal, and oxygen deficient subsurface waters are common in this region (McGill, 1973). Unlike the upwelling zone off Somalia, this region is more "normal" insofar as it is not associated with the strong western boundary currents that characterize the Somali upwelling (Smith and Bottero, 1977). Consequently, mass mortalities may be less common, and perhaps sediments underlying some of the oxygen deficient zones will exhibit a regular (rhythmic) sequence between upwelling and non-upwelling periods.

As stated in the introduction, the days-to-weeks "event scale" nutrient variability, which has received attention during recent upwelling experiments (e.g., Codispoti and Friederich, 1978), is unlikely to leave a recognizable parallel signal in the sediment. However, inter-regional differences in the patterns of short term variability might be preserved. Upwelling systems with well-developed shorter-term variability might be expected to export a higher fraction of their biogenic material to the sediments and/or to the adjoining ocean than systems with a more steady environment (assuming that other factors are similar). This is because variability is one of the factors that can prevent the development of a mature ecosystem sufficiently well organized to consume almost all of the biogenic material produced locally. Short food chains in variable upwelling systems may make them good producers of fish, but these systems are not efficient consumers of the local food supply (Margalef, 1978). We have already suggested how variability at seasonal or lower frequencies might lead to increased sedimentation of biogenic matter (fish kills off Somalia and increased carbon sedimentation off Peru), and some studies also suggest that variability on the days-to-weeks scale may promote sedimentation. For example, Weikert (1977) has suggested that the highly variable nature of shelf break upwelling off northwest Africa may lead to zooplankton mortality. Such an occurrence could affect sedimentation directly (zooplankton remains), or indirectly by enabling a greater export of phytoplankton remains to be transported to the sediments.

Occurrence of hydrogen sulphide formation off Peru (e.g., Dugdale et al., 1977; Doe, 1978; Sorokin, 1978) may represent a type of short term variability (Codispoti, 1981), and such occurrences could cause mass mortalities of higher organisms. During variable upwelling off northwest and southwest Africa, dense nearshore phytoplankton blooms may develop in or near zones of downwelling (Huntsman and Barber, 1977; Hutchings, 1981), and this type of short term variability might contribute to the sedimentation of phytoplankton remains. Conversely, Walsh (1976) suggests that the steady conditions in the equatorial divergence can lead to intense grazing stress which limits the size of phytoplankton populations.

CONCLUSIONS

The above considerations suggest that the following investigations might be both interesting and feasible:

1. A comparison between the biogenic silica content of northwest African and Peruvian sediments over the last ∿ 20,000 years with a view towards determining whether or not results fit with the notion that silicon fractionation between the oceans was less well developed during the last glacial maximum (Codispoti, 1979) might be instructive.

2. Confirmation of Walsh's (1981) suggestion that organic carbon sedimentation off Peru has increased by almost an order of magnitude since the decline of the anchoveta stocks would be very useful. Since primary production has not changed drastically, such a confirmation would demonstrate an important non-linearity in the nutrient-primary production-biogenic matter sedimentation rate relationship: perturbations decrease the efficiency of carbon utilization in the upper portion of the water column until a new "steady state" is achieved (Codispoti and Packard, 1980).

3. If cores can be found off Peru that preserve relatively short term signals, it might be interesting to look for possible effects of El Niños and the massive dinoflagellate blooms of 1917 and 1976. For example, it would be interesting to see if silica deposition decreased and organic carbon storage in the sediments increased as a result of dinoflagellate blooms.

4. Finally, it might be interesting to closely examine the sediments in contact with oxygen deficient water off Arabia for evidence of seasonal and year to year changes in upwelling.

A test of the idea of a positive correlation between short term variability in upwelling regimes and the sedimentation of biogenic matter would also be interesting, but it would require the identification of two upwelling systems that are similar in most respects yet have drastic differences in their short term variability.

ACKNOWLEDGEMENTS

My studies of upwelling systems have been supported by the National Science Foundation under the auspices of the CUEA and INDEX programs (most recently under grant OCE-78-25456), and by the Office of Naval Research (most recently by ONR contract N00014-81-C-0043). Technical assistance was provided by my wife, Codie, and by P. Colby, N. Garfield, P. Oathout, and J. Rollins.

REFERENCES

Berger, W.H., 1970, Biogenous deep-sea sediments: fractionation by deep-sea circulation, Geological Society of America, Bulletin, 81:1385-1402.

Broecker, W.S., 1974, "Chemical Oceanography," Harcourt, Brace, Jovanovich, New York, 214 pp.

Calvert, S.E. and Price, N.B., 1971, Upwelling and nutrient regeneration in the Benguela Current, October, 1968, Deep-Sea Research, 18:505-523.

Caspers, H., 1957, Black Sea and Sea of Azov, in: "Treatise on Marine Ecology and Paleoecology," J.W. Hedgpeth, ed., Ecology 1, Geological Society of America, Washington, 801-889.

Codispoti, L.A., 1979, Arctic Ocean processes in relation to the dissolved silicon content of the Atlantic, Marine Science Communications, 5:361-381.

Codispoti, L.A., 1981, Temporal nutrient variability in three different upwelling regions, in: "Coastal Upwelling," F.A. Richards, ed., Coastal and Estuarine Sciences 1, American Geophysical Union, Washington, 209-220.

Codispoti, L.A., in press, Nitrogen in upwelling systems, in: "Nitrogen in the Marine Environment," E. Carpenter and D. Capone, eds., Academic Press, New York.

Codispoti, L.A. and Friederich, G.E., 1978, Local and mesoscale influences on nutrient variability in the northwest African upwelling region near Babo Corbeiro, Deep-Sea Research, 25: 751-770.

Codispoti, L.A. and Packard T.T., 1980, Denitrification rates in the eastern tropical South Pacific, Journal of Marine Research, 38: 453-477.

Codispoti, L.A. and Richards, F.A., 1976, An analysis of the horizontal regime of denitrification in the eastern tropical North Pacific, Limnology and Oceanography, 21:379-388.

Codispoti, L.A., Dugdale, R.C. and Minas, H.J., 1982, A comparison of the nutrient regimes off northwest Africa, Peru, and Baja California, Conseil Permanent International pour l'Exploration de la Mer, Rapport, 180:177-194.

Cushing, D.H., 1981, The effect of El Niño upon the Peruvian anchoveta stock, in: "Coastal Upwelling," F.A. Richards, ed., Coastal and Estuarine Sciences 1, American Geophysical Union, Washington, 449-457.

Defant, A., 1961, "Physical Oceanograhpy," Vol. 1, Macmillan Co., New York, 729 pp.

Deuser, W.G., Ross, E.H. and Mlodzinska, Z.J., 1978, Evidence for and rate of denitrification in the Arabian Sea, Deep-Sea Research, 25:431-445.

Doe, L.A.E., 1978, A progress and data report on a Canada-Peru study of the Peruvian anchovy and its ecosystem, Project ICANE, Report Series/B1-R-78-6, Bedford Institute of Oceanography, Nova Scotia, Canada, 211 pp.

Dugdale, R.C., 1972, Chemical oceanography and primary productivity in upwelling regions, Geoforum, 11:47-61.

Dugdale, R.C. and Goering, J.J., 1967, Uptake of new and regenerated forms of nitrogen in primary productivity, Limnology and Oceanography, 12:196-206.

Dugdale, R.C., Goering, J.J., Barber, R.T., Smith, R.L., and Packard, T.T., 1977, Denitrification and hydrogen sulfide in the Peru upwelling region during 1976, Deep-Sea Research, 24:601-608.

Emery, K.O., 1954, Source of water in basins off Southern California, Journal of Marine Research, 3:23-45.

Enfield, D.B., 1981, El Niño: Pacific Eastern boundary response to interannual forcing, in: "Resource Management and Environmental Uncertainty: Lessons from Coastal Upwelling Fisheries," M.H. Glantz and J.D. Thompson, eds., John Wiley & Sons, New York, 213-254.

Evans, R.H. and Brown, O.B., 1981, Propagation of thermal fronts in the Somali Current system, Deep-Sea Research, 28A:521-527.

Fahrbach, E. and Meincke, J., 1979, Some observations on the variability of the Cabo Frio upwelling, CUEA Newsletter, 8:13-18.

Foxton, P., 1965, A mass fish mortality on the Somali coast, Deep-Sea Research, 12:17-19.

Friebertshauser, M.A., Codispoti, L.A., Bishop, D.D., Friederich, G.E., and Westhagen, A.A., 1975, JOINT-1 hydrographic station data RV "Atlantis" II cruise 82, Coastal Upwelling Ecosystems Analysis Data Report, 18, 243 pp.

Glantz, M.H. and Thompson, J.D., eds., 1981, "Resource Management and Environmental Uncertainty: Lessons from Coastal Upwelling Fisheries," John Wiley & Sons, New York, 491 pp.

Goldman, J.E., McCarthy, J.J. and Peavy, D.G., 1979, Growth rate influence on the chemical composition of phytoplankton in oceanic waters, Nature, 279:210-215.

Gross, M.G., Gucluer, S.M., Creager, J.S., and Dawson, W.A., 1963, Varved marine sediments in a stagnant fjord, Science, 141: 918-919.

Guillén, O.G. and Calienes, R.Z., 1981, Biological productivity and El Niño, in: "Resource Management and Environmental Uncertainty:

Lessons from Coastal Upwelling Fisheries," M.H. Glantz and J.D. Thompson, eds., John Wiley & Sons, New York, 255-284.

Hood, D.W. and Kelley, J.J., 1976, Evaluation of mean vertical transports in an upwelling system by CO_2 measurements, Marine Science Communications, 2(6):387-411.

Huntsman, S.A. and Barber, R.T., 1977, Primary production off northwest Africa: The relationship to wind and nutrient conditions, Deep-Sea Research, 24:25-34.

Hutchings, L., 1981, The formation of plankton patches in the southern Benguela Current, in: "Coastal Upwelling," F.A. Richards, ed., Coastal and Estuarine Sciences 1, American Geophysical Union, Washington, 496-513.

Hurd, D.C., 1972, Factors affecting solution rate of biogenic opal in seawater, Earth and Planetary Science Letters, 15:411-417.

Hurd, D.C., 1973, Interactions of biogenic opal, sediment and seawater in the Central Equatorial Pacific, Geochemica et Cosmochimica Acta, 37:2257-2282.

Jones, B.H., Jr., 1978, A spatial analysis of the autotrophic response to abiotic forcing in three upwelling ecosystems: Oregon, northwest Africa, and Peru, Coastal Upwelling Ecosystems Analysis Technical Report, 37, 262 pp.

Jones, P.G.W., 1972, The variability of oceanographic observations off the coast of northwest Africa, Deep-Sea Research, 19:5-22.

Lowman, D.M. and Codispoti, L.A., 1973, RV "T.G. Thompson" cruises 035, 037, 046, and 066, University of Washington, Department of Oceanography Technical Report No. 284, 15 pp.

Margalef, R., 1978, What is an upwelling system? in: "Upwelling Ecosystem," R. Boje and M. Tomczak, eds., Springer-Verlag, Berlin, 12-14.

Margalef, R. and Estrada, M., 1981, On upwelling, eutrophic lakes, the primitive biosphere and biological membranes, in: "Coastal Upwelling," F.A. Richards, ed., Coastal and Estuarine Sciences 1, American Geophysical Union, Washington, 522-529.

McGill, D.A., 1973, Light and nutrients in the Indian Ocean, in: "The Biology of the Indian Ocean," B. Zeitzschel, ed., Springer-Verlag, New York, 53-102.

Miller, A.R., Tchernia, P., Charnock, H., and McGill, D.A., 1970, "Mediterranean Sea Atlas," Woods Hole Oceanographic Institution Atlas Series, Vol. III.

Minas, H.J., Codispoti, L.A. and Dugdale, R.C., 1982, Nutrients and primary production in the upwelling region off northwest Africa, Conseil Permanent International pour l'Exploration de la Mer, Rapport, 180:141-176.

Packard, T.T. and Garfield, P., in press, Respiration of the midwater microplankton from the Peru current upwelling system, in: "Productivity of Upwelling Ecosystems," R.T. Barber and M.H. Vinogradov, eds., Elsevier, Amsterdam.

Piper, D.Z. and Codispoti, L.A., 1975, Marine phosphorite deposits and the nitrogen cycle, Science, 188:15-18.

Ramsay, A.T.S., 1977, Sedimentological clues to palaeo-oceanography, in: "Oceanic Micropalaeontology," A.T.S. Ramsay, ed., Academic Press, London, 1371-1453.

Redfield, A.C., Ketchum, B.H. and Richards, F.A., 1963, The influence of organisms on the composition of seawater, in: "The Sea", Vol. 2, M.N. Hill, ed., Interscience, New York, 26-77.

Reid, J.L., 1965, "Intermediate Waters of the Pacific Ocean," Johns Hopkins Press, Baltimore.

Reid, J.L. and Lynn, R.L., 1971, On the influence of the Norwegian-Greenland and Weddell seas upon the bottom waters of the Indian and Pacific oceans, Deep-Sea Research, 18:1063-1088.

Richards, F.A., 1965, Anoxic basins and fjords, in: "Chemical Oceanography," J.P. Riley and G. Skirrow, eds., Academic Press, London, 611-645.

Ryther, J.H. and Dunstan, W.M., 1971, Nitrogen, phosphorus, and eutrophication in the coastal marine enviroment, Science, 171: 1008-1013.

Sayles, M.A., Aagaard, K. and Coachman, L.K., 1979, "Oceanographic Atlas of the Bering Sea Basin," University of Washington Press, Seattle, 158 pp.

Schopf, T.J.M., 1980, "Paleoceanography," Harvard University Press, Cambridge, 341 pp.

Schrader, H., Kelts, K., Curray, J., Moore, D., Aguayo, E., Aubry, M-P., Einsele, G., Fornari, D., Gieskes, J., Guerrero, J., Kastner, M., Lyle, M., Matoba, Y., Molina-Cruz, A., Niemitz, J., Rueda, J., Saunders, A., Simoneit, B., and Vaquier, V., 1980, Laminated diatomaceous sediments from the Guaymas basin slope (central Gulf of California): 250,000 year climate record, Science, 207:1207-1209.

Smith, R.L. and Bottero, J.S., 1977, On upwelling in the Arabian Sea, in: "A Voyage of Discovery," M. Angel, ed., Pergamon Press, Oxford, 291-304.

Smith, S.L. and Codispoti, L.A., 1980, Southwest monsoon of 1979: chemical and biological response of Somali coastal waters, Science, 209:597-600.

Sorokin, Y.I., 1978, Description of primary production and heterotrophic microplankton in the Peruvian upwelling region, Oceanology, 18:62-71.

Suess, E., 1980, Particulate organic carbon flux in the oceans: surface productivity and oxygen utilization, Nature, 288:260-263.

Tsunogai, S., Kusakabe, M., Iizumi, H., Koike, I., and Hattori, A., 1979, Hydrographic features of the deep water of the Bering Sea-the sea of silica, Deep-Sea Reseach, 26:641-660.

Vaccaro, R.F., 1965, Inorganic nitrogen in sea water, in: "Chemical Oceanography," J.P. Riley and G. Skirrow, eds., Academic Press, London, 365-408.

Voituriez, B., 1981, Equatorial upwelling in the eastern Atlantic: problems and paradoxes, in: "Coastal Upwelling," F.A. Richards, ed., Coastal and Estuarine Sciences 1, American Geophysical Union, Washington, 95-106.

Walsh, J.J., 1976, Herbivory as a factor in patterns of nutrient utilization in the sea, Limnology and Oceanography, 21:1-13.

Walsh, J.J., 1981, A carbon budget for overfishing off Peru, Nature, 290:300-304.

Weikert, H., 1977, Copepod carcasses in the upwelling region south of Cap Blanc, N.W. Africa, Marine Biology, 42:351-355.

Whitledge, T.E., 1981, Nitrogen recycling and biological populations in upwelling ecosystems, in: "Coastal Upwelling," F.A. Richards, ed., Coastal and Estuarine Sciences 1, American Geophysical Union, Washington, 257-273.

Whitledge, T.E. and Packard, T.T., 1971, Nutrient excretion by anchovies and zooplankton in Pacific upwelling regions, Investigacion Pesquera, 35:243-250.

Wooster, W.S., Chow, T.J. and Barrett, I., 1965, Nitrite distribution in Peru current waters, Journal of Marine Research, 23:210-221.

Wyrtki, K., 1962, The oxygen minima in relation to ocean circulation, Deep-Sea Research, 9:11-23.

Wroblewski, J.S., 1977, A model of phytoplankton plume formation during variable Oregon upwelling, Journal of Marine Research, 35:357-394.

OXYGEN CONSUMPTION AND DENITRIFICATION BELOW THE PERUVIAN UPWELLING

Theodore T. Packard, Paula C. Garfield and
Louis A. Codispoti

Bigelow Laboratory for Ocean Sciences
West Boothbay Harbor
Maine 04575, U.S.A.

ABSTRACT

Rates of oxygen consumption and denitrification were calculated from the respiratory electron transport activity of deep and intermediate water plankton of the Peru Current upwelling, and compared to rates calculated for the deep water under other oceanic regions. Oxygen consumption off Peru ranged from 1.13 ± 0.85 $\mu\ell\ O_2\ h^{-1}\ \ell^{-1}$ at the sea surface to 0.6 ± 0.4 x 10^{-3} $\mu\ell\ O_2\ h^{-1}\ \ell^{-1}$ at depths below 1000 m. The depth-dependence in oxygen consumption in the Peru upwelling at 15°S was described by the exponential function $R = 1.5\ Z^{-1.03}$ ($r^2 = 0.99$) where R is oxygen consumption in $\mu\ell\ O_2\ h^{-1}\ \ell^{-1}$ and Z is depth in meters. In the intermediate water (100-500 m), oxygen consumption in oxygenated waters and denitrification in anoxic waters decreased seaward from the coastal zone along a gradient from high to low surface productivity.

Denitrification in the southeastern Pacific removed 2.5 x 10^{13} g N yr^{-1} in 1976-1977. This process occurred in the oxygen deficient zone at 100-400 m between 10-25°S, and although the zone extends 1300 km westward into the Pacific, the denitrification rate is most intense in the inner 175 km.

The organic carbon consumed by the O_2 and NO_3^- based respiration in the water column is less than the organic carbon produced by the phytoplankton in the surface waters. The difference of 1.7 g C d^{-1} m^{-2} in the nearshore region and 0.6 g C d^{-1} m^{-2} in the offshore region could be lost to the sediments or exported offshore. Mechanisms by which this organic carbon is transported to the deep sea are discussed.

INTRODUCTION

Oxygen and nitrate consumption in the ocean are not spontaneous processes; they are largely mediated by the respiration of marine organisms. During respiration, organisms use the change in free energy that results from coupling the oxidation of organic matter with the reduction of oxygen, to synthesize biologically-usable high-energy organic compounds. If oxygen is unavailable, certain microbes can use nitrate or nitrite as an alternate electron acceptor in the processes. During these processes, the concentration of organic matter in the seawater decreases and the concentration of CO_2 increases.

Oxygen consumption occurs at all depths and in all regions of the ocean where oxygen is present. It is a vital energy-yielding process employed by most of the organisms in the ocean, from the bacteria and phytoplankton to the zooplankton and nekton. Furthermore, it is a variable that must be quantified before accurate models and budgets of marine ecosystems can be constructed. In a productive euphotic zone, such as one finds in an upwelling area, oxygen consumption is masked by photosynthesis, but below this zone it becomes the dominant process. Oxygen consumption rates in the euphotic zone range from 1 to 20 $\mu\ell\ O_2\ h^{-1}\ \ell^{-1}$. They decrease to 0.05 ± 0.01 $\mu\ell\ O_2\ h^{-1}\ \ell^{-1}$ between the bottom of the euphotic zone and 100 m, and below 100 m they range from 10 to 0.1 x 10^{-3} $\mu\ell\ O_2\ h^{-1}\ \ell^{-1}$ (Packard, 1979). In all cases, oxygen consumption reflects an organism's attempt to replace the energy expended in swimming, feeding, hunting, and growing, and the energy used in basal metabolism. The energy consumed by these processes is produced during the oxidation of carbohydrates, lipids and proteins by the reactions of glycolysis,

$$C_6H_{12}O_6 + 2\ NAD^+ + 2\ P_i + 2\ ADP = 2\ CH_3COCO_2^- + 2\ NADH + 2\ H^+ + 2\ ATP + 2\ H_2O \quad (1)$$

the reactions of the citric acid cycle,

$$CH_3COCO_2^- + 2\ H_2O + FAD + 4\ NAD^+ = 3\ CO_2 + FADH_2 + 4\ NADH + 4\ H^+ \quad (2)$$

and the reactions of the respiratory electron transport system (ETS),

$$NADH + H^+ + FADH_2 + O_2 + 5\ ADP + 5\ P_i = NAD^+ + FAD + 2\ H_2O + 5\ ATP \quad (3)$$

Abbreviations used: NAD^+, oxidized nicotinamide adenine dinucleotide; NADH, reduced nicotinamide adenine dinucleotide; ADP, adenosine diphosphate; ATP, adenosine triphosphate; P_i, inorganic phosphate; $C_6H_{12}O_6$, glucose; $CH_3COCO_2^-$ pyruvate; FAD, flavin adenine dinucleotide (oxidized); $FADH_2$ reduced flavin adenine dinucleotide.

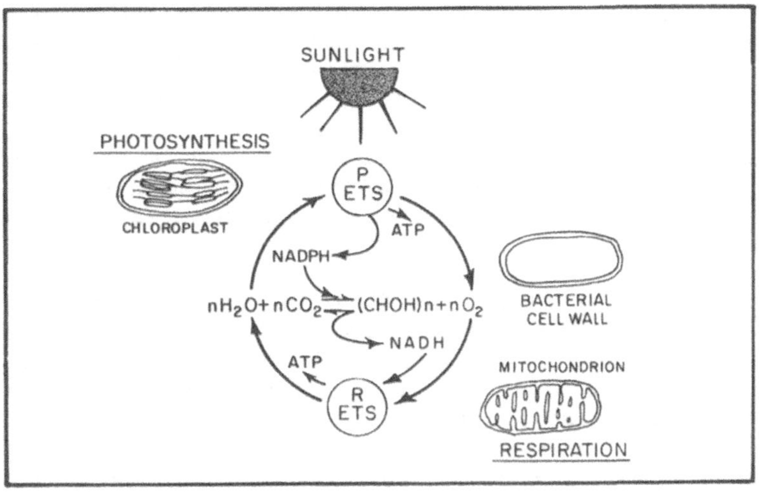

Fig. 1. Respiration and photosynthesis, the two major energy-yielding biological processes in the ocean. In anoxic areas NO_3^-, NO_2^- or SO_4^{-2} take the place of O_2 as respiratory electron acceptors. The electron transport systems (ETS) of both photosynthesis (P) and respiration (R) generate ATP. Photosynthesis occurs in the chloroplasts of the phytoplankton, respiration occurs in bacterial cell walls and in the mitochondria of the phytoplankton, zooplankton and nekton.

The energy produced by these reactions is in the physiologically useful form of adenosine triphosphate (ATP). The overall process is the reverse of photosynthesis and affects seawater by removing particulate organic matter (POM) and O_2, and replacing it with CO_2 (Fig. 1).

Because of the stoichiometry associated with the production of adenosine triphosphate and the metabolism of carbon and oxygen, measurements of oxygen consumption can provide information about the pathways of carbon and energy in oceanic ecosystems. Such measurements are also helpful in understanding many physical and chemical phenomena, such as the ventilation times and circulation patterns of deep water masses and the distribution patterns of many chemical species. The oxygen consumption rates in the deep ocean are difficult to measure because they are low and existing methods for measuring oxygen are not sufficiently sensitive. The rates have to be assessed indirectly from biomass, from advection-diffusion models, from vertical carbon flux measurements, or from the enzymic reactions that control respiratory oxygen consumption. This paper describes the use of the last approach.

The enzymatic processes that control oxygen consumption are essentially the same in all the animals, plants and bacteria (Keilin,

1925; 1966). Oxygen consumption is the terminal reaction of the respiratory electron transport system (ETS) in the mitochondria of plants and animals and in the cell membrane of bacteria (Fig. 2). In both places, the reactions are similar and in both places, the purpose of the oxygen consumption is to generate ATP. The ETS is composed of membrane-bound proteins which, through a series of redox reactions, transfer electrons to oxygen or to some other electron acceptor. This electron transfer and the associated proton translocation generate ATP.

Although oxygen is always the final electron acceptor in aerobic organisms, other electron acceptors are used in nitrate reducing and denitrifying organisms. Fig. 2 shows the organization of the ETS in organisms that can use nitrate, nitrite or nitrous oxide as the terminal electron acceptor in respiration. There are, of course, some differences between the electron transport systems in these different types of organisms. For example, cytochromes b, c, and c_1 in the ETS in aerobes are replaced by the structurally different cytochromes b_1, c_4 and c_5 in the ETS of nitrate reducers and denitrifiers (Doelle, 1969; Payne, 1973a; 1973b; 1976). However, the basic reactions are the same for all organisms.

Denitrification and nitrate reduction are not as widespread as oxygen consumption. They are confined to oxygen deficient fjords and

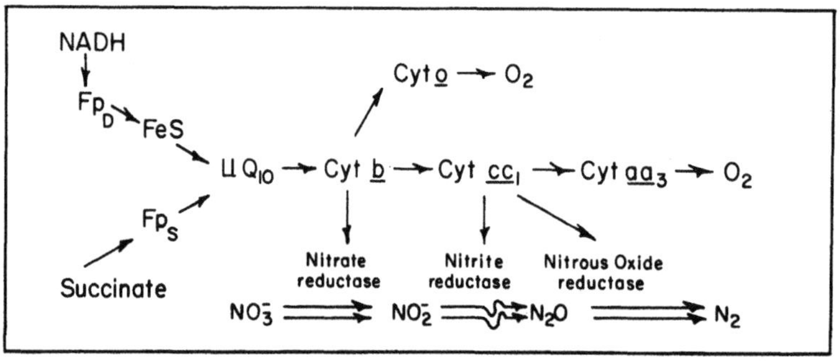

Fig. 2. The electron flow pattern in the respiratory electron transport system in the facultative anaerobe *Paracoccus denitrificans*. The single arrows signify a redox reaction in which the molecule to the left of the arrow is oxidized and the one on the right side is reduced. The double arrows at the bottom represent the reduction steps between NO_3^- and N_2 that are catalyzed by the three enzymes, nitrate reductase, nitrite reductase and nitrous oxide reductase. The wave in the $NO_2^- \rightarrow N_2O$ step represents the possible occurrence of intermediates (hydroxyl amine and/or nitric oxide). Abbreviations: NADH, nicotinamide adenine dinucleotide (reduced); Fp_D, NADH dehydrogenase; Fp_S, succinate dehydrogenase; FeS, non-heme iron; UQ_{10}, ubiquinone; Cyt, cytochrome. The electron flow pattern is modified from Stouthamer (1980).

basins, to the intermediate waters in the southeastern Atlantic, the Arabian Sea and the eastern tropical Pacific, and to oxygen deficient sediments. As with oxygen consumption, the rates of these processes proceed slowly yet have a major effect on the chemistry of seawater. There is no direct method of measuring either nitrate reduction or denitrification, so we measure the activity of the controlling enzymes and calculate the *in situ* rates from the measured enzyme activity.

METHODOLOGY

The basic premise behind this work is that O_2 and NO_3^- utilization and CO_2 and N_2 evolution in the sea are controlled by enzyme reactions in marine organisms. These reactions are stoichiometrically associated with the respiratory electron transport system (ETS) in these organisms, so that by measuring the kinetics of the electron transport in extracts of these organisms one can calculate potential rates of O_2 and NO_3^- utilization as well as CO_2 and N_2 evolution (Fig. 3). For example, if one measures an electron flow of 8 milli-electron equivalents per hour in the respiratory electron transport system extracted from the plankton in an aerobic deep-sea sample, one can calculate from Fig. 3 that the sample's potential oxygen consumption rate was 2 mM/hr or 44.8 ml/hr. If the sample had been taken in the oxygen deficient waters of the secondary nitrite maximum zone, the denitrification rate would be 8/5 mM NO_3^-/hr or 4/5 mM N_2/hr, assuming that the reduction of nitrate goes through nitrite and nitrous oxide all the way to di-nitrogen. The actual *in situ* rate of these reactions can be calculated if the relationship between the potential and the actual rate is understood.

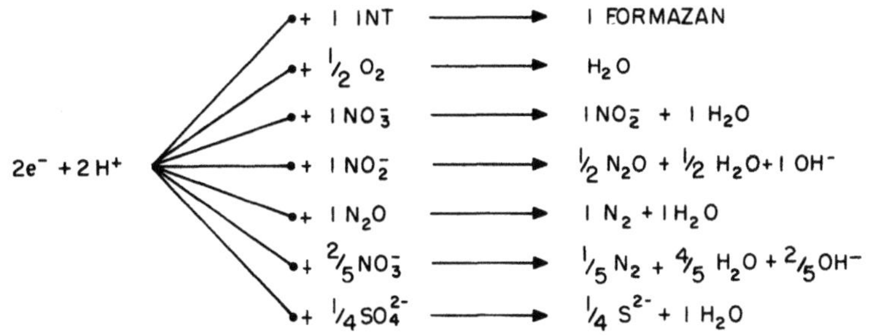

Fig. 3. The reactions of natural respiratory electron acceptors with the reduction products of metabolism (H^+ and e^-) as compared with INT reduction. INT (2-[p-iodophenyl]-3-[p-nitro-phenyl]-5-phenyl tetrazolium chloride is the tetrazolium salt used as the electron acceptor in the ETS assay (see Fig. 4). All the other electron acceptors are used by respiring organisms in the ocean as described by Richards (1965).

To measure the potential rate, we developed an assay based on spectrophotometry that measures respiratory electron transport in mitochondria, microsomes and other biological membranes (Packard, 1969; 1971; Packard, Healy and Richards, 1971; Packard, Harmon and Boucher, 1974; Packard et al., 1977). It measures the combined enzyme activity (maximum reaction rates) of succinate-ubiquinone oxido-reductase (EC. 1.3.99.1), NADH-ubiquinone oxido-reductase (EC. 1.6.99.3) and NADPH-ferricytochrome oxido-reductase (EC. 1.6.2.4; nomenclature of the International Union of Biochemistry). These enzymes control the access to the electron transport systems in bacterial cell walls or in eukaryote mitochondria and microsomes. Their activity is measured by the rate at which they reduce, under optimal conditions, a tetrazolium dye (INT) to its formazan (Fig. 3). This reduction requires two electron equivalents per mole of formazan produced (Fig. 3). Then, from the rate of formazan production, one can calculate, for a sample from oxygen-rich waters, the number of moles of O_2 the electron transport system would have reduced if its flow had not been intercepted by INT. Had the sample been from anoxic, nitrate-bearing, secondary nitrite maximum zones, then the rate of denitrification could be calculated as the number of moles of nitrate reduced and dinitrogen produced (Fig. 3).

The assay has been calibrated against actual oxygen consumption rates for phytoplankton (Kenner and Ahmed, 1975a; 1975b), zooplankton (King and Packard, 1975; Owens and King, 1975), bacteria (Christensen et al., 1980; Packard, Garfield and Martinez, in press) and natural seawater samples (Packard and Williams, 1981). It has also been calibrated against denitrification by Devol (1975), Christensen et al. (1980), and against nitrate reduction by Packard et al. (in press). Since the method has been continually improved, an intercalibration study (Christensen and Packard, 1979) permits comparisons between early and recent results. A flow diagram of the assay procedure is shown in Fig. 4.

The ETS assay has been used to investigate oxygen consumption and/or denitrification in the North Pacific, the north eastern Tropical Pacific, the Costa Rica Dome, the Peru Current upwelling, the central Atlantic, the Arctic Ocean and the Mediterranean Sea. Although the main focus of this paper is on the Peru Current upwelling system, some data from other regions will be used for comparison.

→

Fig. 4. Flow diagram of the extraction-type ETS assay used to determine oxygen consumption and denitrification rates in deep water samples. The substrates used in the assay are nicotinamide adenine dinucleotide (NADH), nicotinomide adenine dinucleotide phosphate (NADPH), and sodium succinate. The electron acceptor (INT) is 2-[p-iodophenyl]-3-[p-nitrophenyl]-5-phenyl tetrazolium chloride.

OXYGEN CONSUMPTION IN SEAWATER

The rate of oxygen consumption in the deep seawater column is not a constant. Using data from the Peru Current upwelling system, one can see that the oxygen consumption rate decreases with depth (Fig. 5); it decreases seaward from the edge of the continental shelf (Fig. 6); and increases under zones of high productivity (Packard and Garfield, in press); it changes with time (Fig. 7); and it displays complex horizontal variations. The rate is so low that it cannot be measured directly without incubation (Bryan, Riley and Williams, 1976) and since incubation techniques give unrealistically high val-

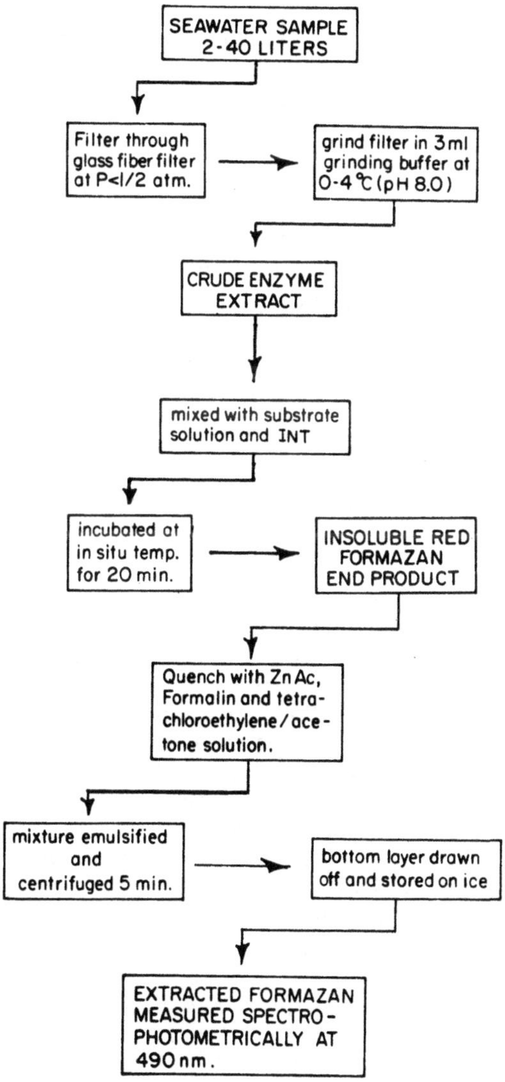

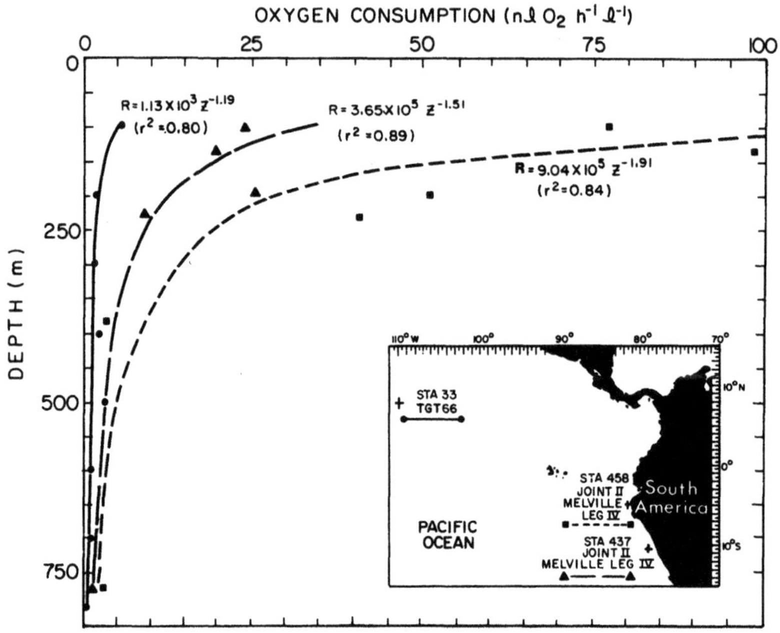

Fig. 5. Comparison of three depth profiles of oxygen consumption. The profile on the left (———) is from the oligotrophic waters in the Northeastern Tropical Pacific; the middle profile (— — —) is from the central part of the Peruvian upwelling system; and the left profile (---------) is from the northern part of the Peruvian upwelling sytem. The integrated oxygen consumption rate between 102 and 195 m off northern Peru (5°S) was 7.4 ml O_2 h^{-1} m^{-2} and only 2.1 ml O_2 h^{-1} m^{-2} for the same depth layer off central Peru (10°S).

ues when applied to deep-sea samples (Seiwell, 1937; Rakestraw, 1947; Skopintsev, 1976), they cannot be used. In spite of the slow rate, oxygen consumption has a significant effect on the distributions of O_2, CO_2, $CaCO_3$, NO_3^-, and PO_4^{-3} in the deep sea (Craig, 1971; Kroopnick, 1974). Before 1970, deep-sea oxygen utilization was calculated from advection-diffusion models (Munk, 1966) or nutrient regeneration (Riley, 1951; Wyrtki, 1962) models. Now it can be calculated from microbial biomass (Holm-Hansen and Pearl, 1972; Williams and Carlucci, 1976), from *in situ* incubations with ^{14}C-labelled substrates (Jannasch et al., 1971; Sorokin, 1972), from helium-tritium dating (Jenkins, 1980), from the rates of particle flux through the water column (Suess, 1980) as well as from a new generation of advection-diffusion models (Craig, 1971; Kroopnick, 1974). From models, from biomass, from dating techniques and from particle flux measurements, the oxygen consumption rates at depths between 1 and 4 km fall in the range of 0.1 to 40 μℓ O_2 yr^{-1} $ℓ^{-1}$. Rates calculated from ETS activity for the waters below 1 km in the Peru Current (Packard and Garfield, in press) and the Costa Rica Dome (Packard et al., 1977) fall into a similar range, i.e., from 0.05 to 6 μℓ O_2 yr^{-1} $ℓ^{1}$.

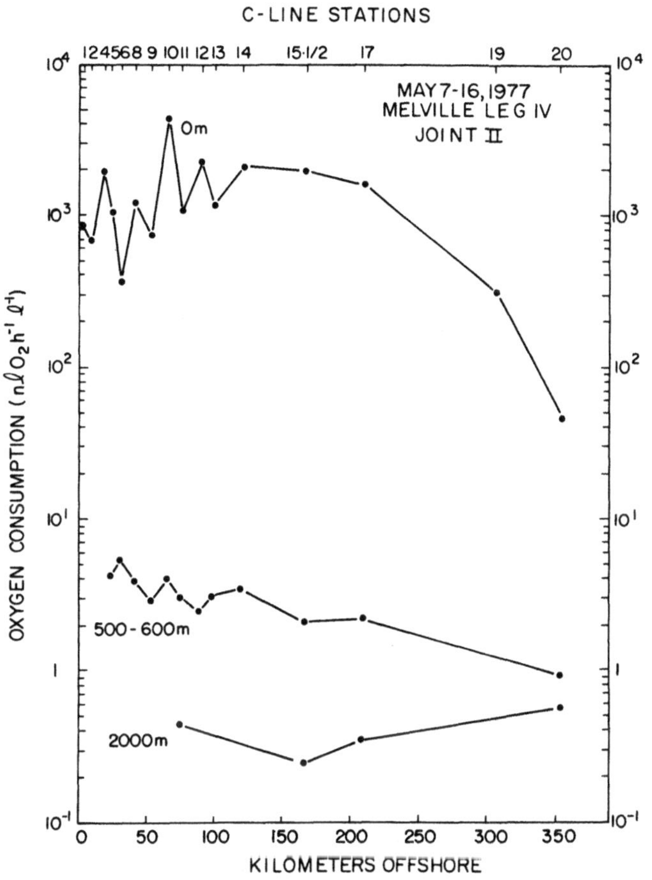

Fig. 6. Oxygen consumption rates in the Peruvian upwelling system (15°S) as a function of distance offshore at three different depth ranges. Each point represents one sample. Oxygen consumption rates were calculated from ETS activities using the equation: oxygen consumption = F • ETS, where F = 0.50 in phytoplankton-dominated surface plankton and F = 0.43 in bacteria-dominated subsurface plankton.

DEPTH PROFILES OF OXYGEN CONSUMPTION

General Features

In May 1977, the oxygen consumption rate in the Peru upwelling region ranged from 1.13 ± 0.85 $\mu\ell\ O_2 h^{-1}\ \ell^{-1}$ at the sea surface to 0.6 ±0.4 x 10^{-3} $\mu\ell\ O_2 h^{-1}\ \ell^{-1}$ at depths below 1000 m. The major decrease occurred in the upper 500 m, where the rate decreased by two orders of magnitude. Between 500 and 1000 m, the rate decreased by another factor of ten and below 1000 m, the rate changed insignificantly (Table 1). The depth profiles off Peru (Fig. 5) are similar

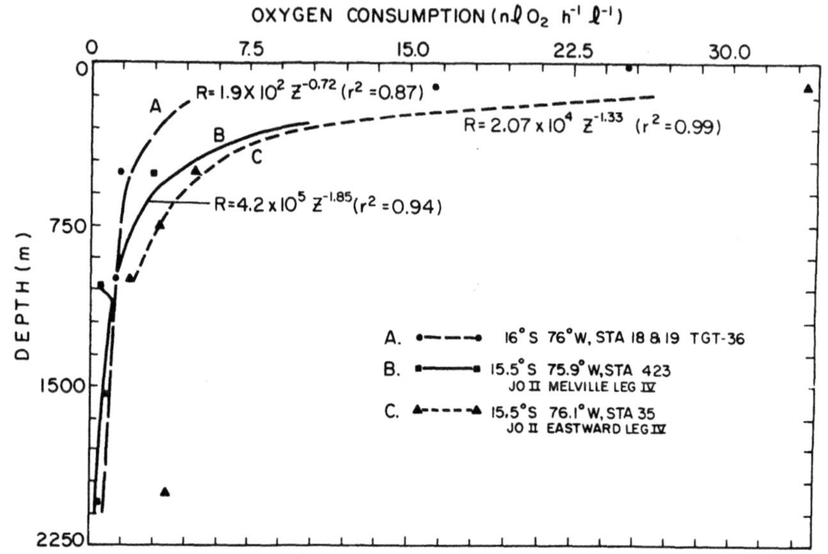

Fig. 7. Variations with time of the oxygen consumption rate at the same part of the southern Peruvian upwelling system. (A) April, 1969; (B) May, 1977; (C) September, 1976.

to the depth profiles in the North Equatorial Pacific (Packard et al., 1971; 1977). Furthermore, they can be predicted from the depth profiles of the particulate organic matter and the depth profiles of organisms that metabolize this organic matter (Menzel, 1967; 1974; King, Devol and Packard, 1978; Garfield, Packard and Codispoti, 1979). However, an important difference between depth profiles of oxygen consumption off Peru and profiles from other oceanic areas is the break in the oxygen consumption rate between 100 and 400 m where denitrification occurs (Fig. 5). If one treats the denitrification zone as a gap in the oxygen consumption profile, rather than a series of null values, then the depth function of oxygen consumption for the Peru upwelling at 15°S (Table 1) is described by the exponential function:

$$R = 1.154\ Z^{-1.03} \quad (r^2 = 0.99) \tag{4}$$

where R is the oxygen consumption rate in $\mu\ell\ O_2\ h^{-1}\ \ell^{-1}$ and Z is depth in meters. In the vicinity of the Costa Rica Dome (9°N 89°W) a similar function,

$$R = 1.63\ Z^{-0.84} \quad (r^2 = 0.98) \tag{5}$$

describes the oxygen consumption depth profile (Packard and Garfield, in press).

Table 1. Summary of the oxygen utilization rates (R) at 15°S in the Peru upwelling system. The data were taken during the RV "Melville" cruise (leg IV, May 1977) of the CUEA expedition, JOINT-II. All rates are reported in units of µℓ O_2 h^{-1} $ℓ^{-1}$. The rates were calculated from ETS activity measurements using the equation R = F x ETS, where F is an experimentally determined ratio between respiration and ETS activity in cultures of phytoplankton or bacteria. In the euphotic zone where phytoplankton dominate, F = 0.50 (Devol and Packard, 1978; Christensen and Packard, 1979). In the deep sea where bacteria are presumed to dominate, F = 0.43 (Christensen et al., 1980).

Depth (m)	Number of Samples	Maximum R	Minimum R	Mean R	Standard Deviation
0	67	4.46	0.030	1.13	0.85
500-600	17	0.0052	0.0007	0.0028	0.0013
1000	8	0.0015	0.0001	0.0005	0.0004
1500	5	0.0008	0.0006	0.0006	0.0001
1750	5	0.0010	0.0002	0.0005	0.0003
2000	6	0.0012	0.0003	0.0006	0.0003

Oxygen Consumption at Mid-Water Depths Off Northern Peru

An outstanding feature of the Peru upwelling at 15°S is the oxygen deficient zone between 100 and 400 m. In this zone, NO_3 and NO_2 are used by microbes in place of oxygen to satisfy their metabolic demand for an electron acceptor. For this reason, oxygen consumption rates cannot be reported in this region between 100 and 400 m (Table 1 and Fig. 5). However, the depth function of the oxygen consumption rate at these depths can be determined in the northern part (5-10°S) of the Peru upwelling where the oxygen concentrations in the oxygen minimum zone are high enough to support aerobic respiration. Between 100 and 250 m, the rates range from 0.02 to 0.1 µℓ O_2 h^{-1} $ℓ^{-1}$, and are several times greater than rates outside the upwelling area. In oligotrophic oceanic waters northwest of the Galapagos Islands, the oxygen consumption rates at comparable depths range from 0.001 to 0.005 µℓ O_2 h^{-1} $ℓ^{-1}$ (Fig. 5). The high rates off northern Peru are likely an important factor in the development of the denitrification zone farther to the south. A major source of water for the large lobe of oxygen-deficient nitrite-rich water off southern Peru (Codispoti and Packard, 1980) is the poleward undercurrent that flows at depths of 100 ± 50 m along the coast of northern Peru (Brockmann et al., 1980). As this water flows south, it receives a heavy load of organic matter falling from the productive surface waters. This material stimulates bacterial growth and, as can be seen in Fig. 5, sustains a high rate of respiration. The oxygen in the undercurrent is depleted by 9°S and subsequent catabolic metabolism is coupled to nitrate reduction and denitrification.

HORIZONTAL VARIATIONS IN THE OXYGEN CONSUMPTION

Large scale variations in deep-sea oxygen consumption are obvious from a comparison of the rates in the abyssal waters of the Peru Trench and the abyssal waters of the central Atlantic (25°N 35°W). Fig. 8 shows that the oxygen consumption rates at 5000 m in the central Atlantic are less than those off Peru by a factor of thirty.

Mesoscale (>100 km) variations at intermediate depths (100-300 m) are obvious from a comparison of the oxygen consumption profiles at 5°S and 10°S in the poleward undercurrent at the shelf edge off northern Peru (Fig. 5). At 5°S the integrated rate between 100 and 200 m is four-fold higher than the comparable rate at 10°S. At 15°S, mesoscale variations were observed of both the oxygen consumption and denitrification rates at intermediate depths (Figs. 6 and 9). The oxygen consumption rate between 500-600 m decreased five-fold and the denitrification rate between 125 and 220 m decreased nearly ten-fold

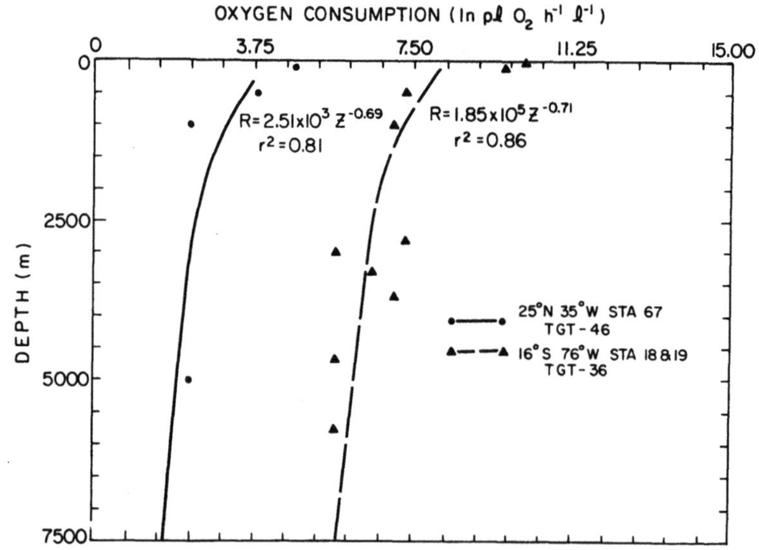

Fig. 8. The decrease in the oxygen consumption rate between the surface and 6000 m at 16°S, 75°35'W in the Peru upwelling system and, for comparison, the decrease in the oxygen consumption rate at 25°N, 35°W in the central North Atlantic. The coefficient of variation is 20%.

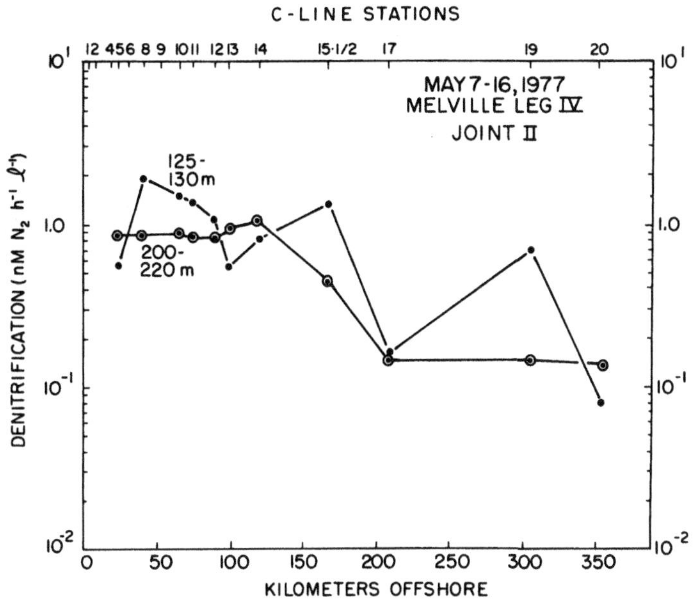

Fig. 9. Denitrification rates for "C-line" stations at 15°S in the Peruvian upwelling system during 7-16 May 1977 as a function of distance offshore. Each point represents a single sample.

along the same offshore transect (Figs. 6 and 9). In contrast to the variations at intermediate depths, variations in the deep sea were small and insignificant off Peru (Fig. 6). At 15°S, the oxygen consumption rate at 2000 m ranged from 0.3 x 10^{-3} to 1.2 x 10^{-3} $\mu\ell$ O_2 h^{-1} ℓ^{-1}, but the variations were not related to distance along the 270 km transect. This does not imply, however, that there is a uniform rate everywhere in the deep sea. Mesoscale variations have been observed under the Costa Rica Dome. The Pinta expedition (Kuntz et al., 1975) looked for a reflection of the high surface productivity of the Costa Rica Dome (Broenkow, 1965) in the horizontal field of the deep-sea oxygen consumption rate. They found that under the statistical center of the Costa Rica Dome (9°N 89°W, Wyrtki, 1964) the deep-sea oxygen consumption rate at 3000 m is more than five-fold higher than comparable rates 160 km to the west.

Small-scale (<100 km) variations have not been investigated in the deep sea, but in mid-water regions (500-600 m) they can be seen in the two-fold decrease within 60 km at the transition between the inshore and offshore zones at 15°S off Peru (Fig. 9). They can also be seen in the five-fold decrease within the same distance at 200 m at 5°S off Peru (Garfield and Packard, 1979).

DENITRIFICATION

Intermediate waters below coastal upwelling regions are often sites of denitrification. This is true for the Peru Current, the Arabian Sea, the eastern tropical North Pacific and the southwest African upwelling systems, but not for the California Current, the Portugese, or the northwest African upwelling systems. In the California Current denitrification does occur in zones where bathymetry restricts circulation such as the Santa Barbara Basin. The factor separating the former cases from the latter is the presence of a subsurface layer of oxygen deficient water. In this layer, the oxygen concentration falls below 0.2 ml O_2 ℓ^{-1} and the marine microbes inhabiting the waters must respond to the changed environment. The microbes are forced to: migrate out of the layer; switch their respiratory chemistry from an oxygen-based to a nitrate-based system; encyst; or die. An individual organism may do any of these, but the net result is a shift from an aerobic population to an anaerobic pop-

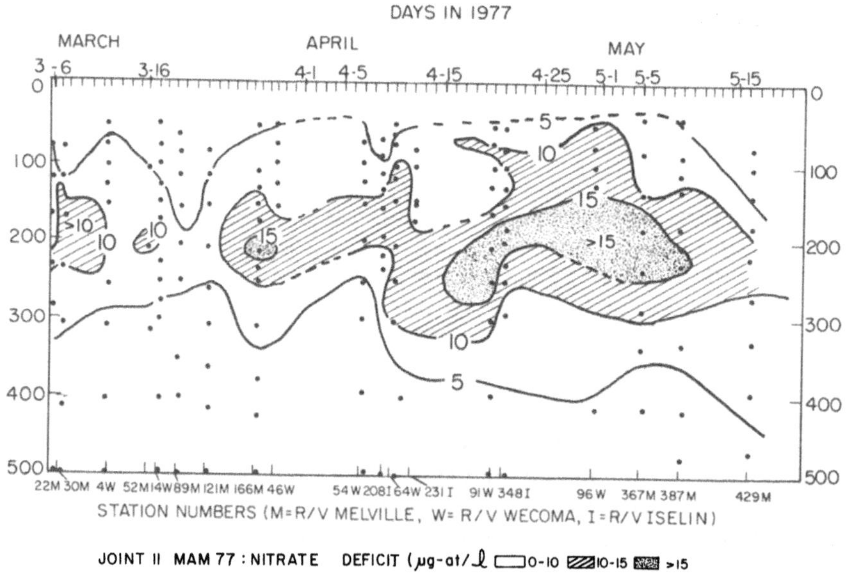

Fig. 10. Variations in the nitrate deficit at the shelf break in the southern part of the Peruvian upwelling system (station C-5, 15°S). The nitrate deficit represents the extent of denitrification in the water column. The data were taken from three vessels during the JOINT II expedition of the CUEA project. Nitrate deficits were calculated from the following formula: $PO_4^{-3} x\ 12.6 - (NO_3^- + NO_2^-) =$ nitrate deficit. Estimates were not made for the upper ∿50 m where such calculations may be unreliable (Codispoti and Packard, 1980).

ulation of denitrifiers. In this process the microbes depend on nitrate as an electron acceptor in place of oxygen. The net product of this type of metabolism is di-nitrogen (N_2) and the net effect on an upwelling system is a loss of combined nitrogen.

The most outstanding example of a denitrification zone occurs in the main secondary nitrite maximum of the Peruvian upwelling system where the NO_3^- can be reduced to zero (Wooster, Chow and Barett, 1965; Fiadeiro and Strickland, 1968; Dugdale, 1972, Dugdale et al., 1977; Packard et al., 1978; Codispoti and Packard, 1980; and Codispoti, this volume). This denitrification zone normally occurs between 100 and 400 m between 10 and 25°S where it can be identified by inspection as a minimum in the depth profile of nitrate and a maximum in the depth profile of nitrite. It extends westward as a triangle with its apex at 15°S, 89°W, 1300 km from the Peruvian coast. The extent of denitrification can be determined by calculating nitrate deficits, which represent the difference between the observable sum of nitrate and nitrite and their estimated concentration just before the onset of denitrification (Fig. 10). In spite of the common occurrence of denitrifying zones in coastal upwelling regions, little is known about their chemistry, their biology and the mechanisms by which they are formed, maintained or dissipated. The Coastal Upwelling Ecosystems Analysis project addressed this problem during the JOINT II expedition to the Peru upwelling system by measuring ETS activity extensively throughout the denitrification zone. Rates of denitrification were calculated by Codispoti and Packard (1980) from ETS activity and are presented here. The rates range from <0.01 to 1 nM N_2 h^{-1}

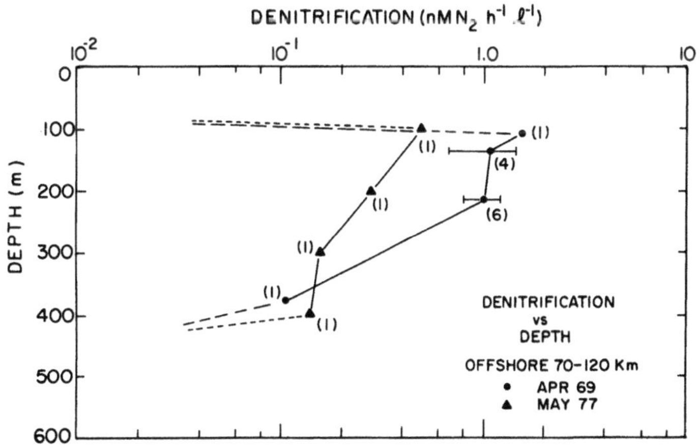

Fig. 11. Denitrification in the oxygen minimum zone of the Peruvian upwelling system at 15°S before and after the 1972-73 El Niño. Since the denitrification zone is bordered above and below by oxygenated water, the denitrification rate is presumed to fall to zero at these borders.

ℓ^{-1}; they decrease offshore and, in the coastal portion of the main secondary nitrite maximum, they appear to have increased between 1969 and 1977 (Figs. 9 and 11).

The denitrification zone is limited at its upper and lower boundaries by oxygenated water, thus the denitrification rate increases from zero at both boundaries to a maximum at the nitrite maximum (Fig. 11). Over the continental slope off Peru where the organic loading is the greatest, the denitrification zone has a thickness of more than 200 m. Offshore, the thickness decreases to 100 m. For discussion, it is convenient to separate the main secondary nitrite maximum (Fig. 12) into three smaller zones (Codispoti and Packard, 1980). The inner one, 175 km wide, is the most active (Fig. 13). The middle one, 125 km wide, and the outer one together contribute only 25% of the total amount of nitrogen lost from the entire denitrification zone.

The processes controlling the horizontal boundaries of the denitrification zone off Peru are not well understood. The northwestern boundary is likely controlled by the onshore flow of oxygenated seawater between 7 and 11°S at the depth of the secondary nitrite maximum (White, 1971, and Tsuchiya, 1975). Along the coast, the oxygenated poleward undercurrent will restrict spreading to the north (Brockman et al., 1980; Pak, Codispoti and Zaneveld, 1980). The southern boundary is probably determined by the Peru offshore current

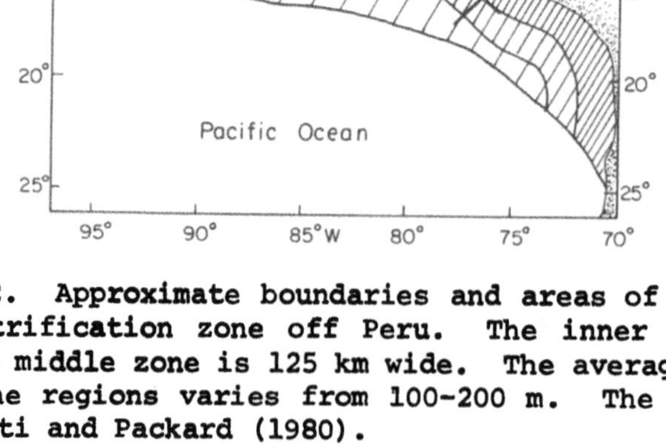

Fig. 12. Approximate boundaries and areas of the three regions of the denitrification zone off Peru. The inner region is 175 km wide and the middle zone is 125 km wide. The average vertical thicknesses of the regions varies from 100-200 m. The figure was taken from Codispoti and Packard (1980).

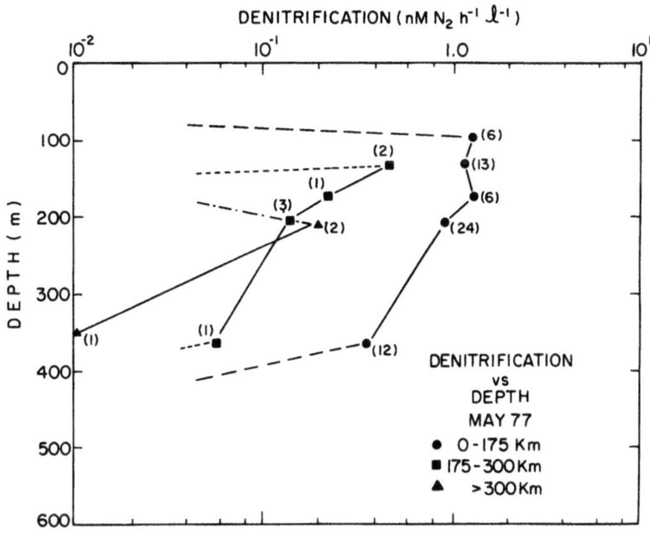

Fig. 13. Comparison of depth profiles of denitrification from three offshore regions. Numbers in parentheses indicate the number of samples used to determine each mean value. All data are from the region of the "C-line" at 15°S off the coast of Peru during May 1977. Dotted lines represent extrapolation of the data to zero denitrification where oxygen concentrations are in excess of 0.2 ml • ℓ^{-1}.

which leaves the coast at 25°S and extends to depths of 700 m (Wyrtki, 1963; 1977).

The previous discussion points out that, in contrast to the oxygen consumption process, denitrification is restricted to relatively small patches in the ocean. Nevertheless, because the availablility of combined nitrogen is so important to biological growth on the planet, any loss of combined nitrogen is important to humanity and should be assessed. We have calculated a loss rate of combined nitrogen of 2.5×10^{13} g N yr^{-1} from the southeastern tropical Pacific during 1976-1977 (Codispoti and Packard, 1980) and most of this removal ($\sim 2 \times 10^{13}$ g N yr^{-1}) occurs within the main secondary nitrite maximum. This removal rate represents ∿25% of the combined nitrogen added to the ocean yearly (Emery, Orr and Rittenberg, 1955; Eriksson, 1959; Codispoti, 1973; Söderlund and Svensson, 1976). Furthermore, this rate may be twice as high as the pre-1972 rate and is comparable to the denitrification in the northeastern Tropical Pacific, the largest marine denitrification site previously identified.

WATER COLUMN CARBON BUDGET FOR THE PERU UPWELLING SYSTEM

From productivity measurements and from our oxygen consumption and denitrification rates, we can construct a simple carbon budget for the water column of the Peru Current System during May 1977 (Fig. 14). To make such a budget one must calculate the gross primary production. We have done this from the net primary production data of Barber et al. (1978) and the respiration data of Setchell and Packard (1978). The resulting gross productivity values are 3.8 g C d^{-1} m^{-2} for the inner 40 km and 1.7 g C d^{-1} m^{-2} for the next 40-km band off the coast. Of this carbon produced by the phytoplankton, 15% is consumed by the microplankton in the euphotic zone and 30% is consumed by the same size group between the bottom of the euphotic zone and 100 m. These percentages were the same for both the nearshore and offshore zones. The macrozooplankton in the upper 200 m of the nearshore zone (or between 0 m and the bottom) consumed another 10%, half of which was consumed in the euphotic zone alone (King and Packard, 1978). We have assumed the offshore macrozooplankton also consumed 10%.

In the secondary nitrite maximum zone the microplankton consumed 3% and 6% in the nearshore and offshore zones, respectively. The deep waters below 350 m consumed 2%. Nekton respiration accounted for only 1% according to Walsh (1981). Thus, 59% of the gross primary productivity is accounted for in the nearshore waters and 64% is accounted for in the offshore waters. The remaining material could be exported from the upwelling system or lost to the sediments where part could be metabolized by the benthic organisms and part could be

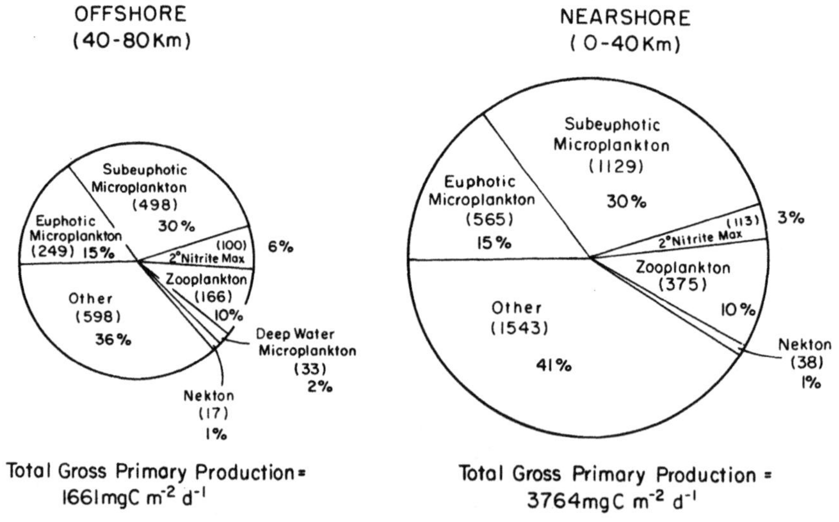

Fig. 14. Carbon budget for the Peru upwelling system during May 1977. Carbon values are in mg C d^{-1} m^{-2}. Total gross primary production was 1.7 g C d^{-1} m^{-2} for the offshore zone (40-80 km) and 3.8 g C d^{-1} m^{-2} for the nearshore zone (0-40 km). Mean depth ranges for the various zones were: euphotic 0-25 m; sub-euphotic 25-150 m; secondary nitrite maximum (2°NO_2 max.) 150-350 m; and deep water 350-bottom. Productivity data were taken from Barber et al., 1978 and Barber and Smith, 1981.

deposited in the sedimentary record. According to the above calculations, 1.7 g C d^{-1} m^{-2} could be lost to the sediments in the nearshore band and 0.6 g C d^{-1} m^{-2} could be lost to the sediments in the offshore band.

MECHANISMS THAT SUPPLY ORGANIC MATTER TO THE DEEP SEA

The processes of oxygen and nitrate consumption in the oceanic water column below the surface layer can only be sustained by importing a supply of oxidizable organic carbon. Since the deep sea does not autotrophically produce organic material except in the vicinity of the hydrothermal vents, the supply must come from the sea surface or from land. From the sea surface the organic carbon can be transported downward: (1) during upwelling-relaxations; (2) at convergence zones; (3) in the rain of particulate matter that sinks into the deep sea; and (4) by vertically migrating fish and zooplankton. From the land, organic carbon is transported to the sea in rivers. This input is negligible, however, off the arid coastal plain of Peru.

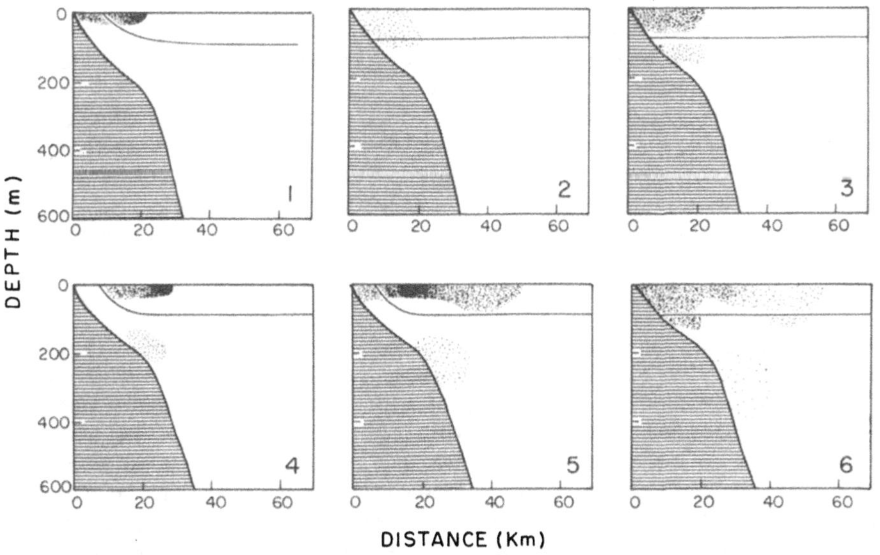

Fig. 15. A mechanism for pumping particulate organic matter into the deep sea that depends on periodic relaxation of coastal upwelling. 1. Coastal upwelling and associated plankton blooms. 2. Relaxation event, isopycnals tend to parallel isobars, plankton entrained and carried below euphotic zone. 3. Moribund plankton starts to sink. 4. Second upwelling event. 5. Extensive bloom develops. 6. Second relaxation occurs, inshore a second batch of entrained plankton is transported out of the euphotic zone.

Upwelling Relaxation

The upwelling-relaxation mechanism was first used to explain the large variations in deep-sea ETS activity, ATP and phaeophytin in the vicinity of the Costa Rica Dome (Packard et al., 1977). Since both upwelling and the plankton productivity associated with the Costa Rica Dome upwelling were observed to fluctuate with time, it was hypothesized that between upwelling events, when the system relaxes and isopycnals become more parallel to isobaric surfaces, plankton could be entrained in a subsiding bolus of water and transported to aphotic sub-surface waters. Once deprived of light, a large fraction of this plankton would become moribund and sink into the deep sea. This moribund plankton would then provide an excellent food source for deep-sea bacteria which would colonize the sinking plankton, oxidize the organic matter and consume oxygen in the process. A similar mechanism (Fig. 15) could occur in coastal upwelling situations between upwelling events when plankton-rich higher density nearshore water sinks along isopycnal surfaces beneath offshore lower density water. Some of this material will provide food for deep-sea fauna and bacteria, some will provide food for the benthos and some will be buried.

Entrainment of Particulate Organic Carbon During Deep-Water Formation

A similar entrainment mechanism will take place in frontal regions where surface waters rich in particulate organic carbon (POC) converge and subside. Once below the euphotic zone, the entrained plankton and organic detritus will sink into intermediate and deep waters and provide a food source for deep-sea bacteria and other organisms. These populations, in turn, will deplete the oxygen supply in the deep waters and replace it with CO_2 (Kroopnick, Margolis and Wong, 1977). This mechanism should occur in all frontal zones.

Cross-Shelf Bottom Flow

Particulate organic matter (POM) may be supplied to intermediate waters off Peru via offshore transport of the bottom nepheloid-layer over the shelf (Pak et al., 1980). Pak and co-workers (1980) have noted that the concentration of particulate matter in the intermediate water shows a connection to the shelf edge. They observed that the particulate matter spreads offshore as a maximum in the depth profiles at the same approximate depth as the shelf break. Offshore transport of water in the bottom nepheloid-layer is consistent with the hydrography of the region. Furthermore, current-meter measurements have shown net offshore transport at several near-bottom locations off the coast of Peru at 15°S. Short-term velocities in this bottom Ekman-layer were reported as high as 5 to 10 cm sec^{-1} (Brink, Allen and Smith, 1978; Brink, Smith and Halpern, 1978; Van Leer and Ross, 1979). Thus, one can easily visualize resuspended shelf sediments rich in POM being injected into offshore intermediate waters.

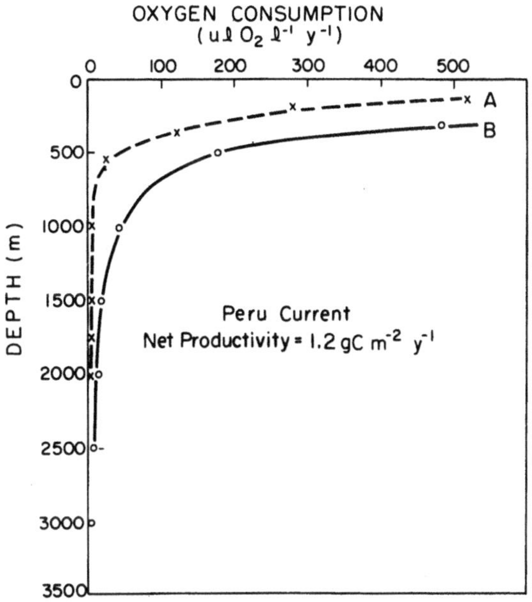

Fig. 16. Oxygen consumption rates in the offshore (D = 40-80 km) region of the Peruvian upwelling system as calculated by two methods. A: this profile was calculated from ETS activity as described in this paper. B: this profile was calculated from sediment trap data using the model of Suess (1980). The Suess (1980) model was based on net primary production data from many sources; consequently, the net production for the offshore region was used in this calculation. The carbon budgets in Fig. 14 are based on gross primary production.

Sedimentation of POC

A more important source of POC for the deep sea in waters distant from the coastal zone is the continuous rain of sedimenting organic particles which falls as dead plankton, fecal pellets or exoskeletons from the surface populations. Suess (1980) has recently analyzed the available data on vertical carbon flux, and has found a linear relationship between the vertical carbon flux and the production of carbon by plankton phytosynthesis in the surface waters. From this relationship, he calculates the rates of mid-water oxygen consumption and the associated respiratory carbon consumption of decaying particles. These rates form exponential depth profiles, similar to ETS-derived oxygen consumption depth profiles. Suess (1980) has generated a family of such profiles for different surface-productivity cases. In the Peruvian upwelling the Suess (1980) model predicts oxygen consumption rates that are much higher than comparable rates calculated from ETS activities (Fig. 16). When used in the water column carbon budget (Fig. 14), these rates would require higher phytoplankton productivity than reported (Barber et al., 1978) and would, furthermore, leave little carbon to be metabolized and/or buried in the sediments. Nevertheless, they do demonstrate the importance of POC sedimentation in sustaining metabolic activity in the deep-sea water column.

Animal Migration

On a daily basis zooplankton and fish migrate up and down in the water column. Riley (1951) and Vinogradov (1955) suggested that these migrations could provide an effective mechanism for transporting surface POC to the deep sea. Most of this material would be in the form of large organisms and fecal material, the oxidation of which would not be measured by any technique based on filtering water samples. This could be one reason for the discrepancy between the rates calculated on the basis of Suess's model (1980) and those based on ETS activity.

ACKNOWLEDGEMENTS

We gratefully acknowledge J. Ammerman, B. Bass, M. Estrada, A. Devol, M. L. Healy, F. D. King, B. R. de Mendiola, and T. Moore who have collaborated with us over the years in making the measurements. We also thank C. Codispoti, M. Colby, A. Levin, and P. Oathout for aid in preparing the manuscript, and J. Rollins for drafting the figures. This work was funded by ONR contract N00014-76-C-0271 and NSF grants OCE 79-19905 and OCE 80-11187 to T. Packard and ONR contract N00014-81-C-0043 to L. Codispoti.

REFERENCES

Barber, R.T. and Smith, R.L., 1981, Coastal upwelling ecosystems, in: "Analysis of Marine Ecosystems," A.R. Longhurst, ed., Academic Press, New York, 31-68.

Barber, R.T., Huntsman, S.A., Kogelschatz, J.E., Smith, W.O., Jones, B.H., and Paul, J.C., 1978, Carbon chlorophyll and light extinction from JOINT II 1976 and 1977, Coastal Upwelling Ecosystems Analysis Data Report No. 49, 476 pp.

Brink, K.H., Allen, J.S. and Smith, R.L., 1978, A study of low-frequency fluctuations near the Peru coast, Journal of Physical Oceanography, 8:1025-1041.

Brink, K.H., Smith, R.L. and Halpern, D., 1978, A compendium of time series measurements from moored instrumentation during the MAM '77 phase of JOINT II, Coastal Upwelling Ecosystems Analysis Technical Report No. 45, Oregon State University, Reference 78-17, 72 pp.

Brockman, C., Fahrbach, E., Huyer, A., and Smith, R.L., 1980, The poleward undercurrent along the Peru coast: 5-15°S, Deep-Sea Research, 27:847-856.

Broenkow, W.W., 1965, The distribution of nutrients in the Costa Rica Dome in the eastern tropical Pacific Ocean, Limnology and Oceanography, 10:40-52.

Bryan, J.R., Riley, J.P. and Williams, P.J.le B., 1976, A winkler procedure for making precise measurements of oxygen concentration for productivity and related studies, Journal of Experimental Marine Biological Ecology, 21:191-197.

Christensen, J.P. and Packard, T.T., 1979, Respiratory electron transport activities in plankton: Comparison of methods, Limnology and Oceanography, 24:576-583.

Christensen, J.P., Owens, T., Devol, A.H., and Packard, T.T., 1980, Respiration and physiological state in marine bacteria, Marine Biology, 55:267-276.

Codispoti, L.A., 1973, "Denitrification in the Eastern Tropical North Pacific Ocean," Ph.D. Thesis, University of Washington, Seattle, 118 pp.

Codispoti, L.A. and Packard, T.T., 1980, Denitrification rates in the eastern tropical South Pacific, Journal of Marine Research, 38(3):453-477.

Craig, H., 1971, The deep metabolism: Oxygen consumption in abyssal ocean water, Journal of Geophysical Research, 76(21):5078-5086.

Devol, A.H., 1975, "Biological Oxidations in Oxic and Anoxic Marine Environments: Rates and Processes," Ph.D. Thesis, University of Washington, Seattle, 208 pp.

Devol, A.H. and Packard, T.T., 1978, Seasonal changes in respiratory enzyme activity and productivity in Lake Washington microplankton, Limnology and Oceanograhy, 23:104-111.

Doelle, H.W., 1969, "Bacterial Metabolism," Academic Press, New York, 486 pp.

Dugdale, R.C., 1972, Chemical oceanography and primary productivity in upwelling regions, Geoforum, 11:47-61.

Dugdale, R.C., Goering, J.J., Barber, R.T., Smith, R.L., and Packard, T.T., 1977, Denitrification and hydrogen sulfide in the Peru upwelling region during 1976, Deep-Sea Research, 24:601-608.

Emery, K.O., Orr, W.L. and Rittenberg, S.C., 1955, Nutrient budgets in the ocean, in: "Essays in the Natural Sciences in Honor of Captain Alan Hancock," University of Southern California, 345 pp.

Ericksson, E., 1959, The circulation of some atmospheric constituents in the sea, in: "The Atmosphere and the Sea in Motion," B. Bolin, ed., Rossby Memorial Volume, Rockefeller Institute, 147-157.

Fiadeiro, M. and Strickland, J.D.H., 1968, Nitrate reduction and the occurrence of a deep nitrate maximum in the ocean off the West Coast of South America, Journal of Marine Research, 26(3): 187-201.

Garfield, P.C. and Packard, T.T., 1979, Biological data from JOINT II RV "Melville", Leg IV, May 1977, Coastal Upwelling Ecosystems Analysis Technical Report No. 53, 186 pp.

Garfield, P.C., Packard, T.T. and Codispoti, L.A., 1979, Particulate protein in the Peru upwelling system, Deep-Sea Research, 26: 623-639.

Holm-Hansen, O. and Pearl, H.W., 1972, The applicability of ATP determination for estimation of microbial biomass and metabolic activity, Symposium on Detritus and its Role in Aquatic Ecosystems, Pallanza, Italy, 23-27 May 1972, U. Melchiorri-Santolini and J.W. Hopton, eds., Memorie dell' Istituto Italiano di Idrobiologia, 29:151-168.

Jannasch, H.W., Eimhjellen, K., Wirsen, C.O., and Farmanfarmaian, A., 1971, Microbial degradation of organic matter in the deep sea, Science, 171:672-675.

Jenkins, W.J., 1980, Tritium and ^{3}He in the Sargasso Sea, Journal of Marine Research, 38(3):533-569.

Keilin, D., 1925, On cytochrome, a respiratory pigment common to animals, yeasts and higher plants, Proceedings of the Royal Society of London on Biological Sciences, 98:312-339.

Keilin, D., 1966, "The History of Cell Respiration and Cytochrome," Cambridge University Press, Cambridge, 416 pp.

Kenner, R.A. and Ahmed, S.I., 1975a, Measurement of electron transport activities in marine phytoplankton, Marine Biology, 33: 119-128.

Kenner, R.A. and Ahmed, S.I., 1975b, Correlation between oxygen utilization and electron transport activity in marine phytoplankton, Marine Biology, 33:129-133.

King, F.D. and Packard, T.T., 1975, Respiration and the activity of the respiratory electron transport system in marine zooplankton, Limnology and Oceanography, 20:849-854.

King, F.D. and Packard, T.T., 1978, Zooplankton respiration, ammonium excretion and protein observations from RV "Cayuse"--legs III and IV, Coastal Upwelling Ecosystems Analysis Data Report No. 44, 16 pp.

King, F.D. and Packard, T.T., 1979, An analysis of chemical, physical, and biological parameters in the vicinity of the Costa Rica Dome in early 1973, Bigelow Laboratory for Ocean Sciences Technical Report No. 2, 64 pp.

King, F.D., Devol, A.H. and Packard, T.T., 1978, On plankton biomass and metabolic activity from the eastern tropical North Pacific, Deep-Sea Research, 25:689-704.

Kroopnick, P.M., 1974, The dissolved O_2-CO_2-^{13}C system in the eastern equatorial Pacific, Deep-Sea Research, 21:211-227.

Kroopnick, P.M., Margolis, S.V. and Wong, C.S., 1977, $\delta^{13}C$ variations in marine carbonate sediments as indicators of the CO_2 balance between the atmosphere and oceans, in: "The Fate of Fossil Fuel CO_2 in the Oceans," N.R. Anderson and A. Malahoff, eds., Plenum Press, New York, 295-321.

Kuntz, D., Packard, T.T., Devol, A., and Anderson, J., 1975, Chemical, physical, and biological observations in the vicinity of the Costa Rica Dome (January-February 1973), University of Washington, Department of Oceanography Technical Report No. 321, 187 pp.

Menzel, D.W., 1967, Particulate organic carbon in the deep sea, Deep-Sea Research, 14:229-238.

Menzel, D.W., 1974, Primary productivity, dissolved and particulate organic matter, and the sites of oxidation of organic matter, in: "Marine Chemistry," E.D. Goldberg, ed., THE SEA, Vol. 5, Wiley and Sons, New York, 659-678.

Munk, W.H., 1966, Abyssal recipes, Deep-Sea Research, 13:707-730.

Owens, T. and King, F.D., 1975, The measurement of respiratory electron-transport-system activity in marine zooplankton, Marine Biology, 30:27-36.

Packard, T.T., 1969, "The Estimation of the Oxygen Utilization Rate in Seawater from the Activity of the Respiratory Electron Transport System in Plankton," Ph.D. Thesis, University of Washington, Seattle, 115 pp.

Packard, T.T., 1971, The measurement of respiratory electron transport activity in marine phytoplankton, Journal of Marine Research, 29:235-344.

Packard, T.T., 1979, Respiration and respiratory electron transport activity in plankton from the Northwest African upwelling area, Journal of Marine Research, 37:711-742.

Packard, T.T. and Williams, P.J.le B., 1981, Rates of respiratory oxygen consumption and electron transport in surface seawater from the Northwest Atlantic, Oceanologica Acta, 4:351-358.

Packard, T.T. and Garfield, P.C., in press, Respiration of the mid-water plankton from the Peru Current upwelling system, in: "Bio-productivity of Upwelling Ecosystems," M. Vinogradov and R.T. Barber, eds., Proceedings of the US/USSR Symposium, P.O. Shirshov Institute of Oceanology, 8-19 October, 1979, Academy of Sciences, U.S.S.R., Moscow.

Packard, T.T., Healy, M.L. and Richards, F.A., 1971, Vertical distribution of the activity of the respiratory electron transport system in marine plankton, Limnology and Oceanography, 16:60-70.

Packard, T.T., Harmon, D. and Boucher, J., 1974, Respiratory electron transport activity in plankton from upwelled waters, Tethys, 6:269-280.

Packard, T.T., Minas, H.J., Owens, T., and Devol, A., 1977, Deep-sea metabolism in the Eastern Tropical North Pacific Ocean, in: "Oceanic Sound Scattering Prediction," N.R. Andersen and B.J. Zahuranec, eds., Plenum Press, New York, 101-116.

Packard, T.T., Dugdale, R.C., Goering, J.J., and Barber, R.T., 1978, Nitrate reductase activity in the subsurface waters of the Peru Current, Journal of Marine Research, 36:59-76.

Packard, T.T., Garfield, P.C. and Martinez, R., in press, Respiration and respiratory enzyme activity in aerobic and anaerobic cultures of the marine bacterium *Pseudomonas perfectomarinus*, Deep-Sea Research.

Pak, H., Codispoti, L.A. and Zaneveld, J.R.V., 1980, On the intermediate particle maxima associated with oxygen-poor water off Western South America, Deep-Sea Research, 27A:783-797.

Payne, W.J., 1973a, Gas chromatographic analysis of denitrification by marine organisms, in: "Estuarine Microbial Ecology," L.H.

Stevenson and R.R. Colwell, eds., University of South Carolina Press, Columbia, 536 pp.

Payne, W.J., 1973b, Reduction of nitrogenous oxides by microorganisms, Bacterial Review, 37:409-452.

Payne, W.J., 1976, Denitrification, Trends in Biochemical Sciences, 1:220-222.

Rakestraw, N.W., 1947, Oxygen consumption in seawater over long periods, Journal of Marine Research, 6:259-263.

Richards, F.A., 1965, Anoxic basins and fjords, in: "Chemical Oceanography," J.P. Riley and G. Skirrow, eds., Academic Press, London, 611-645.

Riley, G.A., 1951, Oxygen, phosphate and nitrate in the Atlantic Ocean, Bulletin of the Bingham Oceanographic Collection, 13: 1-126.

Seiwell, H.R., 1937, Consumption of oxygen in seawater under controlled laboratory conditions, Nature, 140:506-507.

Setchell, F.W. and Packard, T.T., 1978, ETS and nitrate reductase activity in the Peru current March-May 1977 JOINT II RV "Wecoma", Coastal Upwelling Ecosystems Analysis Data Report No. 54, 129 pp.

Skopintsev, B.A., 1976, Oxygen consumption in the deep waters of the ocean, Oceanology, 15:556-560.

Söderlund, R. and Svensson, B.H., 1976, The global nitrogen cycle, in: "Nitrogen, Phosphorus and Sulfur--Global Cycles," SCOPE Report No. 7, Ecological Bulletin, B.H. Svensson and R. Söderlund, eds., Stockholm, 192 pp.

Sorokin, Y.I., 1972, Microbial activity as a biogeochemical factor in the ocean, in: "The Changing Chemistry of the Oceans," D. Dyrssen and D. Jagner, eds., Almquist and Wiksell, Stockholm, 189-204.

Stouthamer, A.H., 1980, Bioenergetic studies on *Paracocus denitrificans*, Trends in Biochemical Science, 5:164-166.

Suess, E., 1980, Particulate organic carbon flux in the oceans--surface productivity and oxygen utilization, Nature, 288(5788): 260-263.

Tsuchiya, M., 1975, Subsurface countercurrents in the eastern equatorial Pacific Ocean, Journal of Marine Research, Supplement, 33:145-175.

Van Leer, J.C. and Ross, A.E., 1979, Velocity and temperature data observed by cyclesonde during JOINT II off the coast of Peru, Coastal Upwelling Ecosystems Analysis Data Report No. 61, Rosensteil School of Marine and Atmospheric Science Data Report No. DR79-2, 34 pp.

Vinogradov, M.E., 1955, Vertical migrations of zooplankton and their importance for the nutrition of abyssal pelagic fauna, Trudy Instituta Okeanologii, 13:71-76.

Walsh, J.J., 1981, A carbon budget for overfishing off Peru, Nature, 290:300-304.

White, W.B., 1971, The westward extension of the low-oxygen distribution in the Pacific Ocean off the west coast of South America, Journal of Geophysical Research, 76:5842-5851.

Williams, P.M. and Carlucci, A.F., 1976, Bacterial utilization of organic matter in the deep sea, Nature, 262:810-811.

Wooster, W.S., Chow, T.J. and Barrett, J., 1965, Nitrate distribution in Peru current waters, Journal of Marine Research, 23: 210-221.

Wyrtki, K., 1962, The oxygen minima in relation to ocean circulation, Deep-Sea Research, 9:11-23.

Wyrtki, K., 1963, The horizontal and vertical field of motion in the Peru Current, Bulletin of Scripps Institution of Oceanography, 8:313-346.

Wyrtki, K., 1964, Upwelling in the Costa Rica Dome, Fishery Bulletin, 63:355-372.

Wyrtki, K., 1977, Advection in the Peru Current as observed by satellite, Journal of Geophysical Research, 82:3939-3944.

EFFECTS OF SOURCE NUTRIENT CONCENTRATIONS AND NUTRIENT REGENERATION ON PRODUCTION OF ORGANIC MATTER IN COASTAL UPWELLING CENTERS

Richard C. Dugdale

University of Southern California
Los Angeles, California 90089, U.S.A.

ABSTRACT

The seawater found at the center of an upwelling plume appears to come from relatively shallow depths. Since these depths correspond approximately to the depths of the pycnocline, small changes in the depth of the source water can result in large changes in the concentrations of nutrients at the surface. The depth of source water varies on seasonal and shorter time scales as a function of both the large-scale circulation and the local wind and topography driven scale, therefore concentration and ratios of nutrients show variations on the same time scales. These two factors are important both in determining biological production rates at upwelling centers and in determining species composition. For example, in Peru weak upwelling results in silicate limitation of primary production with the possibility that poorly silicified diatom frustules are produced at the outside edges of the plume. With strong upwelling, the ratio of silicate to nitrate increases, more of the nitrate is used and 'new production' may increase. These changes are likely to affect the pattern of sedimentation.

INTRODUCTION

Among the primary nutrients required for phytoplankton growth, silicon has a central role in the upwelling production cycle, a consequence of the silicon requirement of diatoms, the fast growing component of the phytoplankton (Barber and Smith, 1981), to construct their outer shells or tests. Dinoflagellates, another important component of upwelling phytoplankton populations, appear to grow more slowly, e.g., when dinoflagellates control an upwelling area, specific nitrate uptake rates are reduced by about an order of magnitude from those of typical diatom dominated systems. The Coastal Upwelling Ecosystems Program (CUEA) carried out a comparative study and the data provide the means to assess the role of dissolved silicon in controlling some aspects of the productivity of these systems.

GROSS EFFECTS OF DISSOLVED SILICON SUPPLY

The role of primary nutrient flux into the euphotic zone in controlling primary production has been described in terms of the nitrogen cycle (Dugdale and Goering, 1967); nitrate uptake is equated to new production, and ammonium uptake (plus urea and one or two other organic nitrogen compounds) is equated to regenerated nitrogen. Epply, Renger and Harrison (1979) have shown that primary production, measured as carbon fixation, rises in exponential fashion as the proportion of nitrate uptake to total nitrogen uptake increases, i.e., the ecosystem is pumped up at an increasing rate by the input of the new primary nutrient, nitrate.

The proportion of new nutrient uptake is measured using nitrate and ammonium labelled with the stable isotope of nitrogen, ^{15}N. The ^{15}N method, as used in the studies cited below, gives nitrate uptake values nearly identical to those indicated by measurements of nitrate reductase activity (Dugdale et al., in prep.), suggesting that the ^{15}N method provides a good measure of nitrate assimilation. Nitrate uptake normalized to particulate cell nitrogen, i.e., specific nitrate uptake, should therefore be proportional to population growth rate. Unfortunately, silicon is not fractionated into a recognizably different chemical compound during regeneration (Dugdale, 1972), and the path of silicon in upwelling productivity cannot be so easily followed with labelled compounds and tracer experiments. Uptake of dissolved silicon has been measured by Nelson, Goering and Boisseau (1981), but these values cannot be equated with new production since they include the uptake of both new and regenerated silicon fractions. For both primary nutrients, it can be seen that the potential new production in an upwelling area or center is the product of upwelling velocities and the concentration of dissolved silicon and/or nitrate in the source waters.

Codispoti, Dugdale and Minas (1982) compared the nutrient concentrations in the northwest Africa, Peru and Baja California upwelling regions. The low nutrient concentrations typical of northwest Africa (about 6 µg-at/ℓ Si, 15 µg-at/ℓ NO_3) were correlated with low new production rates. Primary nutrient concentrations are considerably greater off Baja California and Peru with correspondingly higher new production rates. The high winds characteristic of the northwest Africa upwelling area studied by CUEA resulted in deep mixing; the lower mean light levels for the phytoplankton apparently also affected the primary productivity adversely (Huntsman and Barber, 1977).

EFFECTS ON NUTRIENT GROWTH AND UPTAKE

Dissolved silicon and nitrate uptake by phytoplankton is described by the Michaelis-Menten expression (Dugdale et al., 1981) which gives a saturation curve of uptake vs. concentration. The evidence for dissolved silicon limitation of nitrate uptake, i.e., of new production, is discussed in Dugdale et al. (1981) and includes curves of silicate concentrations vs. specific nitrate uptake, suggestive of Michaelis-Menten kinetics. Other features diagnostic of limiting nutrient conditions, e.g., increasing maximum specific uptake rate and dark uptake of the nutrient in question, have been observed in the Peru upwelling center at 15°S. From published kinetic constants for dissolved silicon uptake, Klim values of 14.3-26.4 µg-at/ℓ Si for field populations, and 4.32-52.7 µg-at/ℓ Si for laboratory populations were calculated (Klim values indicate the concentration of nutrient at which the uptake is reduced to 90% of maximum). It should be noted that kinetic constants for uptake and growth are often different. In two drogue-following experiments in Peru, decreases in specific nitrate uptake were correlated most closely with declining dissolved silicon beginning at about 10-15 µg-at/ℓ Si, which is consistent with the values of Klim and the model of nutrient interactions proposed by Dugdale et al. (1981).

Supporting evidence for nutrient control of specific nitrate uptake rate, and perhaps growth rate as well, is shown in Fig. 1 where dissolved silicon concentration is plotted against $VmaxNO_3$ for all the measurements made in Peru in 1977. Following Eppley (1972), the envelope of these points is shown for Peru, 1977 (JII-77) and also in Fig. 1, the envelopes (but not the data points) for Peru 1976 (JII-76) and northwest Africa (JI). The northwest Africa JI envelope is characterized by low maximum dissolved silicon concentrations and low $VmaxNO_3$ values, the JII-77 envelope by much higher maximum dissolved silicon concentrations and higher values of $VmaxNO_3$. The characteristics of the JII-76 envelope, high minimum dissolved silicon concentrations and low $VmaxNO_3$, are the result of a massive dinoflagellate bloom coinciding with an El Niño (Dugdale et al., 1977). The decreased uptake rates at the high concentration side of the envelope are most reasonably attributed to the effects of strong mixing conditions or to a lack of adaptation time for phytoplankton upwelled with nutrient-rich water (Dugdale et al., in prep.). The high values of dissolved silicon at the left side of the MESCAL 1 envelope, and the high levels of dissolved silicon, generally reflect the lack of uptake resulting from a low diatom and high dinoflagellate population. Taken together, the JI and JII-77 envelopes present a picture that is consistent with a hypothesis of dissolved silicon, or at least of primary nutrient, control of maximum specific nitrate uptake rates in the two coastal upwelling areas.

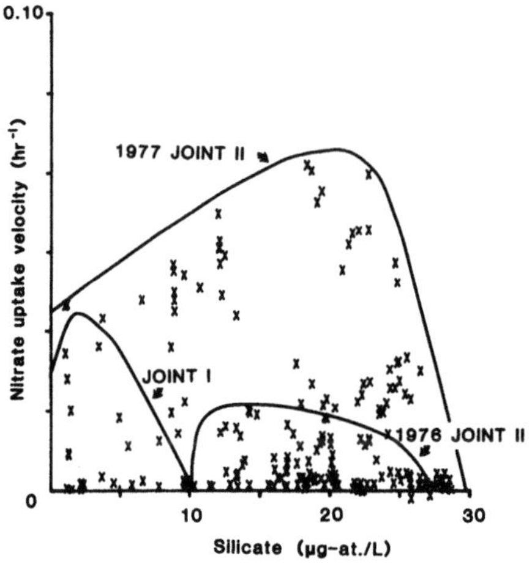

Fig. 1. Values of specific nitrate uptake plotted against silicate concentration for Peru (JII 77); the envelope of similar plots are shown for Northwest Africa (JI) and Peru (JII 76), lower right. Lowest values result from light limitation in screened incubation bottles.

VARIATIONS IN SOURCE WATER, PERU, 1977

The size of the upwelling plume at 15°S increased and the nutrient concentrations at the center changed over the period of the MAM77 study, Table 1. In Fig. 2 it can be seen that dissolved silicon at the upwelling center (location C-1) increased from a low of 17 µg-at./ℓ in mid-March to a high of 31 µg-at/ℓ on April 19, declining to 18 µg-at/ℓ by mid-May. Nitrate, on the other hand, varied little, at about 25 µg-at/ℓ, and actually declined when dissolved silicon reached its maximum. The decline in nitrate with deeper and stronger upwelling, Fig. 2a, is consistent with the nitrate profile in the region which shows a minimum associated with denitrification (Dugdale, 1972; Dugdale et al., 1977; Codispoti and Packard, 1980). The ratio of NO_3/Si, Fig. 2b, declines from 1.3 at the beginning of the experiment to a low of 0.66 on April 19 and increases again to 1.1 by mid-May. The apparent depth of the source water, Fig. 2c, judged from the dissolved silicon sections, increases from 90 m at the beginning of the study to 250 m on April 19 and declines to about 80 m by mid-May. There is, however, a poor correlation between depth of upwelling source water as determined from dissolved silicon isopleths and as determined from sigma-t isopleths, the latter yielding estimates from 30-90 m, in agreement with the depth of inflow estimated from current meter data (Smith, 1981). Redfield, Ketchum and Richards (1963) noted the discrepancy between density and phosphate isopleths along upwelling coasts and attributed it to regeneration and nutrient trapping. It seems likely that for the 15°S system, the

Table 1. C-1 Stations on JII MAM77 15°S Peru.
Concentrations and ratios at 0 m.

Station	Date	Si	NO3	NO3/Si
M48	15 Mar	16.94	21.88	1.29
M61	16 Mar	20.03	24.30	1.21
M93	19 Mar	19.37	26.14	1.35
M125	22 Mar	23.81	25.47*	1.07
M162	27 Mar	22.29	23.36	1.05
I235	12 Apr	24.83	24.88	1.00
I271	16 Apr	27.68	24.36	0.89
I328	19 Apr	31.25	22.67	0.73
I344	20 Apr	30.55	20.61	0.67
M363	5 May	27.61	22.69	0.82
M391	9 May	28.44	22.00	0.77
M409	1[illegible] May	20.76	17.87	0.86
M432	16 May	18.58	20.48	1.10

M=R/V MELVILLE *Data Questionable
I=R/V ISELIN

explanation lies in alongshore inhomogeneity resulting in part from the regeneration of about 10 μg-at/ℓ of dissolved silicon between the equator and 15°S (Friederich and Codispoti, 1981). Water upwelling at 15°S arrives from the north in the poleward undercurrent, but is a mixture of equatorial and subantarctic water at that latitude. On the 26.0 sigma-t level, the subantarctic water has nitrate and dissolved silicon concentrations of about half those of the equatorial water (Friederich and Codispoti, 1981). In March-May 1977, the Subarctic Water exhibited a NO_3/Si ratio of a little greater than 2 at the 26.0 sigma-t level; the corresponding ratio in the Equatorial Water was about 1.7.

In addition to changes in nutrient concentration, the size of the nutrient plume also changed, as shown in Fig. 2d. It can be seen that the distance to the 10 μg-at/ℓ dissolved silicon concentration along the C-line increased from 7 km on 16 March to 46 km on 9 April, remaining at the greater distance for the rest of the study period. The 2.5 μg-at/ℓ distance was often beyond the range of the sections in the later part of the study, sometimes lying beyond 140 km. The changes in both ATP concentration and dissolved silicon at the upwelling center (C-1) and the size of the plume are consistent with the findings of Friederich and Codispoti (1981) that the poleward undercurrent was broader in May, 1977 than in the austral winter months preceeding and that in May, 1977 the water at 15°S on the 26.0 sigma-t surface was composed almost entirely of Equatorial Water.

The appearance of diatoms in the sediments is sometimes a good indicator of high productivity rates and upwelling (Diester-Haass, 1978). Schuette and Schrader (1981) have shown that the three major upwelling centers along the Peru coast are indicated very well by the concentration of diatoms in the surface sediments. The changes in

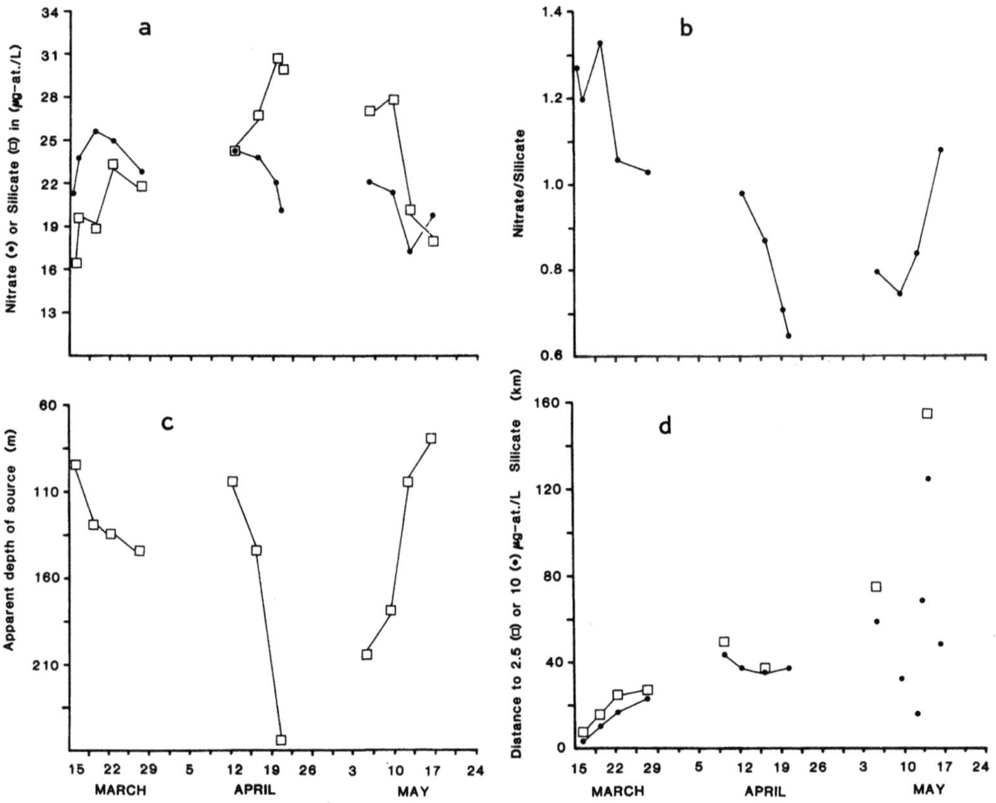

Fig. 2. Variation with time of nutrient parameters in the 15°S Peru upwelling center in 1977; (a) concentration of nitrate (•) and silicate (□); (b) ratio of nitrate/silicate, (c) apparent depth of the source water; (d) size of the upwelling plume measured as distance to the 2.5 and 10 µg-at/ℓ surface silicate concentration.

nutrient concentration at the upwelling center at 15°S appear to influence the primary production cycle and perhaps the sedimentation rates as well. In particular the seasonal increase in dissolved silicon appears to allow significantly higher growth rates as the water drifts away from the center, resulting in populations being able to take up very large amounts of nutrients in a single day (Dugdale et al., in prep.). Under these conditions, the diatom distribution may show a narrower region of higher populations and the sedimentary record might appear more distinct. Another factor influencing the nature of the sedimentary record is the tendency towards weak silicification of diatom tests at low dissolved silicon concentrations (Harrison, Conway and Dudgale, 1976). Early upwelling at the Peru 15°S center could be expected to result in a band of weakly silicified diatoms at the outside end of the plume, although this aspect has not been studied specifically.

Table 2. Distance along the C-line at 15°S along the Peru coast where Si concentrations = 2.5 μg-at /ℓ and 10 μg-at/ℓ and the nitrate concentrations where Si = 2.5 μg-at/ℓ. Data from the CUEA MAM-77 expedition.

Date of Section	C-1 Station	D, Si= 10 μg-at/1 km	D, Si= 2.5 μg-at/1 km	NO3, μg-at/1 at Si=2.5
15-16 Mar	M 48	7	11	10
17-19 Mar	M 93	14	19	15
22 Mar	M125	20	28	10
26-28 Mar	M162	27	30	5
9-10 Apr	I204	47	53	8
12 Apr	I235	40	.50	-
15-16 Apr	I271	38	40	2.5
20-22 Apr	I344	40	46	5
5- 8 May	M363	62	78	5
8- 9 May	M391	35	35	-
12-14 May	M409	19,72,128	158	5
14-16 May	M432	52	108	-

M = R/V MELVILLE
I = R/V ISELIN

SUMMARY AND CONCLUSIONS

The evidence presented suggests dissolved silicon control of phytoplankton processes, including growth rate, in upwelling regions. Furthermore, the changes in dissolved silicon concentration with time are likely to affect the pattern of sedimentation. Since the nitrate concentration remains relatively unchanged with time in the Peru upwelling center, the ratio of nitrate to dissolved silicon changes with time and may result in nutrient limitation switching from dissolved silicon to nitrate as upwelling strengthens in austral autumn. These hypotheses clearly require testing with some carefully designed field experiments.

REFERENCES

Barber, R.T. and Smith, R.L., 1981, Coastal upwelling ecosystems, in: "Analysis of Marine Ecosystems," A.R. Longhurst, ed., Academic Press, 31-68.

Codispoti, L.A. and Packard, T.T., 1980, Denitrification rates in the eastern tropical South Pacific, Journal of Marine Research, 38: 453-477.

Codispoti, L.A., Dugdale, R.C. and Minas, H.J., 1982, A comparison of the nutrient regimes off Northwest Africa, Peru, and Baja California, Conseil Permanente International pour l'Exploration de la Mer, Rapport, 180:5-6.

Diester-Haass, L., 1978, Sediments as indicators of upwelling, in: "Upwelling Ecosystems," R. Boje and M. Tomczak, eds., Springer-Verlag, New York-Heidelberg, 261-281.

Dugdale, R.C., 1972, Chemical oceanography and primary productivity in upwelling regions, Geoforum, 11:47-61.

Dugdale, R.C. and Goering, J.J., 1967, Uptake of new and regenerated forms of nitrogen in primary productivity, Limnology and Oceanography, 12:196-206.

Dugdale, R.C., Goering, J.J., Barber, R.T., Smith, R.L., and Packard, T.T., 1977, Denitrification and hydrogen sulfide in the Peru upwelling region during 1976, Deep-Sea Research, 24:601-608.

Dugdale, R.C., Jones, B.H., Jr., MacIsaac, J.J., and Goering, J.J., 1981, Adaptation of nutrient assimilation, Canadian Bulletin of Fisheries and Aquatic Science, 210:234-250.

Dugdale, R.C., MacIsaac, J.J., Barber, R.T., Blasco, D., and Packard, T.T., in prep., Control of primary production in upwelling centers.

Eppley, R.W., 1972, Temperature and phytoplankton growth in the sea, Fisheries Bulletin (U.S.), 70:1063-1085.

Eppley, R.W., Renger, E.H. and Harrison, W.G., 1979, Nitrate and phytoplankton production in southern California coastal waters, Limnology and Oceanography, 24:483-494.

Friederich, G.E. and Codispoti, L.A., 1981, The effects of mixing and regeneration on the nutrient content of upwelling waters off Peru, in: "Coastal Upwelling," F.A. Richards, ed., Coastal and Estuarine Sciences 1, American Geophysical Union, Washington, 221-227.

Harrison, P.J., Conway, H.L. and Dugdale, R.C., 1976, Marine diatoms grown in chemostats under silicate or ammonia limitation. I. Cellular chemical composition and steady-state growth kinetics of *Skeletonema costatum*, Marine Biology, 35:177-186.

Huntsman, S.A. and Barber, R.T., 1977, Primary production off northwest Africa: the relationship to wind and nutrient conditions, Deep-Sea Research, 24:24-33.

Nelson, D.M., Goering, J.J. and Boisseau, D.W., 1981, Consumption and regeneration of silicic acid in three coastal upwelling systems, in: "Coastal Upwelling," F.A. Richards, ed., Coastal and Estuarine Sciences 1, American Geophysical Union, Washington, 242-256.

Redfield, A.C., Ketchum, B.H. and Richards, F.A., 1963, The influence of organisms on the composition of sea-water, in: "The Sea," Vol. 2, M.N. Hill, ed., Wiley & Sons, New York, 26-77.

Schuette, G. and Schrader, H., 1981, Diatoms in surface sediments: a reflection of coastal upwelling, in: "Coastal Upwelling," F.A. Richards, ed., Coastal and Estuarine Sciences 1, American Geophysical Union, Washington, 372-380.

Smith, R.L., 1981, A comparison of the structure and variability of the flow field in three coastal upwelling regions: Oregon, Northwest Africa and Peru, in: "Coastal Upwelling," F.A. Richards, ed., Coastal and Estuarine Sciences 1, American Geophysical Union, Washington, 107-118.

SKELETAL PLANKTON AND NEKTON IN UPWELLING WATER MASSES OFF NORTHWESTERN SOUTH AMERICA AND NORTHWEST AFRICA

Jörn Thiede

Department of Geology
University of Oslo
Blindern, Oslo 3, Norway

ABSTRACT

In this study we compare the distributions of the skeletal remains of plankton and nekton organisms caught in water samples from the upwelling regions off northwest Africa and northwestern South America. In both regions the near-surface water masses were sampled continuously by means of a seawater pump with an attached filter system effectively collecting all sand-sized material over sampling intervals of limited regional extent. Close to 200 samples were available from each of the regions. The samples were studied for their contents of shells and skeletons. Because the water flow was controlled by means of a water meter, it was possible to calculate relative and absolute concentrations of all components within the sampled size fractions and their regional distributions could thus be determined.

The quantitatively most important components of the collected grain assemblages are diatoms, planktonic and benthic foraminifers, radiolarians, larvae of benthic gastropods and bivalves, pelagic gastropods and fish bones. Samples taken within the coastal water masses of the upwelling regions are often characterized by the massive occurrence of large centric diatoms. Also, benthic foraminifers have frequently been found off South America and are believed to have lived as epi-benthos on drifting algae.

The planktonic foraminifers and their distributional patterns can be related directly to the extent of upwelling water masses. Both off northwest Africa and off South America, the most important planktonic foraminiferal species of the upwelling water masses are *Globigerina bulloides*, *Globigerinita glutinata* and *Globoquadrina dutertrei*. Distributions of planktonic foraminiferal assemblages can be related to the tropical warm water masses, to the eastern boundary currents and to areas with cool upwelled surface waters.

THE COASTAL UPWELLING REGIONS OFF NORTHWESTERN SOUTH AMERICA AND OFF NORTHWEST AFRICA, AND THEIR ECOSYSTEMS

Coastal upwelling develops in regionally well defined areas of the eastern boundary currents under the influence of the trade winds (Smith, 1968, and this volume; Wooster, Bakun and McLain, 1976). The ecosystems which develop in response to the dynamics of the coastal upwelling regime are different from other coastal ecosystems (Barber and Smith, 1981), and it is an interesting question whether or not these differences are also preserved in the geological records of sediments under upwelling regimes. The deposits are usually hemipelagic continental margin sediments (Diester-Haass, 1978) with high proportions of biogenic particles produced by organisms which once lived in the overlying water column and which therefore somehow reflect the hydrography of these waters. To understand the distributional pattern of biogenic sediment particles, one has to develop a basic knowledge of the biogeographic distribution of organisms in the coastal upwelling regimes which once produced these sediment constituents, and quantify the standing stocks of such organisms (Berger, 1970).

Geologically Important Faunal and Floral Components

The geological record preserved in the sediments underlying coastal upwelling regimes is naturally a very selective one because many of the organic components produced close to the sea surface are destroyed or altered on their way to the sea floor or during early diagenesis. An important share of the particle assemblages under coastal upwelling regions (Honjo, Manganini and Cole, 1982; Staresinic et al., this volume) consists of compounds rich in organic carbon (for example, fecal pellets) whose preservation is limited by specific properties of the depositional environment (Suess and Reimers, this volume). The other dominant contribution to the particle assemblages settling to the sea floor comes from organisms which secrete shells or skeletons and which belong to the autotrophic and heterotrophic plankton and nekton. These consist of widely different chemical substances: silica (diatoms, silicoflagellates, radiolarians), calcium carbonate (aragonite: pelagic gastropods and mero-planktonic larvae of benthic gastropods; calcite: coccolithophorids, foraminifers, mero-planktonic larvae of benthic bivalves, of echinoderms and of a few other benthic groups of animals) and calcium phosphate (fish bones). The shells and skeletons of these organisms have reasonably good chances to be embedded in the sea floor although differential solution in the water column, at the sediment water interface as well as in the sediment, will lead to important changes in the state of preservation and in the composition of the biogenic particle assemblages preserved in the geologic record. The presence or absence, the biogeographic distribution, and the regional quantitative distribution of such organisms are important variables which have to be known to understand the geological record of coastal upwelling regimes in modern times as well as in the geologic past.

The groups of organisms listed above have quite different ecologies and belong to widely different levels of the marine food chains or webs whose members are known to be intimately linked ecologically to each other. If only studying the shell- and skeleton-producing members of the floras and faunas which inhabit the surface water masses of coastal upwelling regimes, one has to be keenly aware that only fragments of the marine food chains can be observed. This applies even more to the geological records of such systems. However, it is possible to separate primary producers such as diatoms from secondary producers, such as radiolarians and planktonic foraminifers, from higher members of the marine food webs (molluscs or fishes). It is possible to map zonations of biocoenoses (cf. Jones et al., this volume) which reflect these different levels.

Comparison of Hydrographies of the Upwelling Regimes Off Northwestern South America and Off Northwest Africa

In this study I will try to compare data from two coastal upwelling regimes. After multi-year oceanographic studies, mainly carried out under the CUEA Program (Smith, this volume), it is well known that the upwelling regions off northwest Africa and off northwestern South America (Peru) have some similarities, but that they also show some fundamental differences which have to be considered if comparing plankton and nekton standing stocks. The two coastal upwelling regimes are situated on different hemispheres and important differences between the two include the physiography of their continental margins and the morphology of the adjacent land areas. In northwestern South America a large mountain chain parallels the continental margin, whereas off northwest Africa adjacent to the coastal upwelling area the morphologic gradients are relatively modest. The physiography of the land areas are known to exert an important influence upon the wind regimes which in turn affect coastal upwelling.

The topography of the shelf off northwestern South America and off northwest Africa is widely different. The shelf is 20 km wide, and 150 m deep off Peru, but 50 km wide and only 100 m deep off northwest Africa. The direction of the coast lines is approximately N-S in northwest Africa, but NW-SE in Peru. Differences in the wind regimes and margin morphology affect the circulation patterns in both upwelling regions; for example, the surface layer in which the mean offshore flow occurs is about 35 m thick off northwest Africa, but only 25 m thick off Peru (and therefore less than the thickness of the photic zone). A ubiquitous feature of most modern coastal upwelling regions is a poleward flowing subsurface current which reaches onto the shelf off Peru, but which is mostly confined to the upper slope off northwest Africa. It can be a source for upwelling waters, and it can provide a feedback loop for the upwelling biota (Smith, this volume).

Goals of This Study

Based on the problems outlined in the previous paragraphs, it is then the aim of this study to describe the occurrences and biogeographic distributions of plankton and nekton organisms which secrete skeletal material and which can be expected to produce a geological record in the sediments underlying the coastal upwelling regimes off northwest Africa and off northwestern South America. It is also known, but rarely quantified, that standing stocks of various organisms inhabiting the water masses of coastal upwelling regimes are quantitatively much larger than those of adjacent areas. The quantitative differences in standing stocks (Barber and Smith, 1981) and probably the output of skeletal material as well, can therefore be expected to offer important clues if trying to locate present and former coastal upwelling regimes. Finally, an attempt will be made to evaluate how similarities and differences between two of the major coastal upwelling regimes are reflected in the distributional patterns of the groups of organisms described in this paper.

METHODS

The plankton samples were collected by means of a filter system using pumping installations of the reseach vessels during various cruises to the areas off northwest Africa and northwestern South America. The station points, which appear in the illustrations of this study, represent samples collected by continuously pumping seawater through a set of filters for a number of hours (Figs. 1a and 1b). In both areas, several size fractions were collected and a water meter (without protecting screens) measured the volume of water filtered for each sample. The intake of each ship's pumping systems was located 3-4 m below the actual sea surface. The samples were taken while sailing or on station except when geological instruments were lowered to the sea floor and when the samples were in danger of being contaminated by suspended sediment particles. The individual cruises lasted several weeks in each of the two areas. The ship's speed, even in regions where station work was carried out, exceeded that of the surface water currents. Although somewhat of an oversimplification, it has been assumed for the purpose of this study that the data collected by this method reflect a quasi-synoptic distributional pattern of the plankton and nekton. The problem of plankton patchiness (Boltovskoy, 1971) was minimized either by pumping over extended distances (approx. 2-3 hours while en route) or times (approx. 3-5 hours while on station) even though this probably re-

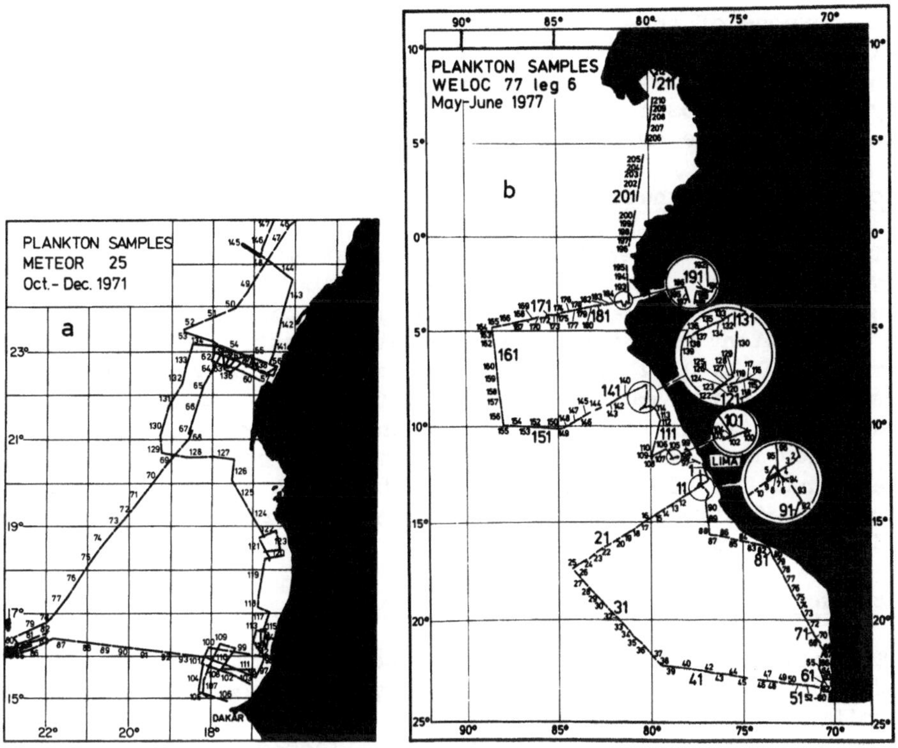

Fig. 1. a) Locations of plankton pump samples taken during cruise "Meteor" 25 (November-December 1971) off northwest Africa. For further details see Thiede, 1975a.

b) Locations of plankton pump samples taken during cruise WELOC 77 (Leg 6, May-June 1977) off northwestern South America.

sulted in a reduction of the spatial resolution of the various, often relatively small or narrow water masses encountered in coastal upwelling regions (Traganza et al., this volume). The largest size fraction of these samples (1.0 mm in diameter) usually did not contain enough particles to yield statistically meaningful results.

The samples were preserved in buffered formalin on board the ships. During the cruise to the northwest African upwelling region, the quality of the samples and of the living plankton and nekton were controlled by means of a plankton microscope (Thiede, 1975a). The laboratory procedures applied to both sets of samples were to separating the shells and skeletal particles from the organic material by ignition (Sachs, Cifelli and Bowen, 1964). The remaining ash was then studied under a binocular stereomicroscope much like a regular micropaleontological sample. The scatter of the quantitative component distributions in adjacent samples has been smoothed by means of a three-point running average (weighted 1-2-1) before contouring.

For the purpose of this comparison of plankton distributions from two upwelling regions, I have aligned the two eastern boundary current regimes along the same direction. Since the two current regimes really flow in opposite directions, the northwest African data had to be plotted with the north direction pointing down.

Collections Off Northwest Africa

The plankton collections off northwest Africa were carried out during cruise no. 25 of RV "Meteor" (Seibold, 1972) in November-December, 1971. The sequences of sieves in the filter set used allowed separation into the following size fractions: >1.0 mm, 1.0-0.25 mm, and 0.25-0.12 mm. About 6 m^3 of seawater were filtered for each sample. After return from the cruise, the samples were washed with distilled water on filter paper and then transferred to a crucible. They were dried in a heater and then reduced to ashes in a muffle furnace at 450°C for about one hour. A total of 177 samples were collected during this cruise. The data from these samples have been described and documented in detail by Thiede, 1975a.

Collections Off Northwestern South America

The plankton samples off northwestern South America were collected during cruise WELOC 77, Leg 6 by RV "Wecoma" of Oregon State University. A set of sieves much like the one in RV "Meteor" was used, but this time the filter system collected all organisms >0.15 mm and, on a separate sieve, >1.0 mm. Most of the samples were preserved with formaldehyde buffered with sodium borate to prevent calcite and aragonite dissolution. A minor set of samples was reserved for geochemical measurements for comparison with sediment samples (Reimers and Suess, this volume); these samples had been preserved with mercuric chloride. A total of 220 samples were collected; after the cruise they were washed with distilled water, freeze-dried and then ignited at approximately 450°C.

DISTRIBUTION OF ORGANIC MATERIAL

The amount of ash produced from the organic material filtered from one m^3 of surface water is a crude measure of the concentrations of plankton and nekton organisms (Fig. 2). The absolute quantities in both upwelling regions are of the same order of magnitude. However, the regional distribution of the material, the gradients between high and low concentrations, and the extent of the upwelling are quite different. The area affected by coastal upwelling extends over >15° of latitude (2°-16°S) off Peru, but only over 8°-10° of latitude (15°-25°N) off northwest Africa. Seaward, the belts of elevated concentrations (>1 mg ash/m^3 surface water) can be traced to about 600 km offshore Peru, whereas a similar belt of high concentrations off northwest Africa is only 200-250 km wide. It is also clear

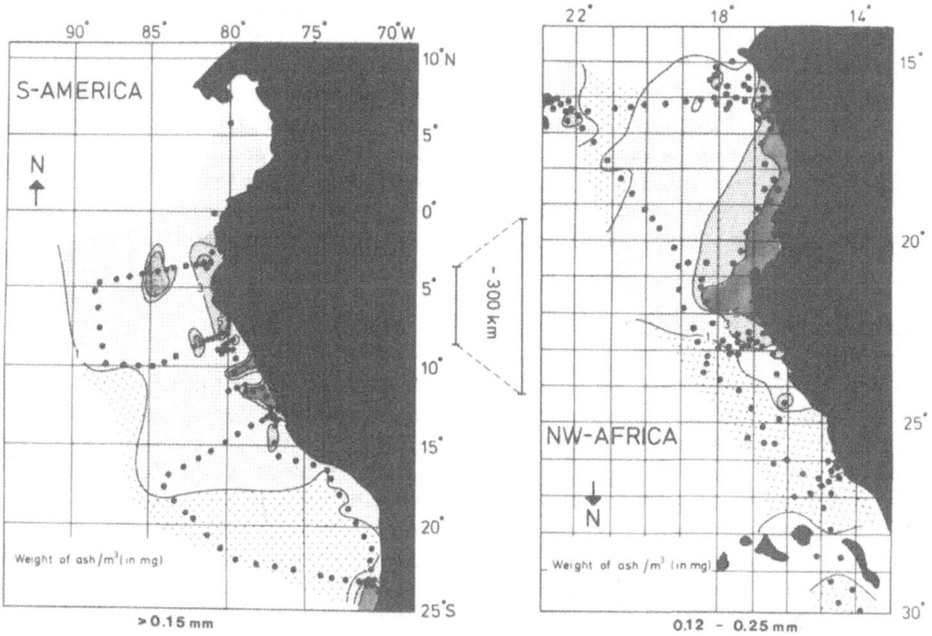

Fig. 2. Distribution of skeletal material (weight of ash in mg per m^3 of pumped water) in surface waters off northwestern South America and off northwest Africa.

that changes in the seasonal intensity and location of coastal upwelling large scale climatic changes and sea level changes (Wooster et al., 1976; Diester-Haass, 1978), could lead to important changes in the width and latitudinal position of these productive regions.

The signal of coastal upwelling off northwest Africa is further complicated by the influence of the Canary and Cape Verde Islands which perturb the surface water currents and which generate upwelling of nutrient-rich subsurface water with high productivities, much the same way as has been observed around the Galapagos Islands (Moore, Heath and Kowsmann, 1973). However, the concentrations of skeletal material around the islands are clearly not as high as in the centers of the coastal upwelling regime.

The distribution of skeletal material is also quite different in the centers of the coastal upwelling regimes off Peru and off northwest Africa. Off northwest Africa, a narrow area of high concentrations can be followed along the entire continental margins from 16°N to 23°N, and concentrations decrease seaward in a fairly regular fashion. Off Peru, however, the highest concentrations of skeletal material are confined to narrow, lobe-shaped regions which seem attached to the continental margin and in location and extent much resemble the distribution of total primary production as compiled from 10 years of measurements by Zuta and Guillen (1970). Isolated water masses with high concentrations of skeletal material off Peru can obviously become detached from the continental margins and drift seaward. If this happens repeatedly over a geological time span, the result could be a wide belt of sediments influenced by coastal upwelling which surround the coastal upwelling centers. The effect of this transport mode would be best developed in eastern boundary currents downstream from the actual coastal upwelling zone.

DISTRIBUTIONS OF SKELETAL MATERIAL OF BENTHIC ORIGIN

The skeletal material in the samples both from the coastal upwelling regions off northwest Africa and off northwestern South America contain a number of components which are not produced by plankton or nekton organisms proper, but by benthic organisms found in these samples because of developmental stages during their life cycles or because of their specialized ecologies. Such material has sporadically been observed in plankton samples before but has rarely been described in a systematic manner (Scheltema, 1971). These components are sometimes quantitatively as important as the plankton and nekton, and they comprise benthic foraminifers and mero-planktonic larvae of benthic organisms such as molluscs, echinoderms, and a number of other unidentified, but always rare, particles. Of these, distributions of benthic foraminifers and of bivalve larvae revealed coherent patterns which seem to be related to the upwelling process (Figs. 3 and 4).

Benthic Foraminifers

Benthic foraminifers were found in the surface waters of both upwelling regions (Fig. 3) but distributions are quite different. Whereas they occurred in low concentrations and in restricted areas off northwest Africa, their concentrations off Peru were an order of magnitude higher and they were found over a wide region of the northern part of that upwelling region. Off northwest Africa, the only area where samples regularly contained benthic foraminifers was north and south of Cape Blanc at 21°N. At the time of sample collection no systematic observations of species compositions of these benthic foraminifers were made (Thiede, 1975a; 1975b), and therefore details have to await further studies.

Off Peru, however, the widely occurring benthic foraminifers were a very conspicuous element of the skeletal assemblages. They consisted entirely of monospecific tests of *Rosalina globularis* which have been reported relatively frequently in plankton samples from other areas (Sliter, 1965; Spindler, 1980). Except during the brief development of a floating chamber before sexual reproduction of the gamonts (the floating chamber supposedly does not exist longer than

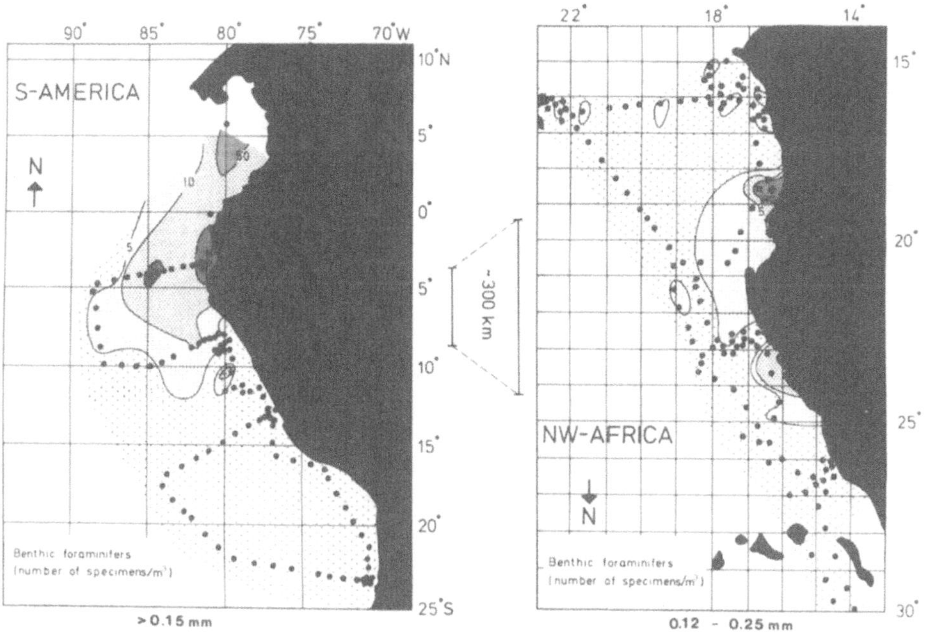

Fig. 3. Quantitative distribution (in number of specimens per m^3 of pumped water) of benthic foraminifers in the surface waters off northwestern South America and off northwest Africa.

18 hours), this species lives epi-benthic in very shallow neritic waters or attached to sea weeds which might float in the oceanic surface water masses (Spindler, 1980). It is therefore interesting to note that the concentrations of this species (Fig. 3) are highest off the mouth of the Gulf of Guayaquil and in the inner Panama Basin. *Rosalina globularis* is very rare or absent south of 9°S off Peru, but a tongue of high abundances extends from the areas of maximum occurrence towards the west and southwest into the open Pacific Ocean, suggesting a considerable seaward transport of surface water masses in the northernmost part of the Peruvian upwelling region. Such species can also be used as indirect indicators of the input of drifting materials which come from the coastal regions and which obviously can travel quite far out into the adjacent ocean basin.

Mero-Planktonic Bivalve Larvae

Mero-planktonic bivalve larvae are a frequent and ubiquitous component of the plankton samples from the surface waters of both upwelling regions (Fig. 4), which are usually living close to the sea surface. It is generally assumed that most of them are produced by shallow water bivalve faunas close to the coast because the size of the larvae increases but their frequency decreases with increasing

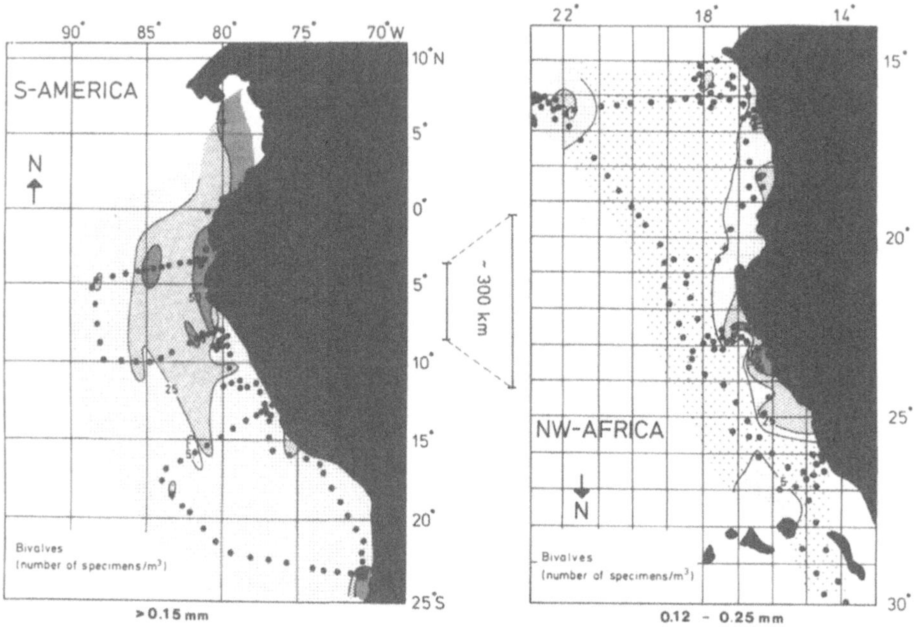

Fig. 4. Quantitative distribution (in number of specimens per m^3 of pumped water) of mero-planktonic bivalve larvae in surface waters off northwestern South America and off northwest Africa.

distance from the coast (Thiede, 1974; 1975a). The larval stages of marine molluscs last only for a few weeks, however some of the gastropod and bivalve larvae are also known to be able to persist drifting for many months and over wide distances across entire ocean basins (Scheltema, 1971).

Concentrations of bivalve larvae of more than 50 specimens per m^3 of filtered water were observed both off Africa and off South America (Fig. 4), but the concentrations decreased rapidly in the surface water masses adjacent to the upwelling region. A continuous, but narrow, belt of high concentrations can be followed along the northwest African continental margin from 16°N to 25°N, but water masses carrying these larvae obviously remain close to the continental margin. The concentrations are higher again around the shelf areas of the Canary and Cape Verde Islands, obviously because of the production of larvae by bivalve faunas indigenous to the shoal areas around the islands. Off South America, bivalve larvae are produced in large quantities in the area north of 5°S but are transported considerably further seaward than off northwest Africa, in part also towards the southwest against the general direction of the Peru Current.

DISTRIBUTIONS OF SILICEOUS PLANKTON

High concentrations of siliceous skeletal material in pelagic and hemipelagic sediments are thought to be confined to and indicative of areas under fertile surface water masses in regions of upwelling (Heath, 1974; Schrader and Baumgartner, this volume). The bulk of the siliceous skeletal material in plankton samples consists of diatom frustules, silicoflagellates and radiolarians. The sampling and laboratory methods used for collecting the data of this study do not provide reliable quantitative data for diatoms and silicoflagellates so that only radiolarian distributions are discussed in detail. Diatoms and silicoflagellates are important as primary producers, however, and as the lowermost links of the marine food web, as discussed in this study.

Diatoms

Diatoms produce the most abundant and characteristic skeletal components found in samples from the innermost part of both coastal upwelling regions (Thiede, 1975a; Schuette and Schrader, 1979). Their abundance, which is highest in the neritic water masses affected by the coastal upwelling, drops rapidly seaward by several orders of magnitude. In particular off Peru, the highest frequencies seem to be confined to areas of relatively persistent upwelling. Large centric diatoms are particularly frequent in the upwelling region off northwest Africa, but in both regions there is also a wide range of other species which can be related to the upwelling, which can be grouped into upwelling assemblages, and which in part are indigenous to the upwelling water masses (Schuette and Schrader, 1979). The sediments under the upwelling regions also contain abundant meroplanktonic diatom species which usually live in the turbulent nearshore waters (Schuette and Schrader, 1979). These document an active transport of such skeletal material across the upper parts of the continental margin of upwelling areas.

Radiolarians

Whereas diatoms as marine algae represent primary producers, radiolarians are protozoans and belong to a higher level of the marine food web. Their distribution in both upwelling areas is therefore considerably different (Fig. 5) from that of diatoms. The quantities of radiolarians in the plankton samples off Peru are higher by a factor of 2-3 than in the northwest Africa upwelling area, and, what seems of particular importance, a basic difference in the pattern of radiolarian distributions exists in each area.

Off northwest Africa, an important decrease of radiolarian standing stocks can be observed within the upwelling region itself. The abundances around the Cape Verde and Canary Islands are slightly higher than in the surrounding open ocean. The highest abundances

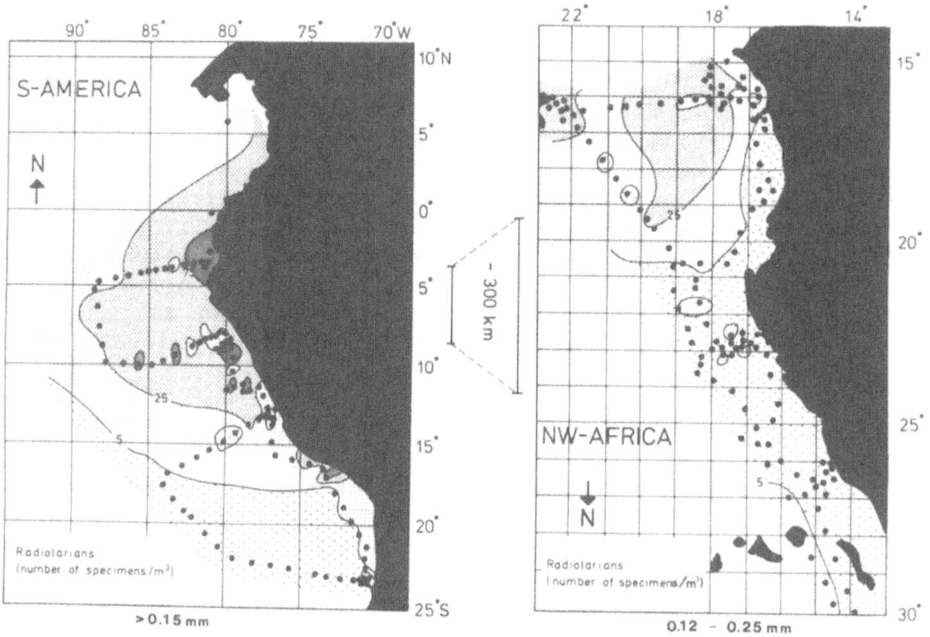

Fig. 5. Quantitative distribution (in number of specimens per m^3 of pumped water) of radiolarians in surface waters off northwestern South America and off northwest Africa.

are confined to the southernmost part of the studied area where surface water temperatures measured during "Meteor" cruise no. 25 (Thiede, 1975a) suggest the presence of water masses warmer than 27°C which protrude northwards, separating the upwelling areas proper, with surface water temperatures of less than 18°C, from the Canary Current, whose surface waters are only 24-26°C and which under the influence of the trade winds turns into a southwesterly direction away from the continental margin.

Off Peru, concentrations of >5 radiolarians/m^3 surface water are found in a wide area north of 17°S covering the entire upwelling area in its widest extent. The highest concentrations are found in samples taken in the outer part of the Gulf of Guayaquil and in a few isolated patches further to the south. The high standing stocks of radiolarians in the surface waters off Peru do not result in high concentrations of radiolarian skeletons in the underlying pelagic and hemipelagic surface sediments at the present time (Molina-Cruz, 1977). It is therefore interesting to note that, despite the differences in standing stocks of both upwelling areas, the sedimentary record of radiolarian distributions is more or less similar for both upwelling regions.

DISTRIBUTIONS OF CALCAREOUS PLANKTON AND NEKTON

Calcareous shells and skeletons are produced by a wide variety of plankton and nekton organisms. Since the distributions of skeletal material produced by benthic organisms have been discussed above, only distributions of the calcitic shell material of planktonic foraminifers and of the aragonitic shells of pelagic gastropods are presented here (Figs. 6 and 7). Within the size fractions studied, these components represent quantitatively some of the most important ones. Whereas planktonic foraminifers as protozoans are from a trophic level comparable to that of radiolarians, pelagic gastropods (in this study only pteropods will be discussed) belong to the heterotrophic nekton and are consequently at a considerably higher level in the marine food web than protozoans are.

Planktonic Foraminifers

Planktonic foraminifers resemble radiolarians in their quantitative distribution. Off Peru, they are relatively rare south of 15°S, but further north a region of large standing stocks (>75 specimens/m^3 surface water) protrudes 400-500 km seaward into the subtropical Pacific Ocean. Off northwest Africa planktonic foraminifers as radiolarians (Fig. 5) are relatively rare in a narrow belt along the

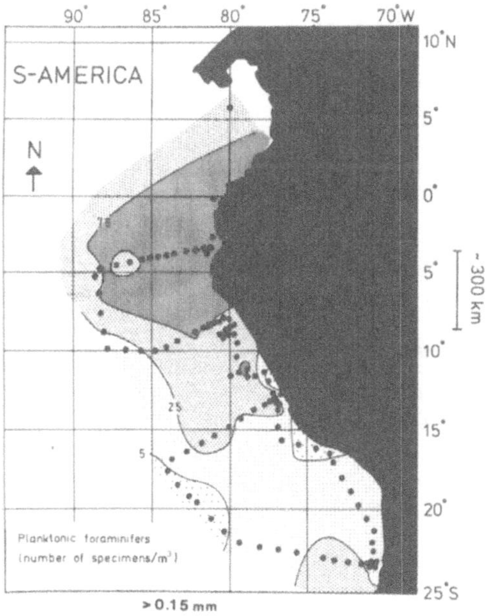

Fig. 6. Quantitative distribution (in number of specimens per m^3 of pumped water) of planktonic foraminifers in surface waters off northwestern South America.

continental margin. Their standing stocks increase in a regular fashion only at a certain distance from the coast. The highest abundances are restricted to a 50-100 km wide belt which can be traced from about 23-25°N to 15°N, but which is limited seaward by water masses with decreasing concentrations of planktonic foraminifers. The frequencies of planktonic foraminifers off Peru as well as off northwest Africa (Thiede, 1975a, Fig. 27e), where more than 100 specimens/m^3 have been observed, are some of the highest observed in the entire world's ocean (Bé and Tolderlund, 1971). Detailed planktonic foraminiferal species and assemblage distributions and their response to coastal upwelling will be discussed below.

Pteropods

Pteropods are active swimmers and therefore able to respond to movements of water masses and to the drift of organisms they are feeding upon in a different way than planktons. Their distribution off Peru and off northwest Africa (Fig. 7) is consequently quite different from most other organisms contributing to the particle assemblages of these samples. Pteropod frequencies off northwest Africa, in particular close to the mouth of the Senegal River and in the broad bight between Cape Blanc and Cape Verde, exceed 200 specimens/

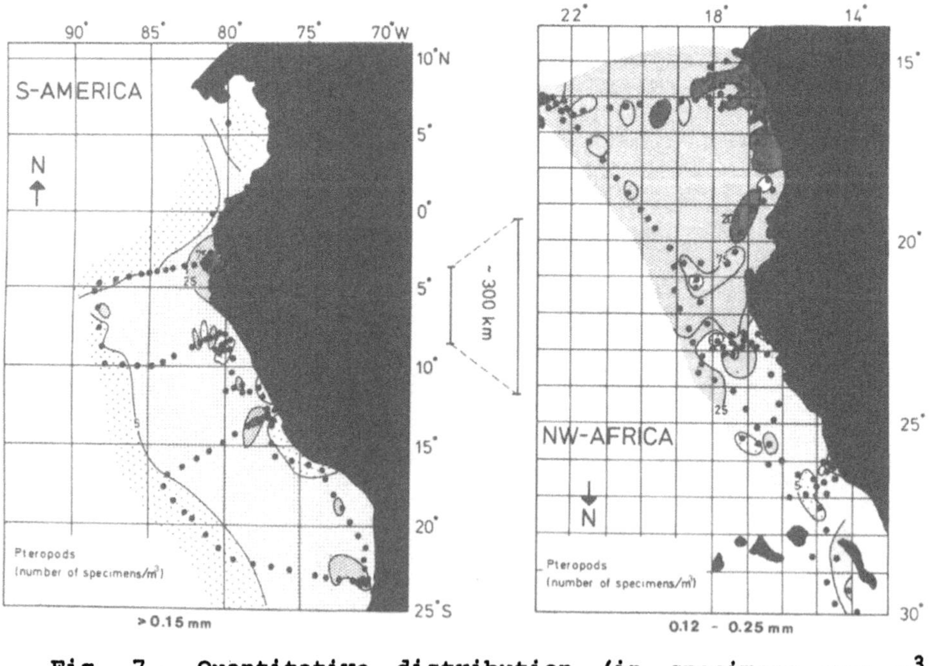

Fig. 7. Quantitative distribution (in specimens per m^3 of pumped water) of pteropods in surface waters off northwestern South America and off northwest Africa.

/m^3 surface water, and standing stocks of 75-200 specimens/m^3 have been found in almost the entire area investigated south of 22°N. Off northwest Africa there is quite obviously a close relationship between the size of the standing stocks and the region where intensive coastal upwelling occurs. Such relationships cannot be established as easily off Peru where concentrations of only >75 specimens/m^3 have been found in a small region in the Gulf of Guayaquil and in one isolated patch of samples off Callao. It seems remarkable that very large standing stocks exist in both coastal upwelling areas in front of river mouths (cf. Sarnthein et al., 1982; Diester-Haass, this volume).

It is also interesting to note that fish bones representing one of the end members of the marine chain are rare in the areas of highest abundances of bulk skeletal material off Peru (Fig. 2), but that their frequencies increase by an order of magnitude in the water masses adjacent to zones of actual coastal upwelling. For geological investigations, it is consequently of the utmost importance in the course of reconstructing former upwelling regimes, to carefully and critically consider the level of the marine food web which has produced the fossils used for such reconstructions.

DISTRIBUTIONS OF PLANKTONIC FORAMINIFERAL POPULATIONS AND ASSEMBLAGES

In the previous chapter, I have discussed distribution patterns of major component groups of the skeletal material found in the plankton and nekton of the two coastal upwelling areas. In this chapter I will describe distributional patterns of individual species and assemblages of planktonic foraminiferal faunas, which are presently the best known and understood group of the geologically important oceanic plankton (Vincent and Berger, 1981). The occurrence of the individual species selected for presentation is expressed not in terms of their absolute standing stocks, but rather in percent of the total of planktonic foraminiferal species.

The planktonic foraminiferal faunas encountered in these samples consist of 26 different species (Thiede, 1975a) but many have a rather patchy and discontinuous distribution. Four of the important species that are known to reflect a specific response to the process of coastal upwelling have been selected for comparison: *Globigerina bulloides* (Fig. 8), *Globoquadrina dutertrei*/synonym: *Neogloboquadrina dutertrei* (Fig. 9), *Globigerinoides ruber* (Fig. 10), and *Globorotalia menardii* (Fig. 11).

Of the other species not illustrated, *Globigerinita glutinata* should especially be mentioned because it occurs relatively frequent close to both upwelling areas yet its ecology is poorly known (Bé, 1977). In contrast to the upwelling area off northwest Africa (Thiede, 1975a, Fig. 13), *G. glutinata* is rare in the region of actual coastal upwelling off Peru but is frequent just outside of it.

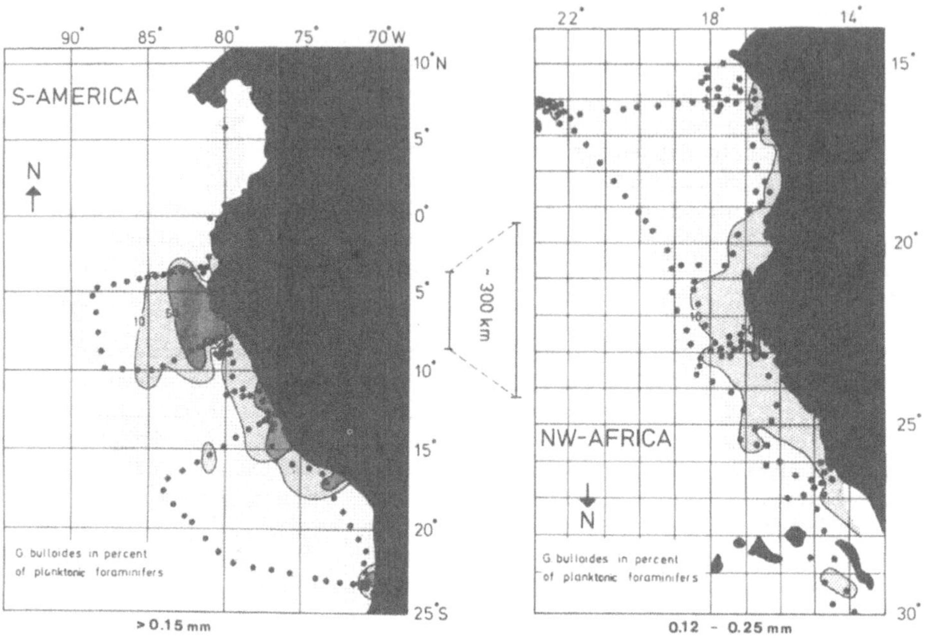

Fig. 8. Distribution of *Globigerina bulloides* (in percent of planktonic foraminiferal faunas) in surface waters off northwestern South America and off northwest Africa.

Species Distributions of Planktonic Foraminifers

G. bulloides is considered to be a species confined mainly to temperate-subpolar surface water masses, but it is also one of the most successful and widely distributed species of planktonic foraminifers (Bé, 1977) which has populated a wide variety of other regions of the oceanic surface water masses. It is particularly typical for upwelling areas and for the eastern boundary currents (Bé and Tolderlund, 1971; Zobel, 1971; Thiede, 1975a) and is therfore, one of the most frequent species of planktonic foraminiferal faunas in the two upwelling regions compared here (Fig. 8). Its concentrations exceed 50% of the total planktonic foraminiferal shell assemblages off Peru in a restricted region which is directly affected by upwelling. There seems to be a close relationship between the distribution of this species and the highest concentrations of organic materials in the surface waters (Fig. 2), whereas in Peru Current water masses this species seems to occur rarely. Off northwest Africa, the relationship of this species to coastal upwelling is more complicated than off Peru because *G. bulloides* constitutes an important component of the Canary Current fauna and the impact of coastal upwelling cannot easily be identified. The highest concentrations of this species, however, are observed in a region off Cape Barbas (close to

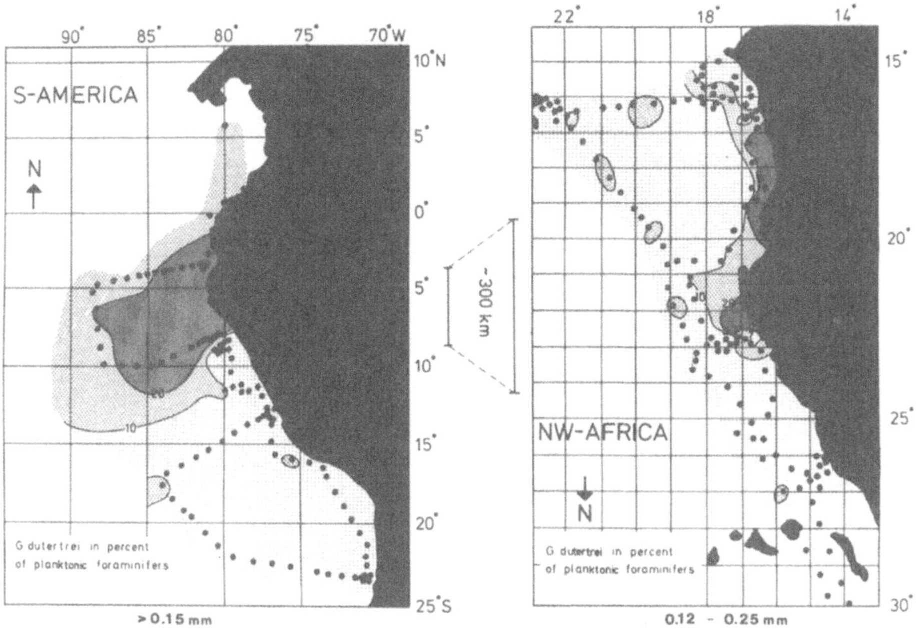

Fig. 9. Distribution of *Globoquadrina dutertrei* (in percent of planktonic foraminiferal faunas) in surface waters off northwestern South America and off northwest Africa.

22°-23°N) which was identified as an area with intense coastal upwelling during the time of sampling.

G. dutertrei, usually classified as a tropical-subtropical species (Bé, 1977), is probably the planktonic foraminiferal species most clearly associated with coastal upwelling. A high abundance of this species (Fig. 9) occurs in both upwelling regions in areas where distributions of other skeletal components also suggest upwelling (Figs. 2 and 8). The distributional patterns, however, suggest important differences between upwelling areas. Off northwest Africa this species achieves its highest concentrations in a narrow belt along the continental margin between 16° and 23°N; off Peru, the area of high concentration during the time of sampling stretches over only 5-6° of latitude but extends much further out into the ocean than off northwest Africa.

G. ruber is the most successful planktonic foraminiferal warm water species (Bé, 1977), and has its peak abundances in subtropical regions. As a species dwelling in the uppermost part of the oceanic water column and frequent in the warm subtropical water masses of both the Canary and Peru Currents, it was, as expected, particularly sensitive to the influence of coastal upwelling (Fig. 10). Off

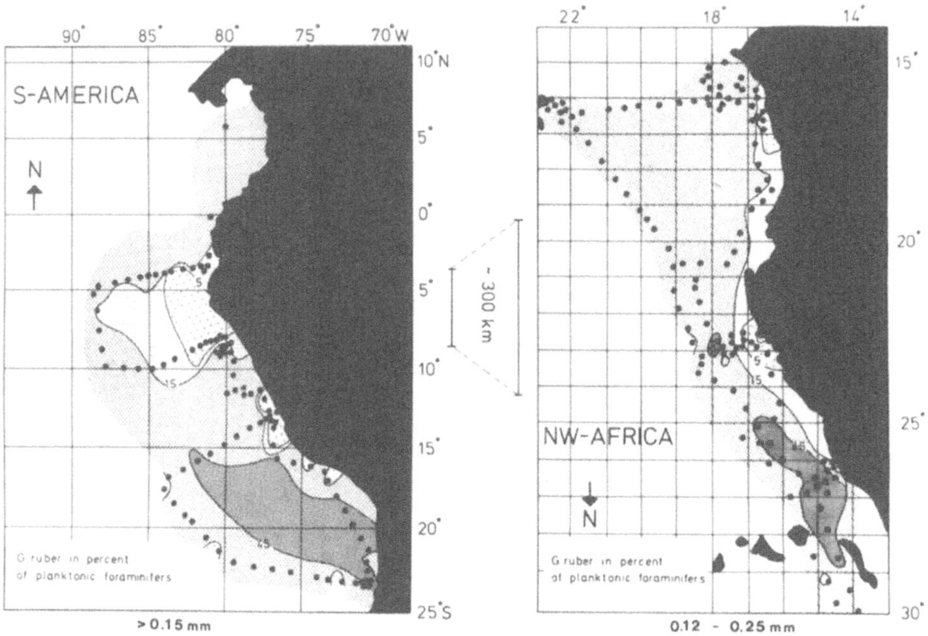

Fig. 10. Distribution of *Globigerinoides ruber*.

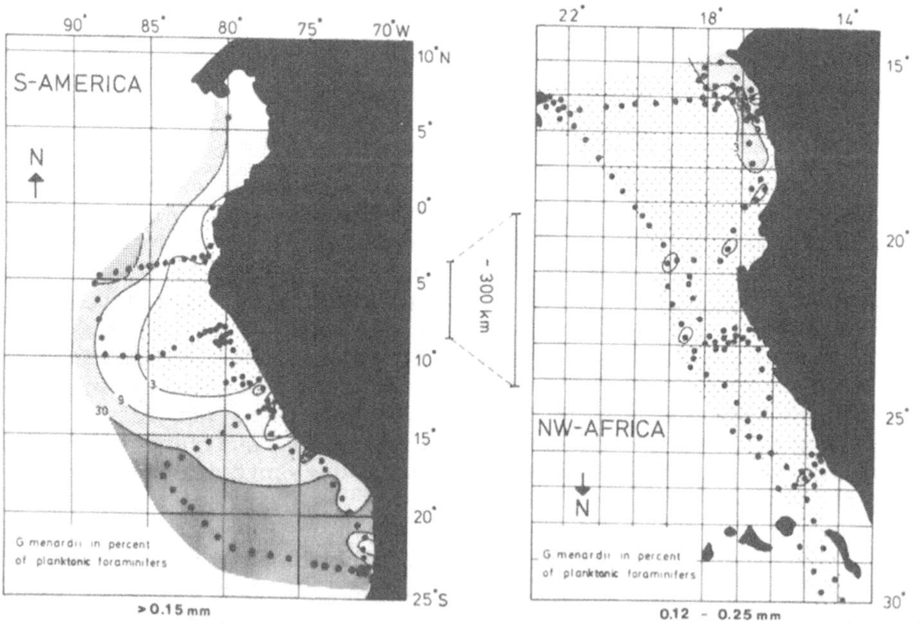

Fig. 11. Distribution of *Globorotalia menardii*.
Each in % of planktonic foraminiferal faunas surface waters off northwestern South America and off northwest Africa, respectively.

northwest Africa it occurs rarely in a long narrow stretch along the continental margin between 16°N and 26°N, with its lowest concentrations at 22°-23°N and at 16°-18°N. Off Peru it is rare in the areas of high productivity at 5°-10°S and 12°-15°S, but the areas of low concentrations can be traced 400-500 km seaward.

G. menardii which is a predominantly tropical species, occurs most frequently in the Atlantic and Indian Ocean, as well as in an isolated area of the eastern tropical Pacific (Bé, 1977). Its distribution is particularly complicated in the two upwelling regions (Fig. 11). Off northwest Africa it occurs frequently in a narrow belt of limited N-S extent (Thiede, 1975a; 1975b), the existence of which seems to be linked to the upwelling of water masses of poleward flowing undercurrent. This interpretation is supported by the fact that *G. menardii* is a deep-dwelling species with its adult stages living in water depths exceeding 100 m (Bé, 1977). Off Peru, however, this species inhabits the surface water masses (Fig. 11) of the subtropical open ocean, but is rare in the surface water masses affected by coastal upwelling. Although an undercurrent is known in the upwelling area off Peru (Smith, this volume), the much less favorable environment of the Peruvian undercurrent, due to the development of the intensive midwater oxygen-minimum, apparently does not permit the lateral displacement of this deep-living planktonic foraminiferal species.

Assemblage Distributions of Planktonic Foraminifers

For the example from northwest Africa, it was attempted not only to map individual species of planktonic foraminifers, but also to apply multivariate statistic techniques (Q-mode factor analysis) to explore the variance of the entire available data matrix of 26 planktonic foraminiferal species (Thiede, 1975b). The planktonic foraminiferal biocoenoses resolve into four important factors or assemblages (Fig. 12), which together account for more than 90% of the total variance of the studied samples. The following 14 species are important contributors to the four factors: *Globigerinoides ruber*, *G. sacculifer*, *G. tenellus*, *Globigerina bulloides*, *G. calida*, *G. falconensis*, *G. quinqueloba*, *Globigerinella aequilateralis*, *Globigerinita glutinata*, *Globoquadrina dutertrei*, *G. pachyderma*, *Pulleniatina obliquiloculata*, *Globoratalia inflata*, and *G. menardii*.

Factor 1 (47% of the total variance, Fig. 12a) with *G. sacculifer* as the dominant species, is confined to the southern and western part of the area studied. It represents a group of species dwelling in tropical and subtropical warm water masses. Factor 2 (18% of the total variance, Fig. 12b) with *G. bulloides* as its dominant species, represents faunas of the Canary Current which reach into the coastal upwelling region. The two remaining factors (Factor 3 accounting for 19% of the total variance, and Factor 4 for 8% of the variance) complement each other. Factor 3 (Fig. 12c) is dominated by *G. ruber* and

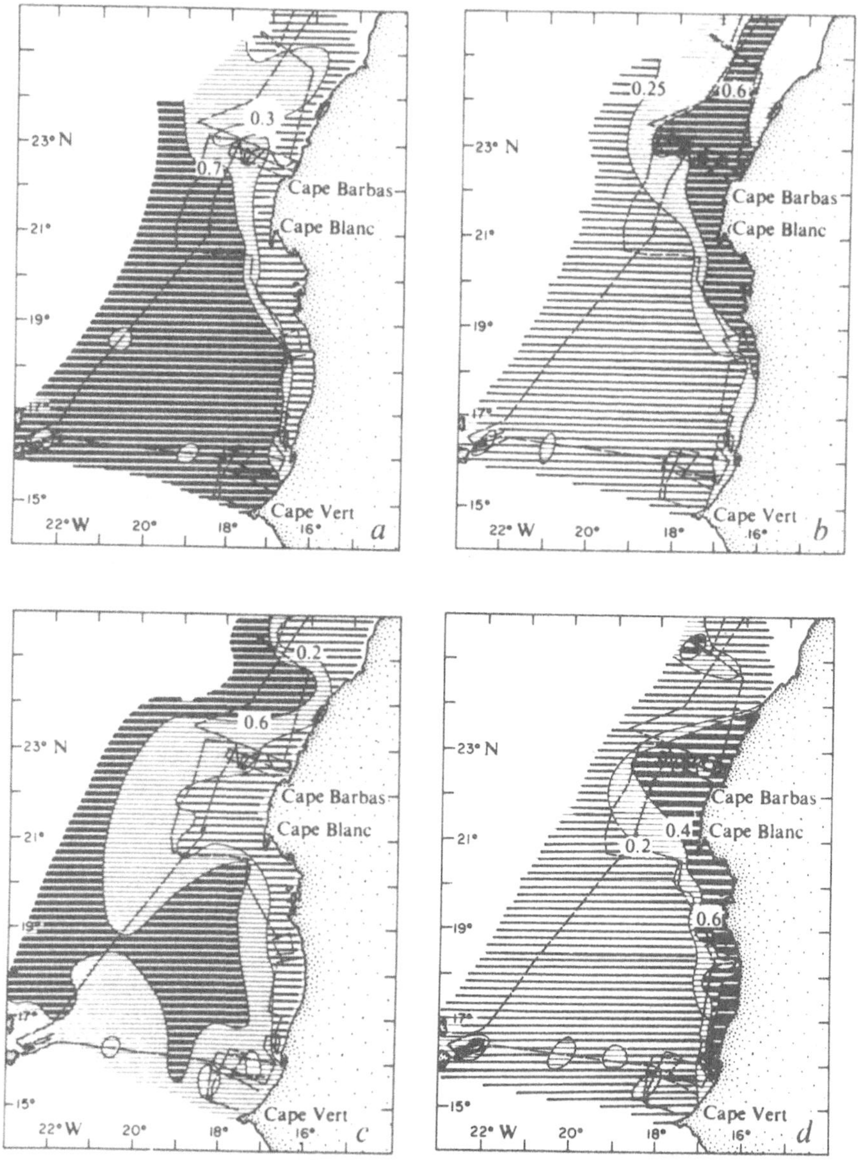

Fig. 12. Distribution of planktonic foraminiferal assemblages off northwest Africa (from Thiede, 1975b). Numbers and associated shading with contours correspond to factor loadings, higher values indicate increasing abundance of respective planktonic foraminiferal assemblages. For further details see text. By permission from Nature, Copyright (c) 1975, Macmillan Jour. Ltd.

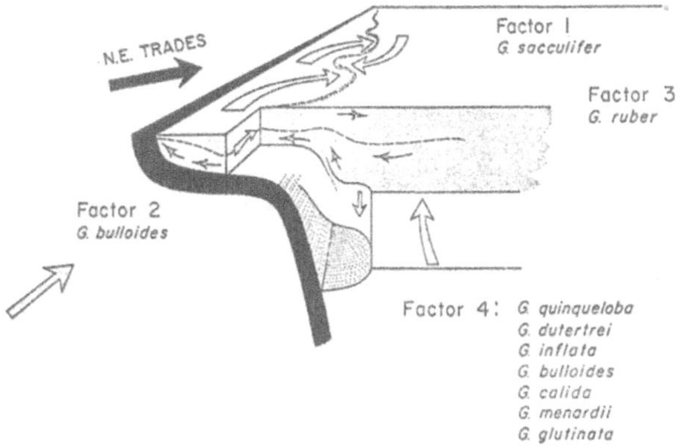

Fig. 13. Schematic relationship of major water masses in the northwest Africa upwelling region and planktonic foraminiferal assemblages inhabiting them. Hydrography simplified after Smith, 1968.

represents assemblages which dwell in the surface waters of the subtropical northeast Atlantic Ocean and which have been displaced from the coastal zone by upwelling. The species dominating Factor 4 (Fig. 12d) are *G. quinqueloba*, *G. dutertrei*, and *G. inflata*. The highest concentrations of this assemblage are found in the bight between Cape Verde (15°N) and Cape Blanc (21°N) in samples with large amounts of algal material. The species dominating the latter factor are all subsurface dwellers, and it is suggested that they represent the planktonic foraminiferal assemblage of the northwest African coastal upwelling region.

In Fig. 13 an attempt was made to relate the distributions of planktonic foraminifers in the surface waters off northwest Africa to the major water masses of this coastal upwelling regime, as conceived by Smith (1968). The distribution of the four planktonic foraminiferal assemblages is well explained by the three-dimensional distribution of major water masses in and around the upwelling area. This reconstruction is also very helpful in understanding the puzzling co-occurrence of the cool to transitional species of Factor 4 with a tropical but deep-dwelling species such as *G. menardii*. The latter appears in this upwelling area (Fig. 11) despite the fact that the surface water temperatures are as much as 10°C lower than in the adjacent open ocean. The explanation might well lie in the above discussed poleward flowing undercurrent, which carries *G. menardii* northward from a region further south where that species occurs more frequently at depth. As the undercurrent flows along the continental slope, it acts as a partial source for the coastal upwelling resulting in the anomalous distribution of this species.

CONCLUSIONS

1. Variations of skeletal components of plankton samples from the coastal upwelling regions off northwest Africa (Mauritania and Senegal) and off northwestern South America (Peru and Chile) are linked in a very specific way to the distribution of the major water masses. They allow mapping of the limits in upwelling areas to pinpoint the zones where upwelling is actually occurring and where water masses are under the indirect influence of the upwelling phenomenon. The two regions display considerable differences both with respect to their N-S and their seaward extensions. It is of particular interest that the Peruvian upwelling can be traced westward into the ocean twice as far as the northwest African upwelling regime.

2. The effect of the poleward flowing undercurrent can clearly be seen in the distribution of planktonic foraminifers off northwest Africa. Off Peru, however, such an influence could not be observed, probably because of the adverse effect of the well developed midwater oxygen minimum zone on foraminiferal ecology.

3. In both areas the zone of actual upwelling was restricted to narrow and well-defined bands close to the coasts where phytoplankton occurred in large abundances. Advection of coastal waters into the adjacent open ocean could be traced by mapping out the frequencies of benthic foraminifers and of mero-planktonic bivalve larvae. It is clear that the seaward advection of such water masses is much more intensive off Peru than off northwest Africa.

4. Islands which are situated in oceanic surface current regimes also induce upwelling. This type can be separated from coastal upwelling by generally lower productivities and certain differences in plankton compositions.

5. Major centers of coastal upwelling seem to be rather stable features which can be traced over many years and which can probably be mapped in bottom sediments.

6. The abundance and distribution of most of the important organisms of oceanic plankton and nekton have been described for both coastal upwelling regions. These groups are composed of siliceous and calcareous skeletal material and contain taxa which are rarely studied in detail. Data on distributions of pelagic pteropods and mero-planktonic mollusc larvae are particularly important when considering the carbonate budget and fluxes in the area.

7. Waters of particularly high bioproduction are found in front of river mouths with considerable fresh water runoff (here the Senegal and Guayaquil rivers). The surface waters outside the areas of such river runoff contain large abundances of nekton organisms (pteropods and fishes).

8. The different skeletal components of the plankton samples represent different levels, but only fragments, of the marine food web. The groups of organisms belonging to similar trophic levels also display considerable similarities in their distribution patterns. A zonation of skeletal material from primary producers, secondary producers and higher trophic levels in and around the actual area of upwelling can be mapped in surface waters. It is likely that such zonations are also important for understanding the geological record of coastal upwelling regimes.

ACKNOWLEDGEMENTS

The collection of samples and data described in this study was made possible during cruises of RV "Meteor", financed by the DFG (German Research Foundation), and of the RV "Wecoma" of Oregon State University, financed by the US-ONR (Office of Naval Research). The support of these two funding agencies and the assistance of numerous colleagues, research assistants and crew members of the two ships, who helped collect and prepare the samples and data, is gratefully acknowledged.

REFERENCES

Barber, R.T. and Smith, R.L., 1981, Coastal upwelling ecosystems, in: "Analysis of Marine Ecosystems," A.R. Longhurst, ed., Academic Press, London, 31-68.

Bé, A.W.H., 1977, An ecological, zoogeographic and taxonomic review of Recent planktonic foraminifera, in: "Oceanic Micropaleontology," A.T.S. Ramsay, ed., Academic Press, London, 1-100.

Bé, A.W.H. and Tolderlund, D.S., 1971, Distribution and ecology of living planktonic foraminifera in surface waters of the Atlantic and Indian Oceans, in: "The Micropaleontology of Oceans," B.M. Funnell and W.R. Riedel, eds., Cambridge University Press, Cambridge, 105-149.

Berger, W.H., 1970, Planktonic foraminifera: Differential production and expatriation off Baja California, Limnology and Oceanography, 15:183-204.

Boltovskoy, E., 1971, Patchiness in the distribution of planktonic foraminifera, in: "Proceedings 2nd Planktonic Conference," A. Fericcani, ed., Roma, 1:107-115.

Diester-Haass, L., 1978, Sediments as indicators of upwelling, in: "Upwelling Ecosystems," R. Boje and M. Tomczak, eds., Springer, Berlin, 261-281.

Heath, G.R., 1974, Dissolved silica in deep-sea sediments, Society of Economic Paleontologists and Mineralogists, Special Publications, 20:77-93.

Honjo, S., Manganini, S.J. and Cole, J.J., 1982, Sedimentation of biogenic matter in the deep ocean, Deep-Sea Research, 29:609-625.

Molina-Cruz, A., 1977, Radiolarian assemblages and their relationship to the oceanography of the subtropical southeastern Pacific, Marine Micropaleontology, 2:315-352.

Moore, T.C., Heath, G.R. and Kowsmann, R.O., 1973, Biogenic sediments of the Panama Basin, The Journal of Geology, 81:458-472.

Sachs, K.N., Cifelli, R. and Bowen, V.T., 1964, Ignition to concentrate shelled organisms in plankton samples, Deep-Sea Research, 11:621-622.

Sarnthein, M., Thiede, J., Plaumann, U., Erlenkeuser, H., Fütterer, D., Koopmann, B., Lange, H., and Seibold, E., 1982, Atmospheric and oceanic circulation patterns off northwest Africa during the past 25 million years, in: "Geology of the Northwest Africa Continental Margin," U. von Rad, K. Hinz, M. Sarnthein, and E. Seibold, eds., Springer, Berlin, 545-604.

Scheltema, R.S., 1971, The dispersal of larvae of shoal-water benthic invertebrate species over long distances by ocean currents, in: "Proceedings 4th European Marine Biology Symposium," D.J. Crisp, ed., Cambridge, 7-28.

Schuette, G. and Schrader, H., 1979, Diatom taphocoenoses in the coastal upwelling area off western South America, Nova Hedwigia, Beiheft, 64:359-378.

Seibold, E., 1972, Cruise 25/1971 of R.V. "Meteor": Continental margin of West Africa - General report and preliminary results, "Meteor"Forschungs-Ergebnisse, C10:17-28.

Sliter, W.V., 1965, Laboratory experiments on the life cycle and ecologic controls of *Rosalina globularis* d'Orbigny, Journal of Protozoology, 12:210-215.

Smith, R.L., 1968, Upwelling, Oceanography and Marine Biology, Annual Reviews, 6:11-46.

Spindler, M., 1980, The pelagic gulfweed *Sargassum natans* as a habitat for the benthic foraminifera *Planorbulina acervalis* and *Rosalina globularis*, Neues Jahrbuch für Geologie und Paläontologie, Monatshefte, 9:569-580.

Thiede, J., 1974, Marine bivalves: Distribution of mero-planktonic shell-bearing larvae in eastern North Atlantic surface waters, Palaeogeography, Palaeoclimatology, Palaeoecology, 15:267-290.

Thiede, J., 1975a, Shell- and skeleton-producing plankton and nekton in the eastern North Atlantic Ocean, "Meteor"Forschungs-Ergebnisse, C20:33-79.

Thiede, J., 1975b, Distribution of foraminifera in surface waters of a coastal upwelling area, Nature, 253:712-714.

Vincent, E. and Berger, W.H., 1981, Planktonic foraminifera and their use in paleoceanography, in: "The Oceanic Lithosphere," C. Emiliani, ed., THE SEA, Vol. 7, Wiley and Sons, New York, 1025-1119.

Wooster, W.S., Bakun, A. and McLain, D.R., 1976, The seasonal upwelling cycle along the eastern boundary of the North Atlantic, Journal of Marine Research, 34:131-141.

Zobel, B., 1971, Foraminifera from plankton tows, Arabian Sea: Areal distribution as influenced by ocean water masses, Proceedings 2nd Planktonic Conference (Rome 1970), 2:1323-1334.

Zuta, S. and Guillén, O., 1970, Oceanografia de las aguas costeras del Peru, Boletin del Instituto del Mar Peru, 2:157-324.

SEASONAL VARIATION IN PARTICULATE FLUX IN AN OFFSHORE AREA ADJACENT TO COASTAL UPWELLING

Kathy Fischer, Jack Dymond, Chris Moser, Dave Murray, and Anne Matherne

School of Oceanography, Oregon State University
Corvallis, Oregon 97331, U.S.A.

ABSTRACT

Successive sediment trap deployments were made over a two-year period at a site 360 km west of Cape Mendocino (39.5°N, 128°W). In the final deployment, the shallowest trap was equipped with a sample changer which provided three consecutive samples during the spring and summer of 1981 in order to study possible seasonal variations due to the influence of coastal upwelling. Biogenic flux data point to increasing productivity at the site from mid-March through the end of August, 1981. The ratio of opal flux to carbonate flux increases from spring to a mid-summer maximum. Radiolarians dominate the spring opal flux, whereas diatoms are the major contributors to opal flux in the mid and late summer. Changes in the distributions of three silicoflagellate species suggest colder temperatures occur in mid to late summer. Upwelling at the latitude of our mooring is greatest in midsummer. Although upwelling occurs near the coast, colder water, possibly with higher nutrient levels, can be advected rapidly offshore in the form of surface plumes. Satellite images from April through July, 1981, reveal at least three plumes crossing our area during the period June 21-July 29. These cold water plumes extended from Cape Mendocino to the mooring location. The occurrence of such plume events can explain the flux patterns and changes in other indicators recorded during the spring-summer period by our multiple sampling sediment trap.

INTRODUCTION

The enhancement of upwelling south of capes and points along the California coast is well known (Reid, Roden and Wyllie, 1958) and can easily be observed using satellite infrared imagery. Colder water upwelled along the coast is seen in satellite images as regions of lighter shading (Fig.1). Eddies or long plumes extending more than 300 km from the coast may form as the colder water moves seaward and mixes with the California Current. Recent studies (Traganza, Conrad and Breaker, 1981; Traganza et al., this volume) have measured *in situ* temperatures, nutrients, and factors relating to biomass. The results show that for an eddy located close to its source, nutrient values are relatively high and correlate well with temperature. Also, biomass peaks are found along the sharpest gradients in temperature and nutrient concentrations. Nutrient values for upwelled water will depend on the source of the water and the extent of mixing and nutrient utilization that has taken place. Consequently, high nutrient values may not always be associated with low temperatures.

Traganza et al. (1981) found that for one plume extending more than 250 km out to sea, nutrients and temperature were still correlated, though not as strongly, and biomass peaks were still found along the edge of the feature. This particular plume was shallow (<50m) at its seaward end and appeared to be supplied with cold, nutrient-rich water from the upwelling center near the coast. Temperature differences of ∿2°C were measured at the extreme seaward edge of the plume, ∿250 km from the coast. Speeds of 18-41 cm/sec (Breaker and Gilliland, 1981) have been estimated from successive satellite images for cold water features along the northern California coast. Thus, in a few days a plume can travel 100 km or more out to sea.

Examination of satellite images reveals that these plumes recur frequently during the upwelling season in areas to the southwest of coastal projections, notably at Cape Mendocino, Point Arena, and Point Sur along the northern and central California coast. The effects of these plumes of cold water, which may have relatively higher nutrient concentrations, upon sedimentation hundreds of kilometers offshore have not been examined. Productivity pulses may arise due to the episodic input of nutrients. If these bursts in biogenic production are rapidly transported to the bottom and remain temporarily "unprocessed" by slowly responding bioturbating organisms, transitory chemical imprints may affect the sediments. An example might be the mobilization of metals through reduction, with subsequent loss to seawater. Another possible effect of recurrent plumes on the sediments results from the input of mixed plankton assemblages along the transition zone between the colder, and possibly more nutrient-rich, plume water and the warmer water surrounding it. Traganza et al. (1981 and this volume) found peaks in biomass on the warm water side

→

Fig. 1. Satellite image for July 8, 1981. The general mooring location (39.5°N, 128°W) is circled and appears to the left of the Cape Mendocino-Point Arena section of the California coast (satellite image from L. Breaker, National Environmental Satellite Service).

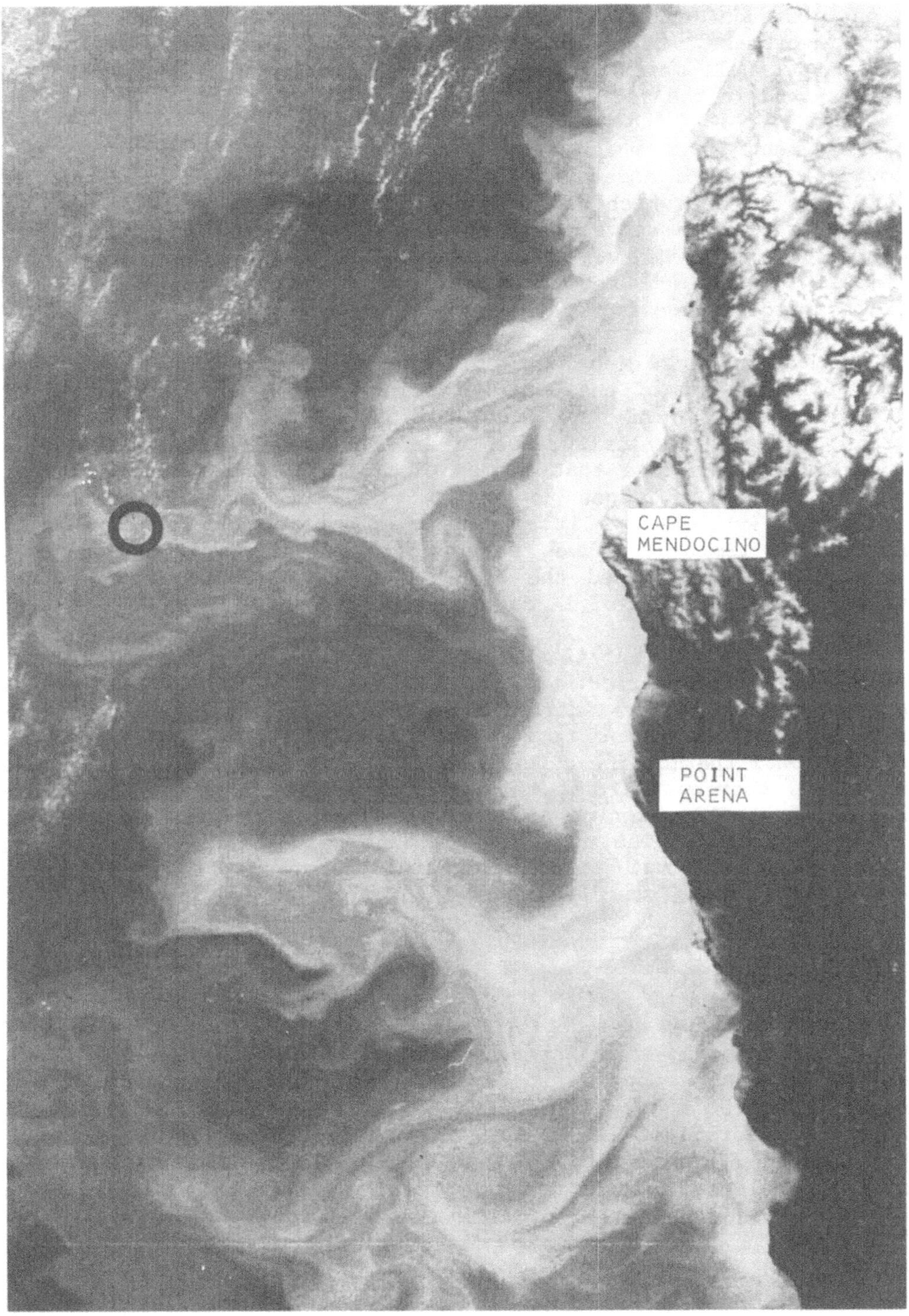
CAPE
MENDOCINO
POINT
ARENA

of this transition zone. Species which are able to grow most rapidly in the presence of higher nutrient levels likely produce the observed peaks in biomass. Depending on many factors (water depth, sedimentation rate, shell material and form), the shells of these species may be preserved in the sediments.

Sediment traps which sample successive, well-defined intervals provide a means of assessing the influence on settling material of cold water plumes which consistently pass through an area. The trap information can be correlated with satellite images for the deployment period and, whenever possible, with *in situ* temperature, nutrient, and biomass measurements. Sediment cores can then be examined for indicators suggested by the trap studies.

We report here trap data collected over two years in an area about 360 km west of the Cape Mendocino-Point Arena section of the California coast. At this latitude, upwelling reaches its peak during June and July as surface winds increase (Reid et al., 1958). It is during this time that plumes of upwelled water frequently are seen in satellite images to extend offshore. From the satellite images that we examined, it appears that our mooring location is at the seaward limit of plume travel at this latitude. Consequently, few plume events crossed the area. Within a 40-day period three events lasting a total of 14 days were observed.

Changes in sedimentation arising from plume events must be separated from longer period seasonal changes. In our study area, the sea surface temperature increases by about 5°C from January to September (Karlin, Heath and Levi, 1982), accompanied by the development of a seasonal thermocline. Inasmuch as a thermocline forms a barrier to mixing, thereby limiting the input of nutrients to surface waters from below, continued high productivity throughout the summer would not be predicted. The expected seasonal pattern would be a spring bloom, followed by much lower productivity levels during the remainder of the summer (Russell-Hunter, 1970).

DESCRIPTION OF THE EXPERIMENT

Sediment traps were deployed at 39°30'N, 128°W, 360 km west of the Cape Mendocino-Point Arena section of the California coast. Table 1 shows the depths and deployment intervals for the sediment traps. The 4050 m trap represents an average sample for the fall, winter, and spring of 1979-1980. The period of maximum upwelling at this latitude (expected to be June-July; Reid et al., 1958) is not completely sampled by this trap. The 2985 m trap deployed during 1980-81 sampled the period of late summer through the winter. The 1235 m, 2985 m, and 3785 m traps cover the time of maximum spring and early to mid summer productivity during 1981. The 1235 m trap collected three sequential samples during the deployment. The sampling

Table 1. Total and biogenic fluxes for traps deployed over a two-year period at 39.5°N, 128°W. Water depth is 4250 m.

Deployment	Trap Depth	Interval	Days	Flux in $mg/cm^2/yr$ Total	CH_2O[a]	Opal	$CaCO_3$	% Biogenic Flux[b] of Total Flux	Key to Figs. 2&3
1	4050 m	8/12/79-6/25/80	318	1.70	0.23	0.57	0.38	69	A
2	2985 m	6/26/80-3/9/81	256	0.56	0.08	0.17	0.14	70	B
3	1235 m-1	3/13/81-5/31/81	79	1.14	0.17	0.44	0.50	97	C
	-2	6/1/81-8/18/81	79	3.56	0.35	1.35	0.60	65	D
	-3	8/19/81-8/29/81	11	6.79	0.75	1.90	1.48	61	E
	1235 m-average			2.64	0.29	0.96	0.62	71	F
	2985 m	3/13/81-8/29/81	169	4.15	0.42	1.56	0.97	71	G
	3785 m	3/13/81-8/29/81	169	3.19	0.35	1.07	0.65	65	H
measured productivity[c] - 0.8 $mgc/m^3/hr$									

[a] Calculated from organic carbon.
[b] Equal to (CH_2O flux + opal flux + $CaCO_3$ flux).
[c] Malone, 1971, for early August, "surface". This value equates to 184 mg $CH_2O/cm^2/yr$ if the measured productivity represents an average for the entire 105 m-deep photic zone.

intervals correspond to spring, early-mid summer, and an 11-day period at the end of August. The two lower traps collected average samples for the entire deployment period March 13-August 29, 1981.

The sediment traps used in these deployments were developed from the basic design of Andrew Soutar (Soutar et al., 1977). Our trap is a single fiberglass cone with a mouth area of 0.5 m^2 and a height-to-diameter ratio of 2:1. A honeycomb baffle which is 5 cm deep and has 1-cm^2 cell openings covers the mouth of the trap. The timer used for sequential sampling (1235 m trap) was checked after the deployment to verify proper functioning. We estimate the bulk flux errors, including weighing and processing errors, timing interval error, and mouth area measurement error, to be less than ±5%. The absolute accuracy of sediment trap fluxes is not known for certain, but intercomparison studies of various trap designs suggest that traps of widely varying shapes collect within a factor of two of each other (Dymond et al., 1981; Spencer, Honjo and Brewer, 1982).

A mixture of four antibiotics in a silica gel matrix (Dymond et al., 1981) was used as a bacterial inhibitor during the first two deployments. Sodium azide was the poison used for the traps in the third deployment. A comparison study of these two methods of bacterial inhibition, as well as formalin, show that the antibiotics reduce bacterial activity greatly, though not as well as azide or formalin. Azide was chosen for the third deployment because chemical effects were less pronounced than for formalin (which mobilizes Mn) or the antibiotic mix (which may add organic C and Si). The conclusions of the comparison study indicate that the use of the two bactericides would not significantly affect the trap compositions examined in this study (Powell and Fischer, 1982).

METHODS

Material collected was sieved at 1 mm. The <1 mm fraction was wet split into four equal parts. One split was preserved with buffered formalin and retained for microscopic particle description. The other three splits were centrifuged to reduce their volumes, rinsed free of salt with pH neutralized double distilled water, and freeze dried. The weight of the remaining wet split was estimated from the weights of the three dried splits. Data reported here pertain to the <1 mm fraction.

Carbonate carbon and organic carbon (±2% precision) were measured directly as evolved CO_2 by successive treatments with phosphoric acid and a potassium dichromate-sulfuric acid solution (Weliky, 1982). Major elements were determined by atomic absorption (Fukui, 1976). Opal was calculated using the following equation.

$$\text{Opal} = 2.61 \times Si_{total} - 2.61 \times (Si/Al)_{detrital} \times Al_{total}$$

The coefficient, 2.61, is the inverse of the weight fraction of silicon in opal (van Bennekom and van der Gaast, 1976). A detrital Si/Al value of three was assumed. This value was found in downcore sediments of cores from this area which contain only minor amounts of opaline tests.

A subsample of the wet split from each trap was used to prepare slides for microscopic studies. Each of the subsamples was centrifuged for 20 minutes at 1500 RPM to reduce water volume. The concentrated sample was treated with H_2O_2 (brought to pH 8.5 with NH_4OH) overnight to remove most of the organic matter. The resulting material was sieved at 150 µm. The >150 µm fraction was rinsed (double distilled water brought to pH8-9 with NH_4OH), dried at 60°C, and placed in foraminiferal counting trays for examination. The <150 µm fraction was left to settle for 24 hours, and then excess water was carefully siphoned off. The sample was stirred by hand, and an aliquot was taken immediately with an automatic pipette. The aliquot was dispensed evenly over a water covered slide and air dried. Two to four slides were prepared for each trap sample.

Satellite images were supplied by L.C. Breaker and R.P. Gilliland, National Environmental Satellite Service, for 33 days during the period from April 26 through July 31, 1981. Clouds, which are opaque in the infrared band used to view water temperature differences, sometimes obscured our mooring area. Often, when this occurred, the Cape Mendocino coastal area was still visible, and the start of plume formation or the base of an existing plume could still be observed. The best data regarding the presence or absence of plumes in our mooring site is for the intervals April 22-May 12, May 31-July 13, and July 27-31, 1981. The April 22-May 12 set of satellite images covers part of the first sampling interval of the 1235 m trap. The satellite images for May 31-July 13 and July 27-31 cover part of the second sampling interval of the 1235 m trap.

RESULTS AND DISCUSSION

Total and biogenic fluxes measured by the traps are presented in Table 1. The average fluxes for the period March 13, 1981, to August 29, 1981, at all water depths are considerably higher than those for the previous two deployment periods which include winter months. The average flux over the year from mid-August, 1979, to mid-August, 1980, was 1.89 $mg/cm^2/yr$ (using the 4050 m and the 3785 m traps). For the following year it was 1.63 $mg/cm^2/yr$ (using the two 2985 m traps).

The average fluxes in the two lower traps (2985 m and 3785 m) for the third deployment are greater than the average flux for the 1235 m trap. Increases in flux with depth have been found in other areas (Honjo, Manganini and Cole, 1982; K. Fischer, unpublished data). Repackaging of material by zooplankton, aggregation into "marine snow", or lateral transport of particle-rich waters have been suggested as processes which can account for increased flux at mid-water depths. For traps within ∿500 m of the bottom (the 3785 m and the 4050 m traps), lateral transport of resuspended sediments is a likely source of material.

The fall-winter flux can be calculated for the first trap deployment at 4050 m by using the average flux from the 3785 m trap to calculate the spring and late summer contribution. This contribution was subtracted from the 4050 m trap average to yield a fall-winter average for this area. The result is 0.78 $mg/cm^2/yr$, nearly an order of magnitude lower than the maximum flux measured during the summer months. Since some of this flux may result from resuspended sediments, near-surface fluxes may be even lower.

The total and biogenic fluxes increased with each successive sampling interval from mid-March through August, 1981. The ratio of opal flux to carbonate flux increased to a mid-summer maximum during this same period (Table 2). Although the opal content of the trapped material increased over the spring-summer sampling periods, the ratio of organic matter to total shell material remained fairly constant. A plot of CH_2O, calculated from organic carbon, vs. ($CaCO_3$ + opal) as molar fluxes reveals the good correlation (Fig. 2). The correlations of organic flux with either carbonate flux or opal flux alone are not as strong. All organisms which produce carbonate or opaline tests also produce organic matter. Therefore, it is reasonable to expect the correlation of total organic matter to total shell material to be better than the correlation of organic matter to either carbonate or opal alone. This would not hold true if significant preferential dissolution of either opal or carbonate had occurred. If there is significant contribution of organic matter from species which do not form tests, a positive intercept could occur.

Table 2. Flux ratios of the three biogenic components for the early spring and summer trap samples and for a trap collecting for almost an entire year

Trap Depth	Deployment Period	Flux Ratios CH_2O	Opal	$CaCO_3$	$Opal/CaCO_3$
1235 m-1	3/13/81-5/31/81	1	2.6	2.9	0.9
-2	6/1/81-8/18/81	1	3.9	1.7	2.3
-3	8/19/81-8/29/81	1	2.5	2.0	1.3
4050 m	8/12/79-6/25/80	1	2.5	1.7	1.5

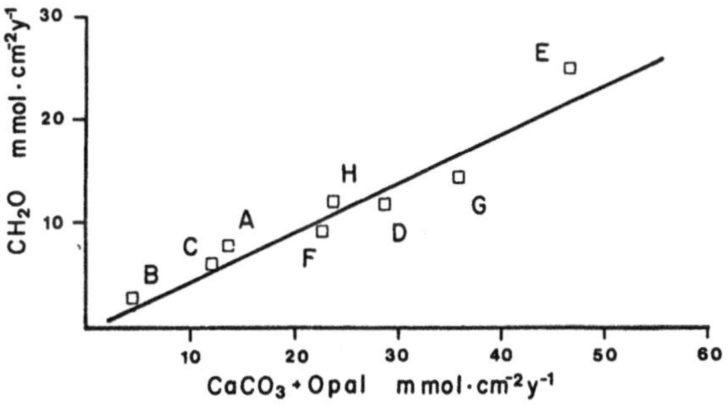

Fig. 2. Plot of molar flux of organic matter (as CH_2O) versus the total molar flux of skeletal material for all sediment traps. The letters refer to traps indicated in Table 1. The slope is 0.48, with a y-intercept of -0.05. The coefficient of determination, r^2, is 0.93.

Fig. 3 is a plot of the relative proportions of three components, biogenic material (the sum of CH_2O, opal, and $CaCO_3$), continental detritus represented by Al, and a resuspended component represented by Mn, which is strongly enriched in surface sediment relative to the shallowest traps. There are other possible sources of Mn. The continental slope is a region where the decay of organic material results in the remobilization of Mn to seawater. The solubilized Mn could precipitate in the oxygenated bottom waters or adsorb onto particles and then be transported seaward before settling. Also, a contribution from hydrothermal sources of Mn cannot be ruled out since the Gorda Ridge is only about 100 km to the north.

The March 13-May 31 sample (1235-1) is predominantly biogenic (Fig. 3). Although the two subsequent sampling periods had higher biogenic fluxes (Table 1), the bulk compositions of the recovered material exhibit increased detrital influence. The source of detritus may be coastal rivers. The Eel River north of Cape Mendocino is a major source of suspended sediment to the nearshore area; however ∿90% of the suspended load is introduced during December through March (Griggs and Hein, 1980). More efficient settling due to zooplankton packaging may be an important factor in the increased detrital flux during the summer months (Scheidegger and Krissek, this volume).

Table 3 shows the flux of skeletal material for radiolarians, silicoflagellates, foraminifers, and coccoliths. The radiolarian and silicoflagellate fluxes remain constant or decrease from the spring to the mid-summer collection periods. Because the opal flux increases substantially during this period, the flux of diatoms must increase. This observation is compatible with an influence of upwelling plumes during the midsummer. Malone (1971) found that the standing stock of net plankton (>22 μm fraction, mostly consisting of

Table 3. Flux of skeletal material (no. of particles/m^2/day) for the three spring-summer samples (1235 m-1, -2, -3) and for the 4050 m trap which represents almost an entire year of collection. Diatoms were not counted. $CaCO_3$ and opal fluxes are also given. Skeletal flux error is estimated at +10%, except for coccolith flux which may vary by a factor of 2.

	Flux, in Particles/m^2/day					mg/m^2/day	
Trap Depth	Foraminifers $\times 10^4$	Coccoliths $\times 10^7$	Pteropods	Radiolarians $\times 10^4$	Silicoflagellates $\times 10^5$	$CaCO_3$	Opal
1235 m-1	3.8	5.2	4.9	3.3	6.2	13.7	12.0
-2	3.5	7.5	17.	3.0	4.6	16.4	37.0
-3	1.3	37.	58.	3.6	4.0	40.5	52.0
4050 m	1.6	5.3	1.2	2.7	3.9	10.3	15.7

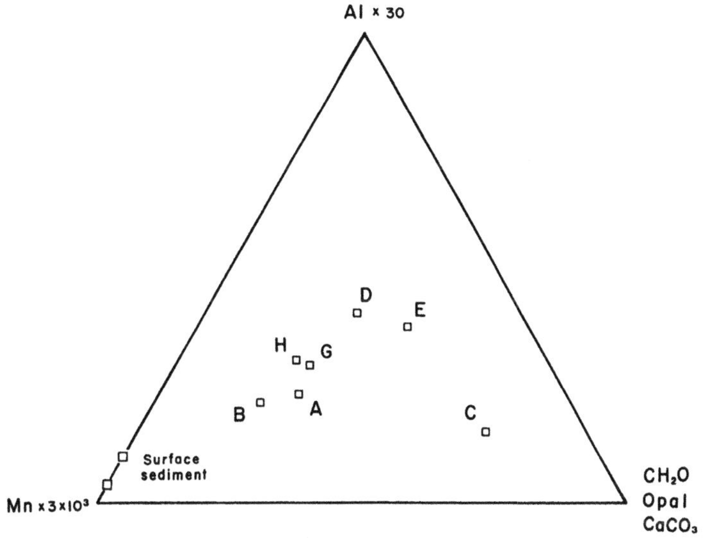

Fig. 3. Partitioning of trap samples and surface sediments into three components represented by Al, Mn, and the sum of CH_2O, opal and $CaCO_3$. To give the three components roughly equal weight, Al concentrations have been multiplied by 30 and Mn concentrations by 3000. Points are shown for surface sediments from two cores taken in the area. The letters refer to traps indicated in Table 1.

chain-forming diatoms) increases in upwelling centers along the California coast if nutrients become more available and vertical water movements prevent sinking. Table 3 shows also that the eleven-day period at the end of August was a time of reduced foraminiferal flux and greatly increased pteropod and coccolith fluxes.

The increasing biogenic flux during the spring is probably a biologic response to increased sunlight. Continued high productivity, however, requires the input of more nutrients. The source of these nutrients, and the resulting increase in biogenic flux observed during June through July, may be due to the addition of nutrients originating in the upwelled water along the coast. Fig. 1 is a satellite image from July 8, 1981, which shows a cold water plume extending over our area. Plumes were seen over the area from June 21-28, July 8-12, and July 29-30, 1981, for a total of ∿14 days out of a 40-day period. This coincides with the time of increased coastal upwelling at this latitude, based on the wind stress pattern (Smith, this volume, Figure 10). It seems reasonable to assume that plume frequency and size are linked to upwelling occurrence, persistence and intensity, although perhaps not in a simple way.

Evidence for cooler waters in the overlying photic zone during the summer trap collection periods comes from an analysis of the

Table 4. Sea surface temperature estimates from silicoflagellate abundances (Poelchau, 1974). Sea surface temperatures from direct measurements are listed for comparison.

Trap Depth	Deployment Period	Sea Surface Temperature, °C			
		Calculated[a] Summer	Calculated[a] Annual	Measurements, Temp.,Month	Mean of Several[c] Measurements, Temp.,Month
1235 m-1	3/13/81-5/31/81	22	20		13°C, Jan
-2	6/1/81-8/18/81	16	14	13°C, Feb[b]	12°C, Apr
-3	8/19/81-8/29/81	17	18	17°C, Aug[d]	16°C, Jul 17°C, Sep
4050 m	8/12/79-6/25/80	*	14		
	surface sediments	14	11		

[a] Calculated from equations given in Poelchau, 1974. See Table 5 of % Relative Abundances.
[b] Heath, 1982, measured 2-13-81.
[c] Naval Weather Service, 1976; see Karlin et al., 1982.
[d] Malone, 1971.
* Since this was almost a full year's deployment, no summer temperature was calculated.

silicoflagellate assemblage compositions in the trap samples (Table 5). Numerous studies (see Poelchau, 1974, for summary) have shown that the relative abundances of silicoflagellate species vary with sea-surface temperatures. Poelchau (1974) developed a set of equations, using the transfer faction technique described by Imbrie and Kipp (1971), to predict annual and summer mean sea-surface temperatures based on the relative abundance of silicoflagellate species in northeast Pacific surface sediments. Exact values of calculated temperature estimates should be used with caution for samples obtained within the water column, but relative changes do reflect temperature variations between samples.

Temperatures were calculated for all samples using Poelchau's equation for determining annual sea-surface temperatures (Table 4). Annual temperatures calculated for seasonal samples collected during the spring and summer are warmer than those for a true annual sample (4050 m sample) and calculated temperatures are slightly warmer than observed values.

Temperatures for successive samples at the 1235 m depth indicate that the warmest period is from March through May. We suggest that more frequent incursions of cold water derived from coastal upwelling plumes during subsequent sampling intervals enhance the abundance of colder water species in the samples. Mean sea-surface temperatures for the seasonal samples were also calculated using Poelchau's summer temperature equation to obtain an estimate more accurately reflecting the temperature during the sampling period. Again, values were highest during the spring sampling period. Those for July and August agree well with the observed temperatures (Table 4).

The top centimeter of box core TT-141-II-12BC, taken near the mooring site (Table 5), was also analyzed. Summer and annual mean sea-surface temperature based on the sediment sample are colder than

Table 5. Percent relative abundances for three silicoflagellate species used to calculate temperatures given in Table 4.

Trap Depth	*Dictyocha messanensis*	*Dictyocha epiodon*	*Distephanus speculum*	*Distephanus octangulus*	Total Counted
1235 m-1	92.1	2.9	5.1	0.0	315
-2	56.6	8.6	34.9	0.0	350
-3	32.1	5.2	62.7	0.0	212
4050 m	66.9	10.4	22.7	0.0	317
Sediment 0-1 cm	62.5	4.5	31.0	2.0	179

observed values or those derived from trap samples. In the North Pacific, the silicoflagellate *Distephanus octangulus* is found in highest abundance in surface sediments of the Bering Sea or the Gulf of Alaska (Poelchau, 1974). Its presence in the sediments but not in trap samples suggests that colder waters were present in this region in the past or that this species is transported to the south by deeper currents. Thus, a colder temperature is obtained from the sediment sample than is observed from recent instrumental data.

CONCLUSIONS

Our biogenic flux and silicoflagellate distributions show steadily increasing productivity from mid-March through August, with the possible incursion of colder water during the middle of the summer. Satellite photographs provide strong evidence that cold water plumes do pass over the area during this time. After the spring phytoplankton bloom and the development and deepening of the summer thermocline, any sudden supply of nutrients would result in a burst of production. Nutrient supply from outside becomes necessary in order for productivity to increase in the presence of an established thermocline.

Our chemical data show a shift in the composition of the settling biogenic material from calcium carbonate to opal over the course of the summer. This opal is mostly in the form of diatoms which are able to rapidly take advantage of a new nutrient supply. Because of the short and intermittent nature of these plume events and the length of our collection periods, compositional changes in our samples may not be as extreme as they would be in samples collected over shorter intervals. Thus, plumes may provide a way for colder, more nutrient-rich water to move seaward for distances of 100-400 km, thereby greatly extending the zone influenced by coastal upwelling.

ACKNOWLEDGEMENTS

We would like to thank the reviewers, Robert Collier, G. Ross Heath, and Mitch Lyle for their suggestions and time. Many thanks to Heidi Powell for nutrient analyses, Jim Robbins for AA analyses, and Bobbi Conard for computer data processing. Special thanks to L.C. Breaker and R.P. Gilliland of the National Environmental Satellite Service, Redwood City, California, for satellite images and to Robert Smith and Jane Huyer who graciously shared them with us. Thanks to Mary Jo Armbrust and Lynn Dickson who typed the manuscript. Portions of this work were supported by Sandia National Laboratories contracts 16-5276 and 73-7808.

REFERENCES

Breaker, L.C. and Gilliland, R.P., 1981, A satellite sequence on upwelling along the California coast, in: "Coastal Upwelling," F.A. Richards, ed., American Geophysical Union, Washington, 87-94.

Dymond, J., Fischer, K., Clauson, M., Cobler, R., Gardner, W., Richardson, M.J., Berger, W., Soutar, A. and Dunbar, R., 1981, A sediment trap intercomparison study in the Santa Barbara Basin, Earth and Planetary Science Letters, 53:409-418.

Fukui, S., 1976, Laboratory techniques used for atomic absorption spectrophotometric analysis of geologic samples, Oregon State University, School of Oceanography, Ref. 76-10.

Griggs, G.B. and Hein, J.R., 1980, Sources, dispersal, and clay mineral composition of fine-grained sediment off central and northern California, Journal of Geology, 88:541-566.

Heath, G.R., 1982, Oceanographic studies through December 1981 at Pacific site W-N, Appendix G in: "Oceanographic Data for Assessing the Sea Disposal of Large Irradiated Structures," D. Talbert, ed., Report SAND 82-1005, Sandia National Laboratories, Albuquerque, 2-10.

Honjo, S., Manganini, S.J. and Cole, J.J., 1982, Sedimentation of biogenic matter in the deep ocean, Deep-Sea Research, 29: 609-625.

Imbrie, J. and Kipp, N.G., 1971, A new micropaleontological method for quantitative paleoclimatology: Application to a late Pleistocene Caribbean core, in: "Late Cenozoic Glacial Ages," K.K. Turekian, ed., Yale University Press, New Haven, 71-181.

Karlin, R., Heath, G.R. and Levi, S., 1982, Summary of historical oceanographic and climatological data for west coast potential disposal sites W-N and W-S, Appendix C in: "Oceanographic Data for Assessing the Sea Disposal of Large Irradiated Structures," D. Talbert, ed., Report SAND 82-1005, Sandia National Laboratories, Albuquerque.

Malone, T.C., 1971, The relative importance of nannoplankton and netplankton as primary producers in the California Current system, Fishery Bulletin, 69(4):799-820.

Poelchau, H.S., 1974, "Holocene Silicoflagellates of the North Pacific: Their Distribution and Use for Paleotemperature Determination," Ph.D. Dissertation, University of California, San Diego, 165 pp.

Powell, H.S. and Fischer, K., 1982, Comparison study of bactericides for sediment traps, American Geophysical Union (Abstracts, Fall Meeting 1982), O52B-0.

Reid, J.L., Jr., Roden, G.I. and Wyllie, J.G., 1958, Studies of the California Current system, California Cooperative Fisheries Investigations Reports, 6:27-56.

Russell-Hunter, W.D., 1970, "Aquatic Productivity: An Introduction to Some Basic Aspects of Biological Oceanography and Limnology," The Macmillan Company, Collier-Macmillan Limited, London, 306 pp.

Soutar, A., Kling, S.A., Crill, P.A. and Duffrin, E., 1977, Monitoring the marine environment through sedimentation, Nature, 266: 136-139.

Spencer, D.W., Honjo, S. and Brewer, P.G., 1982, Panama Basin sediment trap intercomparison experiment: August-November, 1979, American Geophysical Union/American Society of Limnology and Oceanography Meeting Abstracts, EOS, 63(3):45.

Traganza, E.D., Conrad, J.C. and Breaker, L.C., 1981, Satellite observations of a cyclonic upwelling system and giant plume in the California Current, in: "Coastal Upwelling," F.A. Richards, ed., American Geophysical Union, Washington, 228-241.

van Bennekom, A.J. and van der Gaast, S.J., 1976, Possible clay structures in frustules of living diatoms, Geochimica et Cosmochimica Acta, 40:1149-1152.

Weliky, K., 1982, "Clay-organic Associations in Marine Sediments: Carbon, Nitrogen and Amino Acids in the Fine-grained Fractions," M.S. Thesis, Oregon State University, Corvallis, 166 pp.

DOWNWARD TRANSPORT OF PARTICULATE MATTER IN THE PERU COASTAL UPWELLING: ROLE OF THE ANCHOVETA, *ENGRAULIS RINGENS*

Nick Staresinic, John Farrington, Robert B. Gagosian,
C. Hovey Clifford, and Edward M. Hulburt

Woods Hole Oceanographic Institution
Woods Hole, Massachusetts, 02543, U.S.A.

ABSTRACT

Anchoveta fecal material links production and sedimentation of particulate matter, including organic carbon, organic nitrogen and diatom frustules, in the Peruvian coastal upwelling ecosystem. Long-term fluctuations in anchoveta stock could, therefore, affect the quantity and composition of particulate matter input to the benthos and thereby to the sediment record.

INTRODUCTION

Downward transport of surface-produced particulate matter ties pelagic ecosystem processes to benthic material cycles and is currently an active topic of research in biological, chemical and geological oceanography (Reynolds, Wiseman and Gardner, 1980). Sources of marine particulate matter, as well as several pathways by which this material may reach the sediments, are schematically illustrated in Fig. 1. Contributions of carcasses of large fish and whales (Isaacs and Schwartzlose, 1975; Jannasch and Wirsen, 1977) and of parcels of plant detritus such as *Sargassum* (Menzies, Zanefeld and Pratt, 1967) and *Thalassia* (Wolff, 1976) may be locally important in the delivery of particulate matter to depth but are not readily quantified. The more abundant products of lower trophic levels, e.g., phytoplankton detritus (Smayda, 1970) and carcasses (Weikert, 1977), molts (Martin, 1970), and fecal pellets (Osterberg, Carey and Curl, 1963; Turner and Ferrante, 1979) of zooplankton, also contribute to the downward flux of particulate matter in aquatic ecosystems and have been collected with plankton nets (Ferrante and Parker, 1977; Wheeler, 1975), large

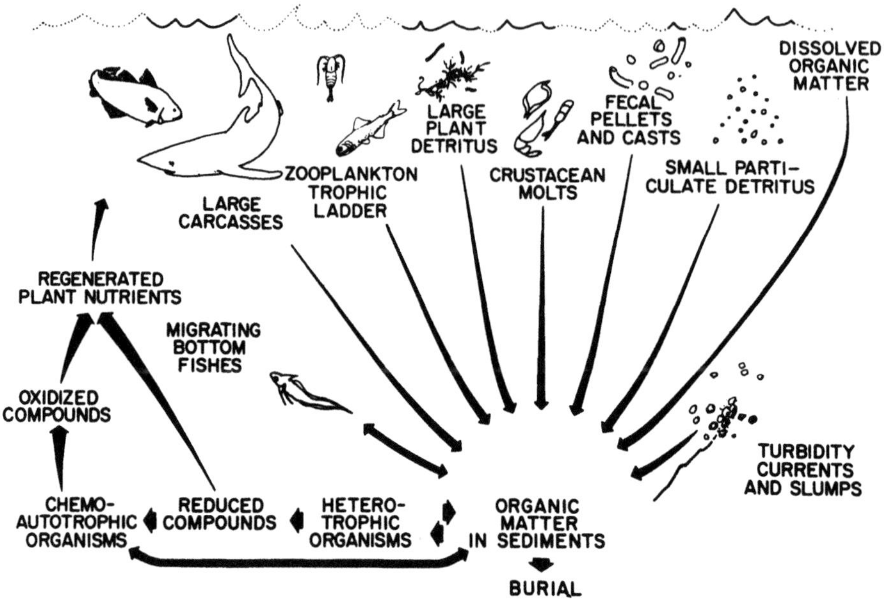

Fig. 1. Some sources of surface-produced particulate matter and pathways by which they may reach the deep sea.

volume *in situ* pumps (Bishop, Ketten and Edmond, 1978) and sediment traps (Moore, 1931; Steele and Baird, 1972; Hargrave and Taguchi, 1978; Honjo, 1978). Of these smaller particles, much attention has focused on zooplankton fecal pellets. The combination of relatively high rates of production (Marshall and Orr, 1955a; Paffenhöfer and Knowles, 1979) and sinking (Smayda, 1969; Small, Fowler and Ünlü, 1979) with high concentrations of biogenic compounds (Volkman, Corner and Eglinton, 1980), various trace elements (Fowler, 1977), certain fallout radionuclides (Cherry et al., 1975; Higgo et al., 1977), and some organic pollutants (Elder and Fowler, 1977; Prahl and Carpenter, 1979), makes them potentially more important in marine biogeochemical cycles than other particles. Zooplankton fecal pellets have commonly been found in sediment trap collections for a variety of marine environments (Steele and Baird, 1972; Honjo, 1978) and sometimes are the dominant identifiable particle collected (Prahl and Carpenter, 1979; Knauer, Martin and Bruland, 1979). However, in high-productivity coastal upwelling areas such as off Peru (Strickland, Eppley and Mendiola, 1969; Zuta and Guillén, 1970; Ryther et al., 1971;), zooplankton are often less important in pelagic ecosystem dynamics than various species of clupeids. During some seasons, zooplankton biomass in the Peruvian upwelling may be only a fraction of a percent of that of the southern anchovy, *Engraulis ringens* (Walsh, 1975). Accordingly, anchoveta may play a relatively greater part in functional aspects of upwelling ecosystems, such as grazing and the regeneration of plant nutrients (Whitledge and Packard, 1971; Whitledge, 1978). Observations from coastal Peru, reported herein, suggest that anchoveta also play an important role in the rapid transport of surface-produced material to depth.

MATERIALS AND METHODS

Free-drifting sediment traps (FST's) (Staresinic, 1978; Staresinic et al., 1978) were used to collect sinking particulate matter in the upwelling system located equatorward of Punta Santa Ana, Peru (ca. 15°S) during austral fall cruises in 1977 ("Wecoma", leg 2) and 1978 ("Knorr" 73-2 and 3). Hydrographic and productivity data for 1977 have been presented by Barber et al. (1978) and by Paul and MacIsaac (1980); those for "Knorr" 73-2 have been reported by Gagosian et al. (1980). A description of the sampling program as well as a more general discussion of the trap data appears in Staresinic (1978; 1982) and in Staresinic et al. (1982). Basically, traps were set at the base of the euphotic zone (11 to 15 m) and at 50 m for periods up to 10 hr. Poisons were not used for relatively short deployments in water about 15°C. Collected material was filtered on precombusted glass-fiber filters and frozen until returned to the laboratory for analysis.

Fragments of anchoveta feces were identified as such by L. Flores (Instituto del Mar del Peru, Callao) by comparison with material observed in large aquaria containing anchoveta. At a 1977 station at which this material was collected, 12 fresh fragments were gently removed prior to filtration and their sinking rates measured aboard ship in a 21-cm graduated cylinder filled with surface seawater of 16°C, 35.03°/₀₀ salinity. Particulate organic carbon and nitrogen content of dried fragments was measured in 1978 on a total of 56 fragments analyzed in batches of 7 to 15 with a Perkin-Elmer CHN analyzer. Protein was measured by the method of Packard and Dirtch (1975).

Diatom species contained in fragments were identified by Hulburt using scanning electron micrographs taken by C.C. Jehanno at the Centre des Faibles Radioactivités, Gif-sur-Yvette, France. Diatom counts were made in eosinophil chambers by light microscopy on aliquots of each of 11 formalin-preserved fragments disaggregated by vortex blending in distilled water. Dilutions were adjusted to insure over 100 individuals of the dominant, *Thallasionema nitzschiodes*, per chamber.

RESULTS AND DISCUSSION

Fragments of the fecal material of the southern anchovy, conspicuous in FST collections from the coastal upwelling area, were cylindrical in shape and up to 1 mm in diameter and 2 mm long, but they were observed underwater as cylindrical strings over 1 cm in

Table 1. Downward flux of anchoveta fecal fragments and their contribution to POC flux at KNORR 73/2 and 3 FST stations.

FST No.[a]	FST Depth (m)	Downward Flux $[m^{-2}(12\ hr)^{-1}]$ Fragments (numbers)	Fragment POC (mg)[b]	Total POC (mg)[c]	Percentage Total POC Flux
4	19	640	11.5	260.4	4.4
5	15	923	16.6	255.6	6.5
6	10	101	1.8	255.6	0.7
7	23	700	12.6	207.6	6.0
8	52	186	3.4	140.4	2.4
9	14	217	4.0	164.4	1.7
10	14	815	14.6	313.2	4.7
11	52	1267	22.8	130.8	17.4
12	11	0	0	203.5	0
13	53	1490	26.9	165.5	16.2
14	11	0	0	343.2	0
15	53	0	0	73.2	0
16	11	0	0	224.4	0
17	53	0	0	208.8	0
18	11	0	0	350.4	0
19	53	516	9.2	111.6	8.3
20	36	0	0	208.2	0
21	53	0	0	66.0	
25	30	2569	46.2	425.0	12.2
26	30	1118	20.1	289.1	6.7

[a] Free Drifting Sediment Trap (FST) deployment information in Staresinic (in prep.).
[b] Calculated with mean POC/fragment = 18.4 μgC (Table 2).
[c] From Staresinic (in prep.).

length. Recovery of the sample from the traps might have broken these strings. Alternatively, the fragments might have been the dominant form present but were not noticed by divers.

No fragments were found near Punta Santa Ana in 1977 when primary productivity was atypically low. In those samples material consisted mainly of the fecal rods of an unidentified animal (Staresinic, 1982). A few were collected, however, during a diatom bloom sampled off Punta Azua (Staresinic, 1978). In 1978, fragments of anchoveta feces were particularly abundant in samples from shelf stations within the region of persistent coastal upwelling. Overall, fluxes varied from 0 to over 2,000 fragments m^{-2} $(12\ hr)^{-1}$ (Table 1). With the exception of FST 13, fragments were not found at stations beyond the shelf break. This is largely consistent with the area distribution of adult anchoveta (Jordan, 1971).

Still-water sinking rates of several fragments collected in 1977 varied from 0.8 to 2.3 cm s^{-1} (691 to 1987 m d^{-1}) and averaged about 1.3 cm s^{-1} (ca. 1100 m d^{-1}). These rates are about an order of magnitude higher than those generally reported for the fecal pellets of zooplanktonic crustacea (Smayda, 1970; Small et al., 1979); they are also about twenty times higher than the sinking rate assigned anchoveta "pseudofeces" in the detrital components of a simulation model of particulate matter dynamics in the upwelling ecosystem (Walsh, 1975). If measured sinking rates apply under *in situ* conditions, they would insure that fragments produced over the shelf would reach the bottom (200 to 400 m) in less than a day.

Analysis of 56 fragments of anchoveta fecal material in groups of 7 to 15 averaged 286 μg (dry), 18.4 μgC, 1.9 μgN, and 35 μg protein per fragment (Table 2). The mean C:N ratio of anchoveta fecal material, 9.7, is higher than that of fresh plankton (ca. 5.4, Parsons, Takahashi and Hargrave, 1979) and similar to that of surface sediments underlying the upwelling area (ca. 9-10, Henrichs, 1980). In terms of bulk composition, an average anchoveta fecal fragment is roughly equivalent to several hundred "typical" adult copepod fecal pellets (Turner, 1977; 1979; Honjo and Roman, 1978; Table 2).

Bulk POC associated with anchoveta fecal fragments accounted for 0 to 17.4% of the total downward POC flux measured with the FST's (Table 1). Higher fragment fluxes in deeper traps reflect anchoveta feeding activity between 15 and 50 m FST's (Staresinic, in prep.). Bulk POC was 6.4% of dry weight. This is lower than values typically reported by zooplankton fecal pellets (Table 1) and may be due to the large amount of siliceous debris contained in the anchoveta material. Microscopic examination of the contents of the fragments revealed two dominant species of planktonic diatoms, the pennate *Thalassionema nitzchioides* (Figs. 3 and 4) and the centric *Thalassiosira eccentrica* (Fig. 2). Mendiola (1971) has reported that both are included in the diet of adult *E. ringens* and that *T. nitzchioides* composed as much as 90% of the gut contents of some of the individuals she examined.

Table 2. Comparison of bulk organic composition and some physical properties of anchoveta fecal fragments and those reported for adult copepod fecal pellets.

	Length (mm)	Width (mm)	Volume ($\mu^3 x 10^5$)	Sinking Rate ($m\ d^{-1}$)	Dry Weight (μg)	Total POC (μg)	PON (μg)	C:N	% POC	Protein[4] (μg BSA)
Anchoveta (1)										
mean:	1.1	1.0	8600	1100	285.6	18.4	1.9	9.7	6.4	35
range:				691-1987	223.0-347.0	14.4-21.1	1.2-2.4	8.7-12.9	6.1-7.0	
Copepod										
Pontella meadi (2)	0.27-0.86	0.06-0.24	10-340	15-153	0.83	0.10	0.02	5.9	12.4	
Calanus finmarchicus (3)	0.60	0.07	230	180-220	10-26					
Acartia clausi (3)	0.20	0.04	2.5	80-150	0.6-4.0	0.10-0.19	0.02-0.04	4.9-6.4		

(1) This paper.
(2) Turner (1977a, 1977b, 1979).
(3) Honjo and Roman (1978).
(4) Bovine serum albumin equivalents (Packard and Dirtch, 1975).

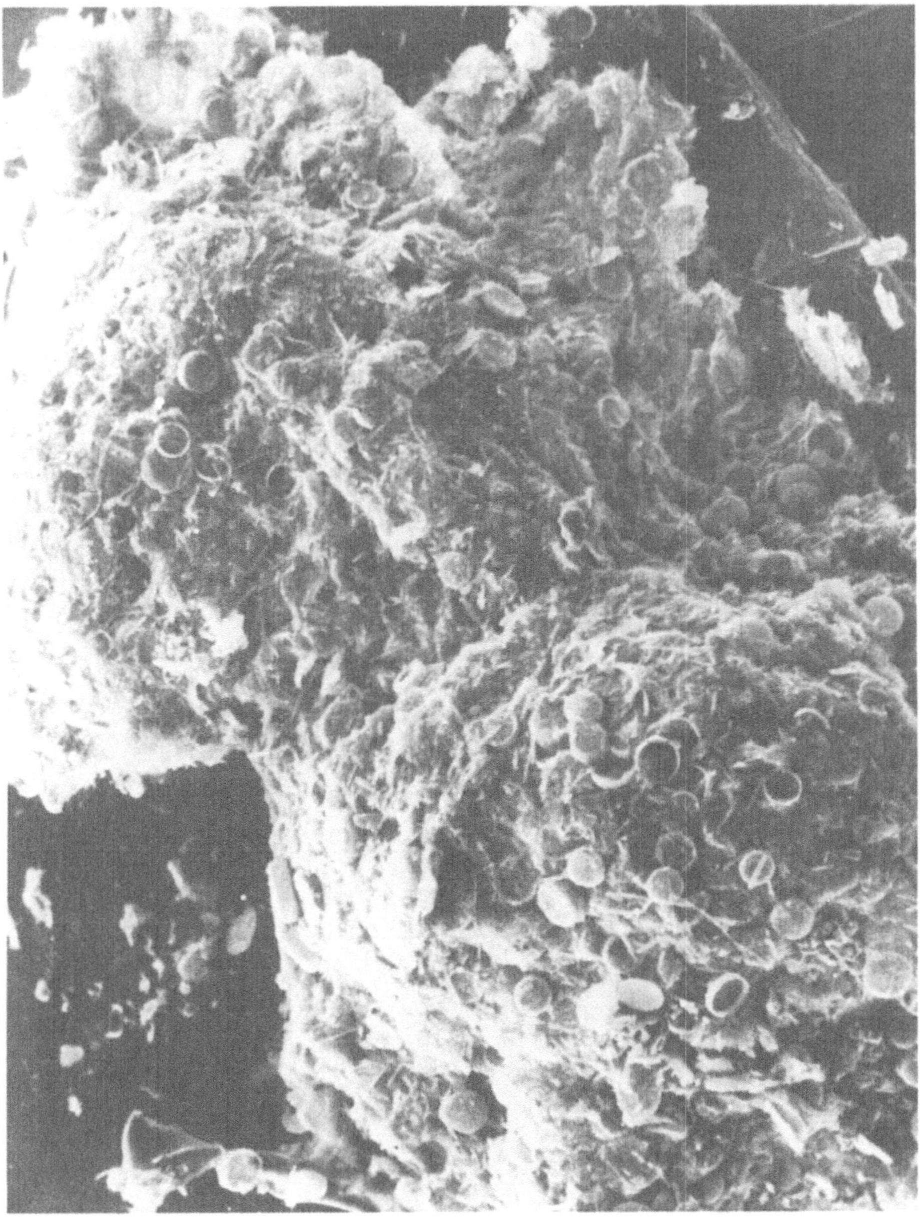

Fig. 2. Scanning electron micrograph of a fecal fragment of the southern anchovy, *Engraulis ringens*, recovered from a sediment trap deployment in the Peru coastal upwelling in 1978. Large centric diatoms, individuals of *Thalassiosira eccentrica*, are about 50 micrometers in diameter. (All photos by C.C. Jehanno, Gif-sur-Yvette.)

Accurate counts of the total number of valves could not be obtained as much of the material in the siliceous fecal fragments was broken. However, valves of *T. nitzschioides* were clearly most abundant in fragments collected in 1978 (Fig. 4); their abundance varied from 414 to 694 x 10^3 and averaged 518 x 10^3 per fragment. Counts for *T. eccentrica*, the next most abundant species, ranged from 8 to 13 x 10^3 and averaged 10 x 10^3 valves per fragment (Table 3). Other species were infrequently encountered. The flux of planktonic diatom valves contained in the fecal fragments of anchoveta ranged from 0 to 14 x 10^8 valves m^{-2} d^{-1} and averaged 8 x 10^8 valves m^{-2} d^{-1} at 50 m (see Tables 1 and 3). Mean flux represents about 10% d^{-1} of the diatom standing-stock expected during the peak of an upwelling bloom (Blasco, 1971) and is of the order of accumulation rates of diatom valves on bottom sediments estimated from two cores taken equatorward of the study area one year earlier (Schuette and Schrader, 1979). This reasonably good agreement may be fortuitous as sedimentation rate estimates from the FST's represent a much shorter time scale than those from sediment cores. The functional relationship between production and sedimentation of sinking particulate matter (Staresinic et al., 1982) suggests that the patchy distribution of diatom remains in the sediment (Schuette and Schrader, 1979) may be in part related to the patchy distribution of anchoveta feeding activity.

Table 3. Abundance of planktonic diatom species in anchoveta fecal material

	Units of 10^3/Fecal Fragment	
	Total Valves	Intact Valves
Thalassionema nitzschioides		
mean	518	254
range	414-694	196-282
Thalassiosira eccentrica		
mean	10	1*
range	8-13	1-3*

* Less than 100 individuals counted per sample.

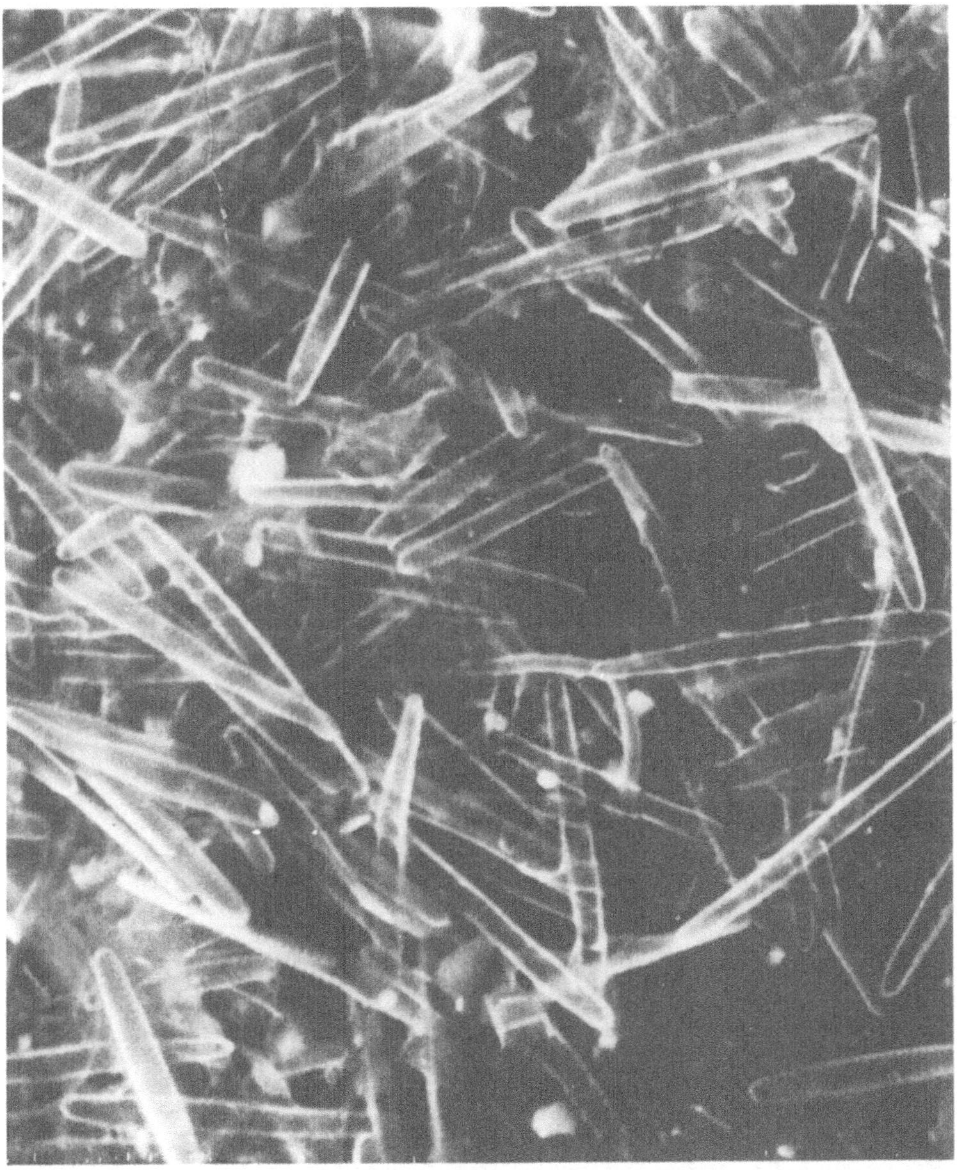

Fig. 3. Section of an anchoveta fecal fragment showing debris of the pennate diatom *Thalassionema nitzschioides*. Each individual is about 40 microns long.

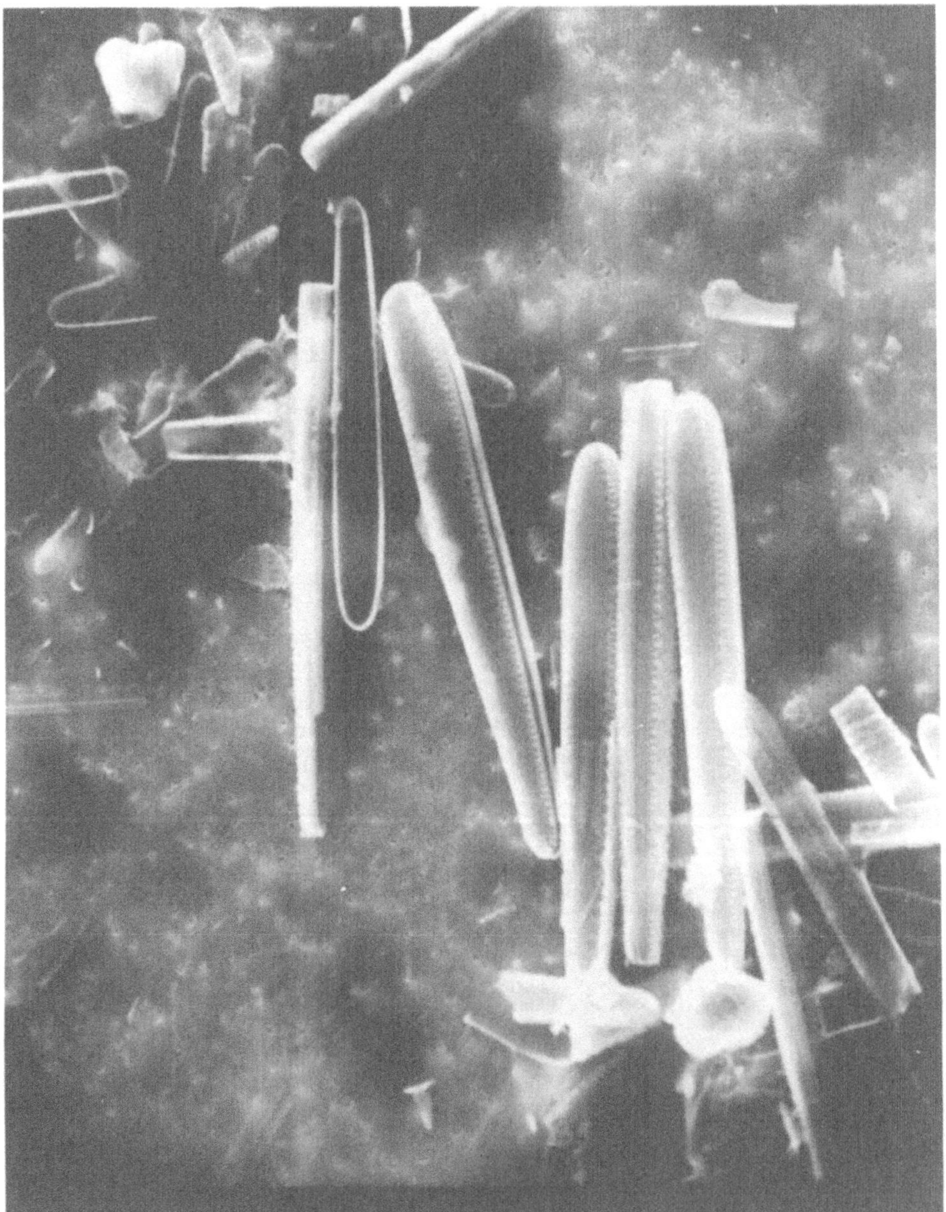

Fig. 4. Close-up of contents of anchoveta fecal fragment showing debris and intact individuals of the pennate planktonic diatom *Thalassionema nitzschioides*.

Of the *T. nitzschioides* present in the fragments, from 196 to 282 x 10^3 per fragment appeared intact and contained visible chloroplasts. Reports of algae passing undigested through the guts of herbivorous zooplankton have been made previously (Marshall and Orr, 1955b; Porter, 1975) and may be related to so-called 'superfluous' feeding at high cell densities (Beklemishev, 1957; 1962). Estimating the volume of *T. nitzschioides* as about 540 micron^3 per cell and converting diatom cell volume to carbon by applying Strathmann's (1967) formula, each intact individual of *T. nitzschioides* could contain about 45 pg carbon. Thus, a mean of 254 x 10^3 intact cells per fragment implies that undigested diatoms could account for as much as 62% of the bulk POC associated with the anchoveta fecal fragments. Undigested cells may also be expected to contribute to the downward flux of a variety of labile organic compounds which would otherwise be degraded by the digestion process (Gagosian et al., this volume). Given the fast sinking rates reported in Table 2, it seems likely that the principal site of remineralization of such compounds would be the benthos. This mode of transport may be an important link in the pathway from production of marine organic matter in the euphotic zone to the formation of petroleum in ancient sediments initially deposited in areas of upwelling (Demaison and Moore, 1980). The relatively short residence time of the fecal matter in the water column, followed by deposition to surface sediments which are primarily anoxic below the upper few centimeters in several areas of the continental shelf and slope off Peru and in other upwelling areas (Demaison and Moore, 1980 and other papers, this volume), should constrain the extent of organic matter transformation prior to deeper burial in sediments. The composition and quantity of organic matter in ancient sediments appears to have a major influence on petroleum formation potential (Tissot and Welte, 1978; Hunt, 1979).

Should some of the phytoplankton remain viable until disaggregation of the fragment, they might eventually be available for resuspension into the water column as part of a "seed" population for a future upwelling event (Tilzer, Paerl and Goldman, 1977). Barber and Smith (1981) have suggested that cross-shelf circulation off Peru would carry sinking individual diatoms to deep water beyond the shelf; their subsequent reinjection into the active surface population would thus be unlikely. Fast sinking anchoveta fecal fragments would insure that at least some of the planktonic diatom population would be retained in shelf sediments. The viability of these cells, however, remains uninvestigated.

These observations clearly suggest the previously unreported importance of anchoveta in linking benthic and pelagic ecosystems in the Peru coastal upwelling through the downward flux of egested particulate matter. Their importance in this process should vary with their abundance. We thus speculate that both the quantity and composition of particulate matter input to the benthos has changed concomitant with the recent dramatic decline of the anchoveta stock off

Peru (Cowles, Barber and Guillén, 1977; Quinn et al., 1978). In particular, we hypothesize that the flux of particulate organic matter to the sediments has decreased since the early 1970's as a result of a relative increase in the contribution of smaller, slower-sinking particles such as phytodetritus and zooplankton fecal pellets. The longer residence time of such particles in the water column could in part account for the increased denitrification noted off coastal Peru by Codispoti and Packard (1980). We recognize that Walsh (1981) has recently submitted an alternate hypothesis. He suggests that the decline in anchoveta stocks off Peru results in less organic matter utilization in the water column by anchoveta thereby making more particulate organic matter available for deposition. We can neither disprove this hypothesis nor prove our hypothesis with the presently available data. Rather, we seek to draw attention to the role of anchoveta feces in the dynamics of the carbon cycle of the Peru upwelling areas.

Schrader (1971) and Honjo (1976) earlier demonstrated the importance of zooplankton fecal pellets in the sedimentation of diatoms and coccolithophores, respectively. The present data indicate that anchoveta fecal matter plays a similar role in the Peruvian coastal upwelling ecosystem. Clupeid species may also function in this manner in other high-productivity ecosystems, such as off northwest Africa (Thiel, 1978) and in the northwestern Adriatic (Piccinetti, 1973).

ACKNOWLEDGEMENTS

We thank J. Volkman, J. Christensen and K. Takahashi for their thoughtful comments. We appreciate the help of R. Chesselet and C.C. Jehanno (Gif-sur-Yvette), L. Flores and C. Delgado (Instituto del Mar del Peru, Callao) and the officers and crew of RV "Knorr", especially J. Cotter.

Financial support was provided by NSF OCE 77-08081, OCE 77-26084, OCE 79-25352 and ONR N00014-79-C-0071. Additional support for Staresinic was kindly made available by the Tai-Ping Foundation and a grant to the Coastal Research Center of the Woods Hole Oceanographic Institution by the Andrew W. Mellon Foundation. Staresinic and Gagosian's participation at this symposium was supported by a grant from NATO.

REFERENCES

Barber, R.T. and Smith, W.O., Jr., 1981, The role of circulation, sinking and vertical migration in physical sorting of phytoplankton in the upwelling center at 15°S, in: "Coastal Upwelling," F.A. Richards, ed., Coastal and Estuarine Sciences 1, American Geophysical Union, Washington, 366-371.

Barber, R.T., Huntsman, S.A., Kogelschatz, J.E., Smith, W.O., Jones, B.H., and Paul, J.C., 1978, Carbon, chlorophyll and light extinction from JOINT-II 1976 and 1977, Coastal Upwelling Ecosystems Analysis Data Report 49, 476 pp.

Beklemishev, K.V., 1957, Superfluous feeding of zooplankton and the question of the sources of food for bottom animals, Trudy Vsesoyunogo Gidrobiologicheskogo Obshchestva, 8:354-358 (in Russian).

Beklemishev, C.V., 1962, Superfluous feeding of marine herbivorous zooplankton, Rapports et Proces-verbaux des Reunions, Conseil Permanent International pour l'Exploration de la Mer, 153:108-153.

Bishop, J.K.B., Ketten, D.R. and Edmond, J.M., 1978, The chemistry, biology and vertical flux of particulate matter from the upper 400 m of the Cape Basin in the southeast Atlantic Ocean. Deep-Sea Research, 25:1121-1161.

Blasco, D., 1971, Composicion y distribucion del fitoplancton en la region del afloramiento de las costas peruanas, Investigacion Pesquera, 35: 61-112.

Cherry, R.D., Fowler, S.W., Beasley, T.M., and Heyraud, M., 1975, Polonium-210: its vertical oceanic transport by zooplankton metabolic activity, Marine Chemistry, 3:105-110.

Codispoti, L.A. and Packard, T.T., 1980, Denitrification rates in the eastern tropical South Pacific, Journal of Marine Research, 38: 453-477.

Cowles, T.J., Barber, R.T. and Guillen, O., 1977, Biological consequences of the 1975 El Niño, Science, 195:285-287.

Demaison, G.J. and Moore, G.T., 1980, Anoxic environments and oil source bed genesis, Organic Geochemistry, 2:9-31.

Elder, D.L. and Fowler, S.W., 1977, Polychlorinated biphenyls: Penetration into the deep ocean by zooplankton fecal pellet transport, Science, 197:459-461.

Ferrante, J.G. and Parker, J.I., 1977, Transport of diatom frustules by copepod fecal pellets to the sediments of Lake Michigan, Limnology and Oceanography, 22:92-98.

Fowler, S.W., 1977, Trace elements in zooplankton particulate products, Nature, 269:51-53.

Gagosian, R.B., Loder, T., Nigrelli, G., Mlodzinska, Z., Love, J., and Kogelschatz, J., 1980, Hydrographic and nutrient data from R/V "Knorr" Cruise 73, Leg 2, February to March, 1978, off the coast of Peru, Woods Hole Oceanographic Institution Technical Report 80-1, 77 pp.

Hargrave, B.T. and Taguchi, S., 1978, Origin of deposited material sedimented in a marine bay, Journal of the Fisheries Research Board, Canada, 35:1604-1613.

Henrichs, S.M., 1980, "Biogeochemistry of Dissolved Free Amino Acids in Marine Sediments," Ph.D. Thesis, Woods Hole Oceanographic Institution/Massachusetts Institute of Technology, 253 pp.

Higgo, J.J.W., Cherry, R.D., Heyraud, M., and Fowler, S.W., 1977, Rapid removal of plutonium from the oceanic surface layer by zooplankton fecal pellets, Nature, 266:623-624.

Honjo, S., 1976, Coccoliths: production, transportation and sedimentation, Marine Micropaleontology, 1:65-79.
Honjo, S., 1978, Sedimentation of material in the Sargasso Sea at a 5367 m deep station, Journal of Marine Research, 36:469-492.
Honjo, S. and Roman, M.R., 1978, Marine copepod fecal pellets: production, preservation and sedimentation, Journal of Marine Research, 36:45-57.
Hunt, J.M., 1979, "Petroleum Geochemistry and Geology," W.H. Freeman and Company, San Francisco, 617 pp.
Isaacs, J. and Schwartzlose, R., 1975, Active animals of the deep sea floor, Scientific American, 233:85-91.
Jannasch, H. and Wirsen, C., 1977, Microbial life in the deep sea, Scientific American, 236:42-52.
Jordan, R.S., 1971, Distribution of anchoveta (*Engraulis ringens* J.) in relation to the environment, Investigacion Pesquera, 35: 113-126.
Knauer, G.A., Martin, J.H. and Bruland, K.W., 1979, Fluxes of particulate carbon, nitrogen and phosphorus in the upper water column of the northeast Pacific, Deep-Sea Research, 26:97-108.
Marshall, S.M. and Orr, A.P., 1955a, "The Biology of a Marine Copepod," Oliver and Boyd, Edinburgh, 188 pp.
Marshall, S.M. and Orr, A.P., 1955b, On the biology of *Calanus finmarchicus*. VIII. Food uptake, assimilation, and excretion in adult and stage V *Calanus*, Journal of the Marine Biology Association, United Kingdom, 34:495-529.
Martin, J.H., 1970, The possible transport of trace metals via moulted copepod exoskeletons, Limnology and Oceanography, 15: 756-761.
Mendiola, B. de Rojas, 1971, Some observations on the feeding of the Peruvian anchoveta, *Engraulis ringens* J. in two regions of the Peruvian coast, in: "Fertility of the Sea," J.D. Costlow, ed., Gordon and Breach, New York, 417-440.
Menzies, R.J., Zanefeld, J.S. and Pratt, R.M., 1967, The distribution and significance of detrital Turtle Grass, *Thalassia testudinata*, on the deep-sea floor off North Carolina, Deep-Sea Research, 14:111-112.
Moore, H.B., 1931, Muds of the Clyde Sea area III, Journal of the Marine Biology Association, United Kingdom, 17:325-358.
Osterberg, C., Carey, A.G. and Curl, H., 1963, Acceleration of sinking rates of radionuclides in the ocean, Nature, 200:1276-1277.
Packard, T.T. and Dirtch, Q., 1975, Particulate protein-nitrogen in North Atlantic surface waters, Marine Biology, 33:347-354.
Paffenhöfer, G.A. and Knowles, S.C., 1979, Ecological implications of fecal pellet size, production and consumption by copepods, Journal of Marine Research, 37:35-49.
Parsons, T.R., Takahashi, M. and Hargrave, B.T., 1979 "Biological Oceanographic Processes," Pergamon Press, New York, 332 pp.
Paul, J.C. and MacIsaac, J.J., 1980, JOINT-II hydrographic data, RV "Wecoma" cruises, 12 March-16 May 1977, Coastal Upwelling Ecosystem Analysis Data Report 51, 145 pp.

Piccinetti, C., 1973, Considerations preliminaires sur les deplacements de l'Anchois (*Engraulis encrascholus* L.) en Haute et Moyenne Adriatique, Rapport du Commission internationale de la Mer Méditerranée, 21(10):763-766.

Porter, K.G., 1975, Viable gut passage of gelatinous green algae ingested by *Daphnia*, Internationaler Verein für Limnologie, Verhandlungen, 19:2840-2850.

Prahl, F.G. and Carpenter, R., 1979, The role of fecal pellets in the sedimentation of polycyclic aromatic hydrocarbons in Dabob Bay, Washington, Geochimica et Cosmochimica Acta, 43:1959-1972.

Quinn, W.H., Zopf, D.O., Short, K.S., and Kuo Yang, R.T.W., 1978, Historical trends and statistics of the southern oscillation, El Niño, and Indonesian droughts, Fishery Bulletin, 76:663-678.

Reynolds, C.S., Wiseman, S.W. and Gardner, W.D., 1980, An annotated bibliography of aquatic sediment traps and trapping methods, Occasional Publication 11: Freshwater Biology Association, 54 pp.

Ryther, J.H., Menzel, D.W., Hulburt, E.M., Lorenzen, C.J., and Corwin, N., 1971, The production and utilization of organic matter in the Peru coastal current, Investigacion Pesquera, 35: 43-59.

Schrader, H.-J., 1971, Fecal pellets: role in sedimentation of pelagic diatoms, Science, 174:55-57.

Schuette, G. and Schrader H., 1979, Diatom taphocoenoses in the coastal upwelling area off western South America, Nova Hedwigia, 64:359-378.

Small, L.F., Fowler, S.W. and Ünlü, M.Y., 1979, Sinking rates of natural copepod fecal pellets, Marine Biology, 51:233-241.

Smayda, T.J., 1969, Some measurements of the sinking rates of fecal pellets, Limnology and Oceanography, 14:621-625.

Smayda, T.J., 1970, The suspension and sinking of phytoplankton in the sea, Oceanography and Marine Biology, Annual Review, 8:353-414.

Staresinic, N., 1978, "The Vertical Flux of Particulate Organic Matter in the Peru Coastal Upwelling as Measured with a Free-drifting Sediment Trap," Ph.D. Thesis, Woods Hole Oceanographic Institution/Massachusetts Institute of Technology, 255 pp.

Staresinic, N., in prep., Downward flux of bulk particulate organic matter in the Peru coastal upwelling, Journal of Marine Research (submitted).

Staresinic, N., Rowe, G., Williams, A.J., III, and Shaughnessy, D., 1978, Measurement of the vertical flux of particulate organic matter with a free-drifting sediment trap, Limnology and Oceanography, 23:559-563.

Staresinic, N., von Bröckel, K., Smodlaka, N., and Clifford, C.H., 1982, A comparison of moored and free-drifting sediment traps of two different designs, Journal of Marine Research, 40:273-292.

Steele, J.H. and Baird, I.E., 1972, Sedimentation of organic matter in a Scottish sea loch, Memorie dell'Instituto Italiano di Idrobiologia, Supplemento, 20:73-80.

Strathmann, R.R., 1967, Estimating the organic content of phytoplankton from cell volume or plasma volume, Limnology and Oceanography, 12:411-418.

Strickland, J.D.H., Eppley, R.W. and Rojas de Mendiola, B. 1969, Phytoplankton populations, nutrients and photosynthesis in Peruvian coastal waters, Boletin Instituto del Mar del Peru, 2:551-582.

Thiel, H., 1978, Benthos in upwelling regions, in: "Upwelling Ecosystems," R. Boje and M. Tomczak, eds., Springer-Verlag, Berlin, 124-138.

Tilzer, M.M., Paerl, H.W. and Goldman, C.R., 1977, Sustained viability of aphotic phytoplankton in Lake Tahoe (California-Nevada), Limnology and Oceanography, 22:84-91.

Tissot, B.P. and Welte, D.H., 1978, "Petroleum Formation and Occurrence," Springer-Verlag, New York, 538 pp.

Turner, J.T., 1977, Sinking rates of fecal pellets from the marine copepod *Pontella meadii*, Marine Biology, 40:249-259.

Turner, J.T., 1979, Microbial attachment to copepod fecal pellets and its possible ecological significance, Transactions, American Microscope Society, 98:131-135.

Turner, J.T. and Ferrante, J.G., 1979, Zooplankton fecal pellets in aquatic ecosystems, Bioscience, 29:670-677.

Volkman, J.K., Corner, E.D.S. and Eglinton, G., 1980, Transformations of biolipids in the marine food web and in underlying bottom sediments, in: "Biogeochimie de la Matiere Organique a l'Interface Eau-sediment Marin," Colloques Internationaux du Centre National de la Recherche Scientifique, Paris, No. 293:185-197.

Walsh, J.J., 1975, A spatial simulation model of the Peru upwelling ecosystem, Deep-Sea Research, 22:201-236.

Walsh, J.J., 1981, A carbon budget for overfishing off Peru, Nature, 2980:300-304.

Weikert, H., 1977, Copepod carcasses in the upwelling region South of Cap Blanc, N.W. Africa, Marine Biology, 42:351-355.

Wheeler, E.H., 1975, Copepod detritus in the deep sea, Limnology and Oceanography, 12:697-701.

Whitledge, T.E., 1978, Regeneration of nitrate by zooplankton and fish in the northwest Arica and Peru upwelling ecosystems, in: "Upwelling Ecosystems," R. Boje and M. Tomczak, eds., Springer-Verlag, Berlin, 90-100.

Whitledge, T.E. and Packard, T.T., 1971, Nutrient excretion by anchoveta and zooplankton in Pacific upwelling regions, Investigacion Pesquera, 35:243-250.

Wolff, T., 1976, Utilization of sea grass in the deep sea, Aquatic Botany, 2:161-174.

Zuta, S. and Guillén, O., 1970, Oceanografia de las aguas costeras del Peru, Boletin Instituo del Mar Peru, 2:157-234.

VERTICAL TRANSPORT AND TRANSFORMATION OF BIOGENIC ORGANIC COMPOUNDS FROM A SEDIMENT TRAP EXPERIMENT OFF THE COAST OF PERU

Robert B. Gagosian, Gale E. Nigrelli and John K. Volkman

Department of Chemistry
Woods Hole Oceanographic Institution
Woods Hole, Massachusetts 02543, U.S.A.

ABSTRACT

Sediment trap experiments using the lipid biomarker approach were undertaken in the Peru coastal upwelling region to answer three main questions concerning the sources, transport and transformation of organic matter in upwelling regimes. 1. How does the temporal and spatial variability of the vertical flux and composition of particulate matter change as it sinks to the sediment surface? 2. How suitable are the various lipids incorporated into the sedimentary record as indicators of the paleoenvironment of deposition? This involves a determination of how much of these compounds survive biological degradation in the water column and surface sediments. 3. What is the relationship between the organic matter composition of sinking particulate material and biological processes in the water column? The composition and vertical fluxes of the steroid class compounds are used as an example of how these questions may be answered.

It was determined that 1.5-6% of the sterols produced by primary production survive water column and surface sediment degradative processes. These values are approximately twice as high as total organic carbon, five times as high as total lipids, and 50% higher than amino acids from primary productivity. A tenfold increase was found in the sterol flux through the water column at night compared with the day. This nighttime increase was probably due to increased zooplankton activity (e.g., fecal pellet, carcass and molt transport). However, compositional differences between day/night, 14 m and 52 m sediment traps were small reflecting overall similar source inputs. Sedimentary sterols contain a strong phytoplankton (diatoms) source signal. These phytoplankton-derived sterols appear to be delivered to the sediment quickly via sinking anchovy fecal pellets. Zooplankton (copepods and euphausiids) molts, carcasses and fecal pellets make significant contributions to the sedimentary sterol pool as well. In the Peru upwelling area, phytoplankton derived sterols appear to be degraded faster than sterols of terrestrial origin suggesting that this terrestrial material is more resistant to degradative processes in the marine environment than is marine derived material.

INTRODUCTION

The use of source marker organic compounds to reconstruct paleoenvironmental conditions of deposition has been an active area of research in organic geochemistry. Applications of this approach to determine an upwelling "chemical signature" in the sedimentary record have been suggested by Brassell and Eglinton (this volume); Gagosian, Volkman and Nigrelli (in press) and Suess and Thiede (this volume). However, to assess the intensity of a phytoplankton productivity event, it is necessary to understand the oceanic processes responsible for (1) the sources and transport of organic material produced in the surface waters delivered to the sediment and (2) the degree of transformation that occurs before the material is buried in the sedimentary record. Time and spatial variability are clearly important considerations.

One approach to determining material fluxes and the nature of particulate material sinking through the water column to the sediments involves the collection of particulate matter in traps deployed at various depths in the water column. Many investigators (Wiebe, Boyd and Winget, 1976; Bishop et al., 1977; 1980; Bishop, Ketten and Edmond, 1978; Honjo, 1978; 1980; Honjo and Roman, 1978; Spencer et al., 1978; Staresinic, 1978; in prep.; Crisp et al., 1979; Knauer, Martin and Bruland, 1979; Prahl and Carpenter, 1979; Rowe and Gardner, 1979; Brewer et al., 1980; Deuser, Ross and Anderson, 1981; Fellows, Karl and Knauer, 1981; Staresinic et al., this volume), using *in situ* water filtering pumps and sediment traps, have shown that a large part of the organic material reaching the sea floor is transported by large particles (>63 μm) not usually sampled by traditional methods using water bottles. Since transport is rapid, an important flux of less degraded and labile, as well as resistant organic material may be transported into deeper water. This is especially important in upwelling areas where a larger proportion of surface primary productivity can reach the sea floor (Bishop et al., 1978). Large, rapidly-sinking (tens to hundreds of meters per day) fecal pellets from zooplankton and fish or aggregates of fecal material appear to be a major mechanism for the transport of labile material to the ocean floor (Schrader, 1971; Elder and Fowler, 1977; Prahl and Carpenter, 1979; Small, Fowler and Onlu, 1979; Turner and

Ferrante, 1979; Honjo, 1980; Staresinic et al., this volume; Gagosian et al., in press).

The transport of organic matter can be traced in general terms by measuring the organic carbon content of sediment trap material (Wiebe et al., 1976; Honjo, 1978; 1980; Knauer et al., 1979; Hinga, Sieburth and Heath, 1979; Deuser and Ross, 1980; Deuser et al., 1981; Staresinic, in prep.). However, knowledge of the distribution of specific "biomarker" lipid class compounds permits a more detailed examination of the sources, transport and transformation processes acting on organic matter in the water column and surface sediments (Crisp et al., 1979; Prahl, Bennett and Carpenter, 1980; Tanoue and Handa, 1980; Wakeham et al., 1980; DeBarr, Farrington and Wakeham, in press; Gagosian, Smith and Nigrelli, 1982; Gagosian et al., in press; Lee and Cronin, 1982; Repeta and Gagosian, 1982; in press; Wakeham, in press; Wakeham, Farrington and Volkman, in press).

Our approach is to compare the lipid distributions of (1) particulate material collected in sediment traps (large particles usually >20 µm), (2) particulate material from bulk seawater samples (<20 µm particles), (3) living phytoplankton and zooplankton, (4) fecal pellets and molts of zooplankton, (5) fecal pellets of fish, and (6) biota living in the sediments with those lipids found in the underlying sediments.

In this manuscript, we report our results for the steroid class compounds in particulate matter collected in the Peru coastal upwelling region using day/night pairs of free drifting sediment traps (FST's) and sediment box cores. We wish to use these results as an example of how the "biomarker" approach can help us to better understand: (1) the temporal and spatial variability of the vertical flux and composition of particulate matter as it sinks to the sediment surface, (2) how much of this material survives biological degradation in the water column and surface sediments before burial in the sedimentary record to determine the suitability of specific lipids as source markers in deeper sediments, and (3) the relationship between the organic matter composition of sinking particulate material and biological processes in the water column. The use of sediment trap experiments to determine biomarker sources in coastal sediments off Peru is discussed in Gagosian et al. (in press) for the steroid class compounds, and in Repeta and Gagosian (in press) for the carotenoid pigments. Other lipid biomarker studies in this region are reported by Volkman et al. (in press) for sedimentary hydrocarbons and ketones, and in Lee and Cronin (1982) and Wakeham et al. (in press) for lipid-soluble amino acids and for fatty acids and their derivatives in sediment trap particulate material.

Although lipids represent only a small portion of biologically produced organic matter in seawater, the lipid class compounds are key biochemicals in marine organisms. They play a major role in energy storage and mobilization, in reproduction, as membrane structural components and in the regulation of metabolic processes. Since these compounds play such a critical role in the life cycle of marine organisms, and many of them are stable for long periods of time, they serve as excellent markers of the biota living above and on the sea floor.

We have chosen one class of the biogenic lipid compounds, the sterols, as an example of this biomarker approach. In addition to those listed above for lipids in general, the steroids were chosen for several other reasons. They are quite stable molecules, many being found in ancient sediments (Simoneit, 1978), and there are several sources of sterols in the ocean (Table 1). Frequently, many of these sources are specific such as the diatoms *Cyclotella nana (Thalassiosira pseudonana)* and *Nitzschia closteria (Cylindrotheca fusiformis)* which contain mainly one sterol (>98%), 24-methylcholesta-5,22E-dien-3β-ol (Rubinstein and Goad, 1974; Orcutt and Patterson, 1975; Ballantine, Lavis and Morris, 1979; Volkman, Eglinton and Corner, 1980b). Dinosterol (4α,23,24R-trimethyl-5α-cholest-22E-en-3β-ol) has only been found in dinoflagellates (Alam et

Table 1. Some sterol biological markers found in marine sediments

Organisms	Sterols Produced
Diatoms, coccolithophorids	(24S)-24-methylcholesta-5,22E-dien-3β-ol
Diatoms	24-methylcholesta-5,24(28)-dien-3β-ol
Crustacea	cholest-5-en-3β-ol*
Crustaceans and phytoplankton	cholesta-5,22E-dien-3β-ol
Dinoflagellates	4α,23,24R-trimethyl-5α-cholest-22E-en-3β-ol (dinosterol)
Coelenterates	22,23-methylene-23,24-dimethylcholest 5-en-3β-ol (gorgosterol)
Land plants	(24R)-24-ethylcholest-5-en-3β-ol (β-sitosterol)

* Although most crustaceans produce cholesterol almost exclusively from sterols, this sterol is also produced to some extent by a wide variety of other organisms.

al., 1979; Boon et al., 1979). Several species of coelenterates have been found to contain gorgosterol as their major sterol (Steudler, Schmitz and Ciereszko, 1977). Because of this specificity of structure, sterols have been used successfully for taxonomic purposes (Nes and McKean, 1977). Sterols have also been used to provide information about the marine or terrestrial origin of sedimentary organic matter (Huang and Meinschein, 1976; Lee, Farrington and Gagosian, 1979). There are very few, if any, species of bacteria that are known to biosynthesize 4-desmethylsterols via the mevalonic acid → squalene → lanosterol sequence (Nes et al., 1980). Thus, the interpretation of sedimentary sterol origins is less complicated than similar studies involving compounds important in bacterial metabolism, such as fatty acids and amino acids.

The steroid hydrocarbon tetracyclic structure forms a relatively stable nucleus which incorporates optically active centers and functional groups such as alcohols (sterols), ketones (stenones and stanones), olefinic linkages (sterenes) and aromatic ring structures, so that we can study the mechanisms of transformation reactions involving these functional groups at a molecular level. The functional groups can undergo oxidation, reduction, structural rearrangement and isomerization reactions. The elucidation of these reaction pathways allows us to trace back from the "transformed" sterol to the original sterol molecule of biological origin. Such structural properties have been used to study biogeochemical and diagenetic processes in Recent marine sediments (Dastillung and Albrecht, 1977; Gagosian and Farrington, 1978; Gagosian et al., 1980a; Taylor, Smith and Gagosian, 1981), ancient sediments (MacKenzie et al., 1982 and references

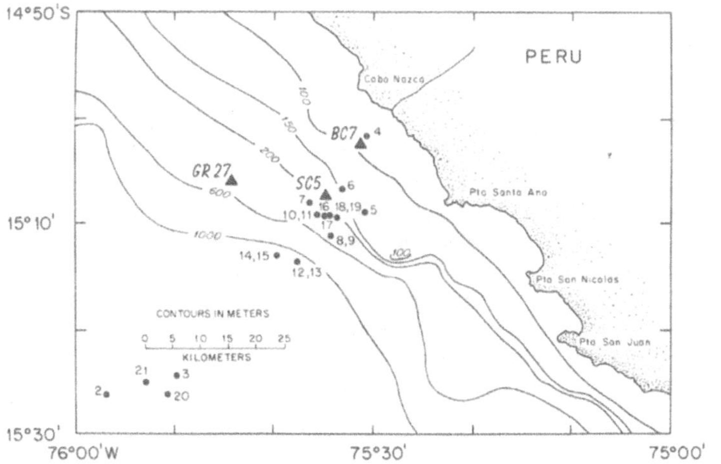

Fig. 1. Locations of sediment samples (•) and sediment trap deployments (▲) during RV "Knorr" cruise 73, leg 2, February to March, 1978 off the Peruvian coast.

therein), and in seawater (Gagosian and Heinzer, 1979). The carbon number and isomer distributions of the transformed sterols present in ancient sediments (e.g., the steranes) have been used in petroleum source-rock/oil correlations (Tissot and Welte, 1978; Seifert and Moldowan, 1978). In addition, the isomerization rates of optically active carbons in the steranes can be applied to the assessment of the extent of thermal maturation of the host sediment (MacKenzie et al., 1982).

EXPERIMENTAL

Sampling and Extraction

Seawater, free drifting sediment trap (FST) and sediment samples were collected during February-March, 1978 on RV "Knorr", cruise 73, leg 2. Detailed hydrographic data for this cruise can be found in Gagosian et al. (1980b). Additional sediment trap samples containing primarily anchoveta fecal pellets were collected in March, 1981 on RV "Atlantis II" cruise 108, leg 3. Locations of the sediment traps and sediments collected on the 1978 cruise are shown in Fig. 1. A schematic of the FST used in this study is shown in Fig. 2. A more detailed description of the FST's (a pair of 41 cm diameter cylinders, 0.26 m^2 total collecting area) is given in Staresinic (in prep.). Day and night trap deployments were set at 14 m (the base of the euphotic zone) and 52 m (below the seasonal thermocline). Details of the sediment samples [SC5 collected with a Soutar corer (0-1 cm) from 15°05.7'S, 75°35.1'W and GR27 collected with a grab sampler (0-2 cm) from 15°05.7'S, 75°43.9'W] can be found in Volkman et al. (in press) and Henrichs (1980). The sediments were organic-rich diatomaceous oozes (∿3-5% organic carbon) located in an oxygen minimum zone (<0.1 ml/l seawater) which either eliminates benthic fauna or restricts it to a few benthic metazoans such as polychaete worms of the family *Ciratulidae*. Filaments of *Thioploca* bacteria (Gallardo, 1977) were conspicuous in each of the surface sediments. A description of the extraction of the sediments and sediment trap material can be found in Volkman et al. (in press) and Wakeham et al. (in press), respectively. Anchoveta fecal pellets were sonically extracted twice with toluene/methanol. Water samples were collected using glass Bodman bottles (Gagosian et al., 1979), transferred to stainless steel and aluminum containers with polypropylene tubing, and extracted with hexane as previously reported (Gagosian and Nigrelli, 1979). Methods used to determine chlorophyll a and particulate organic carbon (POC) can be found in Gagosian et al. (1980b).

Analysis

The lipid extract was fractionated into constituent lipid classes by column chromatography on silica gel (70-230 mesh, deactivated with 5% distilled water). Fourteen fractions were collected by

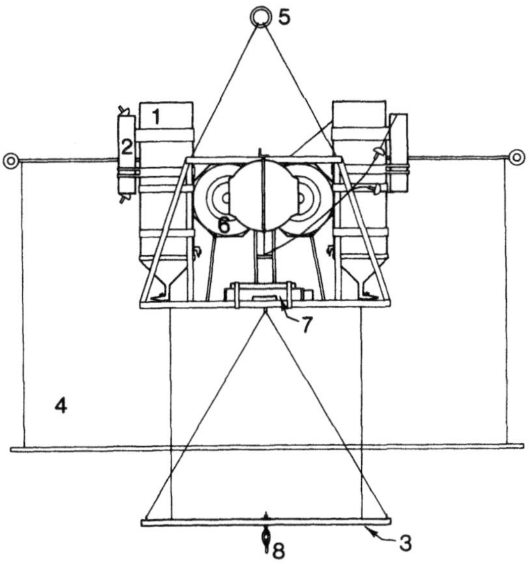

Fig. 2. Schematic diagram of the free-drifting sediment trap (FST). 1-cylindrical PVC sediment traps, 41 cm inside diameter, 2-two 10-liter Niskin bottles, 3-damping plate to damp heave by increasing the system's effective mass, 4-canvas window-shade drogue, 5-bale ring for attachment of surface tether and hoisting, 6-four 17-inch glass flotation spheres with plastic hard hats, 7-two Williams electrolytic timed-release/pingers, 8-disposable weights (Staresinic, in prep.).

elution with mixtures of hexane, toluene, and ethyl acetate. Fraction VII (15% ethyl acetate in hexane) contained the 4-methylsterols and Δ4-3-ketosteroids; fraction VIII (20% ethyl acetate in hexane) contained the 4-desmethylsterols. These fractions were combined for the seawater and sediment trap analyses but kept separate for the sediment analyses. 5α(H)-Cholestane was added as an internal standard. Blanks showed negligible amounts of sterols and steroid ketones.

Sterols were acetylated with acetic anhydride in pyridine and analyzed by high resolution glass capillary gas chromatography on a Carlo Erba Fractovap 2150 gas chromatograph using deactivated (by persilylation; Grob and Grob, 1980) SE-52 coated columns (15 m and 20 m x 0.32 mm i.d.). The helium gas flow was 1.5-2 ml per min. The injector and detector were operated at 300°C. The sediment trap and seawater steryl acetates (1-2 μl) were injected without splitting at 80°C and the oven was then heated to 265°C at 3.5°C/min. The sediment and anchovy fecal pellet sterol acetates were injected at 25°C, and after 1 minute the oven was rapidly heated to 180°C and then to 300°C at 2°C/min. Gas chromatograms of the 4-desmethylsterols of sediment samples GR27 and SC5 are shown in Fig. 3. Quantification of the individual steroids was based on the response each compound gave relative to that of 5α(H)-cholestane (Gagosian and Heinzer, 1979).

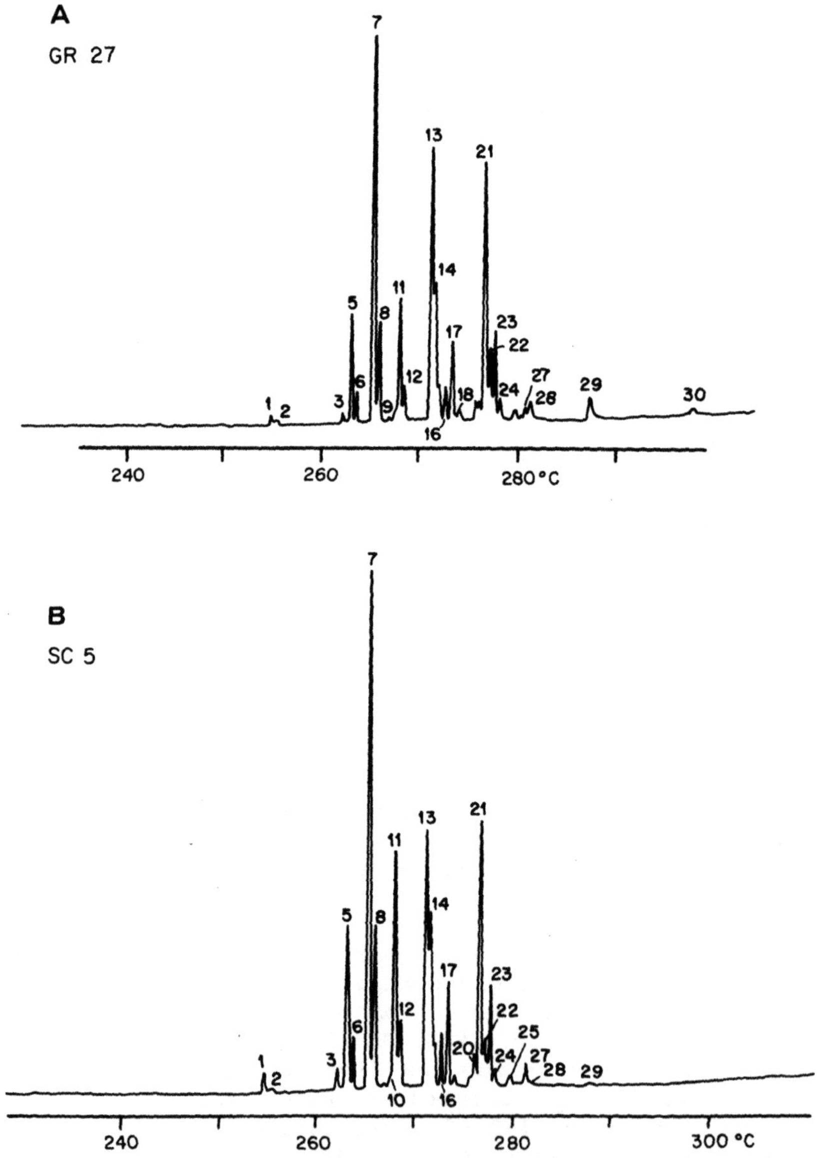

Fig. 3. Glass capillary gas chromatograms of free 4-desmethylsterols (acetates) from surface sediments (a) GR27 and (b) SC5. Peak numbers refer to the structures in Table 5.

Structural identification of the free sterols (as acetates) was based on relative retention times obtained by high resolution glass capillary gas chromatography and comparison of mass spectra obtained by gas chromatography/mass spectrometry (GC/MS) with published mass spectra and mass spectra obtained from authentic standards. Electron impact mass spectra were obtained using a Varian Aerograph 1400 gas chromatograph equipped with an SE-52 glass capillary column (25 m x 0.30 mm i.d.) interfaced with a Finnigan 1015C quadrupole mass spectrometer. The carrier gas (He) flow rate was 2-3 ml/min. The mass spectrometer was scanned linearly from 40-550 or 600 atomic mass units (amu) in 1 second cycles. Data acquisition used a Finnigan INCOS 2300 data system. Further analytical details can be found in Gagosian et al. (1982; in press).

RESULTS AND DISCUSSION

Oceanographic Setting

The study area at 15°S off the Peru coast (Fig. 1) exhibits typical oceanographic parameters for a major upwelling area. Active upwelling was underway during our February-March, 1978, expedition as shown in the sigma-t section from FST 4 to FST 20 locations (Fig. 4). Upwelling extends to about 15 km offshore and appears to be drawn

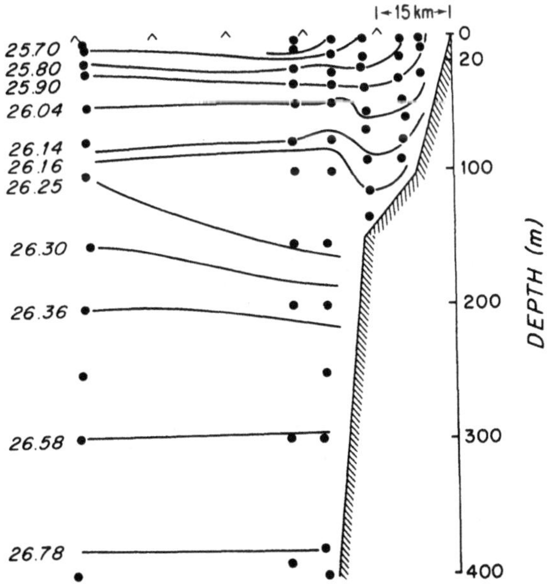

Fig. 4. Sigma-t section perpendicular to the Peruvian coast at about 15°S. Active upwelling, indicated by the intersection of isopycnals with the surface, extended to at least 15 km offshore.

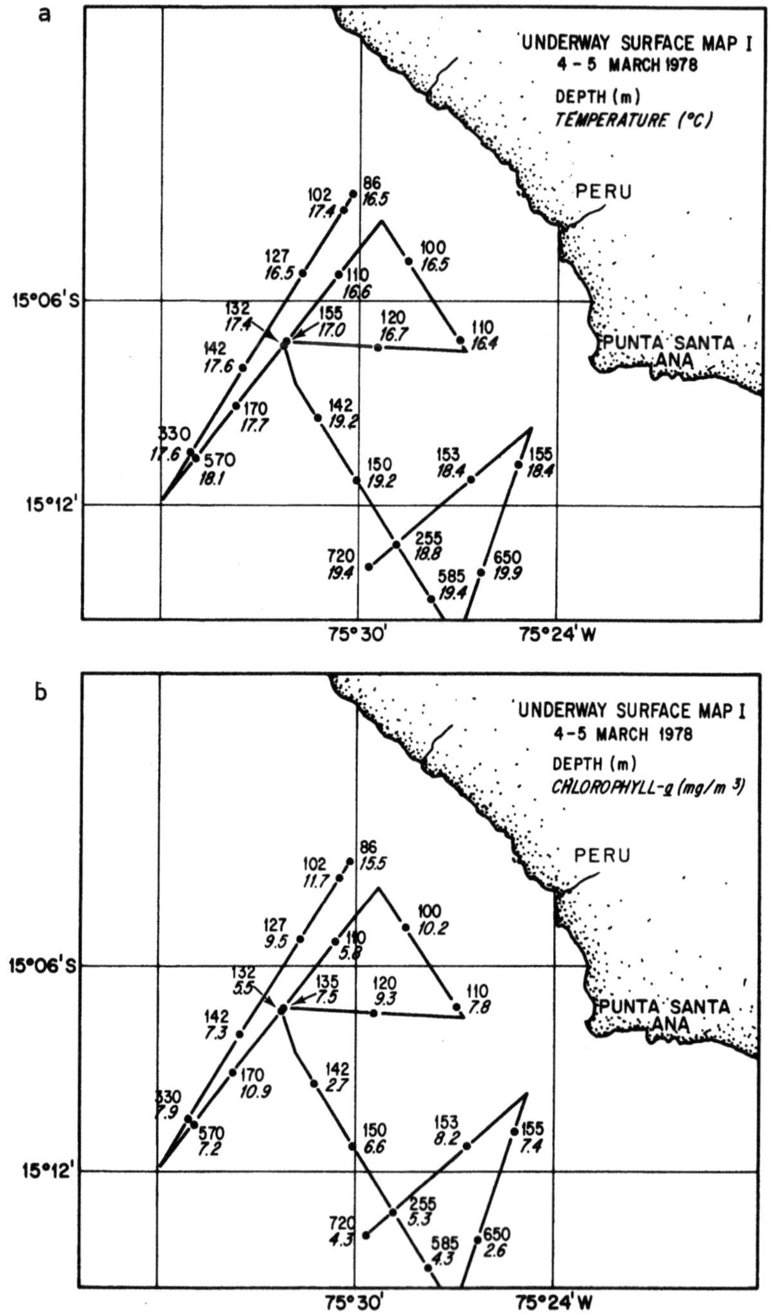

Fig. 5. (a)(b) Underway maps of surface temperature (a) and chlorophyll a (b) for 4-5 March 1978. The upper value at each point gives the depth in meters.

from less than 75 m depth (Smith et al., 1971; Zuta, Rivera and Bustamante, 1978). Brink et al. (1981) have determined that at this depth the mean flow over the shelf during this time of year is poleward. This suggests that equatorial subsurface water was probably the source of the upwelled water.

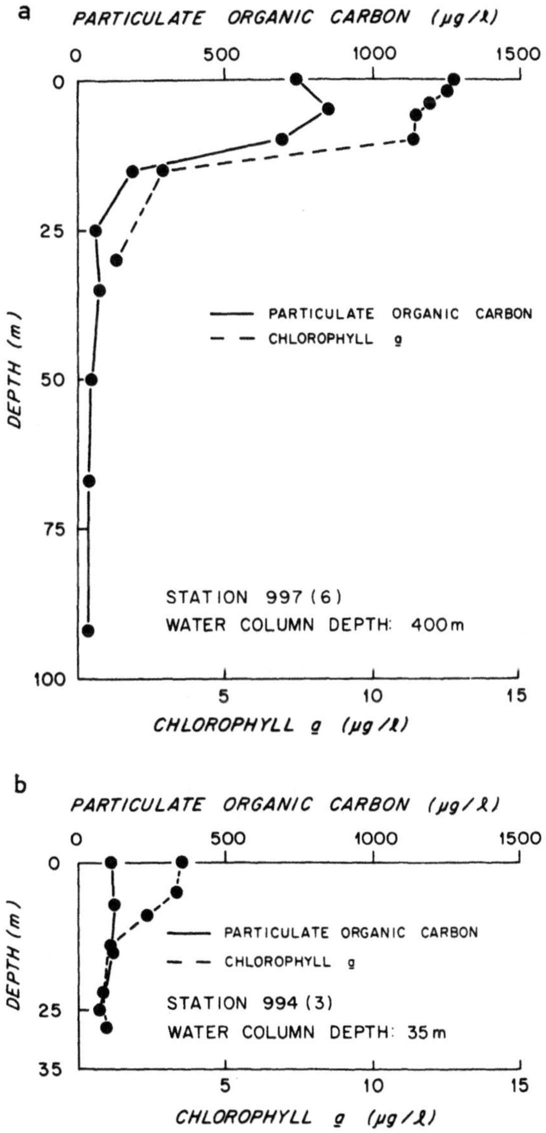

Fig. 6. (a) Particulate organic carbon and chlorophyll a values from the site of FST 8-11 deployments (15°09.9'S; 75°36.0'W).

(b) Particulate organic carbon and chlorophyll a values from a shallow inshore station 22 km northeast of the FST 8-11 deployments (15°04.8'S; 75°25.1'W).

Surface temperature and chlorophyll a underway maps show the gross features of the study area just before FST's 8-11 were deployed (Fig. 5a, 5b). Nearshore surface temperatures increased from 16.5°C to about 17.5°C at 20 km offshore. A sharper southeasterly increase to 20°C was observed offshore of Punta Santa Ana. Surface chlorophyll a varied between 5 to 10 μg/ℓ over most of the study area. Concentrations as high as 27.8 μg/ℓ were recorded at some inshore stations (Gagosian et al., 1980b) but generally were 5-10 μg/ℓ. Chlorophyll a decreased southeasterly toward Punta Santa Ana and reached ca. 2.5 μg/l about 10 to 15 km offshore. Phytoplankton patchiness was indicated by a few scattered high and low chlorophyll a values (Gagosian et al., 1980b). Dense surface patches of dinoflagellates *Gymnodinium sp.* were frequently observed offshore, as earlier reported by Dugdale et al. (1977) but high concentrations of diatoms characterized most shelf station samples. Chlorophyll a: phaeopigment ratios were high (>10) and nutrients were relatively low in the underway study area. Nitrate was less than 2 μM and silicate was approximately 2.5 μM. Overall, the underway map data depict the study area as a system in which upwelled water has been at the surface long enough for phytoplankton to increase their biomass through utilization of available nutrients. Active coastal upwelling was prevalent, as shown by the high phytoplankton primary production which averaged 4 g $C/m^2 \cdot day$ and reached 6 g $C/m^2 \cdot day$ at some inshore stations (Gagosian et al., 1980b). Surface chlorophyll a concentrations in excess of 20 μg/l were common. Fig. 6a shows the high chlorophyll a (∿13 μg/l) and particulate organic carbon (∿750 μg/l) values in the euphotic zone at the deployment location of FST's 8-11. In contrast, Fig. 6b shows the low chlorophyll a (∿3 μg/l) and POC (∿100 μg/l) values from a station 22 km further inshore at a water depth of 35 m. The plankton biomass is quite low due to the lag time for growth in freshly upwelled waters. Diatoms characterized most of the samples except on one occasion (FST 18 and 19 sites) when chlorophyll a values reached in excess of 75 μg/l due to a dense dinoflagellate bloom (J. Kogelschatz, pers. comm.). This high productivity in the Peruvian upwelling area has, until recently, supported a large population of anchoveta (*Engraulis ringens*).

General Characteristics of Sediment Trap Data

The deployment data for FST's 8-11 are shown in Table 2. The mean drift trajectories for all four traps were approximately the same, northwest, with a mean drift speed of 4-12 cm/sec. The mean drift speeds for all traps (FST's 4-21) varied from 5-35 cm/sec (Staresinic, in prep.). Overall, there was no clear relationship between FST speed and deployment depth or wind speed. For example, although FST's 8-11 travelled northwest on 7/8 March, five days later FST's 16-19 from the same area all headed southeast. This type of variability due to current reversals is characteristic of this area (Brink, Halpern and Smith, 1980).

Table 2. Free-drifting sediment trap deployment data

FST Number	Date	Latitude (S)	Longitude (W)	Exposure Period (Local Time)	Deployment Depth (m)	Water Depth (m)	Mean Drift Trajectories (Degrees)	Mean Drift Speed ($cm \cdot s^{-1}$)
8	March 7	15°10.9'	75°34.5'	0840-1722	52	500	338	5
9	March 7	15°10.6'	75°34.6'	0830-1640	14	500	319	12
10	March 7, 8	15°09.3'	75°36.2'	1955-0650	14	400	013	4
11	March 7, 8	15°09.3'	75°36.3'	2105-0716	52	400	327	7

The downward flux of bulk particulate matter, POC, and PON for FST's 4-21 were determined by Staresinic (1982). He found an average of 3.47 g dry mass, 253.3 mgC POC and a 39.1 mgN/m^2•12 hr PON flux through the euphotic zone (∿15 m). The dry mass flux varied very little with sample depth or time of collection. The four FST samples considered in this report (Table 3) exhibit this same pattern. Organic carbon ranged from 3.2 to 9.5% of the dry mass. The mean flux of POC through the euphotic zone was 275 mgC/m^2•12 hr, approximately 11% of the phytoplankton primary productivity. Fluxes through 52 m represented approximately 6% of the primary production. Staresinic (in press) noted a tendency for higher night POC fluxes through the euphotic zone. The POC flux through the euphotic zone at night (FST 10) is about twice the daytime flux (FST 9, Table 3).

The FST's set at the base of the euphotic zone contained large amounts of phytoplankton, mainly diatoms. The mean chlorophyll a flux was 2.5 mg/m^2•12 hr for the euphotic zone traps and averaged

Table 3. Flux of total particulate matter, particulate organic carbon (POC), lipids and sterols as a function of sediment trap depth and time of collection

	Day		Night	
	FST 9 (14 m)	FST 8 (52 m)	FST 10 (14 m)	FST 11 (52 m)
Particle Flux (g/m^2•12 hr)	2.93	2.92	3.34	2.48
POC Flux (mg/m^2•12 hr)	164	140	313	131
Lipid Flux (mg/m^2•12 hr)	13.3	14.9	76.1	24.4
Lipid Flux / POC Flux	0.08	0.11	0.24	0.19
Sterol Flux (mg/m^2•12 hr)	0.71	0.35	6.51	2.26
Sterol Flux / POC Flux	0.0043	0.0025	0.021	0.017
Sterol Flux / Lipid Flux	0.053	0.024	0.086	0.093

0.83 mg/m^2·12 hr for the 52 m FST's. Only 2% of the daily mean standing stock of chlorophyll a in the euphotic zone was collected by the FST's set at the base of the euphotic zone. The corresponding value for the 52 m FST's was about 0.7%. In addition to diatoms, Staresinic et al. (this volume) found intact anchoveta fecal fragments (cylinders about 2 mm long and 1 mm in diameter). These pellets were packed with both the intact and broken remains of planktonic diatoms, *Thalassionema nitzschioides* and *Thalassiosira eccentrica*, both common to the waters and sediments of the Peru coastal upwelling area. The anchoveta fecal material could account for as much as 17% of the total downward flux of POC through 52 m. The FST's also contained phytodetritus, copepod and euphausiid carcasses and molts and zooplankton fecal pellets. The night traps contained greater amounts of the molts than the day traps.

Sterols in Sediment Trap Material

The spatial and temporal variations of particulate organic matter in the upper 50 m of the water column, as reflected by the POC data, are also evident in the concentration data of the specific organic compound classes that make up this material. For example, although total extractable lipid fluxes are approximately the same for day samples (13-15 mg/m^2·12 hr), they are a factor of 2-5 times greater at night (Table 3). Lipids account for only 0.5% of the total particulate flux during the day but make up 1-2% of the particulate matter flux at night. The sample collected at the base of the euphotic zone at night had the highest lipid flux and percent lipid component. Assuming a value of 10-20% lipids (dry weight) in plankton (Orcutt and Patterson, 1975), the lipids found in the 14 m traps represent 4-8% of the lipids produced during primary productivity. This value drops to 2-4% at 52 m (Table 4).

Table 4. Productivity and measured flux values (mg/m^2·day) in Peru coastal waters

	Organic Carbon	Lipids*	Sterols*
Surface Productivity	4300	1100-2100	13-56
14 m	477	89.4	7.2
In % of primary production	11	4-8	13-48
52 m	271	39.3	2.6
In % of primary production	6	2-4	5-20

* See text for calculations.

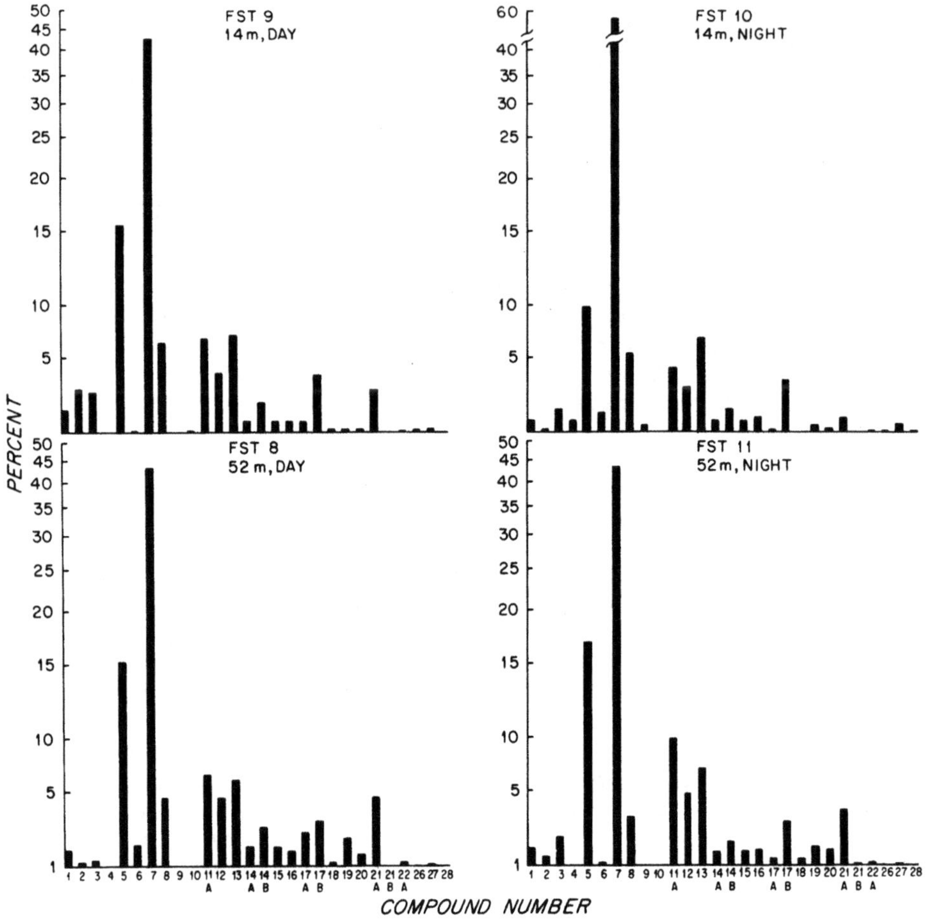

Fig. 7. Sterol distributions for (a) FST 9, 14 m day (b) FST 8, 52 m day (c) FST 10, 14 m night and (d) FST 11, 52 m night. Percent sterols are plotted on a logarithmic scale, compound numbers refer to the structures in Table 5.

The same diurnal pattern observed for lipids is found in the sterol data. The sterol fluxes at night are approximately an order of magnitude higher than the day fluxes in both the 14 m and 52 m traps (Table 3). For the 52 m trap, the sterol flux relative to the lipid flux increased 5-fold at night. Although later changes are observed in total sterol fluxes, compositional differences among the four traps are very small (Fig. 7).

The major source of sterols in the trap material (FST's 8-11) was determined to be copepod and euphausiid fecal pellets, molts, and carcasses. Visual inspection of one sediment trap sample (FST 3) showed that this sample was almost exclusively made up of these copepod components. FST 10 contained mostly euphausiids, *Euphausia mucronata*. The sterol distribution of FST 3 was virtually identical to FST's 8-11. It should be pointed out that the fecal pellets and molts of euphausiids would probably have a similar sterol distribution to those from copepods. Hence, based on sterol data alone, we cannot differentiate between a euphausiid or copepod source. However, studies of wax esters suggest that it may be possible to differentiate between these sources using this class of compounds (S. Wakeham, pers. comm.). An important feature of the sterol distribution in the traps is the predominance of cholest-5-en-3β-ol (7) and cholesta-5,22E-dien-3β-ol (5), both of which are abundant in copepods (Volkman, Corner and Eglinton, 1980a).

In addition to copepod and euphausiid molts and fecal pellets, sediment traps also contained a number of large (∿2 mm length and 1 mm diam.) anchoveta (*Engraulis ringens*) fecal pellets. We were able to analyze intact anchoveta fecal pellets collected on RV "Atlantis" II cruise 108, leg 3, in March, 1981. On this cruise, sediment traps were deployed in the same area as the 1978 cruise. However, almost all of the trap material from 13 deployments was anchoveta fecal pellets, whereas no more than 17% of the particulate organic carbon found in the 1978 sediment trap samples could be accounted for by anchoveta fecal pellets (Staresinic et al., this volume). The main feature of the sterol distribution of the anchoveta fecal pellets is the large amount of 24-methylcholesta-5,24(28)-dien-3β-ol (13). This compound has been found in several diatoms, i.e. *Nitzschia alba* (Kates et al., 1978), *Skeletonema costatum* (Ballantine et al., 1979), and *Chaetoceros simplex calcitrans* (Boutry, Saliot and Barbier, 1979). A culture of cloned *Thalassiosira pseudonana* contained 13 as its major sterol, 81% (J. Volkman, L. Brand, pers. comm.). Indeed, microscopic analysis of the anchoveta fecal pellets showed the pennate diatom *Thalassionema nitzschioides* and the centric diatom *Thalassiosira eccentrica* to be major components. Intact diatoms in the pellets accounted for >50% of the anchoveta pellet POC in 1978 (Staresinic et al., this volume).

Sedimentary Sterols

The sediment sterol distributions are shown in Fig. 8, sterols 5, 7, 11a, 13, 14a, and 21a (Table 5) being the major components. Gas chromatograms for two sediment samples, GR27 and SC5 are shown in Fig. 3.

The major sterol, cholest-5-en-3β-ol (7) undoubtedly comes from zooplankton carcasses, molts and fecal pellets with some contribution (perhaps 20%) from anchoveta fecal pellets and a smaller contribution from phytoplankton (<10%). The sediment trap material (FST 8-11), as well as the copepod molts and fecal pellets (FST 3), was rich in cholest-5-en-3β-ol. Sterol 7 was also present in the sterol distribution (∿10%) of small particles in surface waters suggesting that

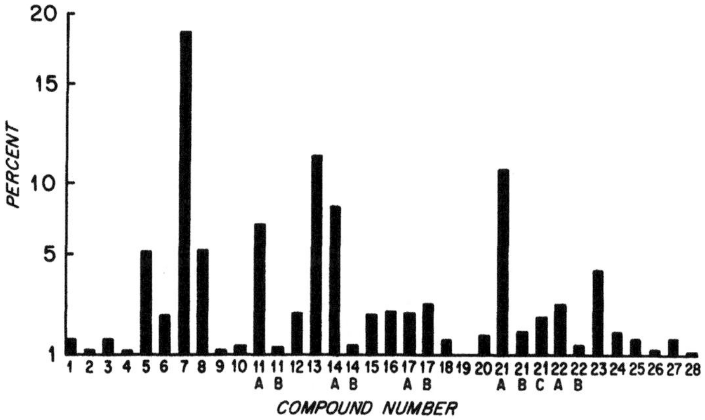

Fig. 8. Sterol distributions for the average of sediment samples GR27 and SC5. Percent sterols are plotted on a logarithmic scale. Compound numbers refer to the structures in Table 5.

phytoplankton may also be a small source of this sterol (Gagosian et al., in press). The source of cholesta-5,22E-dien-3β-ol (5) appears to be from both zooplankton and phytoplankton. This compound was found in high concentrations in water samples taken in the euphotic zone which contained mainly dinoflagellates (Gagosian et al., in press). A few diatoms have also been found to contain this sterol (Volkman et al., 1980b). FST 3, which contained mainly copepod molts and fecal pellets, also had high concentrations of sterol 5.

Anchoveta fecal pellets appear to be a major source of 24-methylcholesta-5,24(28)-dien-3β-ol (13) found in the sediments because this sterol was the major component in the fecal pellets (Gagosian et al., in press). These fecal pellets settle at a rate of ∿1000 m/day as compared with 10-100 m/day for zooplankton fecal pellets and 1-5 m/day for phytoplankton (Staresinic et al., in press). Hence, these fecal pellets can deliver fresh, relatively undegraded material to shelf sediments in a matter of hours. Anchoveta fecal pellets may also be the main mechanism by which 24-methylcholesta-5,22E-dien-3β-ol (11a) is transported to the sediments. As with sterol 13, the source of compound 11a is most likely to be diatoms since many species have been found to contain this compound (Rubinstein and Goad, 1974; Orcutt and Patterson, 1975; Ballantine et al., 1979; Volkman et al., 1980b). A small amount of sterol 11a may also be derived from coccolithophores such as *Emiliania huxleyi* (Volkman et al., 1981), a major non-diatom phytoplankton in coastal Peruvian waters (Ryther et al., 1971).

Because the ultimate source of sterols 11a and 13 in the fecal pellets is undoubtedly phytoplankton, the direct transport of these sterols to the sediment via planktonic settling, rather than incorporation into anchoveta fecal pellets must also be considered. Davies (1975) and Morris (pers. comm.) have observed large diatom mats on the sediment surface of coastal sediments. Evidently, these mats occur when a bloom ends and subsequently sinks to the sediment surface before zooplankton grazing becomes effective. Several of these events could be occurring in Peru upwelling coastal waters as well (Packard, 1977).

Sterols thus far discussed have a clear marine origin. The 24(R) isomer of 24-ethylcholest-5-en-3β-ol (21a) is generally assigned a terrestrial origin. Phytoplankton have also been found to contain a small amount of sterol 21a but its C-24 stereochemistry is unknown. Our gas chromatographic conditions do not allow us to separate these C-24 epimers. The 24(S) epimer of sterol 21a has been found in sponges (Carlson et al., 1978) but sponges are not present in the sediment samples from Peru. As the amount of sterol 21a found in the sediments is much greater than has been found in phytoplankton, it is likely that most of sterol 21a originates from terrestrial sources with some phytoplankton contribution. The major source for another higher plant sterol, 24-methylcholest-5-en-3β-ol (14a), is also most likely terrestrial.

Terrestrial sterols may be transported either by atmospheric aerosols or during river runoff. There are few active rivers near 15°S (Zuta and Guillen, 1970) but during El Niño events (recent examples being 1972 and 1976), when warm water incursions move south along the Peru shelf, high rainfall occurs. Rivers could then bring large amounts of terrestrial debris down the high (300 m) steep continental sandy cliffs into coastal waters where surface currents could eventually move the material further offshore. It is difficult to assess the input of atmospheric transport since no studies have been conducted off this coastline. The strong northwesterly winds cause a constant haze for much of the year so a large amount of this dust is undoubtedly transported over the sea.

Steroid analyses of sediments from another upwelling zone, off southwest Africa (Namibia), show approximately the same levels of total sterols as reported for coastal sediments off Peru, 100 μg/g dry weight sediment (Morris and Calvert, 1977; Wardroper, Maxwell and Morris, 1978; Gagosian et al., 1980a; Lee, Gagosian and Farrington, 1980). The relative sterol distributions however, were quite different in comparing Peruvian with Namibian shelf sediments as well as comparing sterol compositions of sediments from various areas on the Namibian shelf. This is not particularly surprising. Because of the variation in concentrations of oxygen and nutrients and in temperature, the plankton assemblages and microbiological populations on the Peruvian and Namibian shelf are complex spatially and temporally. Although diatoms are usually the predominant phytoplankton in the water overlying these upwelling zones, the quantity and species diversity frequently varies. In addition, large seasonal blooms of dinoflagellates of different species have been recorded (Huntsman et al., 1982). Fluctuations in fish populations are also common. Thus, the different sterol distributions of these areas are indicators of a dynamic and variable system in relation to sedimentary source inputs and microbial transformation. Each sedimentary regime should be considered separately in conjunction with the source organisms inhabiting the overlying water column.

Flux Calculations

Individual sterol fluxes. In order to use certain compounds as biomarkers it is necessary to know how well preserved the compounds are as they pass through the water column and what proportion is buried in the sedimentary record. Using the phytoplankton sterol data from the FST's in conjunction with the sediment sterol data, we can calculate how much of this material is being preserved in the sediments. Hence, in this way we can determine how useful these biogenic compounds are as source biomarkers in the sediments. We will consider two phytoplankton sterols, 24-methylcholesta-5,22E-dien-3β-ol (11a) and 24-methylcholesta-5,24(28)-dien-3β-ol (13) both common in diatoms. The flux of sterol 11a averages 313 and 244 $\mu g/m^2 \cdot day$ through the euphotic zone (14 m) and 52 m depth, respectively. Sterol 13 has an average flux of 486 and 177 $\mu g/m^2 \cdot day$ through the euphotic zone and 52 m, respectively. These calculations show that from 14 to 52 m, phytoplankton sterols 11a and 13 are fairly well preserved. The sterol concentrations of two surface sediment samples, SC5 and GR27 (Fig. 8), taken in the vicinity of FST's 8-11 average 8.2 and 13.1 μg/g dry sediment for sterols 11a and 13, respectively. For several cores in this region, Henrichs (1980) has reported a ^{210}Pb-derived sedimentation rate of approximately 1.0 cm/yr and a water content of from 70-90% for the top 1 cm of sediment. Assuming a water content of 80%, and a wet sediment density of 1.2 g/cm^3, we calculate a sediment accumulation rate of ca. 54 and 86 $\mu g/m^2 \cdot day$, for sterols 11a and 13, respectively.

Thus, the estimated flux of particulate sterols 11a and 13 passing through 52 m in the 500 m water column is 4.5 and 2 times greater than the estimated accumulation rate of these two compounds. Approximately 17% of sterols 11a and 13 which pass through the euphotic zone become buried in the sedimentary record (Table 6).

Both sterols 11a and 13 are derived from marine sources, but sterol 21a (24-ethylcholest-5-en-3β-ol) is usually assigned an origin from terrestrial higher plants (Lee et al., 1979; 1980). Undertaking the same calculations as above for sterol 21a yields a flux of 71 and 95 $\mu g/m^2 \cdot day$ through the euphotic zone (14 m) and 52 m depth, respectively. The sediment accumulation rate of sterol 21a is calculated to be 82 $\mu g/m^2 \cdot day$. Considering the many assumptions involved in this calculation, these numbers are within experimental error. Hence, there is little or no flux decrease with depth for this ter-

Table 5. 4-Desmethylsterols isolated from sediment trap and sediments off the Peruvian coast at 15°S.

GC Peak No.	Identification
1	24-norcholesta-5,22E-dien-3β-ol
2	24-nor-5α-cholest-22E-en-3β-ol
*3	27-nor-24-methylcholesta-5,22E-dien-3β-ol
4	27-nor-24-methyl-5α-cholest-24(28)-en-3β-ol
5	cholesta-5,22E-dien-3β-ol
6	5α-cholest-22E-en-3β-ol
7	cholest-5-en-3β-ol
8	5α-cholestan-3β-ol
9	Not identified
10	cholesta-5,24-dien-3β-ol
11	{ a. 24-methylcholesta-5,22E-dien-3β-ol b. 5α-cholest-7-en-3β-ol (minor)
12	24-methyl-5α-cholest-22E-en-3β-ol
13	24-methylcholesta-5,24(28)-dien-3β-ol
14	{ a. 24-methylcholest-5-en-3β-ol b. 24-methyl-5α-cholest-24(28)-en-3β-ol
15	24-methyl-5α-cholestan-3β-ol
16	23,24-dimethylcholesta-5,22E-dien-3β-ol
17	{ a. 24-ethylcholesta-5,22E-dien-3β-ol b. 23,24-dimethyl-5α-cholest-22E-en-3β-ol
18	24-ethyl-5α-cholest-22E-en-3β-ol
20	23,24-dimethylcholest-5-en-3β-ol (4,24-dimethyl-5α-cholestan-3β-ol) in sediment traps
21	{ a. 24-ethylcholest-5-en-3β-ol b. 23,24-dimethyl-5α-cholestan-3β-ol c. 24-ethylcholesta-5,24(28)E-dien-3β-ol
22	{ a. 24-ethyl-5α-cholestan-3β-ol b. 24-ethyl-5α-cholest-24(28)E-en-3β-ol
23	24-ethylcholesta-5,24(28)Z-dien-3β-ol
24	24-ethyl-5α-cholest-24(28)Z-en-3β-ol
25	Not identified
**26	24-isopropylcholest-5-en-3β-ol
**27	24-isopropylcholesta-5,24(28)-dien-3β-ol
**28	24-isopropyl-5α-cholest-24(28)-en-3β-ol

*Includes some cholesta-5,22Z-dien-3β-ol.

**Structural identification is tentative. The C-24 alkyl group may be n-propyl.

Table 6. Flux data for total sterols, 24-methylcholesta-5,22E-dien-3β-ol (11a) 24-methylcholesta-5,24(28)-dien-3β-ol (13) and 24-ethylcholest-5-en-3β-ol (21a)

	(μg/m²·day)		
	14 m	52 m	Sediment Surface
Total Sterols	7220	2610	767
In % of primary productivity	13-48	5-20	1.5-6
24-methylcholesta-5,22E-dien-3β-ol (11a)	313	244	54
24-methylcholesta-5,24(28)-dien-3β-ol (13)	486	177	86
24-ethylcholest-5-en-3β-ol (21a)	70	95	82

restrially derived sterol in contrast to a fivefold decrease for the two marine produced sterols. Wakeham, Farrington and Volkman (in press) have reached the same conclusion for another higher plant lipid, the $C_{24:0}$ fatty acid. They also found a sediment accumulation rate for the $C_{16:0}$ fatty acid, primarily of planktonic origin, of less than 5% of the flux through 52 m. These results suggest that in this depositional environment some biogenic terrestrial material is more protected from degradation than marine derived material. This is reasonable since the terrestrial material is exposed to prolonged degradation during its transport to the sea and hence only the more resistant material will survive. Soil bacteria and fungi rework this material to a large extent and during aerosol transport OH radical reactions in the atmosphere will further attack any labile lipids present. Plants and soils also contain many protective coatings and the organic compounds present in them may be an intricate part of this coating.

The hypothesis that terrestrial organic compounds are protected from degradation more than marine compounds is further supported by the sediment data of core BC7. The ratio of sterol 21a to phytoplankton sterols 11a and 13 increases dramatically with depth suggesting a slower degradation rate for the land plant sterol. The ratio of land plant derived hydrocarbons (n-C_{27}, n-C_{29}, n-C_{31}) also increases with depth relative to marine derived hydrocarbons in this same core (Volkman et al., in press).

The fact that approximately equal concentrations of sterol 21a were found at 14 m, 52 m and the sediment surface at 500 m, suggests that the input of this land plant sterol into the water column occurs at the surface. Atmospheric and riverine input are the two potential pathways for this input. Land plant and soil sterols could be transported to the sea surface on small particles, incorporated into larger particulate matter such as zooplankton and anchovy fecal pellets by the biota, and transported vertically through the water column. In addition, since shelf currents are quite strong in this study area, alongshore transport of suspended sediments, containing riverine terrestrial debris, could deliver this material to surface waters offshore.

Although evidence has been presented to support the hypothesis that the equal flux of sterol 21a at 14 m, 52 m and at the sediment surface is due to surface water input of resistant land plant material, another hypothesis is that this constant flux with depth may be the sum of degradation processes plus shelf sediment transport into various depths of the water column. Degradation rates must equal input rates for sterol 21a at each depth in the water column for this mechanism to be operable. Although this mechanism cannot be ruled out, the increase with depth of the ratio of land plant to marine derived material in Peru coastal and in other upwelling area sediments (Lee et al., 1980) argues against it.

Total sterol, lipid and carbon fluxes. With the FST, sediment and phytoplankton primary production sterol data, it is also possible to calculate how much of the primary production is surviving at various depths in the water column and how much is finally accumulated in the sediments. The assumption here is that most organisms other than phytoplankton in the water column, such as zooplankton and bacteria, obtain their sterols predominantly through their diet and not from *de novo* synthesis. This is a reasonable assumption since very few, if any, bacteria biosynthesize sterols. Off Peru, the major species of zooplankton (euphausiids and copepods) obtain needed sterols through their diet. The mean primary productivity in the study area during the sediment trap deployment periods was 4.3 $gC/m^2 \cdot day$ (Gagosian et al., 1980b). As discussed in Gagosian and Nigrelli (1979), sterol concentrations are typically 3-13 mg sterols/g phytoplankton organic carbon. Hence the rate of sterol production by phytoplankton in the euphotic zone off Peru at 15°S was 13-56 mg $sterols/m^2 \cdot day$ in February-March, 1978. Dividing this range into the sterol flux at the 14 m trap (7.2 $mg/m^2 \cdot day$), we calculate that 13-48% of the sterols produced in the euphotic zone pass through into deeper water. This is over twice the value found for the equatorial North Atlantic (Gagosian et al., 1982). Using the deeper trap (52 m) sterol flux of 2.6 $mg/m^2 \cdot day$, we find that 5-20% of the sterols produced in the euphotic zone appear at this depth. Thus, 8-28% of the sterols passing through the euphotic zone have either been decomposed or solubilized from the particles before they reach 52 m.

Similar calculations using total lipids and organic carbon are shown in Table 4. The percentages of lipids produced by surface primary productivity passing through the 14 m and 52 m depths are approximately half the organic carbon percentages and only 20-25% of the sterol percentages. This data suggests that phytoplankton organic carbon and bulk lipids are not preserved as efficiently as sterols. In addition, the percentages of sterols preserved through these depths are approximately 50% higher than amino acid values (Lee and Cronin, 1982). These results suggest that sterols are less water soluble or are more stable than the majority of the organic compounds produced by phytoplankton reaching the 14 and 52 m traps. This seems reasonable since sterols are major structural components in membranes and must be protected from decomposition and solubilization relative to other organic compounds in cytoplasmic fluids.

Using a sediment sterol accumulation rate of 0.77 $mg/m^2 \cdot day$, we calculate that 1.5-6% of the sterols produced in the euphotic zone are found in the sediment surface (1 cm). Hence, an additional 3.5-14% of the sterols associated with particles passing through the euphotic zone have been decomposed or solubilized between 52 m and accumulation in the sediment surface. Microbiological transformations at the sediment water interface are probably responsible for a major part of this decomposition (Gagosian et al., 1980a). The sediment

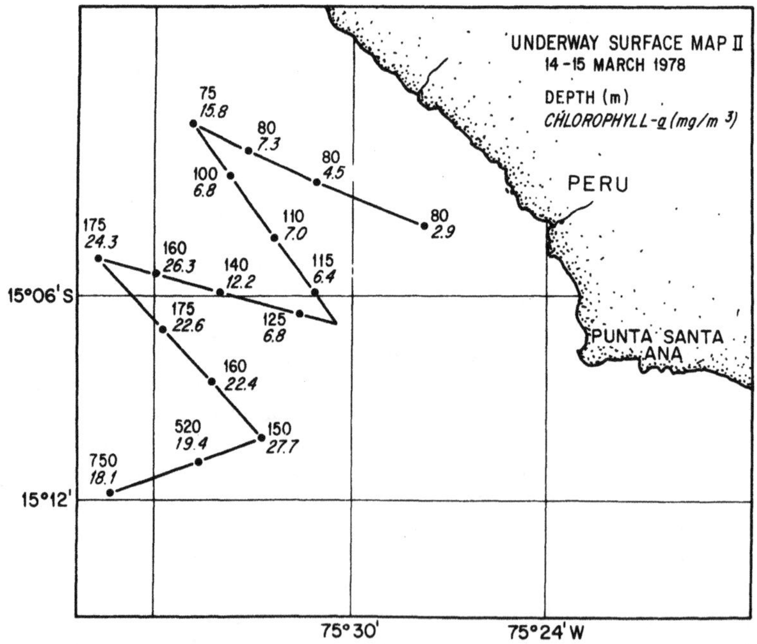

Fig. 9. Underway map of surface chlorophyll a for 14-15 March 1978. The upper value at each point gives the depth in meters.

accumulation value of 1.5-6% is over 20 times that found for the equatorial North Atlantic (Gagosian et al., 1982). This is not particularly surprising considering the difference in water column depth (500 vs. 5000 m) and the much more dynamic situation off Peru indicated by the large difference in primary productivity (4300 vs. 200 $mgC/m^2 \cdot day$). Indeed, as discussed earlier, anchoveta fecal pellets contained many whole diatoms and fragments. This suggests that phytoplankton sterols can be passed through the anchoveta intact and thus reach the sediment surface quickly (∿1000 m/day). For this upwelling zone, it appears that the sterols are sufficiently preserved from their time of formation until their burial in the sediment to allow a direct correlation to be drawn between the organic matter in the sediments with that of the original source organisms.

A major difficulty in verifying the validity of these calculations is the different time scales which the various types of samples represent. For example, sediment traps sample the downward flux of particles through the water column on time scales of hours (this study) to months (Honjo, 1980). On the other hand, surface sediments (1-2 cm depth) represent 1-2 years of accumulation in this area (Henrichs, 1980) and up to several hundred to thousands of years in open ocean areas. The sediments thus integrate hundreds of trap experiments. This short time scale variation can clearly be seen in the tenfold difference in sterol flux at night (FST's 10 and 11) compared with that during the day (FST's 9 and 8) (Table 3).

This variation is also illustrated by the large differences in the amount of anchovy fecal pellets found in the 1978 FST experiment compared with 1981. Typically, phytoplankton blooms and their associated zooplankton grazing events rise and fall in periods of less than one week. An example of these dramatic changes is the comparison of the underway chlorophyll map of March 4-5 (Fig. 5b) with that of March 14-15, 1978 (Fig. 9). The values for the inshore stations are three times higher on March 4-5 than on March 14-15. The opposite is true for the offshore stations. Many differences in these productivity signatures will undoubtedly be left in the surface sediment sampled and changes in anchoveta stocks are expected to track several of these events. In 1978, the maximum amount of FST POC flux that could be accounted for by anchoveta fecal pellets was 17%; in most of the samples it was below 5% (Staresinic et al., this volume). However, in 1981, >90% of the FST POC could be accounted for by anchoveta fecal pellets.

Hence, it is clearly necessary to expand sample coverage both in time and space to approach more fully the accumulation time of the sediments under investigation. Better coverage in the water column, both horizontally and vertically, is needed for sediment trap experiments as well as for the organisms inhabiting the water column at the time these experiments are undertaken. The results reported here are for only one season of the year, the austral fall, when upwelling off the Peruvian coast is well developed. Seasonal changes in particle fluxes have been observed in oligotrophic waters (Deuser and Ross, 1980) and need to be examined in upwelling zones as well.

CONCLUSIONS

(1) A large enough fraction (1.5-6%) of the sterols produced by primary production survive water column and surface sediment degradative processes to make this class of compounds suitable as tracers of the source, transport and transformation processes in the cycling of biogenic organic compounds in the sedimentary deposits of upwelling areas.

(2) Over half (50-90%) of the sterols produced in the euphotic zone are not transported below this depth (14 m). Only 5-20% are transported to depths below 52 m. These values are approximately twice as high as total organic carbon, four times as high as lipids, and 50% higher than amino acids from primary productivity.

(3) In the water column and surface sediments, phytoplankton derived sterols appear to be degraded faster than sterols of terrestrial origin, suggesting that this terrestrial material is more resistant to marine degradative processes than marine derived material in this depositional environment.

(4) There is a tenfold increase in the sterol flux at night compared with the day probably due to increased zooplankton activity (e.g., fecal pellet, carcass and molt transport). The sterol flux relative to the lipid flux for the 52 m trap was 5 times higher at night than during the day. Although these large changes were observed in total sterol fluxes, compositional differences between the day/night 14 m/52 m traps were small reflecting overall similar source inputs.

(5) The identity of the major sterols in the sediments indicate a major phytoplankton contribution, mainly from diatoms. These phytoplankton derived sterols appear to be delivered to the sediment quickly via sinking anchovy fecal pellets. Zooplankton (copepods and euphausiids) molts, carcasses and fecal pellets make significant contributions to the sedimentary sterol pool as well.

ACKNOWLEDGEMENTS

We thank the officers and crew of the RV "Knorr" and RV "Atlantis" II for their efforts at sea. Nick Staresinic, Cindy Lee, John Farrington and Stuart Wakeham made helpful comments on this manuscript. Nelson Frew obtained the mass spectra. This work was funded by grants OCE 77-26084, OCE 79-25352 and OCE 80-18436 from the National Science Foundation, Grant N00014-79-C-0071 from the Office of Naval Research and a WHOI Postdoctoral Scholar Award to J.K.V.

REFERENCES

Alam, M., Sansing, T.B., Busby, E.L., Martinex, D.R., and Ray, S.M., 1979, Dinoflagellate sterols I: Sterol composition of the dinoflagellates of *Gonyaulax* species, Steroids, 33:197-203.

Ballantine, J.A., Lavis, A. and Morris, R.J., 1979, Sterols of the phytoplankton-effects of illumination and growth stage, Phytochemistry, 18:1459-1466.

Bishop, J.K.B., Edmond, J.M., Ketten, D.R., Bacon, M.P., and Silker, W.B., 1977, The chemistry, biology and vertical flux of particulate matter from the upper 400 m of the equatorial Atlantic Ocean, Deep-Sea Research, 24:511-548.

Bishop, J.K.B., Ketten, D.R. and Edmond, J.M., 1978, The chemistry, biology and vertical flux of particulate matter from the upper 400 m of the Cape Basin in the southeast Atlantic, Deep-Sea Research, 25:1121-1162.

Bishop, J.K.B., Collier, R.W., Ketten, D.R., and Edmond, J.M., 1980, The chemistry, biology and vertical flux of particulate matter from the upper 400 m of the Panama Basin, Deep-Sea Research, 27:615-640.

Boon, J.J., Rijpstra, W.I.C., DeLange, F., DeLeeuw, J.W., Yoshioka, M., and Shimizu, Y., 1979, The Black Sea sterol - a molecular fossil for dinoflagellate blooms, Nature, 277:125-217.

Boutry, J.L., Saliot, A. and Barbier, M., 1979, The diversity of marine sterols and the role of algal bio-masses; from facts to hypothesis, Experentia, 35:1541-1543.

Brewer, P.G., Nozaki, Y., Spencer, D.W., and Fleer, A.P., 1980, Sediment trap experiments in the deep North Atlantic: isotopic and elemental fluxes, Journal of Marine Research, 38:703-728.

Brink, K.H., Halpern, D. and Smith, R.L., 1980, Circulation in the Peruvian upwelling system near 15°S, Journal of Geophysical Research, 85:4036-4048.

Brink, K.H., Jones, B.H., Van Leer, J., Mooers, C.N.K., Stuart, D.W., Stevenson, M.R., Dugdale, R.C., and Heburn, G.W., 1981, Physical and biological structure and variability in an upwelling center off Peru near 15°S during March, 1977, in: "Coastal Upwelling," F.A. Richards, ed., Coastal Estuarine Sciences 1, American Geophysical Union, Washington, 473-495.

Carlson, R.M.K., Popov, S., Massey, I., Delseth, C., Ayanoglu, E., Varkony, T.H., and Djerassi, C.J., 1978, Minor and trace sterols in marine invertebrates. VI. Occurrence and possible origins of sterols possessing unusually short hydrocarbon side chains, Bioorganic Chemistry, 7:453-479.

Crisp, T.P., Brenner, S., Venkatesan, M.I., Ruth, E., and Kaplan, I.R., 1979, Organic chemical characterization of sediment trap particulates from San Nicholas, Santa Barbara, Santa Monica, and San Pedro Basins, California, Geochimica et Cosmochimica Acta, 43: 1791-1801.

Dastillung, M. and Albrecht, P., 1977, Δ^2-Sterenes as diagenetic intermediates in sediments, Nature, 269:678-679.

Davies, J.M., 1975, Energy flow through the benthos in a Scottish sea loch, Marine Biology, 31:353-362.

DeBaar, H.J.W., Farrington, J.W. and Wakeham, S.G., in press, Sediment trap experiment in the equatorial Atlantic Ocean: Vertical flux of fatty acids, regeneration and oxygen consumption, Journal of Marine Research.

Deuser, W.G. and Ross, E.H., 1980, Seasonal change in the flux of organic carbon to the deep Sargasso Sea, Nature, 283:364-365.

Deuser, W.G., Ross, E.H. and Anderson, R.F., 1981, Seasonality in the supply of sediment to the deep Sargasso Sea and implications for the rapid transfer of matter to the deep ocean, Deep-Sea Research, 28:495-505.

Dugdale, R.C., Goering, J.J., Barber, R.T., Smith, R.L., and Packard, T.T., 1977, Denitrification and hydrogen sulfide in the Peru upwelling region during 1976, Deep-Sea Research, 24:601-608.

Elder, D.L. and Fowler, S.W., 1977, Polychlorinated biphenyls: Penetration into the deep ocean by zooplankton fecal pellet transport, Science, 197:459-461.

Fellows, D.A., Karl, D.M. and Knauer, G.A., 1981, Large particle fluxes and the vertical transport of living carbon in the upper 1500 meters of the northeast Pacific Ocean, Deep-Sea Research, 28:921-936.

Gagosian, R.B. and Farrington, J.W., 1978, Sterenes in surface sediments from the southwest African shelf and slope, Geochimica et Cosmochimica Acta, 47:1091-1101.

Gagosian, R.B. and Heinzer, F., 1979, Stenols and stanols in the oxic and anoxic waters of the Black Sea, Geochimica et Cosmochimica Acta, 43:471-486.

Gagosian, R.B. and Nigrelli, G., 1979, The transport and budget of sterols in the western North Atlantic Ocean, Limnology and Oceanography, 24:838-849.

Gagosian, R.B., Dean, J.P., Jr., Hamblin, R., and Zafiriou, O.C., 1979, A versatile, interchangeable chamber seawater sampler, Limnology and Oceanography, 24:583-588.

Gagosian, R.B., Smith, S.O., Lee, C.L., Farrington, J.W., and Frew, N.M., 1980a, Steroid transformations in Recent marine sediments, in: "Advances in Organic Geochemistry, 1979," A.G. Douglas and J.R. Maxwell, eds., Pergamon Press, Oxford, 407-419.

Gagosian, R.B., Loder, T., Nigrelli, G., Mlodzinska, Z., Love, J., and Kogelschatz, J., 1980b, Hydrographic and nutrient data for R/V "Knorr" Cruise 73, Leg 2 - February-March, 1978 - off the coast of Peru, Woods Hole Oceanographic Institution Technical Report, WHOI-80, 77 pp.

Gagosian, R.B., Smith, S.O. and Nigrelli, G.E., 1982, Vertical transport of steroid alcohols and ketones measured in a sediment trap experiment in the equatorial Atlantic Ocean, Geochimica et Cosmochimica Acta, 46:1163-1172.

Gagosian, R.B., Volkman, J.K. and Nigrelli, G.E., in press, The use of sediment traps to determine sterol sources in coastal sediments off Peru, in: "Advances in Organic Geochemistry, 1981," M. Bjorøy, ed., Wiley and Sons, New York.

Gallardo, V., 1977, Large benthic microbial communities in sulphide biota under Peru-Chile subsurface countercurrent, Nature, 288: 331-332.

Grob, K. and Grob, G., 1980, Deactivation of glass capillary columns by persilylation, Part 3. Extending the wetability by bonding phenyl groups to the glass surface, Journal of High Resolution Chromatography and Chromatography Communications, 3:197-198.

Henrichs, S.M., 1980, "Biogeochemistry of Dissolved Free Amino Acids in Marine Sediments," Ph.D. Dissertation, Massachusetts Institute of Technology/Woods Hole Oceanographic Institution Joint Program, WHOI Technical Report 80-39, 253 pp.

Hinga, K.R., Sieburth, M.McN. and Heath, G.R., 1979, The supply and use of organic material at the deep-sea floor, Journal of Marine Research, 37:557-579.

Honjo, S., 1978, Sedimentation of material in the Sargasso Sea at a 5367 m deep station, Journal of Marine Research, 36:469-492.

Honjo, S., 1980, Material fluxes and modes of sedimentation in the mesopelagic and bathypelagic zones, Journal of Marine Research, 38:53-97.

Honjo, S. and Roman, M.R., 1978, Marine copepod fecal pellets: Production, preservation and sedimentation, Journal of Marine Research, 36:45-57.

Huang, W.Y. and Meinschein, W.G., 1976, Sterols as source indicators of organic materials in sediments, Geochimica et Cosmochimica Acta, 40:323-330.

Huntsman, S.A., Brink, K.H., Barber, R.T., and Blasco, D., 1981, The role of circulation and stability in controlling the relative abundance of dinoflagellates and diatoms over the Peru shelf, in: "Coastal Upwelling," F.A. Richards, ed., Coastal and Estuarine Sciences 1, American Geophysical Union, Washington, 357-365.

Kates, M., Tremblay, P., Anderson, R., and Volcani, B.E., 1978, Identification of the free and conjugated sterol in a non-photosynthetic diatom, *Nitzschia alba*, as 24-methylenecholesterol, Lipids, 13:34-41.

Knauer, G.A., Martin, J.H. and Bruland, K.W., 1979, Fluxes of particulate carbon, nitrogen and phosphorous in the upper water column of the northeast Pacific, Deep-Sea Research, 26:97-108.

Lee, C. and Cronin, C., 1982, The vertical flux of particulate organic nitrogen in the sea: decomposition of amino acids in the Peru upwelling area and the equatorial Atlantic, Journal of Marine Research, 40:227-251.

Lee, C., Farrington, J.W. and Gagosian, R.B., 1979, Sterol geochemistry of sediments from the western North Atlantic Ocean and adjacent coastal areas, Geochimica et Cosmochimica Acta, 43:35-46.

Lee, C.L., Gagosian, R.B. and Farrington, J.W., 1980, Geochemistry of sterols in sediments from the Black Sea and the southwest African shelf and slope, Organic Geochemistry, 2:103-113.

MacKenzie, A.S., Brassell, S.C., Eglinton, G. and Maxwell, J.R., 1982, Chemical fossils - the geological fate of steroids, Science, 217:491-504.

Morris, R.J. and Calvert, S.E., 1977, Geochemical studies of organic-rich sediments from the Namibian shelf - 1. The organic fractions, in: "A Voyage of Discovery: Deacon Anniversary Volume," M. Angel, ed., Deep-Sea Research Supplement, Pergamon Press, Oxford, 647-665.

Nes, W.R. and McKean, M.L., 1977, "Biochemistry of Steroids and Other Isopentenoids", University Press, Baltimore, 690 pp.

Nes, W.R., Adler, J.H., Frasinel, C., Nes, W.D., Young, M., and Joseph, J.M., 1980, The independence of photosynthesis and aerobiosis from sterol biosynthesis in bacteria, Phytochemistry, 19:1439-1443.

Orcutt, D.M. and Patterson, G.W., 1975, Sterol, fatty acid and elemental composition of diatoms grown in chemically defined media, Comparative Biochemistry and Physiology, 50B:579-583.

Packard, T.T., 1977, The injection of particulate organic matter into the deep sea by a relaxation of oceanic upwelling, CUEA Newsletter, 6(4):46-47.

Prahl, F.G. and Carpenter, R., 1979, The role of zooplankton fecal pellets in the sedimentation of polycyclic aromatic hydrocarbons in Dabob Bay, Washington, Geochimica et Cosmochimica Acta, 43: 1959-1972.

Prahl, F.G., Bennett, J.T. and Carpenter, R., 1980, The early diagenesis of aliphatic hydrocarbons and organic matter in sedimentary particulates from Dabob Bay, Washington, Geochimica et Cosmochimica Acta, 44:1967-1976.

Repeta, D.J. and Gagosian, R.B., 1982, Carotenoid transformations in coastal marine waters, Nature, 245:51-54.

Repeta, D.J. and Gagosian, R.B., in press, Cartenoid transformation products in the upwelled waters off the Peruvian coast: sediment trap, seawater particulate matter, and zooplankton fecal pellet analyses, in: "Advances of Organic Geochemistry, 1981," M. Bjorøy, ed., Wiley and Sons, New York.

Rowe, G.T. and Gardner, W., 1979, Sedimentation rates in the slope water of the northwest Atlantic Ocean measured directly with sediment traps, Journal of Marine Research, 37:581-599.

Rubinstein, I. and Goad, L.J., 1974, Occurrence of 24S-24-methylcholesta-5,22E-dien-3β-ol in the diatom *Phaeodactylum tricornutum*, Phytochemistry, 13:485-487.

Ryther, J.H., Menzel, D.W., Hulburt, E.M., Lorenzen, C.J., and Corwin, N., 1971, The production and utilization of organic matter in the Peru coastal current, Investigacion Pesquera, 35(4): 43-59.

Schrader, H.-J., 1971, Fecal pellets: Role of sedimentation of pelagic diatoms, Science, 174:55-57.

Seifert, W.K. and Moldowan, J.M., 1978, Applications of steranes, terpanes, and monoaromatics to the maturation, migration and source of crude oils, Geochimica et Cosmochimica Acta, 42:77-96.

Simoneit, B.R.T., 1978, The organic chemistry of marine sediments, in: "Chemical Oceanography," 2nd Edition, vol. 7, J.P. Riley and R. Chester, eds., Academic Press, New York, 233-311.

Small, L.F., Fowler, S.W. and Onlu, M.Y., 1979, Sinking rates of natural copepod fecal pellets, Marine Biology, 51:233-241.

Smith, R.L., Enfield, B., Hopkins, T.S., and Pillsbury, R.D., 1971, The circulation in an upwelling ecosystem: The PISCO cruise, Investigacion Pesquera, 35(1):9-24.

Spencer, D.W., Brewer, P.G., Fleer, A., Honjo, S., Krishnaswami, S., and Nozaki, Y., 1978, Chemical fluxes from a sediment trap experiment in the deep Sargasso Sea, Journal of Marine Research, 36:493-523.

Staresinic, N., 1978, The Vertical Flux of Particulate Organic Matter in the Peru Upwelling as Measured with a Free-drifting Sediment Trap," Ph.D. Dissertation, Woods Hole Oceanographic Institution/ Massachusetts Institute of Technology Joint Program in Biological Oceanography, 255 pp.

Staresinic, N., in press, Downward flux of bulk particulate organic matter in the Peru coastal upwelling, Journal of Marine Research.

Steudler, P.A., Schmitz, F.J. and Ciereszko, L.S., 1977, Chemistry of coelenterates. Sterol composition of some predator-prey pairs on coral reefs, Comparative Biochemistry and Physiology, 56B: 385-392.

Tanoue, E. and Handa, N., 1980, Vertical transport of organic materials in the northern North Pacific as determined by sediment trap experiments. Part 1. Fatty acid composition, Journal of the Oceanographic Society of Japan, 36:231-245.

Taylor, C.D., Smith, S.O. and Gagosian, R.B., 1981, Use of microbial enrichments for the study of the anaerobic degradation of cholesterol, Geochimica et Cosmochimica Acta, 45:2161-2168.

Tissot, B.P. and Welte, D.H., 1978, "Petroleum Formation and Occurrence", Springer-Verlag, Berlin, 538 pp.

Turner, J.T. and Ferrante, J.G., 1979, Zooplankton fecal pellets in aquatic ecosystems, Bioscience, 29:670-677.

Volkman, J.K., Corner, E.D.S., and Eglinton, G., 1980a, Transformations of biolipids in the marine food web and in underlying bottom sediments, in: "Biogéochemie de la Matiére Organique a l'Interface Eau Sédiment Marin, Collogues Internationaux du Centre National Recherche Scientifique, No. 293, Paris, 185-197.

Volkman, J.K., Eglinton, G. and Corner, E.D.S., 1980b, Sterols and fatty acids of the marine diatom *Biddulphia sinensis*, Phytochemistry, 19:1809-1813.

Volkman, J.K., Smith, D.J., Eglinton, G., Forsberg, T.E.V., and Corner, E.D.S., 1981, Sterol and fatty acid composition of four marine haptophycean algae, Journal of the Marine Biological Association, United Kingdom, 61:509-527.

Volkman, J.K., Farrington, J.W., Gagosian, R.B., and Wakeham, S.G., in press, Lipid composition of coastal marine sediments from the Peru upwelling region, in: "Advances in Organic Geochemistry, 1981," M. Bjorøy, ed., Wiley and Sons, New York.

Wakeham, S.G., in press, Sources and fates of organic matter from a sediment trap experiment in the equatorial north Atlantic: wax esters, steryl esters, triglycerides and glyceryl ethers, Geochimica et Cosmochimica Acta.

Wakeham, S.G., Farrington, J.W., Gagosian, R.B., Lee, C., DeBaar, H., Nigrelli, G.E., Tripp, B.W., Smith, S.O., and Frew, N.M., 1980, Fluxes of organic matter from a sediment trap experiment in the equatorial Atlantic Ocean, Nature, 286:798-800.

Wakeham, S.G., Farrington, J.W. and Volkman, J.K., in press, Fatty acids, wax esters, triglycerides and glyceryl ethers associated with particles collected in sediment traps in the Peru upwelling, in: "Advances in Organic Geochemistry, 1981," M. Bjorøy, ed., Wiley and Sons, New York.

Wardroper, A.M.K., Maxwell, J.R. and Morris, R.J., 1978, Sterols of a diatomaceous ooze from Walvis Bay, Steroids, 32:203-221.

Wiebe, P.H., Boyd, S.H. and Winget, C., 1976, Particulate matter sinking to the deep sea floor at 2000 m in the Tongue of the Ocean, Bahamas, with a description of a new sedimentation trap, Journal of Marine Research, 34:341-354.

Zuta, S. and Guillen, O., 1970, Oceanografia de las aguas costeras del Peru, Boletin, Instituto del Mar del Peru, 2(5):157-324.

Zuta, S., Rivera, T. and Bustamente, A., 1978, Hydrological aspects of the main upwelling areas off Peru, in: "Upwelling Ecosystems," R. Boje and M. Tomczak, eds., Springer-Verlag, Berlin, 235-260.

RELATIONSHIPS BETWEEN THE CHEMICAL COMPOSITION OF PARTICULATE ORGANIC MATTER AND PHYTOPLANKTON DISTRIBUTIONS IN RECENTLY UPWELLED WATERS OFF PERU

Kevin G. Sellner
Academy of Natural Sciences, Benedict Estuarine Research Laboratory, Benedict, Maryland 20612, U.S.A.

Peter Hendrikson
Institut für Meereskunde
23 Kiel, Federal Republic of Germany

Naomi Ochoa
Instituto del Mar del Peru (IMARPE)
Callao, Peru

ABSTRACT

The offshore advection of the nutrient-rich, recently upwelled surface layer off Peru was accompanied by changing phytoplankton assemblages and declining nutrient concentrations. The succession of large, chain-forming diatoms in nutrient-rich inshore waters to an assemblage of microflagellates, monads, small chain-forming diatoms, ciliates, and dinoflagellates in offshore moving waters resulted in a gradient in the chemical composition of the particulate and dissolved organic matter (POM and DOM) in the euphotic zone over the Peruvian slope and shelf. The "advanced" assemblages had higher proportions of lipid and lower protein and water soluble carbohydrate within the POM, high C/chlorophyll a and C/N ratios and low C/ATP ratios. Primary productivity was low, ranging from 0.2-1.2 $gC{\cdot}m^{-2}{\cdot}d^{-1}$. The more recently upwelled waters were typified by large chain-forming diatoms, e.g., *Chaetoceros* spp. and *Schroderella delicatula*, and higher proportions of protein and water soluble carbohydrate in the POM. Lipid fractions, C/N and C/chlorophyll a ratios declined while C/ATP ratios increased relative to offshore assemblages. Primary productivity exceeded 6 $gC{\cdot}m^{-2}{\cdot}d^{-1}$. Extracellular dissolved saccharide concentrations also increased substantially inshore. These data indicate diatom-dominated upwelling waters have a unique chemical signature in total organic matter and composition, facilitating maintenance of an upwelling facies of organic-rich sediments, a high benthic oxygen demand and anaerobic conditions through sedimentation of the POM.

INTRODUCTION

The flora of the shallow euphotic zone of upwelling regions is generally dominated by diatoms and occasionally dinoflagellates with the former algae producing silicon-rich skeletons that might provide an upwelling signature to the underlying sediments (e.g., see Diester-Haass, 1978). However, the rain of particulate silicates as diatom frustules in fecal matter (Staresinic et al., this volume) is only one characteristic of upwelling regions. Another, perhaps equally important, feature is the supply of organic matter as particulate and dissolved materials. The large influx of organic matter from the highly productive euphotic zone of upwelling areas results in high sediment oxygen demand and anaerobic conditions in bottom waters (Reimers and Suess, this volume). These benthic traits facilitate development of a sedimentary environment perhaps unique to upwelling areas.

The supply of nutrient-rich bottom waters to shallow, lighted depths in upwelling areas is not constant and varies with the strength of local winds (see Barber and Smith, 1981, for review). This variability coupled with net transport of surface waters and associated plankton offshore results in spatial heterogeneity in the available nutrients and changes in phytoplankton assemblages. Phytoplankton patchiness in the types and numbers of species present as well as productivity should therefore be accompanied by marked differences in chemical composition of the particulate matter (Hendrikson et al., 1982), production of extracellular metabolites (Sellner, 1981) and composition of the herbivore community. In October-November, 1977, phytoplankton composition, biomass and productivity were determined in the upwelling region off Peru. Collection of these data was accompanied by measurements of chemical composition of the particulate and dissolved organic pools resulting in characterization of distinct associations of specific community assemblages, productivities and suites of organic compounds.

MATERIALS AND METHODS

The present study was conducted off the Peruvian coast (Fig. 1) in October-December, 1977, as a part of the ICANE program (Investigation Cooperativa de la Anchoveta y su Ecosistema-Canada, Peru). Water samples were collected in hydrocasts at depths to which 100, 25, 11, and 1% of incident light penetrated as determined from Secchi disc depths. Ten hydrocasts were also obtained at one shelf position (9°20'S, 78°51'W) for nine consecutive days. An offshore-onshore

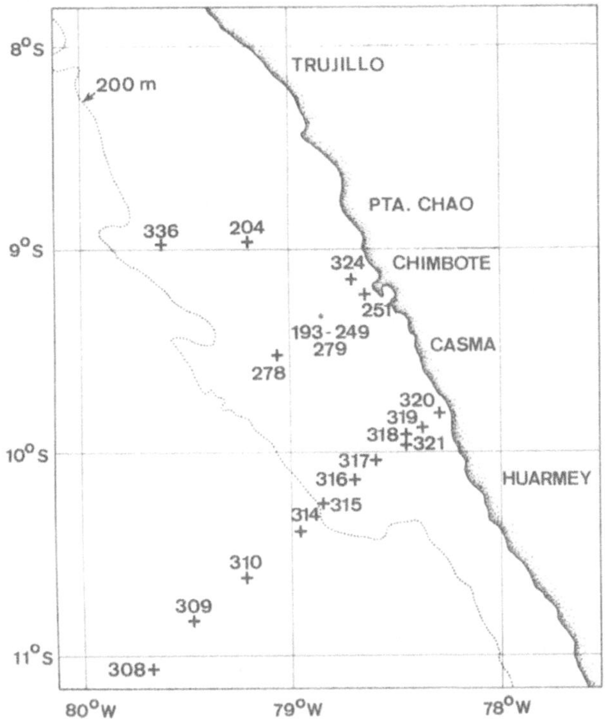

Fig. 1. Station locations in the ICANE oceanographic program, November-December 1977. * represents the location of the sampling site occupied over a 9-day period (from Hendrikson et al., 1982).

transect perpendicular to the shelf was conducted from stations 308-321; data are presented for stations 308, 310, 314, 316, and 319.

Water samples were collected from 12-liter Niskin bottles after inverting several times. For determination of chemical composition of particulate matter, 0.5-liter samples were filtered through pre-combusted glass-fiber filters (Whatman GF/C) and frozen. ATP was extracted in 5-ml Tris buffer (Holm-Hansen and Booth, 1966) with extracts frozen until transport to the laboratory. After drying at 60°C and treating with four drops of 0.1 N HCl, particulate organic carbon (POC) and nitrogen (PON) concentrations were determined on a Hewlett-Packard Model 185 CHN analyzer. After 24-hr hydrolysis in 1-ml 1 N NaOH, protein was measured according to Lowry et al. (1951) using bovine albumin as a standard. Lipid concentrations in chloroform/methanol extracts were recorded for the particulate matter (Zoller and Kirsch, 1962; Smetacek and Hendrikson, 1979) with a conversion of 1.5 times the carbon content (obtained from extracts of net plankton collected on the cruise) for lipid levels. Water-soluble carbohydrate and residual carbohydrate were determined as in

Smetacek and Hendrikson (1979), using glucose standards. The sum of lipid, protein and carbohydrate (weight) was employed as an estimate of particulate organic matter (POM).

Phytoplankton productivity was determined in 1-liter Pyrex bottles receiving 20 μCi $NaH^{14}CO_3$ and incubated 10-48 hr at simulated *in situ* light intensities (Sellner, 1981). Fifty-ml subsamples were withdrawn initially, at sundown and sunrise and filtered through 0.45 μm Millipore filters. After fuming the filters over concentrated HCl for ≥30 sec, the filters were placed in Aquasol or Aquasol II (New England Nuclear) for liquid scintillation counting. Carbon fixation rates were calculated from Vollenweider (1974). Water-column productivity ($gC \cdot m^{-2}$) was computed for each station using trapezoidal integration.

Changes in total dissolved saccharides (DSAC) were determined according to Burney and Sieburth (1977). Technical modifications of these procedures are presented elsewhere (Sellner, 1981). Carbon in saccharides was assumed to be 40%.

Chlorophyll a concentrations were determined on aceton-extracted particulate samples (SCOR-UNESCO, 1966). Nutrient concentrations (nitrate, nitrite, ammonium, orthophosphate) were monitored with automatic colorimetric methods (Whitledge et al., 1980).

Identification and enumeration of the phytoplankton species were performed using the inverted microscope technique (Utermöhl, 1958) on 100-ml samples preserved with buffered formalin (2%). Phytoplankton carbon was calculated from cell volumes (Smetacek and Hendrikson, 1979).

RESULTS

Offshore-Onshore Transect

The phytoplankton assemblages at the offshore stations 308 and 310 were dominated by dinoflagellates, small diatoms, coccolithophorids and unidentified microflagellates (Table 1). Cell densities were low, $1.5 \times 10^5\ \ell^{-1}$ and $0.3 \times 10^5\ \ell^{-1}$, at the two stations, respectively. *Nitzschia closterium* and *Skeletonema costatum* dominated the assemblage at station 308 while *N. closterium* and monads comprised the largest fraction of cells at station 310 (Table 1); overall, diatoms comprised 23% and 44% of total cells at the two stations. Diatom carbon within the euphotic zone formed 80% and 83% of total phytoplankton carbon at the two stations (Fig. 2a). Major contributors to phytoplanktonic carbon were *Thalassiothrix longissima* at station 308 and at the 25% I_o depth, the dinoflagellates *Gyrodinium* sp. and *Prorocentrum gracile* at station 310.

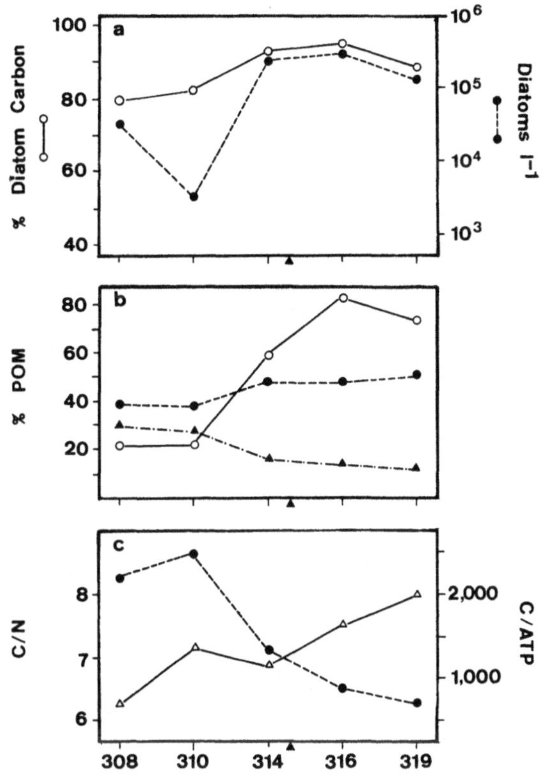

Fig. 2. Phytoplankton and composition of the particulate organic matter (POM) in the offshore-onshore transect, Station 308-319. Δ represents shelf break.

a. Numbers of diatoms ℓ^{-1} at the 25% I_o depth and integrated diatom carbon/total phytoplankton carbon x 100.

b. Mean euphotic zone composition of the POM, including protein (-----), water soluble carbohydrates/total carbohydrate (———) and lipid fractions (-·-·-·-).

c. The mean euphotic zone ratios of carbon/nitrogen (C/N, by weight, -----) and carbon/ATP (———), mg C·µg $ATP^{-1}\cdot m^{-3}$).

The chemical composition of the POM revealed relatively low protein and water-soluble carbohydrate fractions, forming approximately 39% of the POM and 22% of the total carbohydrate for the two stations; lipid formed approximately 30% of the POM (Fig. 2b). The mean euphotic zone C/N ratios in three offshore stations were relatively high, at 8.2 and 8.6, respectively (Fig. 2c), while the C/ATP ratios were 761 and 1418 mgC·µg$ATP^{-1}\cdot m^{-3}$ (Fig. 2c).

Stations 314 and 316 were located on either side of the shelf break and were characterized by increasing contributions of diatoms in the phytoplankton assemblage (Fig. 2a). Diatom densities were 3.9 x 10^5 and 5.7 x 10^5 ℓ^{-1} at the two stations, forming 93% and 95% of

Table 1. Principal phytoplankton species in offshore-onshore transect, November 1977 (ICANE)

Station	Cells/Liter x 10^{-3} at 25% I_0 Depth	Numerically Dominant Species	Major Contributors to Phytoplankton Carbon
308	147.61	*Nitzschia closterium* *Skeletonema costatum*	*Thalassiothrix longissima*
310	33.18	*N. closterium*, monads	*Gyrodinium* sp. *Prorocentrum gracile*
314	424.56	*Chaetoceros compressus*	*Chaetoceros compressus*
316	596.32	*Chaetoceros socialis* *Bacteriastrum* sp.	*Bacteriastrum* sp.
319	191.50	*Chaetoceros* spp.	*Ceratium furca* *Schroderella delicatula*

the phytoplankton carbon (Fig. 2a). Large chain-forming diatoms (*Chaetoceros compressus*, *C. socialis* and *Bacteriastrum* sp.) dominated densities and phytoplankton carbon (Table 1). The shift in species composition was accompanied by different composition of the POM as well (Fig. 2b). Protein increased to 46% of the POM while water-soluble carbohydrate increased to 60% and 82% of total particulate carbohydrate at 314 and 316, respectively. Lipid concentrations, as a percentage of total POM, declined from offshore levels to approximately 15%. The mean euphotic zone C/N ratio also declined from offshore values to 7.0 and 6.5 at the two stations, while the C/ATP ratios ranged from 1111 to 1703 mgC·µg $ATP^{-1} \cdot m^{-3}$ (Fig. 2c). From station 308 to 314, carbon fixation increased from 0.2 to 6.2 gC $m^{-2} \cdot d^{-1}$.

In the most inshore station (319), diatoms formed 95% of total cell numbers (Table 1) and 89% of all algal carbon (Fig. 2a) with *Chaetoceros* spp. becoming numerical dominants. However, *Ceratium furca* and *Schroderella delicatula* were the major contributors to phytoplankton carbon. The POM was characterized by 50% protein and 13% lipid while 73% of total saccharide was observed in the water soluble fraction (Fig. 2b). The mean euphotic zone C/N ratio was 6.3 and the C/ATP ratio of 2008 was the highest recorded in the transect (Fig. 2c). Concentrations of inorganic nutrients, including nitrate, orthophosphate and silicate, were high in the offshore stations, declined at the shelf break and increased inshore (Table 2).

Table 2. Mean euphotic zone concentrations (mg-at $\cdot m^{-3}$) of inorganic nutrients in the cross-shelf transect and the mid-shelf 9-day station, ICANE 1977. Sampling depth (m) in parentheses.

	Nutrient		
	NO_3-N	$Si(OH)_4$-Si	PO_4-P
308(16)	10.39	6.76	0.90
310(16)	13.02	7.48	1.12
314(10)	1.29	2.00	0.33
316(9)	4.01	2.37	0.72
319(10)	12.74	1.88	1.54
193(3)	4.39	0.81	0.80
197(3)	4.40	1.47	0.83
213(3)	3.50	0.88	0.87
222(4)	1.82	0.69	0.77
242(10)	0.95	0.51	0.83
244(10)	1.23	0.42	0.64
248(10)	0.58	0.40	0.63

Fixed Station

During the first four days at the mid-shelf position (stations 193-213), diatoms increased from 5.4×10^3 ℓ^{-1} to 6.0×10^4 ℓ^{-1} and contributed 0-53% of total phytoplankton carbon (Fig. 3a). The numerically dominant species were the ciliate *Mesodinium rubrum* and the coccolithophorid *Emiliana huxleyi* during days 1 and 2; on day 4 (station 213), monads and the diatom *Chaetoceros socialis* became the numerically dominant phytoplankton. The major contributors to phytoplankton carbon were the ciliate *M. rubrum* initially and on day 4, the dinoflagellate *Ceratium tripos* and diatom *C. socialis* (Table 3). For days 5 and 6 (stations 222 and 240), diatoms comprised approximately 53% of total algal carbon while densities were 1.1×10^5 ℓ^{-1} and 4.2×10^4 ℓ^{-1}, respectively (Figure 3a). *C. socialis*, abundant at the deepest depths of the euphotic zone, was a major contributor to cell numbers and with dinoflagellates, to carbon. From day 7-9 (stations 242-279), diatoms formed 72-84% of total phytoplankton carbon (principally *Chaetoceros* spp., *Schroderella delicatula* and the dinoflagellate *Gymnodinium lohmanni*) and densities ranged from 1.5-5.1×10^5 ℓ^{-1} (Fig. 3a; Table 3).

The increasing diatom contribution over the 9-day period was characterized by marked shifts in the chemical composition of the POM (Fig. 3b). While diatoms comprised <53% of the total algal carbon (days 1-6), protein remained relatively constant at 53% of total POM. Lipid concentrations fluctuated between 20-28% of the POM and water

Table 3. Principal phytoplankton species noted at the fixed station, November 14-22, 1977 (ICANE)

Station	Cells/liter x 10^{-3} at 25% I_o Depth	Numerically Dominant Species	Major Contributors to Phytoplankton Carbon
193	9.18	*Mesodinium rubrum*	*Mesodinium rubrum*
197	62.42	*Emiliana huxleyi*	*M. rubrum*, flagellates, dinoflagellates
213	36.87	monads *Chaetoceros socialis*	*Ceratium tripos* *Chaetoceros socialis*
222	101.53	*C. socialis*, monads, flagellates	*C. tripos*
	254.03 (1%I_o)	*C. socialis*	*C. socialis*
240	46.94 126.33 (1%I_o)	flagellates, monads *C. socialis*	dinoflagellates *C. socialis*
242	466.20	*C. socialis*	*C. socialis*
244	955.30	*C. socialis*, *C. compressus*	*C. socialis*, *C. compressus*
246	364.00	*C. socialis*, *C. compressus*	*C. tripos*, *C. socialis*, *Gymnodinium lohmanni*
248	591.32	*C. socialis*, *C. compressus*	*C. socialis*, *C. compressus*
279	521.57	*C. socialis*, *Nitzschia closterium*	*C. socialis*, *Schroderella delicatula*

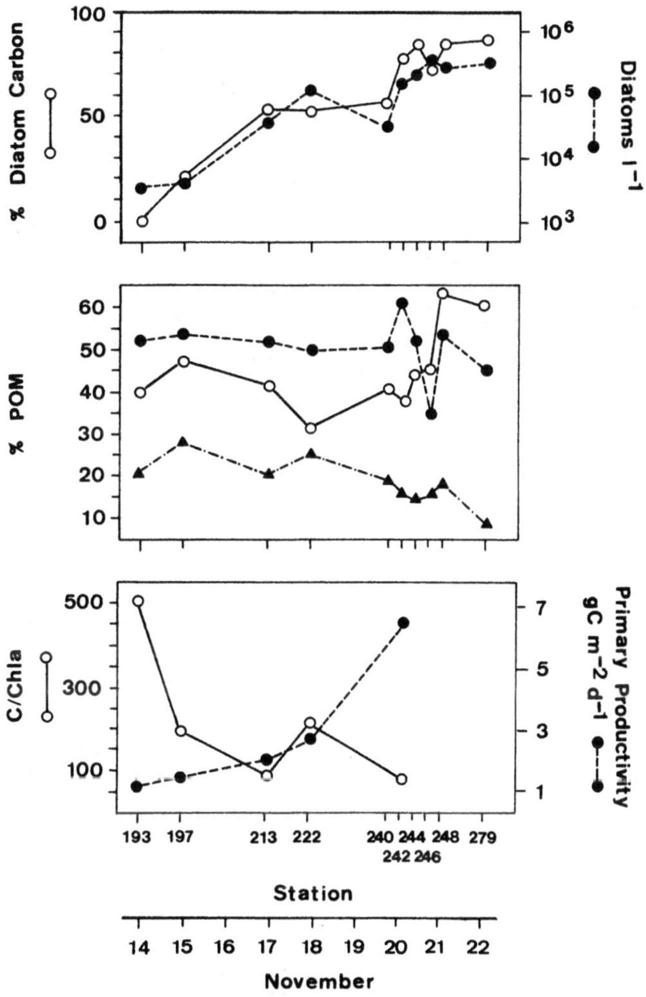

Fig. 3. The changes in phytoplankton and chemical composition of the POM over 9 days at the mid-shelf station.

a. The mean euphotic zone concentrations of diatoms and integrated diatom carbon/total phytoplankton carbon x 100.

b. Mean euphotic zone chemical composition of POM, including protein, water soluble carbohydrate/total carbohydrate and lipid. Symbols as in Fig. 2b.

c. The mean euphotic zone ratios of C/chlorophyll a (———), ($mgC \cdot mg\ chl\ a^{-1} \cdot m^{-3}$) and total primary productivity (------), ($gC \cdot m^{-2} \cdot d^{-1}$).

soluble carbohydrate ranged from 33-47% of total particulate carbohydrate. With the three-fold increase in diatoms from day 6-7 (stations 240-242) and high cell densities through day 9, protein reached 62% of total POM (station 242), water soluble carbohydrate increased from 38% to 65% of total particulate carbohydrate and lipid gradually declined from 20% on day 6 (station 240) to 11% on day 9 (station 279) (Fig. 3b). The C/chlorophyll a ratio continuously declined with increasing diatom-carbon while total productivity was 1.2-2.7 $gC\cdot m^{-2}\cdot d^{-1}$ in days 1-5 and >6.4 $gC\cdot m^{-2}\cdot d^{-1}$ on day 7 (Fig. 3c).

The increases in diatoms and productivity were also accompanied by a large increase in dissolved saccharide carbon (Fig. 4). Initial *in situ* dissolved saccharide concentrations in the euphotic zone increased from 4.84 to 7.96 $gC\cdot m^{-2}$ over the first days. Maximum daily accumulation rates for this pool were noted on day 7. The increase in dissolved saccharide was 1.34 $gC\cdot m^{-2}$ over the first 24 hrs while over the next 12-hr light period the increase was 7.6 $gC\cdot m^{-2}\cdot 12\ h^{-1}$.

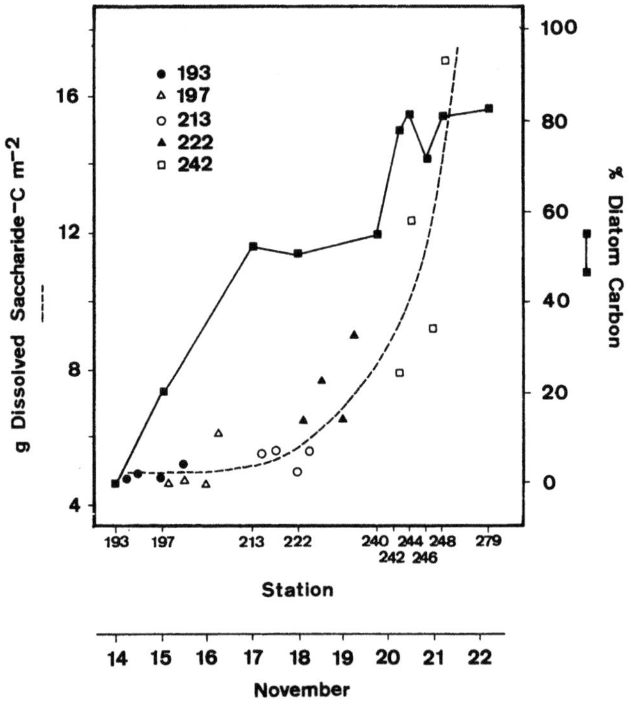

Fig. 4. Concentrations of dissolved saccharide carbon and integrated diatom carbon/total phytoplankton carbon x 100 over 9 days at the mid-shelf station. Each station had four sampling periods (4 symbols) representing concentrations of dissolved saccharide in incubation bottles, initially (*in situ* in text), at sundown, sunrise and sundown of the following day (see Sellner, 1981).

In contrast, the inorganic nutrient concentrations declined, with reductions of 0.81-0.40 μg-at silicate-Si ℓ^{-1}, 0.80-0.63 μg-at PO_4-P ℓ^{-1} and 4.39-0.58 μg-at NO_3-N ℓ^{-1} for days 1-8 (stations 193-248, Table 2).

DISCUSSION

Semi-continuous local winds produce the upwelling of nutrient-rich waters along the Peruvian coast with Ekman flow and wind-driven surface layers moving northwest over the shelf (Smith, this volume; Jones et al., this volume). As the surface light layers move further from the upwelling center, the nutrients in the surface layer will be consumed, the stability of the water column will increase and the species composition of the phytoplankton will change. Each phytoplankton assemblage will have a unique chemical signal dependent on species, mixing and inorganic nutrients present. In the present study, this spatial heterogenity in phytoplankton species and composition of the POM was investigated by making a cross-shelf transect from offshore-inshore and by repeated sampling as water moved past a fixed mid-shelf position.

In general, horizontal patchiness in the distribution of phytoplankton species was associated with marked differences in the chemical composition of the POM. Inshore areas (station 319) and most recently upwelled waters (station 279) were typified by a diatom assemblage, principally *Chaetoceros* spp. and *Schroderella delicatula*, high inorganic nutrient concentrations, a C/N ratio of approximately 6, and protein and lipid comprising roughly 50% and 10% of the POM. In addition, water-soluble carbohydrate formed >60% of the total particulate carbohydrate.

As the phytoplankton assimilated the large reserve of inorganic nutrients, slight changes in composition were observed. *Chaetoceros socialis*, *C. compressus* and *Bacteriastrum* sp. dominated cell numbers and carbon (stations 242-248 and 314-316). Total cell numbers approached 6 x 10^5 ℓ^{-1} with >90% diatoms. Protein and water soluble carbohydrate remained high and lipid low, the C/N ratio remained below 7, and the C/chlorophyll <u>a</u> and C/ATP ratios approximated 75 and 1300, respectively. Water column productivity and the concentration and accumulation of dissolved saccharides were very high with productivity exceeding 6.4 $gC \cdot m^{-2} \cdot d^{-1}$ and dissolved saccharide carbon accumulating at 0.4 $gC \cdot m^{-2} \cdot h^{-1}$.

As nutrients were depleted, metabolism slowed and phytoplankton composition shifted from diatoms to motile forms. From vertical distributions of *C. socialis* observed at station 222 and 240, the diatom evidently settled out and unidentified monads and microflagellates became numerically dominant while dinoflagellates formed the majority of the phytoplankton carbon. Cell densities declined 2-7 fold and productivity declined to 1.2 $gC \cdot m^{-2} \cdot d^{-1}$. Similarly, the quantities of dissolved saccharide were also reduced. The composition of the POM shifted to higher lipid and lower protein; C/N and C/chlorophyll a increased to >7 and >100, respectively. Water soluble carbohydrate declined to 40-50% of the total particulate carbohydrate and C/ATP declined to <1100 $mgC \cdot \mu gATP^{-1} \cdot m^{-3}$. At this time, nutrient regeneration within the water column must have supplied a greater proportion of the available inorganic nutrient, as proposed by Vinogradov and Shushkina (1978).

Finally, phytoplankton species shifted to unidentified flagellates, monads, the diatoms *Nitzschia closterium* and *Skeletonema costatum*, the dinoflagellates *Prorocentrum gracile*, *Gyrodinium* sp. and heterotrophic *Noctiluca*, the ciliate *Mesodinium rubrum* and the coccolithophorid *Emiliana huxleyi* (stations 308-310 and 193-222); cell densities were 7-10 times lower than the maximum densities recorded. Protein was less than 40% and lipids >25% of the POM; water soluble saccharide decreased to <25% of total particulate carbohydrate. The C/N and C/chlorophyll a ratios exceeded 8 and 500, respectively, while C/ATP was <800. Total primary productivity decreased to 0.2-1.2 $gC \cdot m^{-2} \cdot d^{-1}$ and dissolved saccharide concentrations did not change appreciably over the day.

The changes in chemical composition in recent versus aged upwelled water reflect differences in species composition and physiological state of the assemblages. In the advanced assemblages (stations 308-310, 193-222) containing dinoflagellates and unidentified microflagellates, highest lipid levels (as a percentage of POM) were recorded. Parsons and co-workers (1961) reported the chrysophyte *Monochrysis (Pavlova) lutheri* and two dinoflagellates had lipid levels 2 and 3 times higher than four diatoms grown in culture. Lipid concentrations also increase in older diatom cultures (Lewin and Guillard, 1963) as noted in the higher lipid levels observed in the more advanced assemblages at stations 222 and 240, with flagellates in surface waters and *Chaetoceros socialis* at depth. Increasing C/N and C/chlorophyll a ratios in the most advanced assemblages and lower ratios inshore would be expected from results of previous studies (e.g., Sakshaug, 1977; Sakshaug and Holm-Hansen, 1977). However, the increase in C/ATP from advanced to metabolically active assemblages inshore is contrary to the results collected in the Sakshaug studies.

Finally, the increasing importance of diatoms in more recently upwelled waters resulted in significant alterations of the dissolved chemical species in the water column. At the most productive stations (stations 314, 242), highest diatom densities and lowest inorganic nutrient concentrations were recorded. Accompanying nutrient depletions were large increases in dissolved saccharide concentrations, 0.4 $gC \cdot m^{-2} h^{-1}$ at station 242 and 0.5 $gC \cdot m^{-2} \cdot h^{-1}$ (Sellner, 1981) at station 314. In the stations with lower proportions of diatoms (days 1-6, stations 193-240), increases in dissolved saccharide were much lower and nitrate concentrations exceeded 1.8 mg-at $N \cdot m^{-3}$. Myklestad and Haug (1972), Haug, Myklestad and Sakshaug, (1973) and Myklestad (1977) have reported that nitrogen-limited chain forming diatoms (*Chaetoceros* spp.) accumulate and produce large amounts of extracellular saccharide; with abundant nitrogen, however, intracellular and extracellular saccharide concentrations decline. The development of metabolically active diatom populations in upwelled water and their subsequent decline should produce a unique polysaccharide signal over the shelf during upwelling events.

The changes in phytoplankton species and nutrient concentrations noted in the ICANE program correspond to classification schemes developed by Barlow (1980) and Dugdale (pers. comm.) for upwelling areas based on the distribution of nutrients and phytoplankton species from an upwelling center. Both classifications suggest initial phytoplankton colonization inshore. As the surface waters are pushed offshore, concentrations decline, diatoms predominate attaining maximum biomass and productivity within a few days and nutrient concentrations decline. As diatom metabolism slows, dinoflagellates, ciliates, coccolithophorids and small microflagellates eventually replace the sedimenting diatom populations as the advanced phytoplankton assemblage. Productivity and cell numbers decline. These latter assemblages may be the "blue water" assemblages noted by Strickland et al. (1969) and correspond with the heterotrophic plankton communities described by Vinogradov and Shushkina (1978).

The abundance of protein and carbohydrate-rich diatomaceous material inshore and the lipid-enriched "advanced" assemblages with seaward transport should provide suites of compounds perhaps containing individual compounds useful as geochemial markers in sedimented materials. For example, Gagosian et al. (this volume) attribute methylene cholesterol to diatoms and dinesterol to dinoflagellates while other lipids have been offered as biological markers in upwelling sediments (Brassell and Eglinton, this volume). Since anaerobic bottom waters off Peru receive large inputs of organic material in anchovy fecal pellets (Staresinic et al., this volume), anaerobic decomposition products of the large polysaccharides in diatom cell walls and storage products might prove useful as other markers of upwelling in Holocene sediments.

ACKNOWLEDGEMENTS

The authors extend their gratitude to the following individuals for their cooperation in this study: W.G. Harrison, B. Irwin, T. Platt, and S. Oakley for nutrient data, B. Sellner and the crew of the C.S.S. "Baffin" for their cooperation in sample collection and J. Arciprete and M.M. Olson for typing and proofing the manuscript.

REFERENCES

Barber, R.T. and Smith, R.L., 1981, Coastal upwelling ecosystems, in: "Analysis of Marine Ecosystems," A.R. Longhurst, ed., Academic Press, New York, 31-68.

Barlow, R.G., 1980, The biochemical composition of phytoplankton in an upwelling region off South Africa, Journal of Experimental Marine Biology and Ecology, 45:83-93.

Burney, C.M. and Sieburth, J.McN., 1977, Dissolved carbohydrates in seawater II, A spectrophotometric procedure for total carbohydrate analysis and polysaccharide estimation, Marine Chemistry, 5:15-28.

Diester-Haass, L., 1978, Sediments as indicators of upwelling, in: "Upwelling Ecosystems," R. Boje and M. Tomczak, eds., Springer-Verlag, New York, 261-281.

Haug, A., Myklestad, S. and Sakshaug, E., 1973, Studies on the phytoplankton ecology of the Trondheimsfjord. I. The chemical composition of phytoplankton populations, Journal of Experimental Marine Biology and Ecology, 11:15-26.

Hendrikson, P., Sellner, K.G., de Mendiola, B. Rojas, Ochoa, N., and Zimmermann, R., 1982, The composition of particulate organic matter and biomass in the Peruvian upwelling region during ICANE 1977 (Nov. 14-Dec. 2), Journal of Plankton Research, 4:163-186.

Holm-Hansen, O. and Booth, C.R., 1966, The measurement of ATP in the ocean and its ecological significance, Limnology and Oceanography, 11:510-519.

Lewin, J.C. and Guillard, R.R.L., 1963, Diatoms, Annual Review of Microbiology, 17:373-414.

Lowry, O.H., Rosebrough, N.J., Farr, A.L., and Randall, R.J., 1951, Protein measurement with the folinphenol reagent, Journal of Biological Chemistry, 193:265-275.

Myklestad, S., 1977, Production of carbohydrates by marine planktonic diatoms. II. Influence of the N/P ratio in the growth medium on the assimilation ratio, growth rate, and production of cellular and extracellular carbohydrate by *Chaetoceros affinis* var. *willie* (Gran) Hustedt and *Skeletonema costatum* (Grev.) Cleve, Journal of Experimental Marine Biology and Ecology, 29:161-179.

Myklestad, S. and Haug, A., 1972, Production of carbohydrates by the marine diatom *Chaetoceros affinis* var. *willie* (Gran) Hustedt I. Effect of the concentration of nutrients in the culture medium, Journal of Experimental Marine Biology and Ecology, 9:125-136.

Parsons, T.R., Stephens, K. and Strickland, J.D.H., 1961, On the chemical composition of eleven species of marine phytoplankton, Journal of Fisheries Research Board of Canada, 18:1001-1016.

Sakshaug, E., 1977, Limiting nutrients and maximum growth rates for diatoms in Narragansett Bay, Journal of Experimental Marine Biology and Ecology, 28:109-123.

Sakshaug, E. and Holm-Hansen, O., 1977, Chemical composition of *Skeletonema costatum* (Grev.) Cleve and *Pavlova (Monochrysis) lutheri* (Droop) Green as a function of nitrate-, phosphate- and iron-limited growth, Journal of Experimental Marine Biology and Ecology, 29:1-34.

SCOR-UNESCO, 1966, Determination of photosynthetic pigments in sea water, Monographs on Oceanographic Methodology (UNESCO), 1:1-65.

Sellner, K.G., 1981, Primary productivity and the flux of dissolved organic matter in several marine environments, Marine Biology, 65:101-112.

Smetacek, V. and Hendrikson, P., 1979, Composition of particulate organic matter in Kiel Bight in relation to phytoplankton succession, Oceanologica Acta, 2:287-298.

Strickland, J.D.H., Eppley, R.W. and Rojas de Mendiola, B., 1969, Phytoplankton populations, nutrients and photosynthesis in Peruvian coastal waters, Boletin, Instituto del Mar del Peru, 2: 37-45.

Utermöhl, H., 1958, Zur Vervollkommnung der quantitativen Phytoplankton-Methodik, Mitteilungen Internationale Vereinigung für Theoretische und angewandte Limnologie, 9:1-38.

Vinogradov, M.E. and Shushkina, E.A., 1978, Some development patterns of plankton communities in the upwelling areas of the Pacific Ocean, Marine Biology, 48:357-366.

Vollenweider, R.A., 1974, "A Manual on Methods for Measuring Primary Production in Aquatic Environments," IBP Handbook No. 12, Blackwell Scientific Publications, Oxford, 225 pp.

Whitledge, T., Malloy, S., Patton, C., and Wirick, C., 1980, Automated nutrient analysis in seawater, Brookhaven National Laboratory Technical Report, Upton, New York.

Zollner, N. and Kirsch, K., 1962, Über die quantitative Bestimmung von Lipiden (Mikromethode) mittels der vielen Lipoiden (allen bekannten Plasmalipiden) gemeinsamen Sulfo-phospho-vanillin-Reaktion, Zeitschrift für die Gesamte Experimentelle Medizin, 135:545-561.

PARTICULATE GEOCHEMISTRY IN AN AREA OF COASTAL UPWELLING--THE SANTA BARBARA BASIN

Alan M. Shiller

Scripps Institution of Oceanography
University of California, San Diego
La Jolla, California 92093, U.S.A.
Present address:
Department of Earth and Planetary Sciences
Massachusetts Institute of Technology
Cambridge, Massachusetts 02139, U.S.A.

ABSTRACT

The Santa Barbara Basin is a shallow nearshore basin off southern California characterized by seasonal upwelling. Particulates were obtained there by Niskin filtering, and inorganic composition was determined by thin-film x-ray fluorescence spectroscopy. In the mid-water column, resuspension of fine sedimentary material is found to provide the major input of particulates. Upwelling appears to increase the amount of resuspension in the basin. These resuspended sediments may affect coastal productivity.

INTRODUCTION

In studies of the dissolved constituents of seawater, concentration changes are frequently observed that can be attributed to chemical reactions. Generally, these chemical reactions involve a solid phase either at the sea floor or suspended in the water column. In recent years, our knowledge of suspended solids in the ocean (particulate matter) has markedly increased. However, there is still much that remains unknown about the composition, chemistry, distribution, and origin of particulates. The study that is described here is a preliminary attempt to understand the nature of particulate matter in the coastal zone.

The Santa Barbara Basin off southern California is a shallow (590 m) nearshore basin characterized by seasonal upwelling. The waters below sill depth (475 m) are generally low in oxygen and show evidence of nitrate reduction. This basin contains the only known laminated marine sediments in U.S. waters. These sediments constitute a well-preserved, yearly depositional record of use to a wide variety of oceanographers (Soutar et al., 1977). Because the basin has been the subject of many studies providing much supportive information, it was chosen as the site for this particulate study. Samples were collected during four cruises to the basin between April 1978 and July 1980.

EXPERIMENTAL METHODS

The main sampling location was the water column above the zone of laminated sediments in the deepest part of the Santa Barbara Basin (Fig. 1). Particulates were obtained by filtering about a liter of seawater through a 47 mm diameter, 0.4 µm Nuclepore filter. As fecal pellets are only a minor fraction of the particulate standing stock in the basin (Dunbar and Berger, 1981; Shiller, 1982), these samples are primarily composed of fine, individually settling particles. The filters were rinsed with filtered, doubly distilled water and allowed to dry in a desiccator. Major element composition of the particulates was determined by thin-film x-ray fluorescence spectroscopy (Price and Calvert, 1973; Sholkovitz and Price, 1980).

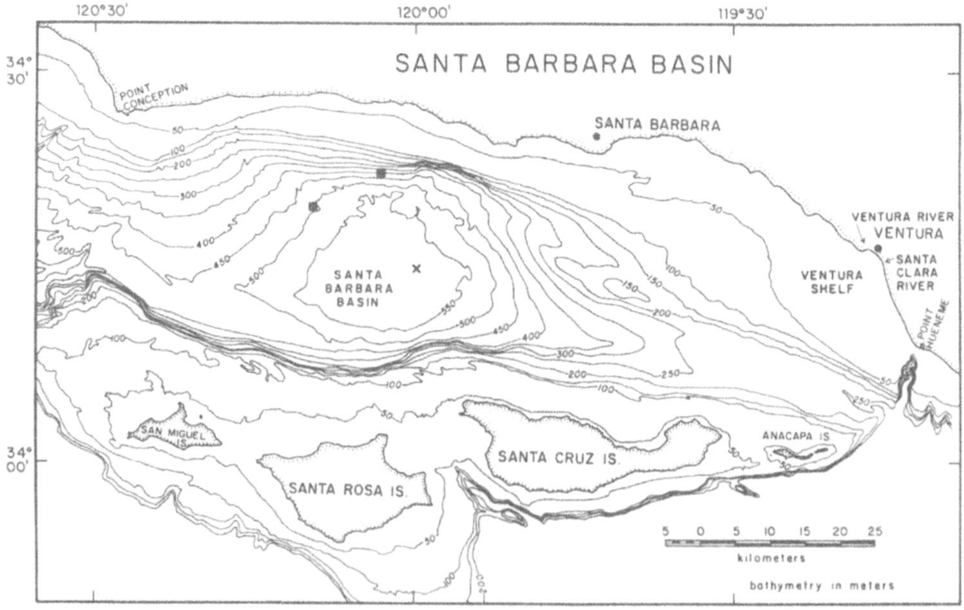

Fig. 1. Bathymetry of the Santa Barbara Basin. The cross shows the main central basin sampling site. Samples were also obtained along the northwest slope at the point marked by the filled boxes.

In addition to the particulate samples, ancillary sediment major element and water column nutrient samples were collected. Surface sediments were obtained by box coring. Bulk sediment and individual size fractions obtained by standard pipette analysis were analyzed by wet chemical methods similar to those of Shapiro (1967) and Shapiro and Brannock (1962). Nutrient samples were analyzed by methods described by Strickland and Parsons (1968). Further details of all sampling and analytical procedures may be found in Shiller (1982).

HYDROGRAPHIC CONDITIONS

Although there are seasonal differences in the current patterns in the Santa Barbara Basin, water generally enters both from the southeast as part of a regional cyclonic eddy and from the northwest as part of the southerly flow of the California Current. Upwelling

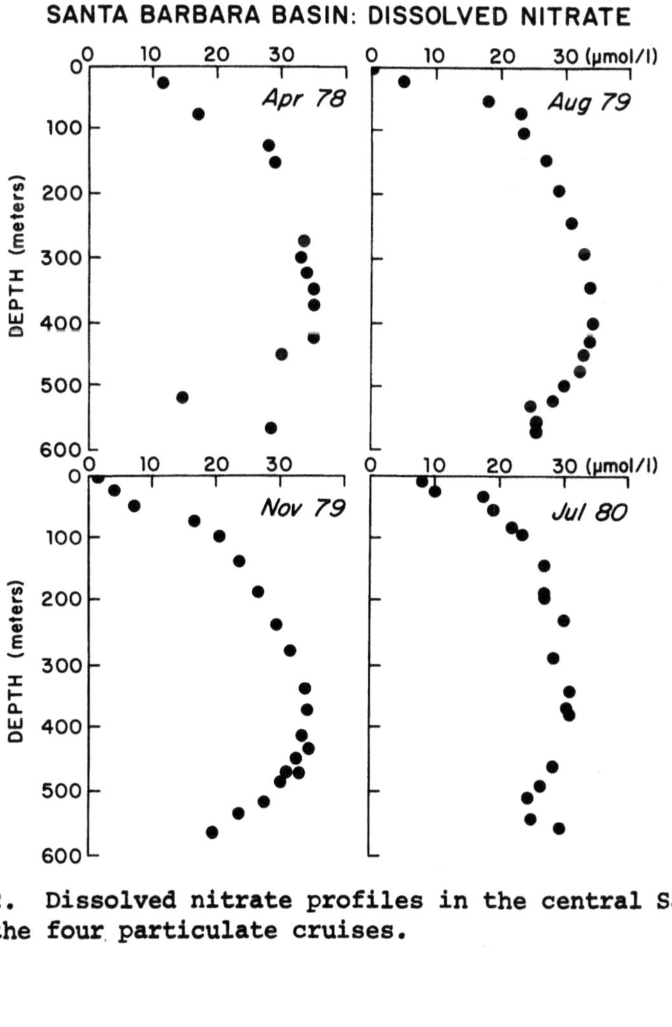

Fig. 2. Dissolved nitrate profiles in the central Santa Barbara Basin for the four particulate cruises.

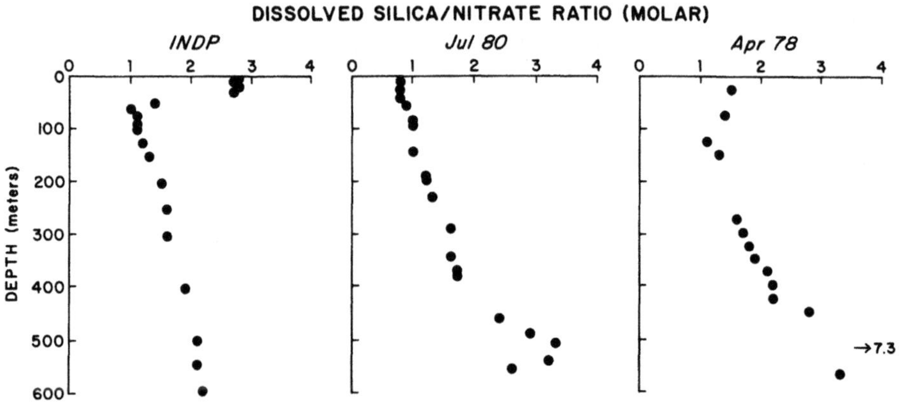

Fig. 3. Dissolved silica/nitrate molar ratio in the California Current (INDP I-1) and the Santa Barbara Basin (July 1980 and April 1978).

occurs in the basin from March to July with strongest intensities usually in May and June. Below the basin sill depth of 475 m, the water is relatively quiescent as evidenced by the occurrence of nitrate reduction. This sub-sill water can be replaced by two mechanisms: complete "flushing", which appears to be associated with periods of intense upwelling (Sholkovitz and Gieskes, 1971) and "spill-over" which replaces only a fraction of the bottom waters. These events are associated with turbidity flows (Sholkovitz and Soutar, 1975). During flushing and spillover, the replacement water apparently enters the basin from the west (Emery, 1954; Sholkovitz and Gieskes, 1971) and is readily distinguished by its higher nitrate and oxygen concentrations.

Fig. 2 shows dissolved nitrate in the water column for the four sampling periods considered in this study. In the sub-sill waters "normal" conditions were present during November 1979, a possible minor spillover is seen in the August 1979 data, and major spillover events are evident during July 1980 and April 1978. In the upper water column, upwelling is evident in the July 1980 profile. The April 1978 profile is difficult to interpret due to the lack of samples in the upper waters. Examination of the dissolved SiO_2/NO_3^- ratio is of use here. Zentara and Kamykowski (1981) have used SiO_2-NO_3^- relationships to study processes in the South Pacific, and Codispotti (this volume) has used the ratio in the Peru upwelling area. Fig. 3 shows SiO_2/NO_3^- profiles for July 1980, April 1978 and station 1 of INDOPAC I (outside the basin, 150 km from shore). The INDOPAC station gives a general California Current profile: ratios are high in the nitrate-depleted surface waters, decrease in the upper hundred meters and then slowly increase again. These ratio changes are due to the more rapid water-column regeneration of nitrate compared to silica. For the July 1980 upwelling profile, surface water values of the ratio are low and increase monotonically to sill depth. Below sill depth, the increase in the ratio is more rapid than at the INDOPAC station due to nitrate reduction in the basin. In the upper waters, the SiO_2/NO_3^- ratios of the April 1978 profile appear intermediate to the ratios of the other two profiles. Thus, seasonal upwelling may have just been starting at the time of the April 1978 cruise.

SANTA BARBARA BASIN SEDIMENTS

As particulates may be considered proto-sedimentary material, sediments from the Santa Barbara Basin were analyzed to provide an interpretive framework for the particulate study. There is, of course, no reason to expect particulate and sedimentary compositions to be in total agreement. Diagenesis will affect sedimentary composition even at the sediment-water interface; and both fecal pellets and near-bottom transport (turbidity currents, etc.) may bring to the sediments material of different composition than the fine particulates collected for this study.

Besides bulk sediment analysis, a slightly more detailed understanding of sediment composition was obtained through the analysis of sediment size fractions from a standard pipette analysis. This method of sedimentary component analysis is somewhat arbitrary, especially since the disaggregation process may break up natural aggregates and fragile biogenic particles. Nonetheless, compositional changes with size should still be apparent. The sediment sample used in this experiment was from the surface of a box core taken at the central

Table 1. Santa Barbara Basin sediment analysis

Cumulative Size (μm)	Al (%)	Mass Ratios Ti/Al	K/Al	Mg/Al	Percent of Total Sediment (%)
>53	5.9	.030	.34	.10	∿2
<44	7.2	.061	.27	.20	98
<22	7.0	.062	.26	.22	88
<16	6.8	.063	.26	.23	77
<8	7.5	.057	.25	.23	61
<4	7.5	.050	.24	.24	50
<2	7.3	.048	.19	.24	41
<1	6.1	.045	.19	.26	31
fine	2.1	.034	.16	.25	3

basin particulate collection site. In this discussion, only the mass ratios of titanium, potassium, and magnesium to aluminum will be considered as these are the least variable in the bulk sediment and least subject to diagenetic change of the sedimentary elemental ratios. It should be noted that, based on ion exchange data for Santa Barbara Basin surface sediment (Bischoff, Clancey and Booth, 1975), exchangeable potassium and magnesium contribute less than 5% of the total amount of these elements in the sediments. Thus, replacement of exchangeable ions during pipette analysis should have little effect on the results reported here.

Table 1 summarizes the results of the sediment analyses. The titanium/aluminum and potassium/aluminum ratios are seen to decrease and the magnesium/aluminum ratio increase with decreasing cumulative size. (The large changes in the >53 μm fraction are of no consequence since this fraction accounts for only ∿2% of the sediment.) In a mineralogical examination of basin sediments, Fleischer (1972) found the large size fractions dominated by quartz, plagioclase, K-feldspar, and amphibole, while the smaller size fractions showed an increasing importance of montmorillonite. Thus, the trends in the ratios of potassium and magnesium to aluminum are consistent with the mineralogy. Increases in titanium/aluminum ratios with increasing particle size have been reported in other sedimentary systems (Hirst, 1962; Price and Calvert, 1973).

As suggested earlier, these results provide an interpretive framework for the particulate data. If fine particulates are representative of total sediment composition, then ratios of titanium, potassium, and magnesium to aluminum of about 0.06, 0.3, and 0.2, respectively, are expected for the particulates. Resuspended material, which presumably is enriched in the finer sedimentary components, should have ratios of titanium, potassium, and magnesium to aluminum of about 0.04, 0.2, and 0.25, respectively. Temporal changes in these ratios may indicate seasonal changes in the predominant sources of fine particulate matter.

SANTA BARBARA BASIN PARTICULATES

Figure 4 shows typical profiles of particulate aluminum, iron, and phosphorus in the central Santa Barbara Basin. Three zones are readily apparent: surface waters, characterized by low aluminum and high phosphorus; the mid-water column, characterized by high aluminum and low phosphorus; and deeper waters, characterized by very high iron and phosphorus. With some variability, all four basin particulate profiles show these features. The surface, mid-water, and high-iron particulates can be easily distinguished on an Al + Fe + P = 100% ternary diagram (Fig. 5). To simplify further discussion, these three types of Santa Barbara Basin particulates may be defined in a somewhat arbitrary manner. All samples taken above 100 m will be

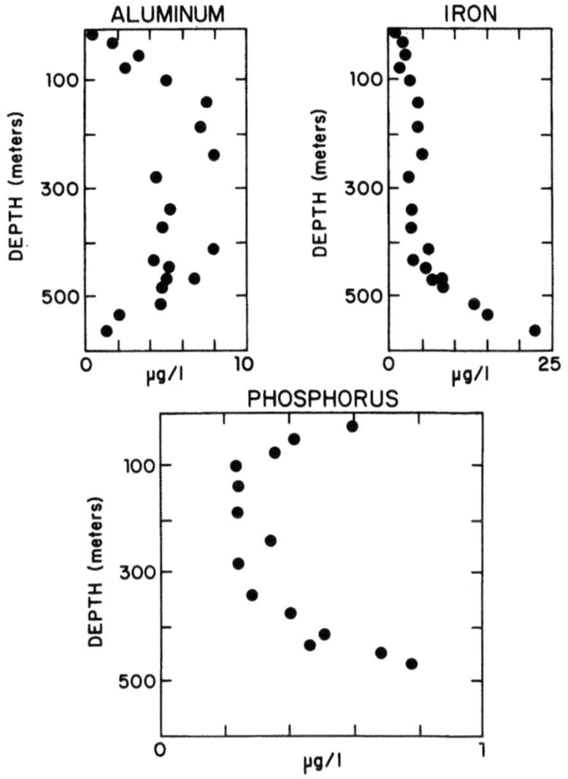

Fig. 4. Particulate composition in the Santa Barbara Basin during November 1979.

considered surface-water samples. All samples taken below 100 m that have Fe/Al ratios greater than 0.7 (in mass units) will be considered high-iron samples. The others (samples below 100 m with Fe/Al <0.7) will be the mid-water particulates. The composition of surface-water particulates is undoubtedly affected by biological productivity, while the high iron apparently results from the transport of a ferric hydroxo-phosphate phase into the basin (Shiller, 1982). In this report, discussion will be confined to the composition of the mid-water particulates.

Let us begin by considering inter-element linear correlations. This type of statistical approach has been used previously in particulate studies (Krishnaswami and Sarin, 1976; Copin-Montegut and Copin-Montegut, 1978; Buat-Menard and Chesselet, 1979) but is not without its hazards: a correlation need not imply a specific causal relationship. Table 2 summarizes the inter-element correlations for the mid-water particulates: S indicates a significant correlation at the 0.1% level, M indicates significance at the 1% level. The M level is chosen such that, with the number of available element

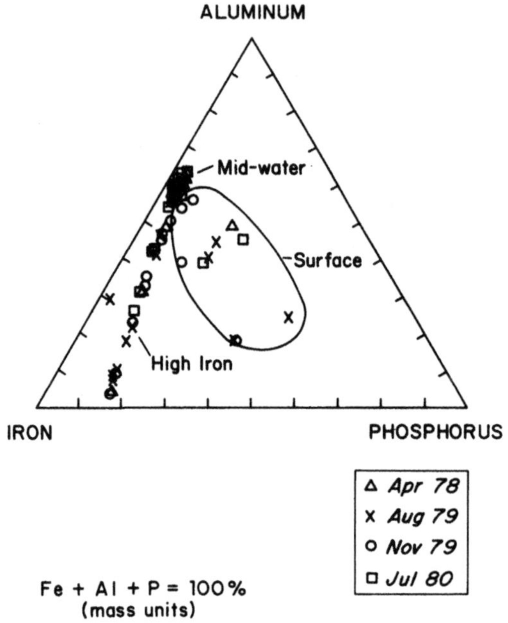

Fig. 5. Ternary composition diagram for Santa Barbara Basin particulates.

pairs, there may be one "significant" correlation in the table due to random chance. In the mid-water column it is found that the elements fall into one large group, which might be called a general sedimentary mix. Some distinction can be made among these elements in that the terrigenous elements (e.g., Al, Fe, Ti) tend to show more significant correlations than the biogenic elements (e.g., Ca, S, P).

Examining the composition of the mid-water particulates (Table 3), it is seen that the particulate Mg/Al, K/Al, and Ti/Al ratios are similar to the fine fraction of the basin sediment rather than to bulk sediment. These ratios, coupled with the general correlation of the major elements, suggest that resuspension of fine sedimentary material accounts for the major input of particles to the mid-water column. Evidence from other workers also suggests that sediment resuspension is an important factor in the basin. Drake (1971), using a series of transmissometer profiles, found tongues of turbidity extending into the basin from the Ventura Shelf; and Dymond et al. (1981) obtained basin nephelometer data that are best explained by particulate advection.

To better understand the source of the mid-water particulates, two stations were occupied along the northwest slope of the basin (Fig. 1) during the July 1980 cruise. A total of four samples within 65 m of the bottom were taken there. Aluminum concentrations for

Table 2. Interelement correlations: mid-water column

	Al	Fe	Ti	K	Si	P	Mn	S	Mg	Ca
Al		S	S	S	S	S	S	S	S	M
Fe	S		S	S	S	S	S	S	S	M
Ti	S	S		S	S	M	S	S	S	S
K	S	S	S		S	S	S	S	M	M
Si	S	S	S	S		M	S	S	M	
P	S	S	M	S	M		M	S		
Mn	S	S	S	S	S	M				
S	S	S	S	S	S	S			M	
Mg	S	S	S	M	M			M		M
Ca	M	M	S	M					M	

S: Significant at $p < 0.001$.
M: Significant at $p < 0.01$.

these samples were higher than for any of the central basin samples. A comparison of the average composition of these four samples and the average mid-water composition for the July 1980 central basin station is shown in Table 3. The ratios of titanium, potassium, and magnesium to aluminum in the basin-slope samples are similar to both the mid-water and fine-fraction sediment ratios. It is interesting to note that for the slope samples, the Ti/Al and K/Al ratios are slightly higher and the Mg/Al ratio slightly lower than the mid-water average. This would indicate that the slope samples are composed of slightly coarser material. This seems likely since the slope samples are closer to the sedimentary source and coarser particles have thus had less time to settle out. That there appears to be only a slight change in the three ratios, though, suggests that very little coarse material is resuspended in the first place. That is, during this observation period at least (July 1980), bottom current shear was not great enough to cause wholesale transport of slope sediments.

The ratios of the biogenic elements (Ca, Si, P, S) to aluminum in Table 3 provide additional source information regarding these elements. Note that all of the biogenic elements are enriched in the mid-water averages when compared to the slope samples. This indicates that the primary flux of material from the surface waters contributes biogenic material to the mid-water column. Some diatoms and forams have been shown to settle as individuals in the basin (Dunbar and Berger, 1981). These freely settling particles could provide the needed extra input. Additionally, the mid-water bacteria and microplankton could contribute to the mid-water samples. Unfortunately, little is known about the mid-water microbiota.

Table 3. Average mid-water particulate compositions in the Santa Barbara Basin in μg/ℓ

	Cruise				
	Central Basin				Slope
	APR78	AUG79	NOV79	JUL80	JUL80
Al	4.8	5.6	6.0	8.3	20.2*
Fe/Al	.59	.62	.62	.59	.58
Ti/Al	.034	.035	.033	.031	.032
K/Al	.21	.21	.21	.21	.23
Si/Al	4.0	5.4	3.6	4.8	4.0
P/Al	.060	.048	.050	.040	.032
S/Al	.057	.052	.062	.055	.036
Mg/Al	.28	.31	--	.31	.28
Ca/Al	.59	.67	.59	.37	.34
$MnAlx10^3$	5.1	5.0	4.9	5.0	4.0

* Average of four samples: 12.3, 14.7, 16.1 and 37.8 μg Al/liter.

Let us finally examine the temporal variability of particulate composition. Elements most closely associated with terrigenous matter (Al, K, Fe, Mn, Ti, Mg) show little variability in their element/aluminum ratios (Table 3). This is consistent with the resuspension origin of the particulates (Feely, Massoth and Landing, 1981). Elements with a substantial biogenic contribution (P, Si, Ca, S) do show some temporal variability, again consistent with an extra primary input of these elements. Increased biogenic contribution, though, does not appear to correspond to the upwelling situation for the July 1980 cruise. This need not be of concern as there is undoubtedly a time lag between the appearance of the hydrographic upwelling signal, the response of the biota, and the transfer of primary biogenic material from the surface waters to the mid-water column. Additionally, upwelling events need not always be associated with increased productivity and vice versa.

One compositional change which does correspond with the July 1980 upwelling event is the increase in the aluminum concentration. From the nitrate profiles it would seem that there was substantial water movement at the time of the July 1980 sampling. Note that the maximum nitrate concentration for this profile is several μmoℓ/ℓ less than the maximum concentrations of the other profiles (Fig. 2). Evidently, upwelling involves the movement of water at all depths in the basin, thereby increasing the amount of resuspended particles in the water column (see Sholkovitz and Gieskes, 1971). For the April 1978 profile, based on the intermediate SiO_2/NO_3^- ratios in the upper water column and the sharpness of the sub-sill nitrate gradient, an upwelling-spillover event appears to have just begun. The low average aluminum concentration suggests that there had not been enough time for newly resuspended material to appear in the mid-water column. Nonetheless, the bottom sample of this profile (characterized by high-nitrate basin replacement water) does have the highest aluminum concentration of all sub-sill particulate samples ($\sim$7 μg/ℓ), thus reinforcing the upwelling-spillover-resuspension connection.

CONCLUSIONS

Previous work has shown how resuspension causes transport of sediment from the local shelf and slope areas to the Santa Barbara Basin (Drake, Fleischer and Kolpack, 1971; Drake, Kolpack and Fischer, 1971) and thereby causes "excess" deposition of ^{210}Pb (Krishnaswami et al., 1973) and ^{228}Th (Moore, Bruland and Michel, 1981). In this work, resuspension has been shown to be the major source of mid-water particles in the basin. Upwelling appears to increase the amount of resuspended matter in the water column.

That the fine particulates are mainly resuspended sediments may have implications for coastal productivity. The resuspended sediments could increase productivity by allowing more particulate nutrients to be regenerated in the water column rather than in the sediments. Zooplankton may also benefit directly by the ingestion of resuspended particulate nutrients. More than just nutrients could be added to the water column in the resuspension process. Boyle, Huested and Jones (1981) have discussed coastal zone trace-metal enrichment in the water column. If resuspended particles are sites of oxidative regeneration, organically bound trace metals may also be released.

ACKNOWLEDGEMENTS

I wish to thank J. Gieskes and B. Price for their help and advice during all phases of this project. This paper was improved following the suggestions of J. Dymond and another anonymous reviewer. L. Jackson helped prepare the manuscript, and figures were drafted by W. Borst. This work was supported by NSF grant OCE78-27376.

REFERENCES

Bischoff, J.L., Clancey, J.J. and Booth, J.S., 1975, Magnesium removal in reducing marine sediments by cation exchange, Geochimica et Cosmochimica Acta, 39:559-568.

Boyle, E.A., Huested, S.S. and Jones, S.P., 1981, On the distribution of copper, nickel, and cadmium in the surface waters of the North Atlantic and North Pacific Ocean, Journal of Geophysical Research, 86:8048-8066.

Buat-Menard, P. and Chesselet, R., 1979, Variable influence of the atmospheric flux on the trace metal chemistry of oceanic suspended matter, Earth and Planetary Science Letters, 42:399-411.

Copin-Montegut, C. and Copin-Montegut, G., 1978, The chemistry of particulate matter from the south Indian and Antarctic Oceans, Deep-Sea Research, 25:911-931.

Drake, D.E., 1971, Suspended sediment and thermal stratification in Santa Barbara Channel, California, Deep-Sea Research, 18:763-769.

Drake, D.E., Fleischer, P. and Kolpack, R.L., 1971, Transport and deposition of flood sediment, Santa Barbara Channel, California, in: "Biological and Oceanographic Survey of the Santa Barbara Channel Oil Spill 1969-1970," Vol. II, R.L. Kolpack, ed., Allan Hancock Foundation, 181-217.

Drake, D.E., Kolpack, R.L. and Fischer, P.J., 1971, Sediment transport on the Santa Barbara-Oxnard Shelf, Santa Barbara Channel, California, in: "Shelf Sediment Transport," D.J.P. Swift, D.B. Duane, and O.H. Pilkey, eds., Dowden, Hutchinson and Ross, 307-331.

Dunbar, R.B. and Berger, W.H., 1981, Fecal pellet flux to modern bottom sediment of Santa Barbara Basin (California) based on sediment trapping, Geological Society of America, Bulletin, 92: 212-218.

Dymond, J., Fischer, K., Clauson, M., Cobler, R., Gardner, W., Richardson, M.J., Berger, W.H., Soutar, A., and Dunbar, R., 1981, A sediment trap inter-comparison study in the Santa Barbara Basin, Earth and Planetary Science Letters, 53:409-418.

Emery, K.O., 1954, Source of water in basins off southern California, Journal of Marine Research, 13:1-21.

Feely, R.A., Massoth, G.J. and Landing, W.M., 1981, Major- and trace-element composition of suspended matter in the north-east Gulf of Alaska: relationships with major sources, Marine Chemistry, 10:431-453.

Fleischer, P., 1972, Mineralogy and sedimentation history, Santa Barbara Basin, California, Journal of Sedimentary Petrology, 43:49-59.

Hirst, D.M., 1962, The geochemistry of modern sediments from the Gulf of Paria. The relationship between mineralogy and the major elements, Geochimica et Cosmochimica Acta, 26:309-334.

Krishnaswami, S., Lal, D., Amin, B.S., and Soutar, A., 1973, Geochronological studies in Santa Barbara Basin: ^{55}Fe as a unique trace for particulate settling, Limnology and Oceanography, 18: 763-770.

Krishnaswami, S. and Sarin, M.M., 1976, Atlantic surface particulates: composition, settling rates and dissolution in the deep sea, Earth and Planetary Science Letters, 32:430-440.

Moore, W.S., Bruland, K.W. and Michel, J., 1981, Fluxes of uranium and thorium series isotopes in the Santa Barbara Basin, Earth and Planetary Science Letters, 53:391-399.

Price, N.B. and Calvert, S.E., 1973, A study of the geochemistry of suspended particulate matter in coastal waters, Marine Chemistry, 1:169-190.

Shapiro, L., 1967, Rapid analysis of rocks and interals by a single-solution method, U.S. Geological Survey Professional Paper 575-B:B187-B191.

Shapiro, L. and Brannock, W.W., 1962, Rapid analysis of silicate, carbonate and phosphate rocks, U.S. Geological Survey, Bulletin, 1144-A:1-56.

Shiller, A.M., 1982, "The Geochemistry of Particulate Major Elements in the Santa Barbara Basin and Observations on the Calcium Carbonate-Carbon Dioxide System in the Ocean," Ph.D. Dissertation, University of California, San Diego, 197 pp.

Sholkovitz, E.R. and Gieskes, J.M., 1971, A physical-chemical study of the flushing of the Santa Barbara Basin, Limnology and Oceanography, 16:479-489.

Sholkovitz, E.R. and Price, N.B., 1980, The major-element chemistry of suspended matter in the Amazon Estuary, Geochimica et Cosmochimica Acta, 44:163-171.

Sholkovitz, E.R. and Soutar, A., 1975, Changes in the composition of the bottom water of the Santa Barbara Basin: effect of turbidity currents, Deep-Sea Research, 22:13-21.

Soutar, A., Kling, S.A., Crill, P.A., Dufferin, E., and Bruland, K.W., 1977, Monitoring the marine environment through sedimentation, Nature, 266:136-139.

Strickland, J.D.H. and Parsons, T.R., 1968, A manual of sea-water analysis, Bulletin of the Fisheries Research Board of Canada, 167:311 pp.

Zentara, S.-J. and Kamykowski, D., 1981, Geographic variations in the relationship between silicic acid and nitrate in the South Pacific Ocean, Deep-Sea Research, 28:455-466.

ZOOPLANKTON AND NEKTON: NATURAL BARRIERS TO THE SEAWARD TRANSPORT OF SUSPENDED TERRIGENOUS PARTICLES OFF PERU

Kenneth F. Scheidegger and Lawrence A. Krissek*

School of Oceanography
Oregon State University
Corvallis, Oregon 97331, U.S.A.
*Present Address:
Department of Geology
Ohio State University
Columbis, Ohio 43210, U.S.A.

ABSTRACT

Some of the strongest biogenic signals of coastal upwelling reported in Recent marine sediments (e.g., organic carbon content >20 wt. %) are found in a mud facies on the inner continental margin off Peru, particularly between 10° and 14°S. Terrigenous sediments found there are anomalously fine grained and have overall grain-size distributions which are similar to those of deep-sea sediments collected seaward of the Peru Trench. To explain the observed dispersal patterns of fine silts and clays on the inner continental margin, rapid vertical transport of large aggregates (fecal pellets) from a wind-driven surface layer is required.

For the Peru upwelling zone, zooplankton ingest phytoplankton cells and occur in sufficient numbers to remove the total suspended particulate content from the surface layer each day. In addition, large populations of anchoveta feed on the zooplankton and produce fecal matter which can dominate the flux of particles from the surface layer to the seafloor. Sediment budget calculations for the mud facies indicate that most of the terrigenous sediment from adjacent continental fluvial sources is trapped in the inner continental margin environments each year by the combined efforts of zooplankton and anchoveta. Phytoplankton abundances appear to increase in response to seasonal fluctuations in upwelling intensity, a process ultimately controlled by annual fluctuations in the intensity of low-level trade winds. Zooplankton and nekton populations increase as the availability of food increases, and their peak abundances coincide with the

December to April period of peak discharge from coastal rivers. This places the largest biological filtering capability of the year at the time of maximum terrigenous sediment input, and may explain why so little terrigenous material escapes from the continental margin.

The preservation of biogenous and fine-grained terrigenous particles at the sediment-water interface of the mud facies is ensured by a combination of weak-to-negligible bottom currents and oxygen-deficient bottom waters. The former prevents winnowing of fine-grained particles from the sedimentary components which reach the sea floor; the latter inhibits the development of a large benthic fauna which might rework the deposits and thereby destroy the strong signals of coastal upwelling.

INTRODUCTION

Sedimentologists, micropaleontologists and geochemists have examined sediments beneath coastal upwelling centers (e.g., Peru, Namibia and northwest Africa) to find key indicators that reflect the long-term and anomalously high productivity of the overlying water column. High organic carbon and phosphate contents, high biogenous opal contents, concentrations of fish debris, key diatom species, and various lipids and amino acids are among the suggested signals (DeMaster, 1981; Schuette and Schrader, 1981; DeVries and Pearcy, 1982; Brassell and Eglinton; Staresinic et al.; Reimers and Suess; all this volume). A major problem, however, is that abundances or accumulation rates of such components can vary markedly among upwelling systems or even in a localized area. As Krissek and Scheidegger (this volume) have stressed, deposition and preservation of sediments bearing strong coastal upwelling signals can only occur if productivity is high, bottom water remains oxygen poor, bottom currents are negligible, and the accumulation rate of terrigenous components is minimal.

Among recognized eastern-boundary-current upwelling centers, the Peru system is emerging as one of the best examples of a sedimentary environment where signals of primary productivity are well preserved (e.g., Krissek, Scheidegger and Kulm, 1980; Reimers and Suess, this volume). Off Peru, sediments strongly influenced by upwelling are found in an upper slope mud lens between 10° and 14°S where a mid-water oxygen minimum layer impinges on the sea floor. These sediments exhibit extremely high values or abundances of key indicators. In addition, terrigenous sediments found in this mud lens are anomalously fine grained for upper continental deposits, and they resemble adjacent deep-sea deposits in overall grain-size distributions (Krissek et al., 1980; Scheidegger and Krissek, 1982). Such deposits deserve special attention because they may provide clues that will enable us to better understand upwelling and its imprint on the sediment record.

In this paper we examine the salient factors that appear responsible for the formation and preservation of the upper slope mud lens off Peru. We initially review observations on the nature, distribution and characteristics of the sediments in relation to known centers of upwelling and to the ambient current regime. Our goal is to provide a synthesis of the factors and processes recognized from various disciplines of oceanography which pertain to the origin of this unique deposit. Sedimentary budgets are then developed for the mud lens, and they show that much of the fine-grained terrigenous sediment supplied by adjacent continental sources is deposited in this environment. We show that the locations, concentrations and annual fluctuations of various biological communities in the surface layer of the water column over the mud lens combine to make biological feeding activity a natural "filter" or barrier to the seaward transport of suspended terrigenous particulate matter in the wind-driven surface layer. We suggest that terrigenous and biogenous particles are incorporated into fecal matter of zooplankton and nekton and settle rapidly to the oxygen-deficient sedimentary environment of the mid-shelf to upper slope off Peru. They are preserved by this environment and create a strong biological imprint on a developing sedimentary record.

PHYSICAL OCEANOGRAPHY OF THE PERU UPWELLING ZONE: INFLUENCE ON THE BIOTA

Brockmann et al. (1980) and Smith (1981) have recently summarized available data on currents and upwelling along the Peru coast. They note that surface currents along the Peru continental margin are driven toward the northwest (equatorward) by persistent southeasterly trade winds (Fig. 1; Smith, 1981). Such currents typically flow at velocities of 10-30 km/day (Zuta and Guillén, 1970). Because of wind-stress-induced Ekman transport, the surface layer (<25 m in depth) is driven away from the coast, thus causing nutrient-rich water to be upwelled from depths of 100 m or less and creating conditions favorable for high primary production.

Beneath the surface layer, a poleward undercurrent, fed by oxygen-deficient and nutrient-rich water of the South Equatorial countercurrent, flows at velocities of 5-20 cm/sec typically at water depths between 50 and 150 m (Fig. 1; Brockmann et al., 1980). This current is found over the mid-shelf to upper slope. Limited data exist on near-bottom current velocities along the upper Peru continental margin where the poleward undercurrent may interact with the sea floor. Brockmann et al. (1980) note much stonger poleward undercurrent velocities at 200 m near 5°S than at 12° and 15°S (Fig. 1). Currents measured at greater depth were commonly less than 5 cm/sec and variable in direction.

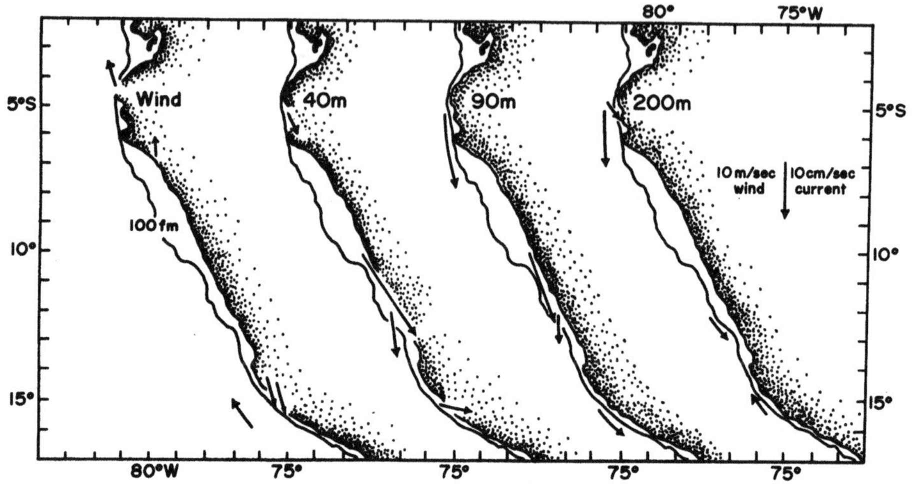

Fig. 1. Mean wind and current vectors at different locations along the Peru coast for a 38-day period (2 April to 10 May 1977). The edge of the continental shelf is indicated by the 100 fm contour. The wind is generally equatorward at all locations, and the currents at about 40 and 90 m are all poleward. Data and illustration are from Brockman et al. (1980). By permission from Deep-Sea Res., Copyright (c) 1980, Pergamon Press, Ltd.

Oxygen concentrations and the topography of the oxygen minimum zone in such a physical oceanographic setting are crucial to both the distribution of zooplankton and nekton (Jordan, 1971; Boyd, Smith and Cowles, 1980; Smith et al., 1981a) and to the preservation of sedimentary components on the inner continental margin off Peru (Krissek and Scheidegger, this volume). Off northern Peru near 5°S, the oxygen minimum is typically located near 400 m and occupies a rather limited depth range, whereas near 15°S oxygen-deficient water extends from about 100 to more than 400 m (Fig. 2; Zuta and Guillén, 1970; Brockmann et al., 1980). From these observations and from data compiled by Zuta and Guillén (1970), it appears that mid-shelf to upper slope sedimentary environments are bathed in oxygen-depleted water and this results in a near absence of benthic fauna (Soutar, Johnson and Baumgartner, 1982). A marked decrease in percent oxygen saturation also coincides with the top of the poleward undercurrent at a depth of <50 m (see Fig. 2), and this dramatic change in oxygen availability confines the distribution of zooplankton and nekton to the more oxygenated surface waters (Jordan, 1971; Boyd et al., 1980; Smith et al., 1981a).

Surface primary productivity values in the Peru coastal upwelling zone are among the highest measured in coastal waters, commonly exceeding 1.0 $gC/m^2/day$ (Fig. 3; Zuta and Guillén, 1970). Values of 0.5 $gC/m^2/day$ or greater are normally encountered within about 100 km

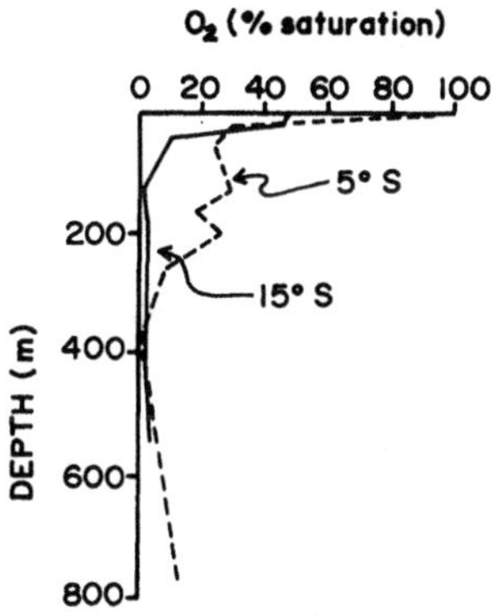

Fig. 2. Profiles of dissolved oxygen (percent of saturation) over the upper continental slope at 5°S (dashed) and 15°S (solid). Data and illustration modified from Brockman et al. (1980).

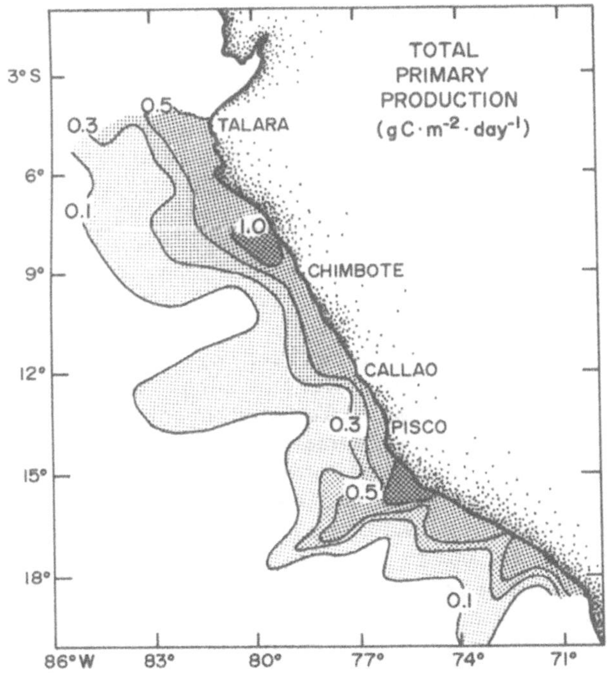

Fig. 3. The distribution of total primary production over the Peru continental margin as compiled by Zuta and Guillén (1970) from measurements of surface water during the years 1960-1970.

from shore. Such conditions prevail during "normal" years, but high productivity values can be dramatically reduced with the penetration of warm equatorial surface waters during El Niño events (Zuta, Rivera and Bustamente, 1978).

DISTRIBUTION AND NATURE OF THE MUD LENS

Available 3.5 kHz seismic records from the Peru continental margin have been examined (Krissek et al., 1980), and they provide insight into the distribution and thickness of an upper slope mud lens and its shelf equivalent which are associated with the upwelling regime. Between 10° and about 14°S, such mud deposits are found between a water depth of 500 m and inner shelf depths of perhaps 50 m (Fig. 4). At both extremes in depth the muds pinch out (Fig. 4). The mud facies can be traced to the north and south of the 10°-14°S region, but in both cases it appears to be confined to shelf depths of <200 m (Fig. 4). The facies is more than 100 km wide to the north of 10°S where the shelf is wide, but it narrows dramatically to the south to widths of less than 20 km (Fig. 4). Marked increases in the steepness of the slope appear to limit the seaward extent of the muds, and wave base and associated turbulence limit its landward extent. Characteristic maximum thicknesses of the muds are about 22 m between 8° and 10°S, 38 m between 10° and 15°S, and near 15 m near 16°S.

Grain-size analyses have been performed on surface sediment samples collected from the mud lens and surrounding areas, and they indicate that the sediments deposited within the mud facies are anomalously fine grained for a continental margin setting (Krissek et al., 1980). Many of the key textural characteristics of the sediments are apparent in Fig. 5 where we present histograms of the grain-size data for samples taken along transects across the mud lens at 11° and 13°S. Samples collected from the mud lens typically have >60% clay (>8 phi), silt fractions characterized by significant quantities of medium to very fine silt (5-8 phi), a near absence of coarse silt (4-5 phi), and a small but variable sand fraction which consists dominantly of biogenic calcium carbonate (Krissek and Scheidegger, this volume). Terrigenous particles are dominant in the 4- to 6.5-phi size range, and terrigenous particles mixed with diatom frustules and fragments are subequal in abundance in the finer fractions. Seaward of the mud lens and to water depths of at least 1500 m, upper slope sediments are much coarser; they typically have low clay contents (<40%), silt fractions dominated by coarse- to medium-grained terrigenous silts, and high sand contents. At mid to lower slope depth and continuing to sites located seaward of the Peru trench, the expected seaward fining of sediments is observed (Fig. 5; Krissek et al., 1980; Scheidegger and Krissek, 1982). This includes an increase in the percentage of clay, a shift in mean silt grain-size toward fine silt, and a near absence of particles in the coarse silt size

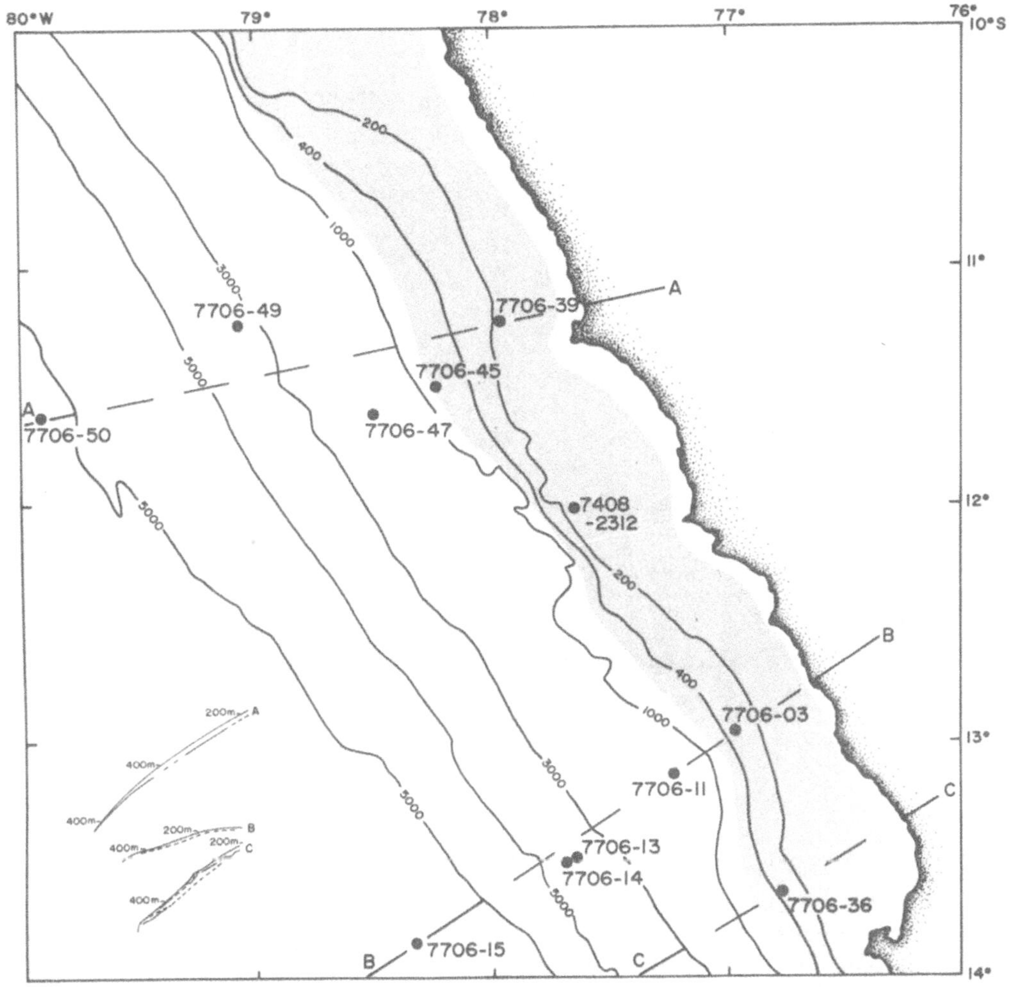

Fig. 4. Map showing the distribution of the mud facies on the upper Peruvian continental margin as determined from 3.5 kHz seismic records. Selected line drawings of records across the mud facies are shown. Except for Scripps core 7408-1909, which occurs at 14°39'S and at a water depth of 183 m, the locations of all cores discussed in this report are shown.

range (4-5 phi). We thus find sediments on the lower slope and in adjacent deep-sea environments with grain-size distributions which closely resemble those deposited on the upper slope within the mud lens.

A lack of bottom current measurements along the two transects makes it difficult to fully assess the significance of the textural

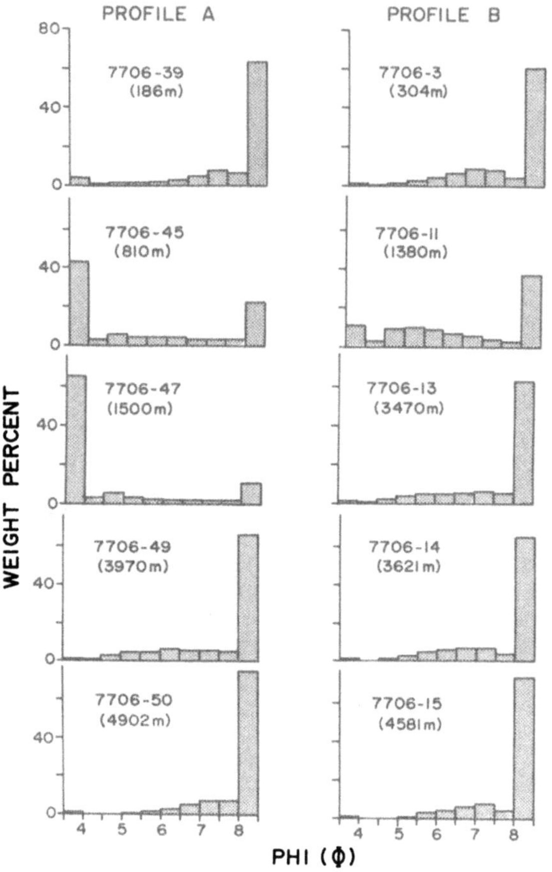

Fig. 5. Histograms of grain-size data obtained on 10 surface sediment samples from profiles A and B (see Fig. 4 for core locations). Far left and far right portions of each histogram correspond to all material coarser that 4 phi (62 μm) and finer than 8 phi (4 μm), respectively. Note presence of coarser sediments seaward of the inner continental mud facies cores (cores 7706-3 and 7706-39), and the similarity in overall grain-size distributions of samples 7706-50 and 7706-15, located seaward of the Peru Trench, and the two mud facies samples.

data. However, Brockmann et al. (1980) note that poleward undercurrent velocities at 200 m near 5°S are much stronger than at 12° and 15°S (Fig. 1). These observations are consistent with the very fine-grained sediments deposited at 11° and 13°S (Fig. 5) and the tendency for upper slope sediments to be coarser north of the mud lens (Krissek et al., 1980). The grain-size data and available bottom current observations point to the presence of tranquil near-bottom conditions associated with the mud lens.

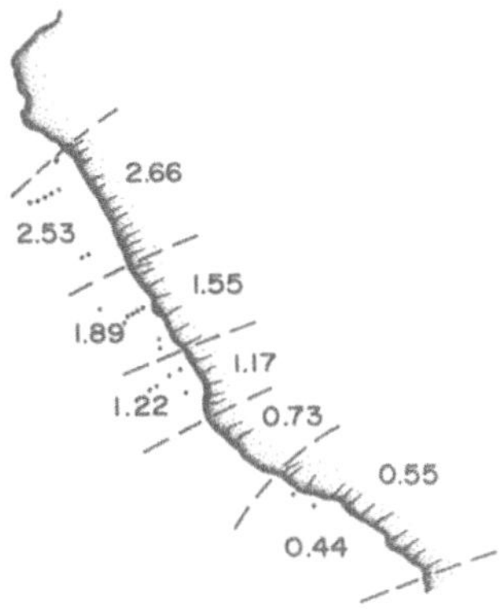

Fig. 6. Sketch map showing the average quartz/plagioclase feldspar ratios found in <4 μm material collected from cores taken on the Peru continental margin (modified from Scheidegger and Krissek, 1982). Offshore provinces with average values of 1.22 and 1.89 correspond to the 10° to 14°S area shown in Fig. 4.

SOURCES AND DISPERSAL PATTERNS OF TERRIGENOUS SEDIMENTARY COMPONENTS

Scheidegger and Krissek (1982) have used latitudinal differences in the quartz to plagioclase feldspar ratios of material derived from the Peru coastal sediment sources to document dispersal patterns of fine-grained terrigenous sedimentary components to continental margin and neighboring deep-sea depositional environments. Mineral data from the Peru continental margin are summarized in Fig. 6. Quartz to plagioclase feldspar ratios of silt- and clay-sized components on the slope mirror comparable values for sediments transported by adjacent coastal streams. These data point to pronounced offshore transport from distinctive continental sediment sources. If dispersal patterns for neighboring deep-sea environments are considered (Scheidegger and Krissek, 1982), it is clear that there is also a significant northwesterly component to the transport directions. The relationship of the physical oceanographic setting of the area (Fig. 1) to the observed dispersal patterns of fine-grained terrigenous components (Fig. 6) suggests that the wind-driven surface currents and resulting Ekman transport are responsible for transporting particles to the northwest.

CHARACTERISTICS DEPENDENT ON HIGH PRODUCTIVITY AND THE O_2 MINIMUM

Some of the distinctive indicators of pronounced coastal upwelling include high organic carbon and phosphate abundances and high biogenic opal concentrations in the underlying sediments (Diester-Haass, 1978). For the upper slope mud facies off Peru, organic car-

Table 1. Chemical compositions of representative surface sediment samples from the mud facies off Peru (see Fig. 4 for core locations). All values are expressed in weight percent except for P-org and PO_4 (soluble) which are in ppm.

	7706-3	7706-36	7706-39
SiO_2	45.27	37.72	57.84
Al_2O_3	11.62	7.08	7.70
Fe_2O_3	3.11	1.98	2.50
MgO	1.75	1.63	1.44
CaO	2.75	2.48	2.09
Na_2O	1.41	0.88	0.74
K_2O	1.77	1.05	1.04
TiO_2	0.43	2.37	0.26
MnO_2	0.04	0.03	0.03
$CaCO_3$	2.89	2.01	1.64
C-org	14.52	21.16	12.54
P-org	632	1120	707
PO_4	3638	2879	1105

bon values commonly exceed 10 wt.%, with maximum >20 wt.%, while total phosphate ranges from about 2000 to 4000 ppm (Table 1). Biogenic silica contents in samples recovered from the mud facies typically range from 16 to 19 wt.% (determined by Na_2CO_3 leach; DeMaster, 1981). The relation,

$$SiO_2(\text{opal}) = SiO_2(\text{total}) - 3\ Al_2O_3(\text{total}),$$

provides an independent estimate of biogenic silica (Leinen, 1977), and yields an average biogenic silica content of 20.5% for the samples listed in Table 1. Combining the organic carbon, phosphate, opal, and carbonate (Table 1) contents, we find that as much as 40% of the sediments in the mud facies is biogenic material. Because of the low specific gravity of fibrous organic matter and opaline silica (0.2 g/cm^3; Heath, 1974), the resultant sediment is volumetrically dominated by biogenous components (Reimers and Suess, this volume) and has low dry bulk densities and related anomalous geotechnical properties (Busch and Keller, 1982).

Krissek et al. (1980) and Reimers and Suess (this volume) have shown that the magnitude of the signals of coastal upwelling vary latitudinally along the Peru margin and are most intense between 11° and about 15°S in the mud facies where the O_2-minimum impinges on the sea floor. The strength of the O_2-minimum also intensifies to the south and reaches a maximum in this area (Brockmann et al., 1980) due to continual oxygen utilization by bacteria and animals in waters of

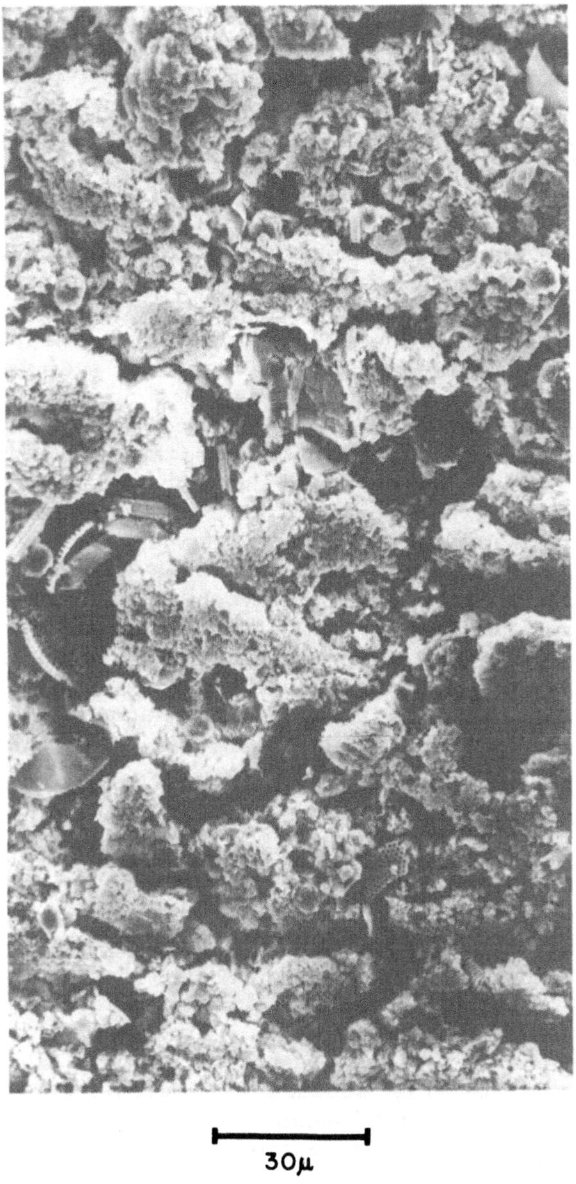

Fig. 7. Scanning electron photomicrograph of sediment fabric from core 7706-41 (2 cm level) taken in the organic-rich surface sediments of the Peru margin. Core 7706-41 is located between cores 7706-39 and 7706-45 at a water depth of 411 m. Note the presence of organo-mineral aggregates with dimensions of 20 to 80 µm. Much of the terrigenous components found in cores from the mud facies is incorporated into such aggregates. By permission from C.E. Reimers, Copyright (c) 1981.

the poleward undercurrent (Barber and Smith, 1981). Under these low oxygen bottom conditions, the organic matter undergoes limited oxidation (Reimers and Suess, this volume). Therefore, there is little opportunity for oxidative degradation of biogenous components as they settle through the water column; once on the bottom, these same waters inhibit the development of benthic biological communities that could ingest and destroy such organic-rich detritus (Soutar et al., 1982).

A key observation which helps to explain the origin of the anomalous fine-grained, organic-rich mud facies is the presence of ubiquitous fecal pellets and aggregates in its surface sediments (Krissek et al., 1980; Reimers and Suess, this volume). The pellets and related aggregates are commonly up to 80 µm long and perhaps 20-30 µm in width (see Fig. 7), although DeVries has noted that the >500 µm fraction of some of the sediment samples contains much larger fecal pellets. The volumes of the smaller aggregates fall in the range of 10^5 to 10^3 μm^3 and may thus represent the fecal pellets of small copepods (Small, Fowler and Ünlü, 1979). The ones larger than 500 µm are grooved and rod-shaped and have been tentatively identified as being produced by galatheid crabs (DeVries, 1979; Soutar et al., 1982). Because such crabs require oxygen levels above what is commonly observed over the mud lens, Soutar and co-workers (1982) speculate that every few years oxygen levels may become high enough to support a temporary population of such organisms. At such times, the crabs would be expected to ingest smaller aggregates of copepods, loosely bound fecal pellets of anchoveta, or other biogenous or terrigenous debris which may be on the bottom.

ACCUMULATION RATES AND SEDIMENT BUDGET CALCULATIONS FOR THE MUD FACIES

Because many key compositional, physical and time-dependent characteristics are known for the mud facies, it is possible to estimate how rapidly particular sedimentary components are accumulating. In addition, for some components (e.g., fine-grained terrigenous sediments) it is possible to estimate how much material is entering the marine environment. In this section we address the problem of how efficiently fine-grained terrigenous sediments are trapped in the shelf to upper slope mud facies located between 10° and 14°S.

Sedimentation rates are available for six cores located in the mud facies from 11°15'S to 14°39'S (compare Fig. 4 and Table 2; DeMaster, 1979; Koide and Goldberg, 1981; Reimers and Suess, this volume). The dated cores appear representative geographically and cover water depths between 186 and 411 m; we find no apparent correlation between sedimentation rates and either water depth or geographical location. The observed sedimentation rates vary by an order of magnitude (30-340 cm/1000 y), although higher rates appear

Table 2. Data used to estimate accumulation rates of terrigenous components in six cores from the mud lens off Peru. Accumulation rate of terrigenous components represents the difference between the bulk and sum of organic C, $CaCO_3$ and opal accumulation rates. Values in parentheses were calculated according to assumptions noted below. Data are from Reimers and Suess (this volume) unless otherwise noted.

Core	Latitude	Longitude	Water Depth (m)	Sedimentation Rates (cm/1000 y)	Accumulation Rates ($g/cm^2/1000$ y)				
					Bulk	C-org	$CaCO_3$	Opal	Terrigenous Components
7706-04	12°58.9'S	76°58.0'W	325	30	9	1.3	0.2	(0.5)[f]	7.0
7706-36	13°37.3'S	76°50.5'W	370	50[a]	11	1.6	0.2	0.6[a]	8.6
7706-39	11°15.1'S	77°57.4'W	186	160[a]	28	3.3	0.6	4.8[a]	19.3
7706-41	11°20.6'S	78°07.0'W	411	160	33	6.3	3.4	(5.6)[g]	15.3
7408-2312	12°02.1'S	77°43.0'W	200	340[b]	(75)[c]	(8.0)[d]	(1.1)[e]	12.7[a]	53.2
7408-1909	14°38.9'S	76°10.3'W	366	320[b]	(70)[c]	(8.0)[d]	(1.1)[e]	16.8[a]	44.1

[a] = DeMaster, 1979.
[b] = Koide and Goldberg, 1981.
[c] = Assumed dry bulk density of 0.22 g/cm^3 (Busch and Keller, 1982).
[d] = Based on assumption maximum accumulation rates of organic C of 9-12 $g/cm^2/1000$ y occur near 15°S where sedimentation rates in excess of 1000 cm/1000 y are found (Reimers and Suess, this volume).
[e] = Average of cores with 7706 numbers.
[f] = Assumed same opal content (5%) as core 7706-36.
[g] = Assumed same opal content (17%) as core 7706-39.

more typical (Table 2). In this regard, Heinrichs (1980) indicates that sedimentation rates in the mud facies are 1100 and 1200 cm/1000 y at 15°04'S and 15°09'S, respectively.

For the six cores within the study area, dry bulk densities, bulk accumulation rates, and accumulation rates of organic carbon, $CaCO_3$ and opal are known or can be closely approximated. From these data it is possible to use the differences between the bulk and total biogenic accumulation rates to obtain a meaningful estimate of the terrigenous component accumulation rate at each site (see Table 2). Our estimates indicate that accumulation rates of terrigenous sediment components range from 10 to about 50 $g/cm^2/1000$ y.

The known distribution of the mud facies between 10° and 14°S (Fig. 4) and our accumulation rate estimates of terrigenous components can be used to calculate the mass of fine-grained terrigenous sediment deposited annually within the mud facies. We calculated the area of the mud facies between 10° and 14°S and water depths of 50 and 500 m to be 25.3×10^{13} cm^2. Multiplying this area by accumulation rates of 10, 25 and 50 $g/cm^2/1000$ y, we estimate the total mass flux for low, mean and high accumulation rates to be 2.5×10^{12}, 6.3×10^{12} and 12.6×10^{12} g/y, respectively.

From previous discussion of the mineralogy of these fine-grained terrigenous sediments (see Fig. 6), we are confident that they have been derived from the neighboring small coastal streams. Zuta and Guillen (1970) present data on the monthly and total yearly river runoff values for the 10 major rivers between 10° and 14°S over approximately a 20-year interval; total annual fluvial discharge calculated from these data is 7.6×10^9 m^3/y. Krissek and Scheidegger (this volume) have considered the problem of estimating the amount of suspended terrigenous sediments transported by rivers in such arid areas. They suggest that values of 0.3 g/ℓ and 1.0 g/ℓ are reasonable but conservative for low and high water discharges, respectively. We assume that half of the total annual discharge occurred at each of these concentration levels, and calculated a total annual fine-grained sediment supply to the marine environment of 4.9×10^{12} g/y. Such a value compares favorably with a mean value of 6.3×10^{12} g/y calculated from accumulation rates of terrigenous components, and suggests that the mud lens is a major site for the deposition of fine-grained terrigenous sediment.

BIOLOGICAL CONTROLS ON DEPOSITION OF FINE-GRAINED TERRIGENOUS SEDIMENTS

McCave (1975) has drawn attention to the paradox that deposition in pelagic environments is controlled by the settling of large particles with fast settling velocities when only much finer grained particles (<2 μm) are generally observed in suspension. This is especially relevant to the case where very fine eolian silt and clay-sized particles (with settling velocities in water on the order of 10^{-3} to 10^{-4} cm/s) settle rapidly enough to the seafloor to preserve wind patterns although ambient current velocities may be several orders of magnitude higher than the settling rates (Scheidegger and Krissek, 1982). As Schrader (1971), McCave (1975) and Honjo and Roman (1978) have documented, such situations require that zooplankton and nekton "package" such suspended particles into fecal pellets as a part of their normal feeding activities. Additional work by Deuser (1982) indicates that organisms can remove inorganic debris from the surface layer very efficiently. There is little doubt that much of the mass flux of material from the surface layer of the water column to pelagic sedimentary environments is controlled by this process.

For hemipelagic sedimentary environments, particularly those strongly influenced by coastal upwelling, the importance of such mechanisms is less clear. Most studies which have used sediment traps to examine the importance of fecal pellet transport have focused on deeper water environments (e.g., Honjo and Roman, 1978; Honjo, 1980), although Dunbar and Berger (1981) have documented the importance of the process for Santa Barbara Basin, and Staresinic (1982) has stressed its importance for non-terrigenous components for the Peru upwelling system. However, for continental margin environments there are still a number of added problems that require examination before such a process can be fully assessed:

1. There is a paucity of data concerning the particle size distribution of suspended terrigenous sediments in water over continental margin sedimentary environments.

2. The connection between the feeding activity of zooplankton and nekton and their ability to feed on terrigenous particles of different sizes and concentrations is tenuous.

3. The important standing stocks of zooplankton and nekton which can ingest such components may fluctuate annually in response to the availability of food. As a result, they may not be able to efficiently "package" all of the terrigenous particles which pass through the margin system, particularly if such components are introduced into the marine environment during periods of low standing stock or in quantities which exceed the filtering capacity of the existing zooplankton and nekton populations.

Even though such unknowns remain, there can be little doubt that fecal pellet transport also occurs in hemipelagic environments. The main question concerns the importance of fecal pellet transport relative to other depositional processes. In this and subsequent sections, we present arguments which point to the importance of fecal pellet transport in the Peru upwelling zone.

Particle Sizes of Suspended Terrigenous Sediment Particles off Peru

Little is known about the relationship of suspended terrigenous components in waters over the Peru continental shelf and slope to sediments on the underlying sea floor. Assuming that suspensions of terrigenous particles introduced into Peru coastal waters behave like their counterparts in better studied areas, coarser suspended components (i.e., >30 μm) should settle out largely in the littoral zone or contribute to the development of a bottom nepheloid layer on the shelf (e.g., Harlett and Kulm, 1973; McCave, 1979). Near-bottom currents on the Peru shelf are generally weak (<10 cm/s), favoring the retention of such sediment in inner continental margin sedimentary environments. The remaining suspended particles would be expected to move seaward as part of a surface plume (Pak, Beardsley and Park, 1970). From a study of the Zaire River plume, Eisma and Kalf (pers. comm.) provide additional insight into the possible characteristics of suspended particles tens of kilometers from a river mouth. At such distances the particles are commonly less than 18 μm, although coarser grains may occasionally be found. They further suggest that flocculation has a minimal effect on the overall particle-size distribution of the suspended material and appears to be confined to <1 μm particles which may form fine silt-size terrigenous aggregates (Whitehouse, Jeffrey and Debbrecht, 1958). However, recent work by Avnimelech, Troeger and Reed (1982) suggests the added possibility that algae and clay-size particles may flocculate together to produce silt-size aggregates in the marine environment.

For the Peru margin a surface layer of about 25 m (Brockmann et al., 1980) is advected offshore and to the northwest at 10-30 km/day (Zuta and Guillen, 1970). Using a mean offshore transport rate of 20 km/day, and neglecting the influence of the turbulence expected within a wind-driven surface layer, it can be shown that particles as coarse as 20 μm (settling rate of 10^{-2} cm/sec) should remain in suspension until they are well beyond the known limits of the mud lens some 50 km offshore. Similar considerations suggest that it would take nearly 300 days for particles <2 μm (10^{-4} cm/sec) to settle through just the 25 m of the surface layer. Since particles <4 μm make up more than 60% of bottom sediments in the mud facies off Peru (Fig. 5), accelerated vertical transport is as necessary in this setting as it is in more pelagic sedimentary environments.

Feeding Activities of Zooplankton and Nekton

Before biologically mediated sediment packaging can be proposed as the cause of accelerated vertical transport in the Peru upwelling system, several points must be examined. These include whether zooplankton and nekton can ingest medium silt- to clay-sized particles as part of their routine feeding activities; and if so, do the organisms occur in sufficient concentrations to make an impact on the accumulating sedimentary record. It is beyond the scope of this paper to address the feeding activities of the many groups which inhabit the surface layer off Peru. Rather, we will address two representatives of the plankton (*Acartia tonsa*, a calanoid copepod; *Salpa fusiformis*, a gelatinous member of the macrozooplankton) and one of the nekton (*Engraulis ringens*, the common Peruvian anchoveta). As noted below, both calanoid copepods and anchoveta are known to be ubiquitous in waters off Peru, whereas the distribution and abundance of gelatinous zooplankton is much less well-known . Our goal is to draw attention to potential roles of such groups in continental margin sedimentation.

Calanoid copepods. *A. tonsa* and other calanoid copepods were initially believed to use their second maxillae as filters or sieves (Cannon, 1928; Loundes, 1935). The "filter" of *Acartia* mainly consists of a basal part with four endites and a five-segment endopod, all having long setae (Marshall, 1973). The distance between setae in the resulting network gives some idea of the particle size this filter can retain (Nival and Nival, 1976). In most species the pore size varies from 2 to 30 μm (Marshall, 1973; Nival and Nival, 1976), and seems to be independent of the size of the adult copepod (Marshall, 1973).

The manner in which copepods use their filtering mechanisms to gather food particles is unclear (see Poulet and Marsot, 1980; Alcaraz, Paffenhöfer and Strickler, 1980). In some cases they may simply move their maxillae indiscriminately as sieves and gather food particles that are retained on the setae (Nival and Nival, 1976); in other instances, they may use chemoreceptors to sense if suspended particles are in fact food before attempting to gather them (Poulet and Marsot, 1980; Alcaraz et al., 1980). Because to our knowledge no studies have addressed how copepods react to the presence of suspended terrigenous particles when feeding, we cannot definitely assess whether such particles would be avoided. However, it is well known that clay minerals will adsorb dissolved organic molecules (see Whitehouse et al., 1958) and that clay particles will flocculate with algae (Avnimelech et al., 1982). This should be particularly true off Peru where dissolved organics and primary productivity are high. It is further known that such resulting particles are inhabited by bacteria and are thus considered as food sources (Odum and de la Cruz, 1967). We speculate that this may be the case for copepods which rely on chemoreceptors; their chemoreceptor systems may not necessarily reject suspended terrigenous particles or algae-clay aggregates as nonfood items. It is also known that copepods will ingest foreign matter, such as crude oil (Conover, 1971) and red clay (Paffenhöfer, 1972).

Many studies have been made on the filtering efficiency or clearing rates of different species of copepods at different ages.

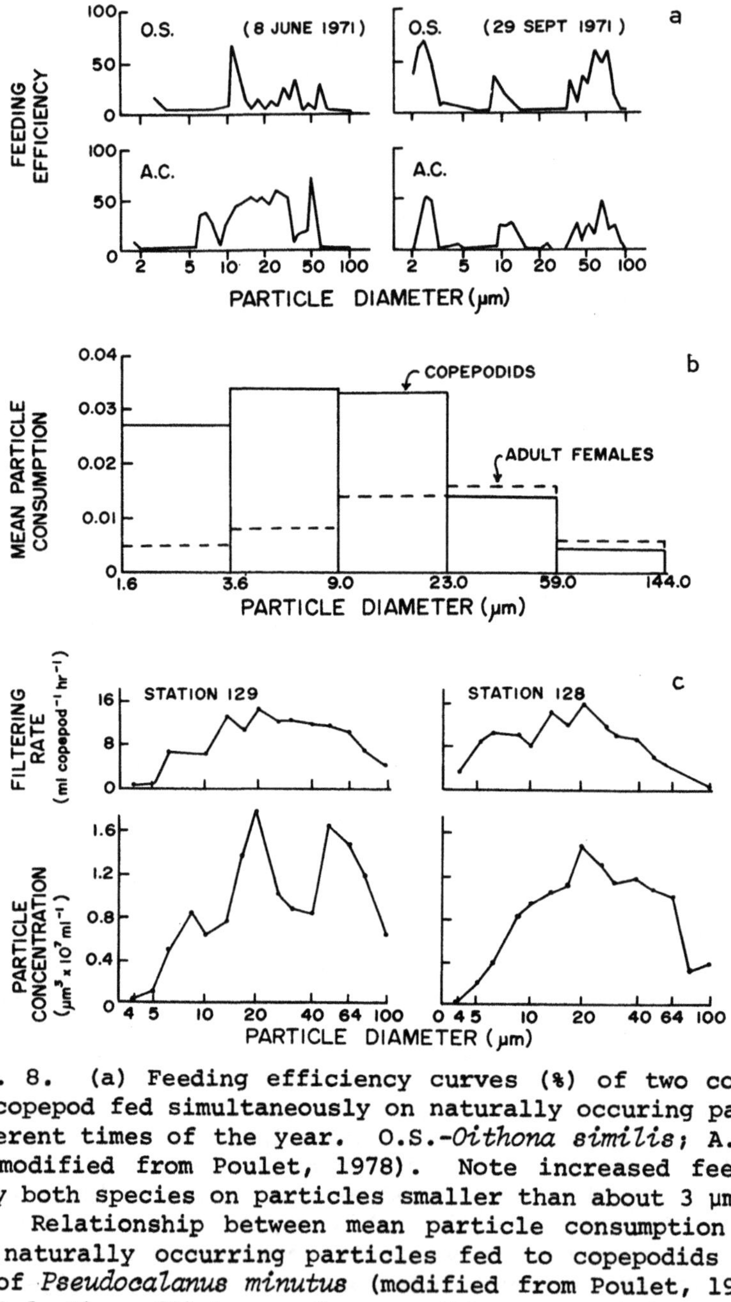

Fig. 8. (a) Feeding efficiency curves (%) of two coastal species of copepod fed simultaneously on naturally occuring particles at two different times of the year. O.S.-*Oithona similis*; A.C.-*Acartia clausi* (modified from Poulet, 1978). Note increased feeding efficiency by both species on particles smaller than about 3 μm.

(b) Relationship between mean particle consumption and grain size of naturally occurring particles fed to copepodids and adult females of *Pseudocalanus minutus* (modified from Poulet, 1977). Note increased feeding efficiency of copepodids on particles smaller than 23 μm.

(c) Relationship between filtering rate and particle size distribution for *Calanoides carinatus*. Particle size is given in spherical diameter. Modified from Schnack and Elbrächter (1981).

In Fig. 8a we present data from Poulet (1978) on the feeding efficiency of two nearshore species of marine copepods on naturally occurring particles (cells) in seawater; in Fig. 8b we show data from Poulet (1977) on the grazing behavior of copepodids and adults on similar particles; and in Fig. 8c we show typical filtering rates of copepod *Calanoides carinatus* (Schnack and Elbrächter, 1981). Several interesting observations can be made from these data:

1. Copepods feed on a wide range of particle sizes, and the relative importance of different size groups varies during the year. In particular, the feeding efficiency of adult copepods of both species on 2-3-μm diameter particles can be very high;

2. Copepodids may consume very fine particles (1.58-15.0 μm) very efficiently compared to adults (Fig. 8b), and the young may thus play a major overall role in the consumption of fine particles due to their high grazing ingestion capability (Paffenhöfer, 1971);

3. Filtering rates of adult copepods may be of the order of 5-10 mℓs/hr (Fig. 8c).

Since *Acartia tonsa* can attain abundances of >400,000/m^2 over the outer shelf and upper slope off Peru, and since they are confined largely to the surface layer (Bueno, 1981) because of the presence of the oxygen minimum, it can be calculated that one copepod of this species alone is present in about every 125 mℓ of water. With filtering (clearing) rates such as those shown in Fig. 8c and the realization that it will take perhaps 2-5 days for the surface layer to be driven offshore, it becomes apparent that waters of the surface layer may have been "cleared" a few times before reaching the seaward edge of the mud facies. In this regard, Paffenhöfer (1982) suggests that calanoid copepods and copepodids could ingest close to 100% of the average daily productivity in the Peru upwelling zone.

Gelatinous zooplankton. In an excellent study, Silver and Bruland (1981) have studied the filtering efficiency of a salp (*Salpa fusiformis*) and a pteropod (*Corolla spectabilis*). Both organisms use mucoid membranes to collect suspended particles. Salps pump water through internal mucus sheets (Madin, 1974) which serve as filters and retain even bacteria-sized particles (Gilmer, 1974; Harbison and Gilmer, 1976). By comparison, *C. spectabilis* uses external webs with adhesive properties to collect food particles. Water passes through such webs as they settle through the water column, but it does not actively pump water as salps do (Gilmer, 1972; 1974). The webs of *C. spectabilis* can also recover very fine particles.

In Fig. 9 we present data from Silver and Bruland (1981) on the distribution of various size categories of phytoplankton in water of

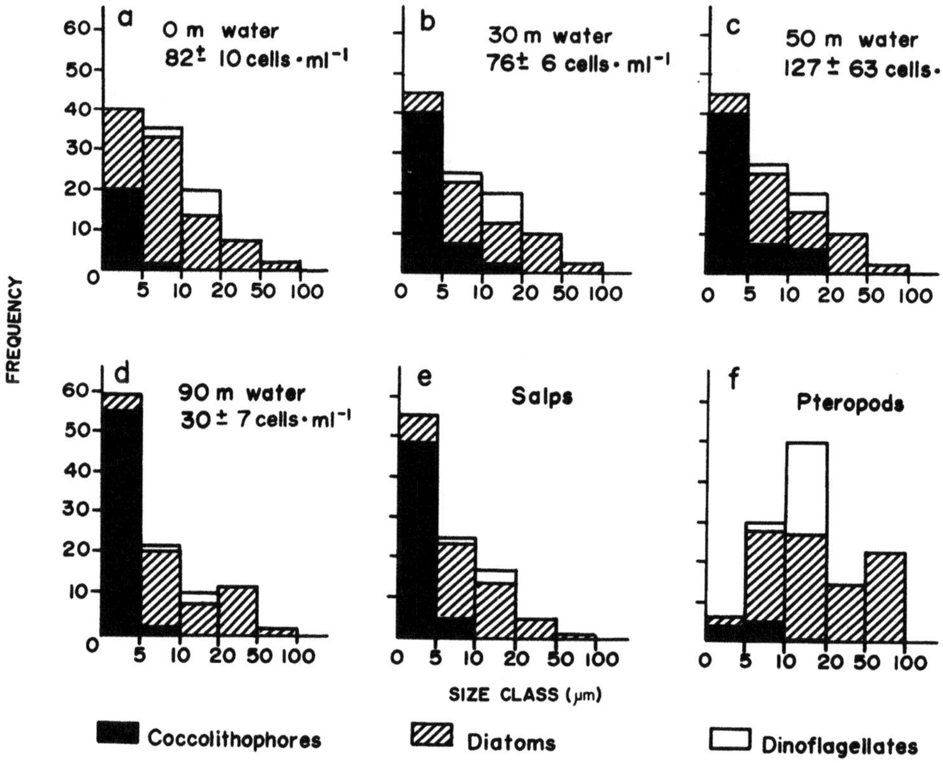

Fig. 9. Frequency (%) distribution of phytoplankton in various size categories. (a)-(d) Water samples within mixed layer, mean values ± standard errors; (e) fecal pellets of salps (*Salpa fusiformis* and *Pegea socia*); (f) fecal pellets of pteropods (*Corolla spectabilis*). From Silver and Bruland (1981). Note dominance of <5 μm material in fecal pellets of salps.

various depths off California and, for comparison, the distribution of such components in fecal pellets of salps and pteropods. Such data clearly indicate that mucoid filter feeders can readily retain particles <5 μm. Unfortunately, little is known about the abundance of such organisms within the surface layer off Peru. It follows, though, that mucoid filter feeding members of the zooplankton would also be very important in nonselectively filtering suspended particulate matter from the water column. In this regard Dunbar and Berger (1981) have suggested that at least 60% and possibly as much as 90% of the trapped matter recovered from a sediment trap deployed at 341 m depth in the Santa Barbara Basin came from the fecal pellets of salps, euphausiids, and copepods.

Peruvian anchoveta. The Peruvian anchoveta, *Engraulis ringens*, is ubiquitous off Peru and is responsible for sustaining one of the world's most famous fisheries, with millions of tons harvested annually during the last two decades (Cushing, 1981). *E. ringens* inhabits surface waters to a depth of about 40 m (Jordan, 1971) over essentially the same area occupied by the mud lens (Fig. 4). It commonly spawns in the months of August to March with peaks in September and January; the young enter the fishery in December through May (see Fig. 10; Paulik, 1970). Young anchoveta feed on phytoplankton, but with increasing size they switch to zooplankton (Paulik, 1970; Cushing, 1975; 1978). Although young anchoveta might be expected to ingest fine-grained suspended terrigenous particles while feeding on similar sized diatoms, their role in sedimentation may be more secondary. That is, as members of a higher trophic level, they may feed primarily on zooplankton which may have already ingested terrigenous components.

The relationships between the abundances of anchoveta and zooplankton in the upwelling zone off Peru have been addressed by Walsh et al. (1971) and Whitledge (1978). Whitledge (1978) has calculated that anchoveta are present in abundances of 0.5 fish (3.6 g)/m^3, and Walsh et al. (1971) suggest that the mean zooplankton biomass ranges from 39.1 to 46.1 mg/m^3 during late summer. Cushing's (1978) work on small fish indicates that anchoveta may consume 1-5% of their body weight per day with 5% probably being closer to the actual amount. Assuming that these values are reasonable first approximations, we find that, if one anchovy were to consume all of the zooplankton in an average cubic meter, this would amount to 1.1 to 1.3% of its body weight. Although there is uncertainty in the numbers, it appears that anchoveta could be expected to rather effectively remove copepods and other zooplankton from the water column. In this regard, Cushing (1975) notes that individual anchoveta can consume 1000 calanus copepods per day.

Since they feed on zooplankton and/or phytoplankton in the upwelling zones, the anchoveta and their fast-settling fecal pellets have been considered the primary mechanism for deposition of organic matter on the sea floor (von Bröckel, 1981; Staresinic, in prep.). Using floating sediment traps which recovered dominantly anchoveta fecal pellets at a depth of 30 m, von Bröckel (1981) showed that an average of 13% of the mean daily primary production was incorporated into fast settling aggregates. For the Peru upwelling zone, total annual primary productivity has been calculated to be 300 to 500 g C/m^2/yr (Ryther, 1969; Rowe, 1971). Combining the two results, von Bröckel (1981) calculated that between 39 and 65 g C/m^2 should reach the sea floor each year, or 3.9 to 6.5 g C/cm^2/1000 yrs. This compares favorably with the average accumulation rate of 4.8 g C/cm^2/1000 yrs. that can be calculated from the data in Table 2. Comparable flux estimates are not available for terrigenous components known to be present in the fecal pellets (N. Staresinic, pers. comm.). However, if one assumes that suspended terrigenous sediment in the surface layer is removed each day even more efficiently than the 13%

value given by von Bröckel (i.e., that some organic carbon is used in body growth and respiration by organisms) and, if it takes a few days for the waters to move to the seaward edge of the mud facies, then it becomes obvious that such a mechanism would serve as an effective barrier to the seaward transport of such components as well.

Additional evidence supporting the mechanism outlined above includes the observations by Scheidegger and Krissek (1982) and Thiede (this volume) that ashed residues of zooplankton recovered from the surface layer over the continental margin contain silt and clay-sized terrigenous mineral components. These components were presumably in the guts of copepods and other zooplankton. In addition, Staresinic (pers. comm.) has observed terrigenous particles in fecal pellets of anchoveta. Finally, Reimers (1982) has shown that remnants of fecal pellets are ubiquitous in the upper slope mud lens sediments and consist of diatoms and terrigenous components.

DISCUSSION

Although the feeding activity of zooplankton and nekton offer a mechanism for incorporating fine-grained terrigenous particles into rapidly settling aggregates, it is still possible that maximum abundances of organisms may not coincide with the time when most terrigenous sediments are introduced into the marine environment. For example, the eastern Gulf of California and the Oregon upwelling regimes have significant seasonal upwelling which is totally out of phase with the times when terrigenous sediment enters the marine environment (Smith, 1981; Donegan and Schrader, 1982; Krissek and Scheidegger, this volume). Under those circumstances, biological activity might remove fewer suspended terrigenous particles from the water column.

For the Peru continental margin the maximum discharge of coastal rivers between 10° and 14°S occurs in the summer months of January, February, March, and April (Fig. 10; Zuta and Guillen, 1970). This marked seasonal discharge is caused by snowmelt and rainfall along the western flank of the Andes. Thus, from June through November little terrigenous sediment would be expected to be introduced into the marine environment.

Persistent southeasterly trade winds are known to cause the Ekman transport of surface waters both offshore and equatorward off Peru (Smith, 1981). However, the relationships between annual fluctuations in the trade winds in the southeastern tropical Pacific and upwelling intensity, or between upwelling intensity and primary productivity, are less well understood. From satellite data on the movement of clouds, Enfield (1981a) has described an annual cycle in southeastern tropical Pacific low-level wind speeds north of 15°S. He finds that, although the winds remain constant in direction and

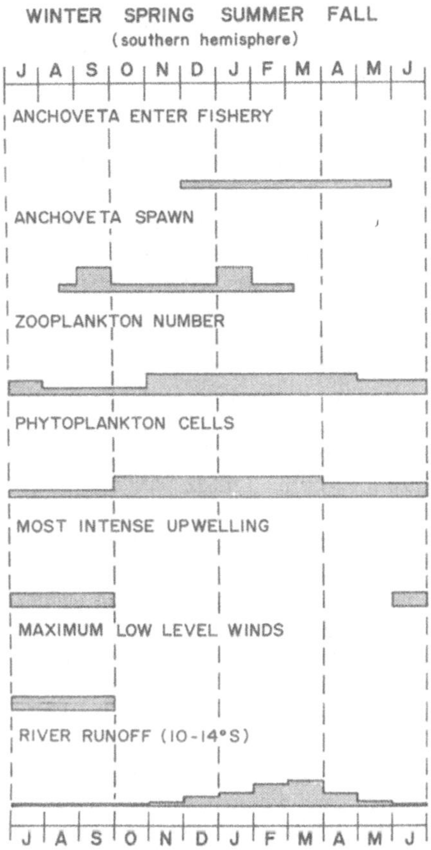

Fig. 10. Relationship between river runoff (10-14°S) and atmospheric, physical oceanographic and biological factors influencing the Peru upwelling regime. Variations in heights of bars indicate qualitative differences in parameters described in text; lengths correspond to times when parameters are most important. Data sources are as follows: river runoff (10-14°S), (Zuta and Guillén, 1970); low level winds (Enfield, 1981a); most intense upwelling (Zuta and Guillén, 1970; Zuta et al., 1978); phytoplankton cells (Rojas de Mendiola, 1981); zooplankton numbers (Cushing, 1975); anchoveta spawn (Cushing, 1981); and anchoveta enter fishery (Paulik, 1970; Cushing, 1981). Note relationship between time of maximum river runoff (fluvial sediment discharge) and times when zooplankton and nekton are believed to be most prevalent.

are favorable for upwelling throughout the year, they typically reach a maximum in the southern hemisphere winter (July-September; see Fig. 10) and a minimum about six months later (January-March). Although land-sea thermal forcing of coastal winds is responsible for more complex secondary variations near Lima (Enfield, 1981a; 1981b), annual fluctuations in the intensity of the trade winds should control much of the coastal upwelling along the inner Peru continental margin. In this regard, Zuta and Guillen (1970) and Zuta et al. (1978) suggest that maximum coastal upwelling occurs from May until November (Fig. 10), a time period which coincides with the maximum in winter trade wind intensities noted by Enfield (1981a).

Annual fluctuations in phytoplankton abundances along the entire Peru coast have been summarized by Rojas de Mendiola (1981) from a study of more than 2000 samples taken over a 10-year period. Although phytoplankton cell densities are generally high along the coast of Peru, Rojas de Mendiola (1981) recognized interesting spatial and temporal variability. Four persistent and intense upwelling centers were recognized at Pimentel (∿6°S), Chimbote (∿9°S), Callao (∿12°S) and Tambo de Mora-Pisco (∿15°S). At each area and at intervening areas, maximum numbers of phytoplankton cells per liter were noted in October through March, intermediate numbers in April through June, and low values in July through September (Fig. 10). At first glance, the near inverse correlation between the period of most intense upwelling and fluctuations in the numbers of phytoplankton cells seems puzzling. However, as Cushing (1975) notes, it is the rate at which water upwells that is important. If it is too fast (e.g., >1 m/day), such as might be expected during the period of most intense winds in July, August, and September (Fig. 10), lower primary production would be anticipated.

In response to seasonally varying phytoplankton abundances, zooplankton and other organisms which feed on such plant material are expected to become more abundant as the availability of food increases (Cushing, 1975). For the Peru upwelling zone, however, very little is known about annual fluctuations in zooplankton abundances. In Fig. 10 we have assumed that the relative abundances of zooplankton follow the phytoplankton abundances with an estimated one-month lag time characteristic of low latitude areas (Cushing, 1975). This scenario predicts that an increased zooplankton biomass is associated with a greater food supply and, perhaps more importantly, with the mid-spring to mid-fall time period when most terrigenous particles are introduced into the marine environment (Fig. 10).

Additional indirect evidence on the seasonal fluctuations of zooplankton comes from work by Cushing (1981) on the reproductive cycle of the anchoveta. Adults typically spawn from late winter until mid-summer with peaks in the months of September and January (Cushing, 1981). The diet of anchoveta larvae initially is zooplankton nauplii or phytoplankton. As they mature and enter the fishery months later (Fig. 10), however, their diet switches over to zooplankton (Paulik, 1970; Cushing, 1981). Thus, the annual cycle of the anchoveta would appear to be well positioned in time to take advantage of maximum phytoplankton and zooplankton abundances.

It becomes clear from these studies that in the Peru upwelling system there are annual fluctuations in trade wind intensity, upwelling intensity, and abundances of phytoplankton, zooplankton, and nekton. River runoff data also point to an annual fluctuation in the nonbiogenic sediment flux into the marine environment. The striking feature of the data presented in Fig. 10, however, is that peak abundances of phytoplankton, zooplankton and nekton occur when terrigenous sediment components are expected to be most prevalent in the surface layer. This situation favors the efficient removal of suspended terrigenous particles from the water column over the inner continental margin as a consequence of feeding activity.

Although fecal pellet transport appears to be essential to the formation of the mud facies off Peru, the nature of the current regime and the oxygen content of the waters are equally important. Large fecal pellets produced by anchoveta, with settling velocities on the order of 1000 m/day, settle rapidly to the shelf and upper slope environments and should not be significantly affected by horizontal advection. Small fecal pellets of juvenile anchoveta and copepods (settling rates 10-200 m/day; Small et al., 1979), and individual uningested suspended terrigenous particles coarser than about 10 µm (settling rates >10 m/day; Shepard, 1963), however, may be influenced by lateral transport during settling. For the Peru upwelling zone, this does not mean net transport offshore; rather, with settling rates >10 m/day and surface layer transport offshore of 10-30 km/day (Zuta and Guillén, 1970), such slow settling aggregates or particles would settle into the poleward undercurrent. Since the undercurrent provides water for coastal upwelling (Smith, 1981), the aggregates or particles may then be transported back onshore. Only fine particles with settling rates perhaps much less than 10 m/day and uninfluenced by biological packaging may be expected to be transported permanently seaward of the upwelling system by surface currents. Zooplankton and nekton off Peru also take advantage of the opposing directions of the surface layer flow and the poleward undercurrents to remain in essentially the same geographical area (Smith, Boyd and Lane, 1981b) by simply controlling their vertical position in the water column. The near absence of bottom currents also plays an important role in the retention of fecal pellets and other fine-grained particulates on the inner continental margin. Similarly, the O_2-deficient waters of the undercurrent inhibit the chemical oxidative alteration of the organic-rich sedimentary components and favor the preservation of fecal pellets at the sediment-water interface (Reimers and Suess, this volume). These same oxygen-deficient waters confine most nekton and zooplankton and their resulting feeding activities to the surface layer or the uppermost portion of the poleward undercurrent.

In summary, we suggest that between 10° and 14°S the many physical, chemical, biological, and geological processes described above interact to favor the deposition and preservation of a mud facies with a very strong signal of coastal upwelling. Many crucial factors (e.g., the persistence of upwelling; the influence of the O_2- minimum on benthic fauna, nekton and zooplankton; the near absence of bottom currents; the minimal influx of terrigenous sediments relative to biogenous ones; and the synchroneity of seasonal fluctuations in plant and animal life in the surface layer and their relationships to seasonal river runoff patterns) interact to produce, deposit and preserve, virtually intact, sedimentary components formed in one of the most productive upwelling regimes known. Outside of the area of focus (e.g., seaward or to the north), this optimum combination of factors does not occur. Thus, we conclude that, although near-surface biological activity may be as crucial in hemipelagic as in pelagic sedimentary environments in bringing about rapid vertical transport of terrigenous and biogenous components, processes acting on the particles at the sediment-water interface are equally effective in determining the type of coastal upwelling signal that is eventually preserved within the sedimentary record.

ACKNOWLEDGEMENTS

This study has benefited considerably from discussions we have had with several of our colleagues. These include E. Suess who made available a number of pertinent papers and provided some needed stimulus; H. Schrader and C. Miller, who reviewed an earlier draft of this manuscript and made a number of helpful comments; and V. Kulm, L. Small and H. Pak who provided some key references and took part in helpful discussions. Many of the chemical, mineral and textural analyses discussed in the text were performed by M. Hower. Portions of this study have been supported by the Office of Naval Research through contract N00014-76-C-0067.

REFERENCES

Alcaraz, M., Paffenhöfer, G.-A. and Strickler, J.R., 1980, Catching the algae: A first account of visual observations on filter-feeding calanoids, in: "Evolution and Ecology of Zooplankton Communities," W.C. Kerfoot, ed., University Press of New England, Hanover, New Hampshire, 241-248.

Avnimelech, T., Troeger, B.W. and Reed, L.W., 1982, Mutual flocculation of algae and clay: Evidence and implications, Science, 216:63-65.

Barber, R.T. and Smith, R.L., 1981, Coastal upwelling ecosystems, in: "Analysis of Marine Ecosystems," A.R. Longhurst, ed., Academic Press, London, 31-68.

Boyd, C.M., Smith, S.L. and Cowles, T.J., 1980, Grazing patterns of copepods in the upwelling system off Peru, Limnology and Oceanography, 25:583-596.

Brockmann, C., Fahrbach, E., Huyer, A., and Smith, R.L., 1980, The poleward undercurrent along the Peru coast: 5 to 15°S, Deep-Sea Research, 27A:847-856.
Bueno, H.S., 1981, The zooplankton in an upwelling area off Peru, in: "Coastal Upwelling," F.A. Richards, ed., Coastal and Estuarine Sciences 1, American Geophysical Union, Washington, 411-416.
Busch, W.H. and Keller, G.H., 1982, The physical properties of Peru-Chile continental margin sediments--The influence of coastal upwelling on sediment properties, Journal of Sedimentary Petrology, 51:705-719.
Cannon, H.G., 1928, On the feeding mechanism of the copepods *Calanus finmarchicus* and *Diaptomus gracilis*, British Journal of Experimental Biology, 6:131-144.
Conover, R.J., 1971, Some relations between zooplankton and bunker C oil in Chedabucto Bay following the wreck of the tanker Arrow, Journal of the Fisheries Research Board of Canada, 28:1327-1330.
Cushing, D.H., 1975, "Marine Ecology and Fisheries," Cambridge University Press, London, 278 pp.
Cushing, D.H., 1978, Upper trophic levels in upwelling areas, in: "Upwelling Ecosystems," R. Boje and M. Tomczak, eds., Springer-Verlag, Berlin, 101-110.
Cushing, D.H., 1981, The effect of El Niño upon the Peruvian anchoveta stock, in: "Coastal Upwelling," F.A. Richards, ed., Coastal and Estuarine Science 1, American Geophysical Union, Washington, 449-457.
DeMaster, D.J., 1979, "The Marine Budgets of Silica and ^{32}Si," Ph.D. Thesis, Yale University, New Haven, 308 pp.
DeMaster, D.J., 1981, The supply and accumulation of silica in the marine environment, Geochimica et Cosmochimica Acta, 45:1715-1732.
Deuser, W.G., 1982, Temporal variations in deep-sea particle flux and relation to surface-water suspended matter, EOS Transactions American Geophysical Union, 63:45.
DeVries, T.J., 1979, "Nekton Remains, Diatoms, and Holocene Upwelling off Peru," M.S. Thesis, Oregon State University, Corvallis, 85 pp.
DeVries, T.J. and Pearcy, W.G., 1982, Fish debris in sediments of the upwelling zone off coastal Peru: a late Quaternary record, Deep-Sea Research, 28:87-109.
Diester-Haass, L., 1978, Sediments as indicators of upwelling, in: "Upwelling Ecosystems," R. Boje and M. Tomczak, eds., Springer-Verlag, New York, 261-281.
Donegan, D. and Schrader, H., 1982, Biogenic and abiogenic components of laminated hemipelagic sediments in the central Gulf of California, Marine Geology, 48:215-237.
Dunbar, R.B. and Berger, W.H., 1981, Fecal pellet flux to modern bottom sediment of Santa Barbara Basin (California) based on sediment trapping, Geological Society of America, Bulletin, 92:212-218.

Enfield, D.B., 1981a, Annual and nonseasonal variability of monthly low-level wind fields over the southeastern tropical Pacific, Monthly Weather Review, 109:2177-2190.

Enfield, D.B., 1981b, Thermally driven wind variability in the planetary boundary layer above Lima, Peru, Journal of Geophysical Research, 86:2005-2016.

Gilmer, R.W., 1972, Free-floating mucus webs: a novel feeding adaptation for the open ocean, Science, 176:1239-1240.

Gilmer, R.W., 1974, Some aspects of feeding in thecosomatous pteropod molluscs, Journal of Experimental Marine Biology and Ecology, 15:127-144.

Harbison, G.R. and Gilmer, R.W., 1976, The feeding rates of the pelagic tunicate *Pegea confederata* and two other salps, Limnology and Oceanography, 21:517-520.

Harlett, J.C. and Kulm, L.D., 1973, Suspended sediment transport on the northern Oregon continental shelf, Geological Society of America, Bulletin, 84:3815-3826.

Heath, G.R., 1974, Dissolved silica in deep-sea sediments, in: "Studies in Paleo-Oceanography," W.W. Hay, ed., Society of Economic Petrologists and Mineralogists, Special Publication, 20: 77-93.

Heinrichs, S.M., 1980, "Biogeochemistry of Dissolved Free Amino Acids in Marine Sediments," Ph.D. Dissertation, Woods Hole Oceanographic Institution, WHOI-80-39, Woods Hole, 253 pp.

Honjo, S. and Roman, M.R., 1978, Marine copepod fecal pellets: production, preservation and sedimentation, Journal of Marine Research, 36:45-57.

Honjo, S., 1980, Material fluxes and modes of sedimentation in the mesopelagic and bathypelagic zones, Journal of Marine Research, 38:53-97.

Jordan, R., 1971, Distribution of anchoveta (*Engraulis ringens J.*) in relation to the environments, Investigacion Pesquera, 35:113-126.

Koide, M. and Goldberg, E.D., 1981, Transuranic nuclides in coastal marine sediments off Peru, Earth and Planetary Science Letters, 57:263-277.

Krissek, L.A., Scheidegger, K.F. and Kulm, L.D., 1980, Surface sediments of the Peru-Chile continental margin and the Nazca Plate, Geological Society of America, Bulletin, 91:321-331.

Leinen, M., 1977, A normative calculation technique for determining opal in deep-sea sediments, Geochimica et Cosmochimica Acta, 40:671-676.

Loundes, A.G., 1935, The swimming and feeding of certain calanoid copepods, Proceedings, Zoological Society London, 1935:687-715.

Madin, L.P., 1974, Field observations on the feeding behavior of salps (Tunicata:Thaliacea), Marine Biology, 25:143-147.

Marshall, S.M., 1973, Respiration and feeding in copepods, Advances in Marine Biology, 11:57-120.

McCave, I.N., 1975, Vertical flux of particles in the ocean, Deep-Sea Research, 22:491-502.

McCave, I.N., 1979, Suspended material over the central Oregon continental shelf in May 1974: I. concentrations of organic and inorganic components, Journal of Sedimentary Petrology, 49:1181-1194.

Nival, P. and Nival, S., 1976, Particle retention efficiencies of an herbivorous copepod, *Acartia clausi* (adult and copepodite stages): effects on grazing, Limnology and Oceanography, 21: 24-38.

Odum, E.P. and de la Cruz, A.A., 1967, Particulate organic detritus in a Georgia salt marsh estuarine ecosystem, in: "Estuaries," G.H. Lauff, ed., American Association for the Advancement of Science, Washington, 383-388.

Paffenhöfer, G.A., 1971, Grazing and ingestion rates of nauplii, copepodids and adults of the marine planktonic copepod *Calanus helgolandicus*, Marine Biology, 11:286-298.

Paffenhöfer, G.A., 1972, The effects of suspended "red mud" on mortality, body weight, and growth of the marine planktonic copepod, *Calanus helgolandicus*, Water Air Soil Pollution, 1:314-321.

Paffenhöfer, G.A., 1982, Grazing by copepods in the Peru upwelling, Deep-Sea Research, 29:145-147.

Pak, H., Beardsley, G.F., Jr. and Park, P.K., 1970, The Columbia River as a source of marine light-scattering particles, Journal of Geophysical Research, 75:4570-4578.

Paulik, G.J., 1970, The anchoveta fishery of Peru, Center for Quantitative Science in Forestry, Fisheries and Wildlife, Seattle, University of Washington, Paper 13, 79 pp.

Poulet, S.A., 1977, Grazing of marine copepod developmental stages on naturally occurring particles, Journal of the Fisheries Research Board of Canada, 34:2381-2387.

Poulet, S.A., 1978, Comparison between five coexisting species of marine copepods feeding on naturally occurring particulate matter, Limnology and Oceanography, 23:1126-1143.

Poulet, S.A. and Marsot, P., 1980, Chemosensory feeding and food-gathering by omnivorous marine copepods, in: "Evolution and Ecology of Zooplankton Communities," W.C. Kerfoot, ed., University Press of New England, Hanover, New Hampshire, 198-218.

Reimers, C.E., 1982, Organic matter in anoxic sediments off central Peru: Relations of porosity, microbial decomposition, and deformation properties, Marine Geology, 46:175-197.

Rojas de Mendiola, B., 1981, Seasonal plankton distribution along the Peruvian coast, in: "Coastal Upwelling," F.A. Richards, ed., Coastal and Estuarine Sciences 1, American Geophysical Union, Washington, 348-356.

Rowe, G.T., 1971, Benthic biomass in the Pisco, Peru upwelling, Investigacion Pesquera, 35:127-135.

Ryther, J.H., 1969, Photosynthesis and fish production in the sea, Science, 166:72-76.

Scheidegger, K.F. and Krissek, L.A., 1982, Dispersal and deposition of eolian and fluvial sediments off Peru and northern Chile, Geological Society of America, Bulletin, 93:150-162.

Schnack, S.B. and Elbrächter, M., 1981, On the food of calanoid copepods from the northwest African upwelling region, in: "Coastal Upwelling," F.A. Richards, ed., Coastal and Estuarine Sciences 1, American Geophysical Union, Washington, 433-439.

Schrader, H., 1971, Fecal pellets: Role in sedimentation of pelagic diatoms, Science, 174:55-57.

Schuette, G. and Schrader, H., 1981, Diatoms in surface sediments: A reflection of coastal upwelling, in: "Coastal Upwelling," F.A. Richards, ed., Coastal and Estuarine Sciences 1, American Geophysical Union, Washington, 372-380.

Shepard, F.P., 1963, "Submarine Geology," Harper and Row, New York, Evanston and London, Second edition, 557 pp.

Silver, M.W. and Bruland, K.W., 1981, Differential feeding and fecal pellet composition of salps and pteropods, and the possible origin of the deep-water flora and olive-green "cells," Marine Biology, 62:263-273.

Small, L.F., Fowler, S.W. and Unlu, M.Y., 1979, Sinking of natural copepod fecal pellets, Marine Biology, 51:233-241.

Smith, R.L., 1981, A comparison of the structure and variability of the flow field in three coastal upwelling regions: Oregon, Northwest Africa and Peru, in: "Coastal Upwelling," F.A. Richards, ed., Coastal and Estuarine Sciences 1, American Geophysical Union, Washington, 107-118.

Smith, S.L., Brink, K.H., Santander, H., Cowles, T.J., and Huyer, A., 1981a, The effect of advection on variations in zooplankton at a single location near Cabo Nazca, Peru, in: "Coastal Upwelling," F.A. Richards, ed., Coastal and Estuarine Sciences 1, American Geophysical Union, Washington, 400-410.

Smith, S.L., Boyd, C.M. and Lane, P.V.Z., 1981b, Short-term variation in the vertical distribution of small copepods off the coast of northern Peru, in: "Coastal Upwelling," F.A. Richards, ed., Coastal and Estuarine Sciences 1, American Geophysical Union, Washington, 417-426.

Soutar, A., Johnson, S.R. and Baumgartner, T.R., 1982, In search of modern depositional analogs to the Monterey Formation, in: "The Monterey Formation and Related Siliceous Rocks of California," G.E. Garrison and R.G. Douglas, eds., Society of Economic Paleontologists and Mineralogists, Tulsa, 123-147.

Staresinic, N., in prep., Downward flux of bulk particulate organic matter in the Peru coastal upwelling, Journal of Marine Research (submitted).

von Bröckel, K., 1981, A note on short-term production and sedimentation in the upwelling region off Peru, in: "Coastal Upwelling," F.A. Richards, ed., Coastal and Estuarine Sciences 1, American Geophysical Union, Washington, 291-297.

Walsh, J.J., Kelley, J.C., Dugdale, R.C., and Frost, B.W., 1971, Gross features of the Peruvian upwelling system with special reference to possible diel variation, Investigacion Pesquera, 35:25-42.

Whitehouse, U., Jeffrey, L.M. and Debbrecht, J.D., 1958, Differential settling tendencies of clay minerals in saline waters, in:

"Seventh National Conference Clays and Clay Minerals," Pergamon Press, London-New York, 1-79.

Whitledge, T.E., 1978, Regeneration of nitrogen by zooplankton and fish in the northwest Africa and Peru upwelling ecosystems, in: "Upwelling Ecosystems," R. Boje and M. Tomczak, eds., Springer-Verlag, New York, 90-100.

Zuta, S. and Guillen, O., 1970, Oceanografia de las aguas costeras del Peru, Instituto del Mar del Peru, Boletin, 2:161-323.

Zuta, S., Rivera, T. and Bustamente, A., 1978, Hydrologic aspects of the main upwelling areas of Peru, in: "Upwelling Ecosystems," R. Boje and M. Tomczak, eds., Springer-Verlag, New York, 235-256.

GEOCHEMISTRY OF COASTAL UPWELLING SYSTEMS

GEOCHEMISTRY OF NAMIBIAN SHELF SEDIMENTS

Stephen E. Calvert

Department of Oceanography
University of British Columbia
Vancouver, British Columbia, Canada, V6T 1W5

N. Brian Price

Grant Institute of Geology
University of Edinburgh
Edinburgh, United Kingdom, EH9 3JW

ABSTRACT

The modern sediments accumulating on the Namibian shelf between 20° and 26°S latitude are composed predominantly of biogenous material: organic matter, diatomaceous silica and calcium carbonate. A belt of anoxic, organic-rich diatom ooze, containing fish debris and Recent and pre-Recent phosphorite, occurs in a coastal zone up to 70 km wide off Walvis Bay in water depths shallower than 140 m. A blanket of foraminiferal and skeletal calcarenites, containing reworked phosphorite, and calcareous muds covers the central and outer shelf. The diatom ooze is a modern equilibrium facies reflecting the intensity of coastal upwelling in the Benguela Current; sedimentation rates range between 30 and 120 cm/10^3 yr. Seaward of the diatom ooze, benthic and planktonic carbonate is mixed with a reworked, shelly transgressive facies on the middle part of the shelf, the sediment becoming richer in foraminifera and more fine-grained towards the shelf edge, where sedimentation rates are on the order of 5 cm/10^3 years.

The major- and minor-element geochemistry of the surface sediments reflects the complex intermixture of several sedimentary components (terrigenous, biogenous and authigenic) and can be used to identify several important diagenetic processes and to examine the extent of minor metal enrichment in the anoxic, diatom ooze. Five components are responsible for most of the variability in composition: calcium carbonate (Ca, Sr, CO_2), terrigenous debris (Si, Al,

Ti, K, Rb, Zr), organic matter and opal (Si, C, S, Cu, Mo, Ni, Pb, Zn), phosphorite (P, Sr, U, Y) and glauconite (Fe, K, Rb). Barium and iodine are distributed independently of these components, Ba increasing in concentration away from the diatom ooze towards the slope and I being enriched in oxidized, relatively organic-rich surface sediments of the outer shelf and upper slope.

The distributions of Cu, Mo, Ni, Pb and Zn are very similar to the distribution of organic matter. Considering the overwhelmingly biogenous nature of the sediment and the low concentrations of metals in recent analyses of bulk plankton, we suspect that these metals are enriched in the diatom ooze diagenetically, although the mechanism of enrichment is unknown. The organic-rich diatom ooze, which contains modern phosphorite concretions, represents the sediment facies produced as a consequence of coastal upwelling. The presence of diatoms signifies an upwelling environment, high organic matter contents signify a relatively shallow-water environment and/or an environment of deposition adjacent to an arid hinterland. The occurrence of phosphorite signifies an environment where extensive diagenetic recycling occurs and where dissolved interstitial phosphate levels are elevated by a high rate of accumulation of unaltered planktonic organic material and fish debris. The co-occurrence of these three components serves to identify a coastal upwelling environment; each of them alone, however, is insufficiently diagnostic.

INTRODUCTION

Recent sediments accumulating beneath areas of coastal upwelling are known to have higher concentrations of organic material compared with other nearshore areas. Off California, for example, sediments in the nearshore basins contain up to 6% organic carbon (Emery, 1960), while off Peru, where primary production is much more intense, a 30- to 50-km-wide belt of organic-rich mud, containing up to 21% organic carbon (Krissek, Scheidegger, and Kulm, 1980), lies on the upper part of the slope in water depths between 110 and 680 m.

On the Namibian continental margin, a similar facies is present; the organic- and diatom-rich mud, containing up to 22.3% organic carbon, lies on the inner part of a wide and deep shelf platform (Senin, 1968; Avilov and Gershanovich, 1970; Calvert and Price, 1970). Such unusual sediments are produced in an environment where planktonic organic and skeletal material suffers minimal dilution by terrigenous sediment, because the coastal region is arid and permanent rivers are absent, and where the deposited organic material has had a short transit from the ocean surface to the sediment.

While the distribution of modern organic-rich sediments on mid-latitude continental margins characterized by coastal upwelling is known in broad outline, relatively little information is available on the geochemistry of the deposits. The following questions can usefully be asked: Does the basic geochemical and mineralogical make-up of such sediments provide a distinction from other nearshore sediments? Can modern situations be used to construct analogues of the formation of metalliferous black shales, fossil fish beds and petroleum source rocks (as cogently and imaginatively argued in a series of publications by Brongersma-Sanders, 1948; 1951; 1965; 1966; 1969).

The data discussed in this paper result from the collection of samples and observations during CIRCE expedition of the Scripps Institution of Oceanography on board RV "Argo" in October 1968, supplemented by additional sediment samples obtained from the University of Cape Town. During a six-day period we carried out a reflection seismic, bathymetric, hydrographic and sediment survey of the shelf area off Walvis Bay (Fig. 1A). The data have previously been used to provide descriptions of the upwelling and nutrient regeneration in the Benguela Current (Calvert and Price, 1971a), the bathymetry and shallow structure of the shelf (van Andel and Calvert, 1971), some aspects of the nature and geochemistry of the surface sediments (Calvert and Price, 1970; 1971b; Price and Calvert 1973; 1977; Morris and Calvert, 1977; Calvert and Morris, 1977) and the age and composition of phosphorites (Veeh, Calvert and Price, 1974; Price and

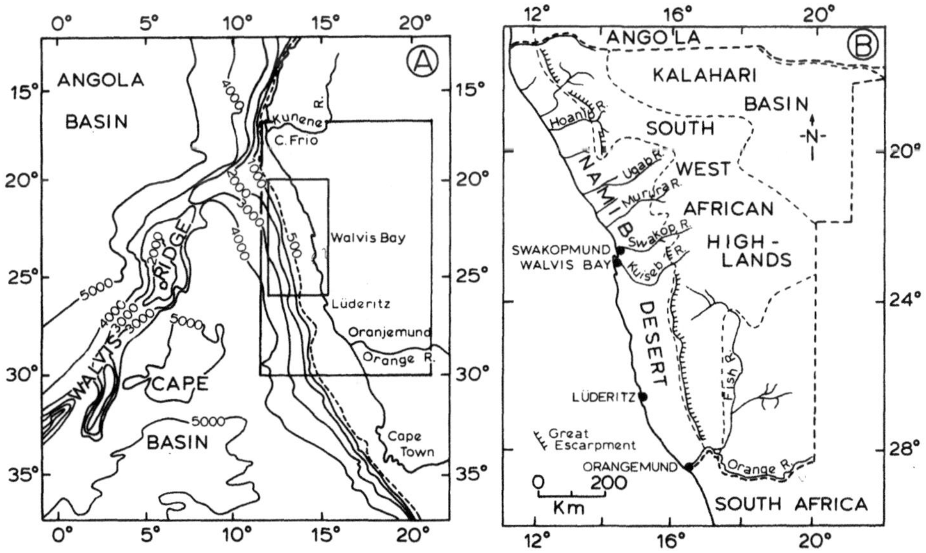

Fig. 1. (A) Location of the area investigated (smaller rectangle) with relation to the physiography of the southeast Atlantic. Isobaths in meters. Larger rectangle delineates area of Fig. 1B.

(B) Physiographic divisions of Namibia with surface drainage. From Logan (1960) and Haughton (1969).

Calvert, 1978). The purpose of this paper is to describe the modern sediment facies on the shelf, paying particular attention to the major- and minor-element bulk compositions in order to examine the distinctive composition of the various sediment facies, to examine the extent of trace metal enrichment in the organic-rich sediments and to identify diagenetic mineral phases.

SAMPLING AND ANALYTICAL METHODS

Sediment samples were collected during CIRCE expedition by open-barrel gravity coring; the cores were stored frozen in butyrate liners or other appropriate containers and preserved in the same state until sampled in Edinburgh. Additional surface samples were

Table 1. Summary of analytical methods

First Quarter Core	Second Quarter Core
Dispersed in H_2O; heated with H_2O_2; washed with distilled water and wet sieved at 62 μm.	Dried at 110°C; ground to 200 mesh.
1. Coarse Fraction (>62 μm) Dried at 110°C, sieved at 2 mm 1a. Coarse Fraction (>2 mm) Size distribution by sieving 1b. Fine Fraction (0.062-2 mm) Size distribution by sedimentation tower[a] Sieved at 0.25 mm, grain types in 0.062-0.25 mm fraction enumerated by point counting[b] 2. Fine fraction (<62 μm) Size distribution by falling-drop method[c]	1. Total and carbonate carbon by Leco analyzer; organic C by difference[d] 2. Opal by infra-red absorption spectrophotometry[e] 3. Major elements by X-ray emission[f] 4. Minor elements by X-ray emission[g] 5. Sulphur and chlorine by X-ray emission using diluted powder pellets[h]

[a] Zeigler et al. (1960)
[b] Shepard and Moore (1954)
[c] Moum (1965)
[d] Kolpack and Bell (1968)
[e] Chester and Elderfield (1968)
[f] Norrish and Hutton (1969)
[g] Reynolds (1963)
[h] The chlorine values were used to correct bulk compositions to a salt-free basis, assuming standard seawater composition.

kindly provided by A. O. Fuller of the University of Cape Town from gravity cores collected by RV "Sardinops" in 1961; these were received air dried but were otherwise untreated. The locations of the samples are shown in the subsequent sediment distribution maps.

In the laboratory, the cores were sectioned longitudinally and one half was preserved at 3°C. The other half was sectioned again, the two quarters being used for mineralogical and textural analyses and for chemical analyses, respectively. Each quarter was split into 5-cm thick subsamples; only the near-surface section was studied in detail. Duplicate cores from four stations were sampled and prepared for radio-carbon dating (see Veeh et al., 1974). The analytical methods utilized are summarized in Table 1.

PHYSIOGRAPHIC AND OCEANOGRAPHIC SETTING

Namibia forms part of the western boundary of the great African plateau, the latter stretching from southern Africa to the southern border of the Sahara. The plateau ranges from 1000 to 3500 m in elevation above sea level, the outer edge of which forms the highest ground and the watershed between plateau rivers and coastal drainage. The original margin of the plateau was probably coincident with the coastal margin of the continent, but with uplift and headward erosion, the margin has migrated eastwards. This uplifted margin is the Great Escarpment of southern Africa (Wellington, 1955), a somewhat variable feature in terms of altitude and rock structure but a fundamental physiographic feature of the region (Fig. 1B).

In Namibia, the Great Escarpment separates a coastal peneplain, the Namib Desert, from the western plateau region, the southwest African Highlands (Fig. 1B). The Namib Desert stretches approximately 2000 km from the Cape region of South Africa to Angola, extending from the coastline to an altitude of about 700 m. In the south, the desert consists of a coastal strip of sand, the sandveld, about 50 km wide, and an inner zone, the hardveld with bare rock outcrops, stretching to the plateau margin. The sandveld maintains a sparse coverage of succulents which are maintained by the very wet coastal mists due to the influence of the Benguela Current.

The largest extent of sand is found in the area immediately south of Walvis Bay (Fig. 2). An extensive dune field, up to 100 km wide, is present with dune crests up to 130 m high above the intervening troughs. The northern boundary of this sand sea is abruptly marked by the Kuiseb River (Fig. 1B) where periodic floods check the northward migration of the sand (Logan, 1960). A relatively narrow strip of sand between Walvis Bay and Swakopmund is similarly bound on the north by the Swakop River. North of Swakopmund the desert is quite different. Behind a very narrow coastal sand belt, bare rock outcrops consisting of Pre-Cambrian basement, are widespread. Vege-

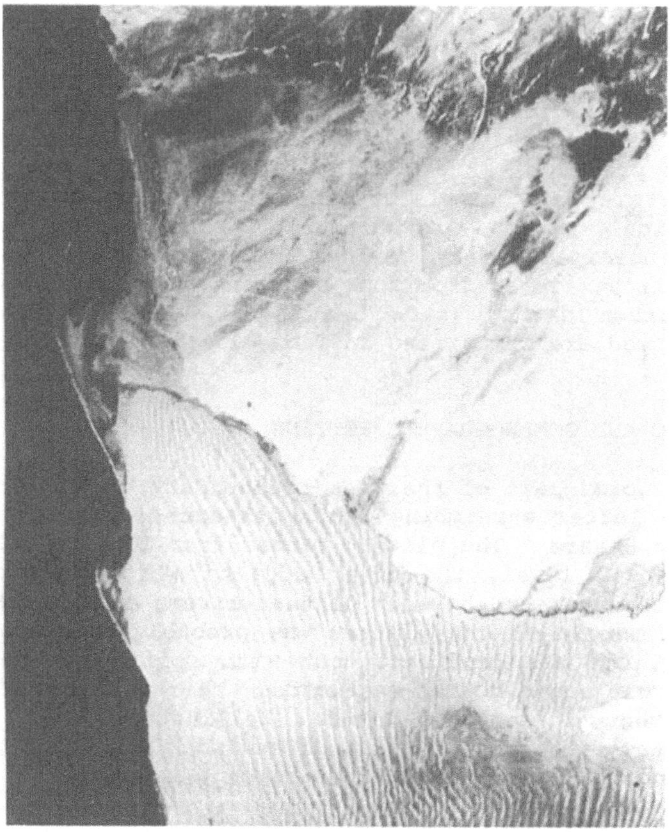

Fig. 2. Gemini photograph of Walvis Bay (located immediately east of the prominent spit in the center of the photograph) and adjacent part of the Namib desert. Prominent dune fields, crests oriented roughly north-south in the southern half, are bounded sharply by the course of the intermittent Kuiseb River. A small area of coastal sand desert north of the Kuiseb River is bounded to the north by the Swakop River.

tation is very sparse and the desolation of the region is well described by the name "Skeleton Coast" given to the area from the Kuenene to the Marura Rivers.

Rainfall in the Namib Desert is everywhere less than 20 cm per annum and the coastal Namib receives less than half this amount. Rainfall increases inland to >50 cm per annum in parts of the interior Kalahari. The only perennial rivers draining Namibia are the Orange and Kuenene Rivers. Many of the other water courses shown in Fig. 1B are intermittent streams containing storm water perhaps once in a decade which seldom reaches the sea.

The Namibian continental margin is a distinctly wide platform, contrasting markedly with the shelf north of Cape Frio where the Walvis Ridge abuts the African continent (Fig. 1A). The basement rocks of the shelf are extensions of the Pre-Cambrian complexes on land with Cretaceous rocks overlying this basement (Simpson, 1971). Prograded and slumped Cenozoic strata smooth the abrupt margin of the underlying Mesozoic terrace.

The continental shelf is 100 to 160 km wide (Fig. 3). Between 23°00'S and 23°30'S and between 24°30'S and the southern part of the CIRCE survey, the shelf has a gently sloping surface and a rounded shelf break. The shelf edge lies at 350-400 m depth. The remaining parts of the shelf are more complex and consist of a narrow inner-shelf platform and an outer, deeper shelf separated by a rather abrupt step at 140-160 m depth (Fig. 3, inset). This intermediate shelf break may be an irregular, cliff-like feature or a relatively rounded break-in-slope. The outer shelf is either a smooth platform, sloping seawards more steeply than the inner platform, or consists of a distinct trough.

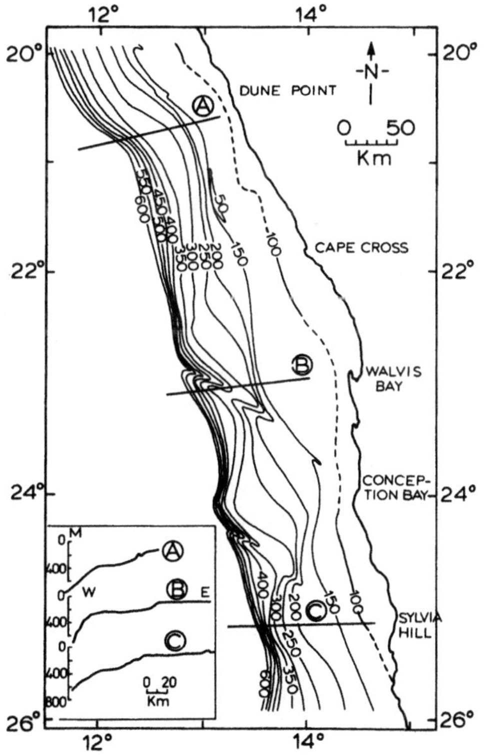

Fig. 3. Bathymetry of the central Namibian continental shelf. Isobaths (meters) drawn from sections obtained during Expedition CIRCE (1968). Inset shows three profiles of the continental shelf indicated by the cross-shelf lines A, B and C.

The shallow substrate over most of the shelf consists of bedded sediment. On the shelf and upper slope, 3.5 kHz echo-sounding records show reflectors sharply truncated by the sea floor. The truncated beds produce minor relief on the shelf; in addition, there are a number of small notches and terraces 5-20 m high. The most widespread of these occurs as a sharp escarpment at about 110 m depth, forming the seaward limit of an inshore zone of a very rugged small-scale relief. This depth is close to the accepted late Pleistocene low stand of sea level (Curray, 1965) so that most of the remaining part of the shelf was submerged at that time. The other topographic notches and terraces may well represent other late Quaternary or earlier sea level positions (van Andel and Calvert, 1971).

The coastal waters of Namibia form the eastern limb of the South Atlantic mid-latitude anticyclonic gyre which determines the surface current system, the Benguela Current (Hart and Currie, 1960). The northward-flowing current is about 200 km wide and relatively cool. The prevailing winds are the southeast trades with average speeds between 10 and 25 knots. These are augmented by coastal winds, a response to diurnal heating and cooling of the land surface, which may blow inland during the day and from the land at night. In the Walvis Bay region, some winter easterlies, the Berg winds, blow with considerable force from the hot interior and carry large quantities of sand and dust out over the coast.

The prevailing wind system drives surface water offshore during most of the year and is consequently responsible for coastal upwelling. During the 1968 CIRCE survey, winds were blowing at 12-23 knots from the south and southeast and cool, low-salinity, oxygen-poor and nutrient-rich water was present along the entire stretch of coastline studied (Calvert and Price, 1971b), representing the water that had been recently upwelled, evidently from depths of 150 to 220 m at the shelf edge. In addition, the bottom water on the shelf was markedly nutrient-enriched --on some profiles more so than the water at the depth of upwelling at the shelf edge-- and oxygen deficient, the oxygen concentrations decreasing towards the coastline. These distributions are a consequence of the regeneration and biological oxygen demand of the settling organic matter produced in the fertile surface waters.

THE SURFACE SEDIMENTS: BULK COMPOSITION AND FACIES

The texture, coarse fraction (0.062-0.25 mm) compositions and the concentrations of organic carbon, carbonate and opal of the surface sediments have been used to construct a surface lithofacies map (Fig. 4). A nearshore belt of diatomaceous organic-rich, silty clay, containing abundant fish debris, the diatom ooze, is present from Conception Bay to the area between Cape Cross and Dune Point, with two smaller patches north of Sylvia Hill and off Dune Point (see also

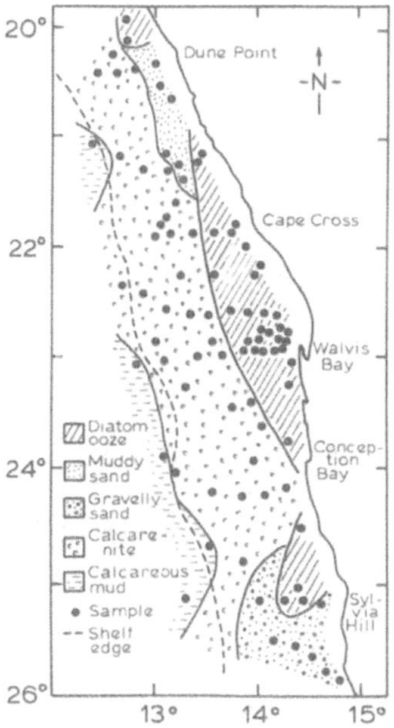

Fig. 4. Distribution of sediment lithofacies delineated by textural, coarse-fraction, carbon, carbonate, and opal analyses.

Bremner, 1980a). This diatom ooze corresponds with the so-called "azoic zone" of Copenhagen (1953), a belt of extremely fluid, olive-green muds containing abundant H_2S and lacking benthos. It is clearly recognized on 3.5 kHz echo-sounding records by the presence of a variable thickness of acoustically transparent sediment below a weak sea floor echo (Fig. 5). The diatom ooze contains abundant opal, less than 25% $CaCO_3$ and high concentrations of organic carbon (Fig. 6). The opal and organic carbon distributions appear to be displaced one from the other. Samples containing high concentrations of opal are found in the area nearest the coastline consistent with the locus of maximum production of diatoms in the nearshore zone (Hart and Currie, 1960), and with the distribution of neritic diatom shells in the sediment (Schuette and Schrader, 1981), whereas the highest carbon contents are present in a north-south belt approximately 30 km from shore. The high accumulation rate of opal in the nearshore region serves to dilute the sediment and thereby lower the overall carbon content. The area occupied by sediments containing more than 5% by weight organic carbon roughly coincides with the inner shelf platform (Fig. 3) apart from the southern part of the shelf where the mid-shelf and upper slope samples are both carbon-rich. A somewhat similar facies occurs beneath the Peru current on the upper part of the slope (Krissek et al., 1980) where the sediments are equally fluid and organic-rich, the latter property evidently imparting unusual physical properties to the deposits (Busch and Keller, 1981).

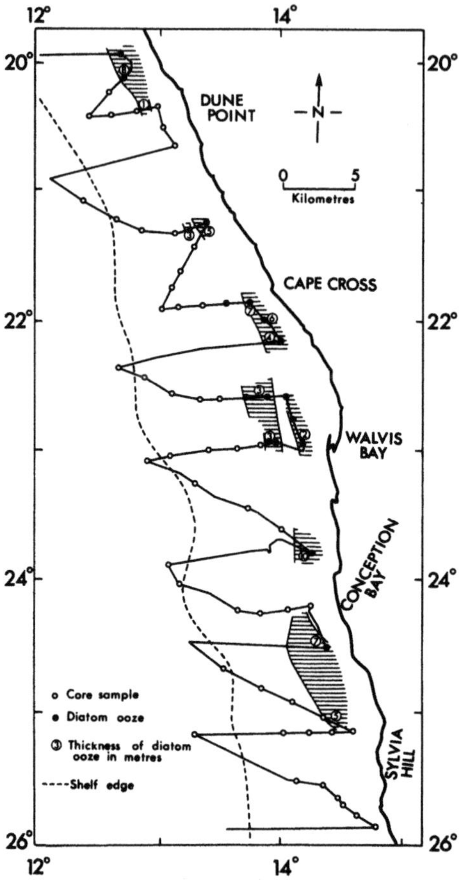

Fig. 5. Location of sampling traverses on the shelf. Areas of acoustically transparent sediment shown by shading with indications of the thickness of ooze obtained from 3.5 kHz echo-sounding records. Data compiled by Tj. H. van Andel.

The areas of the inner shelf off Conception Bay and immediately south of Sylvia Hill and Dune Point are covered by coarse-grained sediments, either gravelly sand (the gravel consisting of schist pebbles) or terrigenous muddy sands, respectively.

The remaining areas of the shelf are covered by carbonate-rich sediments of varying textures. The concentration of carbonate is derived from the CO_2 values in the total sediment, expressed as $CaCO_3$ (Fig. 6), since carbonate phases other than calcite and very minor aragonite are absent. Values reach 88.5% by weight in the outer part of the shelf where the bulk of the carbonate consists of comminuted mollusc fragments and foraminifers in the calcarenite facies. The sediments are quite coarse-grained immediately seaward of the central

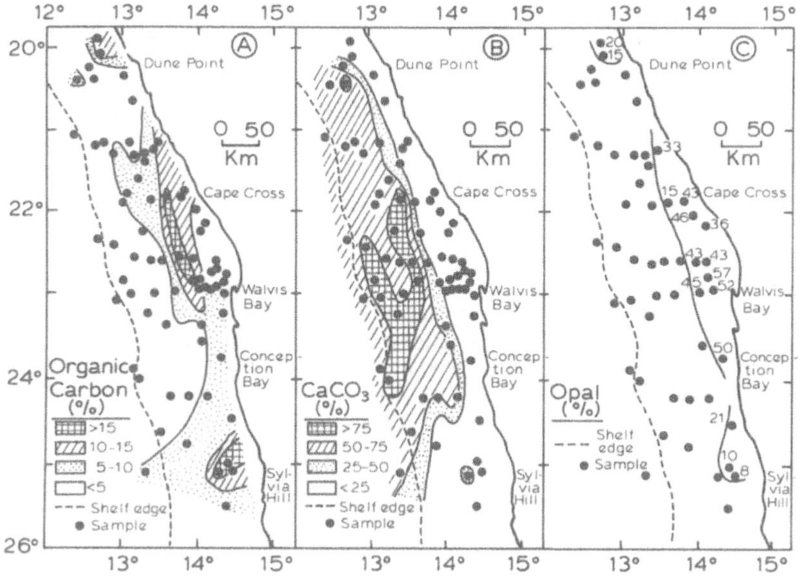

Fig. 6. Distributions of organic carbon, calcium carbonate and opal in the surface sediments.

belt of diatom ooze, but they become finer-grained toward the edge of the shelf, passing into calcareous muds. A component of secondary importance in some of the nearshore gravels and sands and in the calcarenites of the central part of the shelf is represented by phosphorite (Summerhayes et al., 1973). It occurs as cream-yellow concretionary fragments and as black, ovoid pellets (Veeh et al., 1974; Price and Calvert, 1978; Bremner, 1980b).

As discussed earlier, only the inner shelf platform, shallower than about 140 m, could have been exposed during the last Pleistocene low stand of sea level. The diatom ooze which occurs here to a thickness of up to 10 m has therefore accumulated only during the last 3,000 to 15,000 years, consistent with measured uncompacted sedimentation rates ranging from 29 to 120 cm/10^3 yrs. (Veeh et al., 1974). This sediment facies can be identified, therefore, as the modern equilibrium facies of the shelf which reflects the location of the most intense upwelling and concomittant production in the Benguela Current. The sediment on the remaining part of the shelf, which was never exposed, is probably partly relict, as attested by the presence of stained and abraded shell fragments and reworked phosphorite grains, and partly modern, fresh carbonate being worked into old calcareous sands.

GEOCHEMISTRY OF THE SURFACE SEDIMENTS

A total of 52 surface sediment samples were analyzed for major and minor element contents by x-ray emission methods. The analyses have been corrected both for the diluting effect of salt on the weight of the dried samples and for the contributions of Mg, Ca, K, S, and Sr from sea salt to the total compositions and are shown in Table 2.

Major Elements

The major element concentrations reflect the mixture of terrigenous, biogenous and authigenic phases in the sediment and mirror the distribution of the surface sediment facies. Thus, the concentration of Si is highest in a rather narrow nearshore zone (Fig. 7) corresponding with the areas covered by diatom ooze and nearshore terrigenous sands. The concentrations decrease rapidly away from the coast and are very low in the central part of the shelf. Aluminium contents are high in two nearshore areas separated by the belt of diatom ooze off Walvis Bay. Higher Al contents depict the distribution of fine-grained aluminosilicates, mainly clay minerals; they are similarly low in the central shelf.

The relationship between Si and Al, as suggested by Fig. 7, is highly variable (Fig. 8). They are reasonably well correlated in the fine-grained sediments containing small amounts of opal but are much higher, and variably so, in diatom oozes and nearshore sandy sediments.

The distributions of Ti and Fe (Fig. 7) are similar to that of Al; high values are found in the nearshore areas north and south of Walvis Bay and low values are found on the central shelf. The relationship between Fe and Al (Fig. 8B) shows that the Fe/Al ratio is somewhat variable. The presence of glauconite in two outer shelf samples increases the ratio markedly, while the association of Fe with detrital aluminosilicates (predominantly feldspar and mica with minor montmorillonite and kaolinite) and pyrite in the diatom ooze, and with low levels of aluminosilicates alone in the calcareous sediments, produces the scattering of points around the regression line in Fig. 8B. The distribution of K (Fig. 9) again reflects to a major degree the distribution and composition of fine-grained aluminosilicate material; high concentrations are confined to the nearshore areas north and south of Walvis Bay where the highest Al, Ti and Fe values are also found. The relationship between K and Al (Fig. 8C) shows that a ratio close to 0.4 describes all but two samples, which are the same two glauconite-bearing samples distinguished by high Fe/Al ratios in Fig. 8B.

The distribution of Ca in the surface sediments (Fig. 9) largely reflects the distribution of carbonate, which is present mainly as

Table 2. Chemical composition of surface (0-5 cm) sediments of the Namibian shelf. Major elements in weight %; minor elements in ppm. All data salt-corrected. [a]Samples collected during CIRCE Expedition; [b]total iron, [c]organic carbon.

Samp.[a]	Si	Al	Ti	Fe[b]	Ca	Mg	K	P	S	CO_2	C[c]	Ba	Cu	I	Mo	Ni	Pb	Rb	Sr	U	Y	Zn	Zr
148	12.1	1.44	0.16	2.65	20.3	0.8	1.19	6.21	1.66	5.0	4.5	248	41	140	21	75	16	65	1320	114	210	80	174
150	6.5	1.74	0.10	1.00	25.2	0.6	0.59	0.14	0.82	25.2	8.0	1400	19	1990	6	63	8	46	1050	6	21	86	54
152	10.5	2.21	0.15	1.89	14.9	1.0	0.99	1.49	1.94	12.0	15.2	228	95	506	57	152	10	69	772	57	51	152	89
153	11.4	2.39	0.17	1.87	10.9	1.6	1.22	1.49	1.72	8.4	14.3	184	122	566	61	184	15	99	810	84	46	337	107
155	7.6	2.01	0.11	1.16	11.1	0.5	0.59	2.14	1.99	6.6	16.4	186	86	424	60	152	13	66	716	80	36	139	60
157	5.1	0.73	0.07	0.75	24.3	1.2	0.44	9.40	1.61	3.7	6.0	550	35	141	56	268	9	49	2410	360	66	63	63
158	4.6	1.09	0.07	0.79	31.0	0.3	0.47	0.74	0.52	30.8	4.1	234	37	956	5	64	5	32	1100	11	58	42	74
159	25.9	2.17	0.23	1.35	7.7	0.3	0.81	1.44	1.05	4.1	6.7	168	67	146	39	73	11	56	538	45	34	73	247
161	6.4	0.91	0.07	0.57	27.9	0.3	0.35	0.66	0.78	28.5	5.6	86	43	302	22	81	9	22	1140	11	32	65	81
162	2.6	0.64	0.03	1.13	33.3	0.5	0.31	5.64	1.39	21.1	2.5	153	22	148	26	66	8	20	1600	97	128	31	72
163	2.6	0.75	0.03	0.81	31.9	0.4	0.28	3.10	0.91	27.8	3.2	156	28	518	5	57	9	21	1320	57	93	31	47
164	2.6	0.57	0.03	0.49	36.7	0.3	0.20	0.28	0.42	39.0	1.0	52	23	208	5	42	16	16	1220	26	21	31	57
165	5.5	1.55	0.09	0.87	29.4	0.6	0.51	1.11	0.51	30.9	4.5	692	43	1040	5	53	7	43	1090	5	23	48	75
166	24.6	1.19	0.13	1.02	6.9	1.4	0.71	0.70	2.00	3.9	9.5	252	66	179	80	106	6	40	464	73	13	46	40
167	3.6	0.74	0.04	0.82	28.2	0.5	0.33	1.84	0.71	29.2	4.5	106	39	223	21	74	6	27	1290	42	50	42	53
169	3.3	0.86	0.04	0.56	34.0	0.3	0.31	0.70	0.51	34.2	3.0	136	28	576	5	52	10	21	1080	10	37	37	58
170	6.3	1.47	0.09	0.99	28.6	0.3	0.66	0.20	0.31	31.2	3.2	474	32	765	5	65	3	40	1010	5	31	35	81
171	3.7	1.07	0.05	0.66	33.5	0.4	0.35	0.14	0.40	35.0	3.2	171	43	706	5	43	7	27	1070	5	16	32	54
172	3.0	0.90	0.05	0.73	32.0	0.3	0.29	1.22	0.58	30.3	5.9	98	46	760	6	98	6	22	1170	27	51	33	49
173	1.9	0.37	0.02	0.38	35.2	0.3	0.15	0.52	0.49	36.2	2.8	96	16	225	5	32	9	11	1280	11	11	22	43
175	20.0	0.76	0.07	0.78	5.0	1.5	0.41	0.22	1.93	4.5	0.5	122	101	263	61	117	30	61	477	5	5	61	20
176	26.0	0.86	0.07	0.98	2.5	0.4	0.37	0.12	1.00	1.6	8.4	88	61	175	61	79	11	53	289	9	3	53	26
177	29.8	0.90	0.12	0.87	3.2	0.7	0.35	0.28	0.69	3.2	7.8	129	46	144	46	72	10	36	266	5	7	57	22
178	19.9	0.76	0.19	1.07	2.5	0.6	0.28	0.19	0.13	0.8	8.9	19	66	160	56	104	19	28	255	19	13	56	28
179	23.2	0.79	0.07	0.69	5.5	0.8	0.34	0.29	1.67	4.6	13.9	150	63	265	79	110	13	40	395	16	8	47	28
180	5.8	1.59	0.10	0.92	19.9	0.8	0.53	0.22	1.29	18.5	16.2	260	95	484	82	341	14	44	851	27	20	68	68
181	2.0	0.29	0.02	0.02	35.5	0.1	0.11	0.18	0.44	27.4	2.6	107	29	188	5	32	4	11	1310	5	11	16	70
182	2.4	0.88	0.08	0.67	32.6	0.4	0.24	1.09	0.59	32.5	4.4	77	35	686	5	72	6	28	1180	22	52	28	5
183	2.8	0.95	0.10	0.65	32.7	0.5	0.27	0.33	0.39	32.4	4.3	119	32	765	5	65	5	22	1070	6	24	32	49
184	3.0	0.90	0.04	0.45	34.9	0.3	0.22	0.19	0.28	35.3	2.6	128	29	492	5	48	3	21	1040	5	16	27	43
185	10.1	1.95	0.10	4.54	23.0	0.3	2.16	0.12	0.27	22.5	3.2	616	37	807	5	42	4	90	797	5	16	42	27
186	27.7	2.05	0.18	1.58	2.9	0.7	0.67	0.79	1.41	2.3	4.9	193	54	109	28	97	13	77	297	77	15	58	77
187	27.8	1.05	0.08	0.86	3.7	0.5	0.46	1.09	1.44	7.2	6.2	452	57	134	85	106	7	49	424	92	14	49	42
188	20.6	1.28	0.10	1.30	7.5	0.8	0.43	2.49	2.77	0.6	13.6	219	78	227	102	156	13	63	565	63	23	86	47
189	11.8	1.09	0.20	1.09	9.6	0.9	0.44	0.87	1.48	6.8	22.3	142	92	483	142	242	11	50	569	36	24	78	43
190	2.2	0.60	0.04	0.44	32.5	0.2	0.22	0.27	0.63	32.8	6.7	79	96	390	8	85	8	23	1140	11	19	34	39
192	4.3	1.29	0.07	0.84	29.2	0.5	0.44	0.57	0.58	29.1	5.7	177	37	731	5	104	5	37	1030	5	37	37	47
194	4.2	0.87	0.05	0.50	31.2	0.3	0.29	0.17	0.50	32.2	5.5	98	54	423	5	71	6	27	1030	5	13	33	43
195	10.8	1.58	0.09	1.02	20.5	0.7	0.59	4.50	1.15	10.9	9.2	103	55	276	40	147	9	46	1070	92	115	40	98
196	25.0	2.96	0.20	0.20	5.3	0.9	1.11	1.35	1.32	0.1	7.4	163	52	134	35	81	13	111	419	23	20	76	70
197	12.0	1.99	0.13	1.08	18.9	0.8	0.71	1.97	0.42	16.9	9.5	106	59	319	35	154	12	53	851	17	24	53	106
198	30.6	2.27	0.10	0.63	6.9	0.3	1.10	0.10	0.20	6.0	1.7	326	26	128	5	41	10	71	337	5	31	31	77
199	7.1	1.96	0.12	0.99	27.2	0.1	0.60	1.49	0.47	26.9	5.2	169	53	584	5	85	16	48	966	16	16	42	74
200	5.8	1.60	0.10	0.83	30.5	0.4	0.46	0.10	0.45	30.9	3.6	225	34	562	5	51	7	34	933	11	13	34	62
201	6.8	2.04	0.10	1.13	27.8	0.7	0.61	0.11	0.37	30.2	4.3	406	46	1096	8	82	6	44	933	5	19	49	60
202	33.5	2.92	0.31	1.24	3.8	0.7	1.05	0.22	0.51	3.1	3.4	232	28	100	16	63	11	74	232	11	23	32	253
204	29.7	3.01	0.22	1.60	6.1	0.8	1.31	2.32	0.60	4.2	3.7	266	34	96	48	191	13	90	350	5	23	37	218
206	4.8	0.81	0.07	0.71	32.3	0.5	0.34	0.12	0.47	34.2	2.7	109	33	391	5	49	9	27	1650	38	5	22	43
207	7.9	2.56	0.16	1.38	25.3	0.6	0.70	0.39	0.60	24.9	6.6	120	49	459	5	120	11	49	842	22	30	44	66
208	31.6	2.26	0.22	0.81	7.6	0.5	0.97	0.22	0.23	6.0	2.1	246	26	108	15	82	13	62	379	10	26	21	272
209	16.1	3.21	0.21	2.31	9.1	1.2	1.01	0.09	1.35	9.9	12.9	141	77	323	112	225	14	98	492	56	21	70	77
210	19.7	3.13	0.23	2.37	5.1	1.0	1.10	0.13	0.18	3.8	12.2	143	71	286	78	299	13	110	338	10	22	71	65

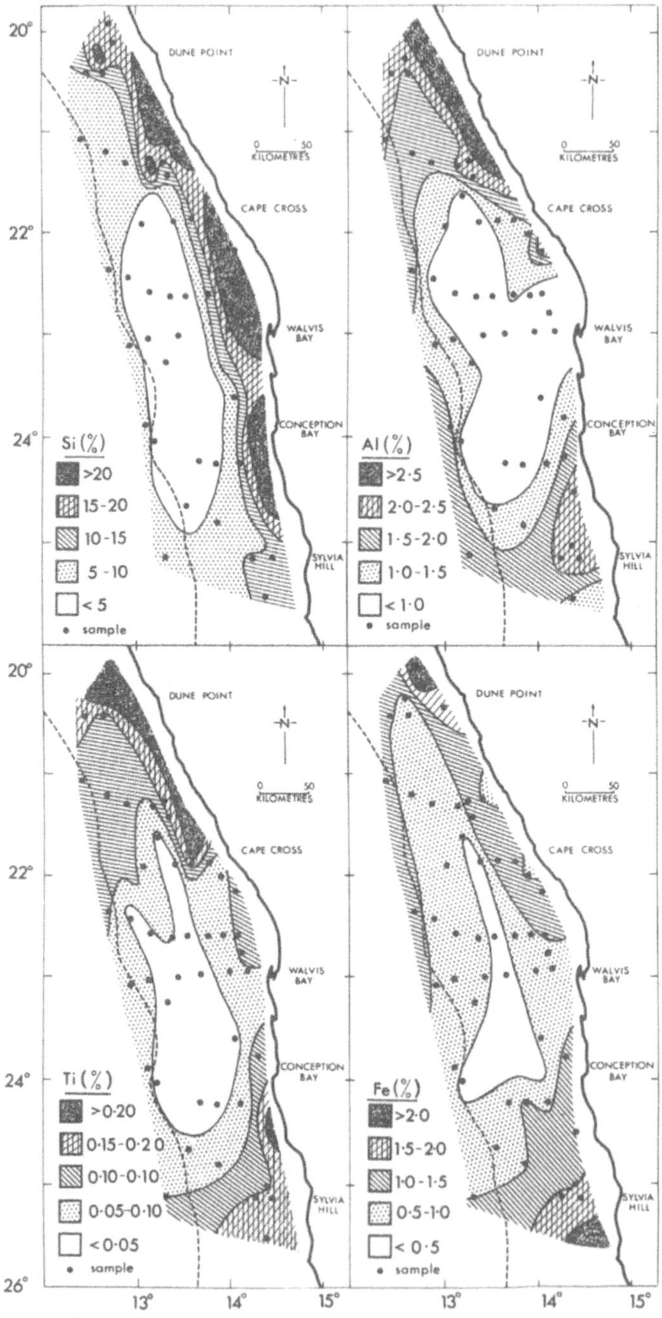

Fig. 7. Distribution of total Si, Al, Ti, and Fe in the surface sediments.

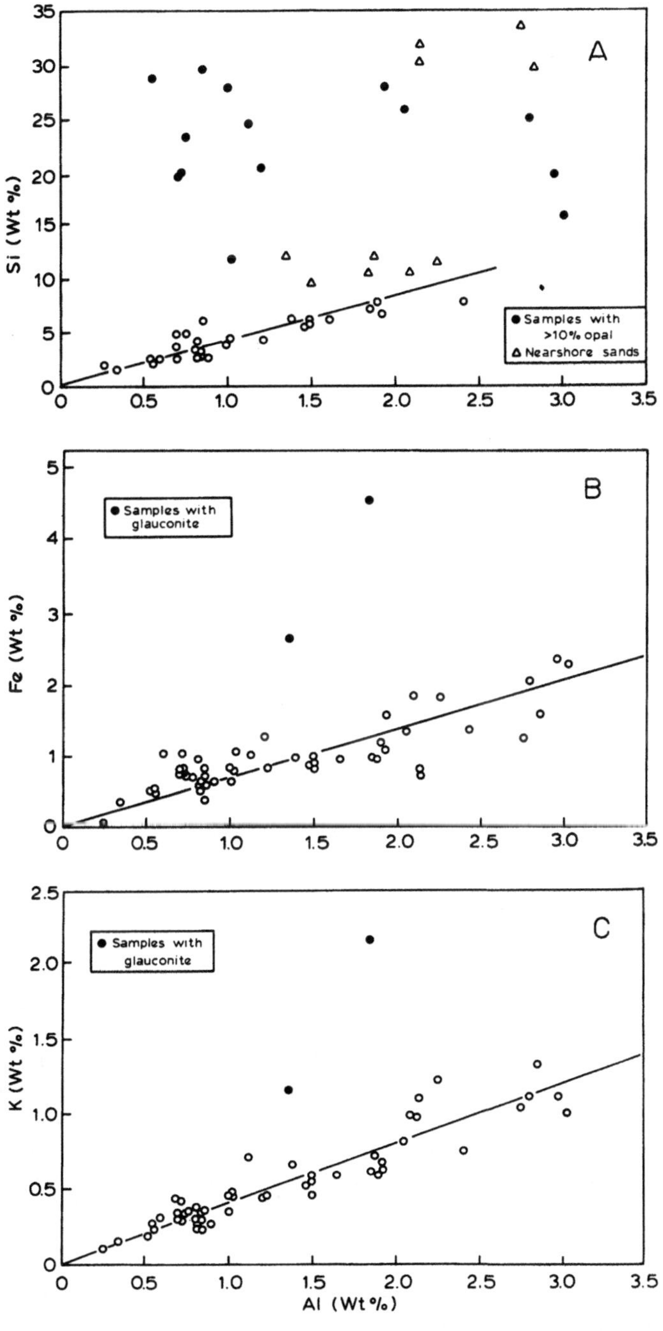

Fig. 8. Relationships between Si, Fe, K, and Al in the surface sediments.

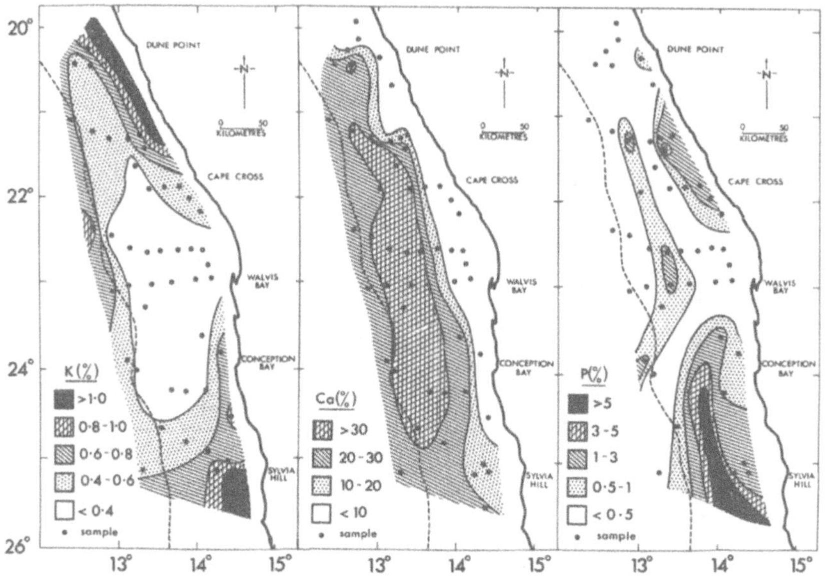

Fig. 9. Distributions of K, Ca and P in the surface sediments.

foraminiferal and macrobenthonic shell debris. Fig. 10 shows that the majority of the samples have Ca and CO_2 contents contributed by ideal $CaCO_3$, with a small amount of Ca not balanced by CO_2 representing the Ca in aluminosilicates. Other samples have Ca contents in excess of those necessary to balance the CO_2 and these are samples containing significant quantities of phosphorite where the CO_2 contents are generally <4%.

The distribution of phosphorite is accurately delineated by the P content (Fig. 9). Maximal values are found in a nearshore and mid-shelf zone in the southern part of the shelf; and moderate concentrations are found in the nearshore areas off Cape Cross and in a narrow mid-shelf zone between Walvis Bay and Cape Cross. Senin (1968), Summerhayes et al. (1973) and Bremner (1980b) have shown that phosphorite, in the form of reworked concretionary and pelletal varieties is widely distributed on the Namibian and South African shelves. The mid-shelf belts of high P content reflect the distribution of this type. On the other hand, the nearshore zone off Cape Cross is one where modern, unconsolidated and friable phosphorite is forming (Baturin, 1969; 1971a; Veeh et al., 1974; Price and Calvert, 1978). In addition, note that the phosphorus contents of many of the organic- and opal-rich oozes have significantly higher levels of total P than many other nearshore muds. It is likely that the bulk samples analyzed here contain small quantities of newly-formed, fine-grained and dispersed phosphorite (Price and Calvert, 1978) as well as fish debris.

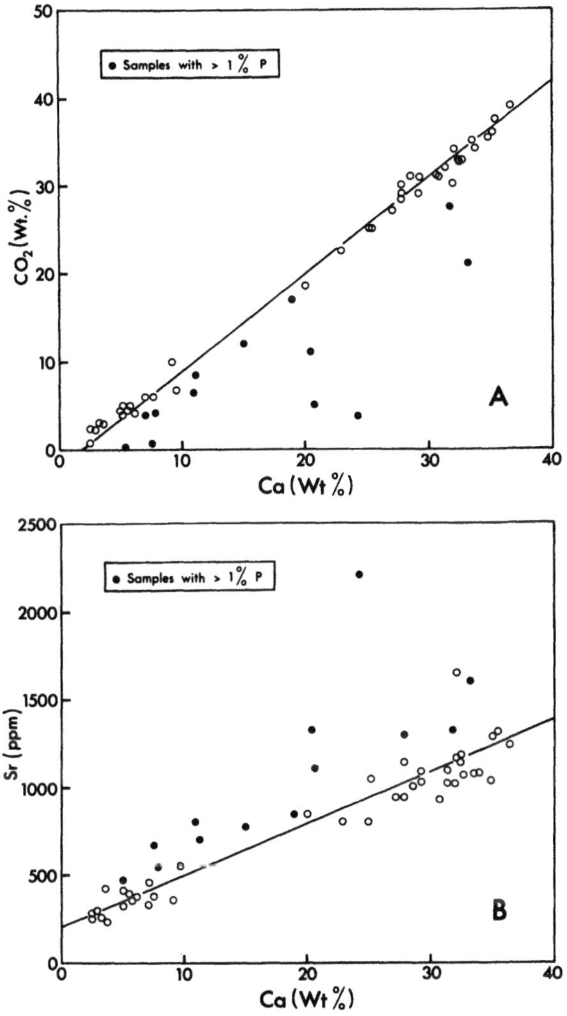

Fig. 10. Relationships between Ca, CO_2 and Sr in the surface sediments.

Minor Elements

The distribution of the minor elements in the surface sediments allows further refinement of the factors controlling the bulk geochemistry of the surface sediments and permits an examination of the degree of enrichment of minor elements, if any, in the distinctive sediment facies occurring on the shelf.

The distribution of Sr (Fig. 11) reflects the distributions of both carbonate and phosphorite; the highest concentrations are found in the southern part of the shelf in a north-south belt where phos-

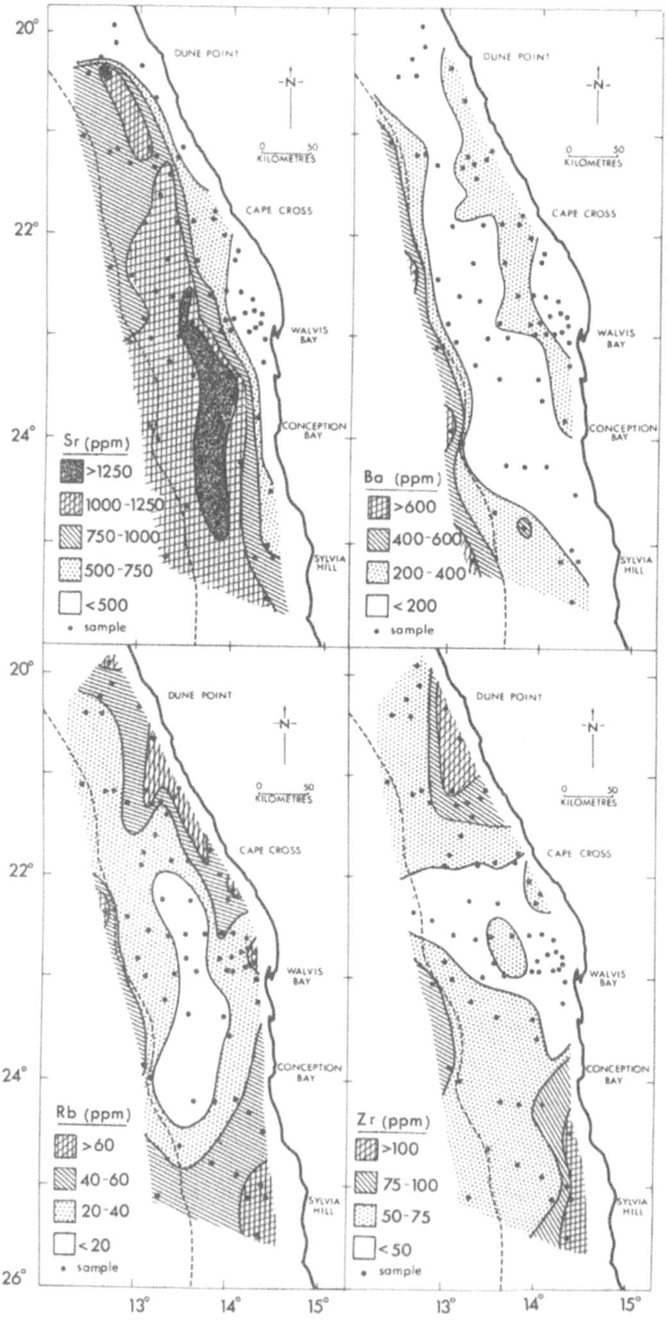

Fig. 11. Distributions of Sr, Ba, Rb, and Zr in the surface sediments.

phorite is abundant. The carbonate is almost entirely calcitic and the levels of Sr observed in the highly calcareous sediment are consistent with the presence of approximately 1400 ppm Sr, typical of foraminiferal calcite (Turekian, 1974). This is illustrated by the relationship between Sr and Ca (Fig. 10) which also shows that Sr is enriched in samples containing significant amounts of phosphorite as indicated by the P contents. This again is consistent with the substitution of Sr for Ca in carbonate fluorapatite (Swaine, 1962; Gulbrandsen, 1966).

The highly calcareous sediments of the central shelf contain very low levels of Ba (Fig. 11). This is consistent with the observations of Turekian and Tausch (1964) that Ba and $CaCO_3$ in the surface sediments of the Atlantic were significantly negatively correlated. Barium levels are a little higher in a nearshore region between Walvis Bay and Dune Point; some of the Ba here might reflect the presence of feldspar in terrigenous muddy sands (see Fig. 4). Much higher levels occur, however, on the upper slope area, an observation also made by Turekian and Tausch (1964).

Brongersma-Sanders (1967) has suggested that some diatoms and in particular some species of the genera *Rhizosolenia* and *Chaetoceras*, concentrate Ba from seawater and thereby act as important conveyors of this metal to bottom sediment. Furthermore, Brongersma-Sanders et al. (1980) report that Ba levels range up to 3660 ppm, with an isolated value of 8% Ba, in samples of the organic-rich diatom ooze off Walvis Bay and argue that this level of enrichment is due to its concentration by diatoms.

Goldberg and Arrhenius (1958) showed that the Ba contents of pelagic sediments are related to surface organic production and Goldberg (1958) found that the Ba content of Bering Sea sediments co-varied with the opal content. The cause of this relationship has not been established although Goldberg (1958) suggested that the element is conveyed to sediments in organic-rich particles and subsequently released in the bottom sediments. On the Namibian shelf, however, the nearshore diatom ooze, according to the data presented here, does not show any Ba enrichment. On the other hand, the concentration levels in the upper slope calcareous ooze are much higher. This might reflect the presence of a different, oceanic plankton and the lower accumulation rate of total sediment allowing a greater degree of degradation of the biogenous conveyor and a higher level of diagenetic enrichment. Suess (pers. comm.) has suggested that microflagellate algae, some species of which are known to contain barite micro-crystals in their storage products (Fresnel, Galle and Gayral, 1979) are the conveyors of Ba to marine sediments, and the data of Dehairs, Chesselet and Jedwab (1980) are consistent with Ba being an element which reflects high, open ocean productivity.

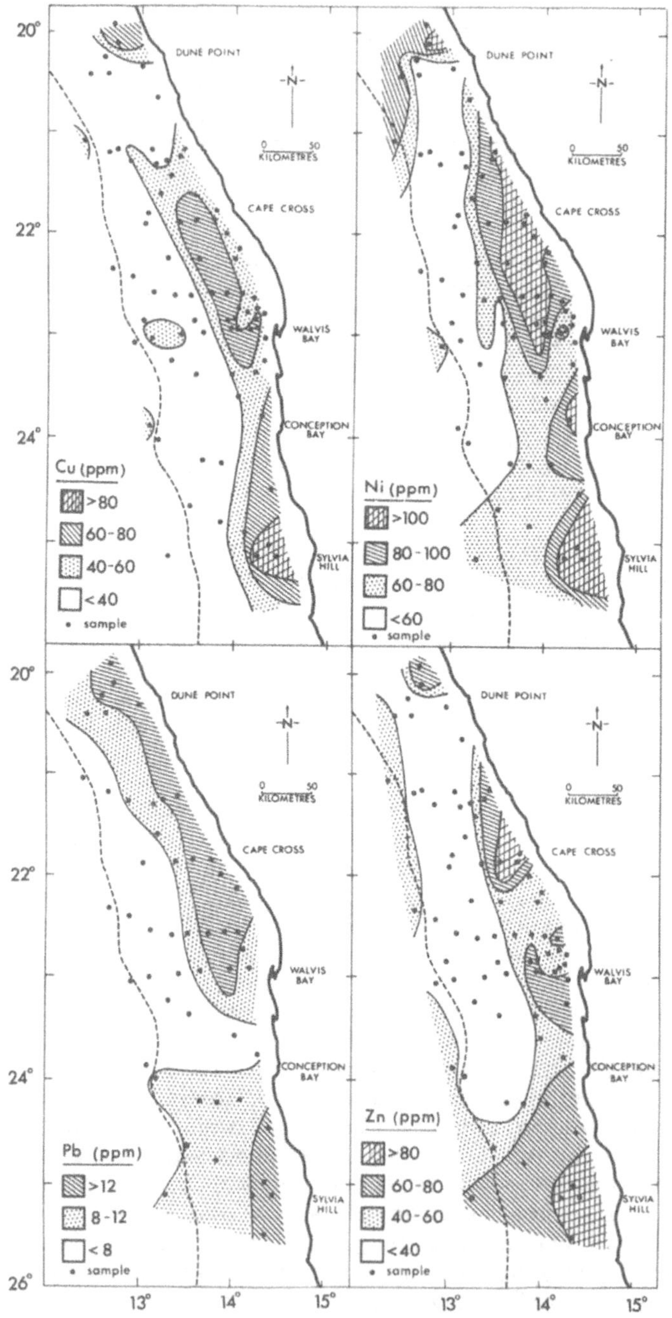

Fig. 12. Distributions of Cu, Ni, Pb, and Zn in the surface sediments.

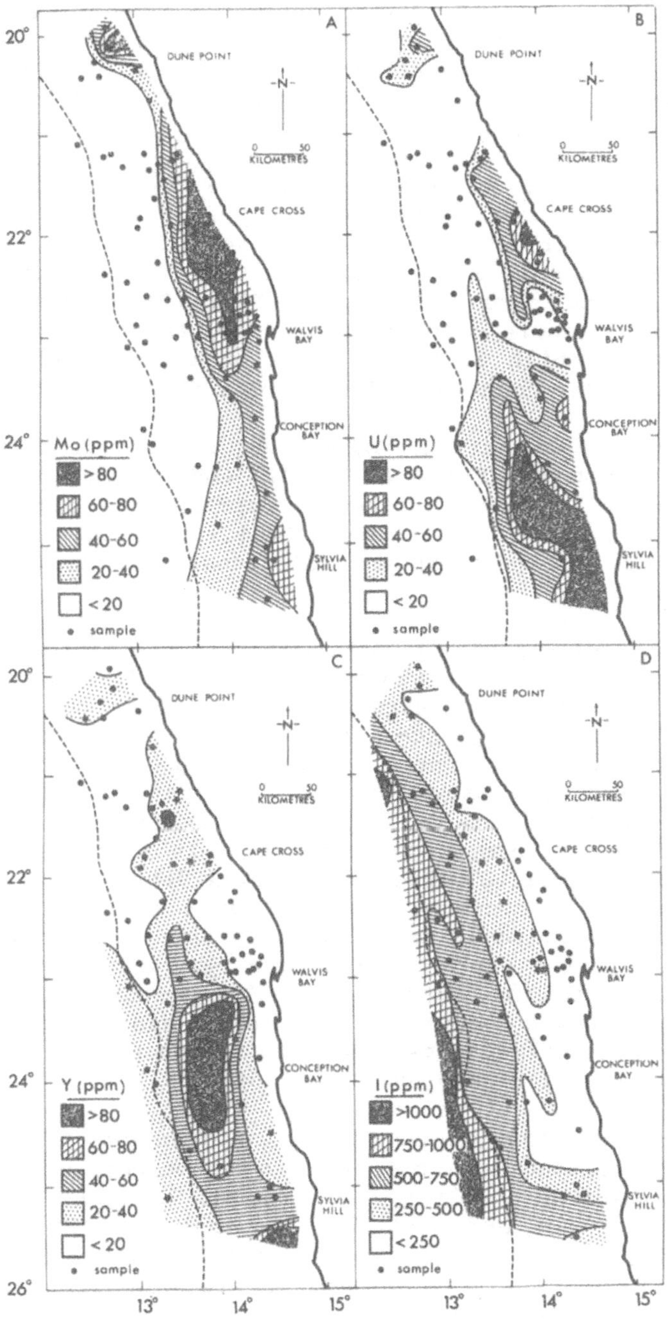

Fig. 13. Distributions of Mo, U, Y, and I in the surface sediments.

The distributions of Rb and Zr in the surface sediments (Fig. 11) reflect almost entirely the distribution of detrital aluminosilicates, as is also clear from other nearshore areas (Hirst, 1962; Wright, 1972). High values are found off Sylvia Hill and Dune Point while the lowest values occur in the calcarenites of the central part of the shelf. These two elements also serve to outline the relative distribution of sand- and clay-sized material. Rubidium occurs in feldspars and in mica, where it substitutes for potassium (Horstmann, 1957), whereas Zr occurs almost entirely in zircon. High Zr/Rb ratios, therefore, reflect the presence of coarse-grained sediments and the ratio decreases in muddy sediments. On the Namibian shelf, the Zr/Rb ratio ranges from 0.3 to 6.4 (Table 2). Lowest values are found in the nearshore muds where the clay fraction (<4 μm) is relatively abundant, while values >3 occur in a mid-shelf belt of sandy sediments which may contain a higher proportion of Namib desert sand.

The distributions of Cu, Mo, Ni, Pb, and Zn are very similar in the surface sediments (Figs. 12 and 13) and can be conveniently considered as a single group. Highest concentrations occur in the diatom ooze off Dune Point, Cape Cross, Walvis Bay, and off Sylvia Hill. Very low values are found in the calcareous sediments of the middle and outer shelf.

The mean contents of the five metals in the diatom ooze are compared with other nearshore sediments in Table 3. Apart from Mo, it can be seen that the concentrations are similar to those in other sediments, including organic-rich, reducing sediments (columns 1-4) and sediments largely composed of terrigenous debris (columns 5-7). However, when the bulk composition of the Namibian shelf sediments is taken into account, some enrichment is evidently present because these sediments are composed almost entirely of biogenous material, either dispersed organic matter or skeletal debris. This can be seen by comparing the Al contents (Table 3) as a reflection of the relative amounts of aluminosilicate debris in the various sediments. In the case of Mo, the concentrations are very much higher than those found in most nearshore sediments; similar concentrations occur however, in other anoxic sediments, and Bertine and Turekian (1973) have argued that such sediments represent important marine sinks for this element as well as for uranium (Veeh, 1967).

The concentrations of Cu, Mo, Ni, and Zn are all significantly positively correlated with the organic carbon contents (see Fig. 4). Similar correlations involving Cu, Mo, Ni, V, and U have been reported in Black Sea and Mediterranean sediments (Kochenov et al., 1965; Baturin, Kochenov and Shimkus, 1967; Volkov and Fomina, 1971; Pilipchuk, 1972). Such relationships are commonly taken to imply that metals are associated with the organic fraction of the sediment (Curtis, 1966), although it is also possible that the correlations are indirect (see Calvert, 1976). It is well established (Trask, 1953; van Andel, 1964) that fine-grained sediments contain higher

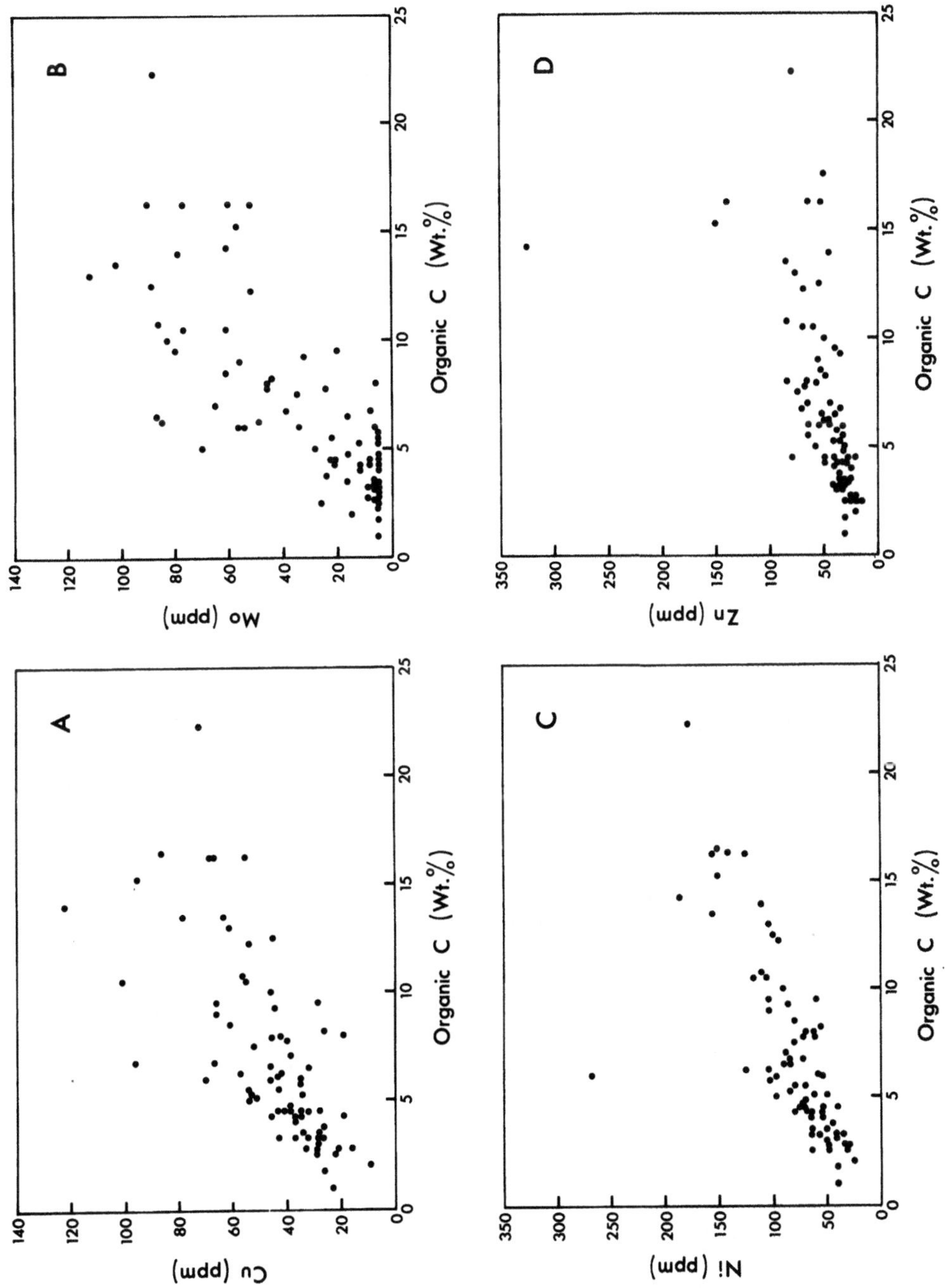

Fig. 14. Relationships between Cu, Mo, Ni, and Zn and organic carbon in the surface sediments.

Table 3. Minor metal (ppm) organic carbon and aluminium (%) contents of some Holocene nearshore sediments

	1	2	3	4	5	6	7
Cu	68	45	64	30	17	32	45
Mo	53	26	--	33	--	--	3
Ni	108	26	146	67	31	38	68
Pb	12	--	51	24	22	2]	20
Zn	68	71	--	147	--	87	95
C	9.35	2.92	2.74	10.67	0.85	1.2	0.1
Aℓ	1.54	4.5	4.5	5.42	8.79	6.05	8.84

1-Namibian shelf diatom ooze (Table 2).
2-Saanich Inlet, British Columbia (Gucluer and Gross, 1964; Gross, 1967).
3-Gulf of California, Mexico.
4-Black Sea sapropels (Glagoleva, 1961; Pilipchuk and Volkov, 1968; Belova, 1970; Lubchenko, 1970).
5-Gulf of Paria clays (Hirst, 1962).
6-Barents Sea (Wright, 1972).
7-Shales (Wedepohl, 1971).

concentrations of organic carbon; such sediments may, therefore, contain fine-grained, metal-bearing components and this is borne out by the analysis of different grain-size fractions of Barents Sea sediments by Wright (1972). The possible role of the organic fraction in controlling the minor metal geochemistry of sediments is discussed later.

Uranium and yttrium have somewhat similar distribution patterns in the surface sediments (Fig. 13), but there are some important differences. Yttrium occurs in highest concentration in the middle shelf phosphatic calcarenites and at intermediate levels in the nearshore terrigenous muds and sandy muds north and south of Walvis Bay. The element occurs in terrigenous aluminosilicate minerals, notably in the heavy mineral fraction, and to a greater extent in phosphorites where it substitutes for Ca in the carbonate-fluorapatite structure. The pattern shown in Fig. 13 therefore reflects the distribution of concentrationary and pelletal phosphorite. Uranium is also present in this phase, but unlike Y is present in relatively high concentration in some organic-rich sediments. Fig. 15 shows the effect of these two hosts on the distribution of U.

The relatively high U values in the surface sediments off Cape Cross (Fig. 13) are due to the presence of modern phosphorite (Veeh

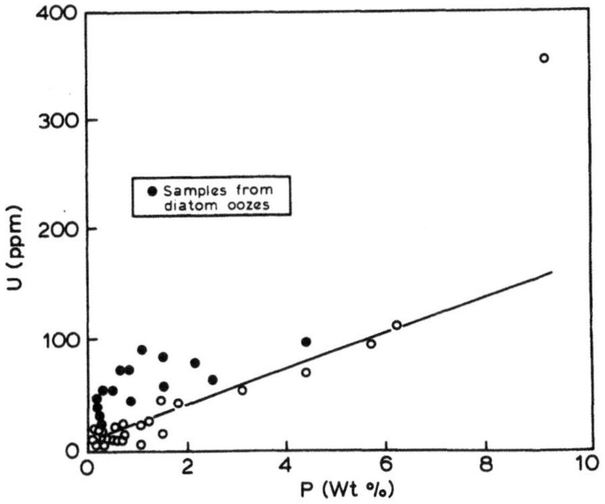

Fig. 15. Relationship between U and P in the surface sediments.

et al., 1974; Price and Calvert, 1978), and because some U is associated with organic matter; this has the effect of increasing the U/P ratio above a value of around 9 x 10^{-4}, typical of the phosphorites from this area. Veeh et al. (1974) have shown that the U in some of the diatom ooze is fairly uniformly distributed within the sediment rather than being concentrated in phosphatic grains. Moreover, the sediment U has a $^{234}U/^{238}U$ ratio close to 1.15, close to the value characteristic of seawater (Thurber, 1962), identifying contemporary seawater as the U source (Veeh et al., 1974).

Kolodny and Kaplan (1973) showed that between 45 and 90% of the U in some organic-rich sediments of an anoxic fjord was authigenic, with a $^{234}U/^{238}U$ ratio of 1.16; more than half of this U was probably incorporated in organic material. They were also able to show that dissolved pore water U also had a $^{234}U/^{238}U$ ratio of 1.15. On the Namibian shelf, U concentrations in the pore waters of the diatom ooze are in the range 1-650 μg ℓ^{-1} (Baturin, 1971b) compared with open ocean values of around 3 μg ℓ^{-1}. These data serve to show that U is highly mobile in marine sediments and that some of it is probably adsorbed by dispersed organic material in such sediments.

Iodine contents of the surface sediments range from 96 to 1990 ppm (Table 2) and include some of the highest values recorded for marine sediments (see Vinogradov, 1939; Shiskina and Pavlova, 1965; Bennett and Manual, 1969; Bojanowski and Paslawska, 1970; Price, Calvert and Jones, 1970). Fig. 13 shows that the highest I values occur in the sediments of the shelf edge and that they decrease towards the coastline. In the sediments of the middle shelf off Walvis Bay, the concentrations of iodine appear somewhat variable because the sediments have widely varying organic carbon contents over small distances (see Fig. 6).

In areas of more uniform modern sedimentation, where the sediments are well oxygenated, I contents are well correlated with organic carbon (Fig. 16A) (Bojanowski and Paslawska, 1970; Price et al., 1970). However, no such relationsip is found on the Namibian shelf (Fig. 16B); I/C ratios vary continuously between the high values typical of the Barents Sea, and these come from outer shelf sediments, and very low values in the nearshore, anoxic diatom oozes.

Price and Calvert (1973) have suggested that the large amount of variation in the I contents of the Namibian shelf sediments is related to the degree of oxygenation of the surface sediments. The

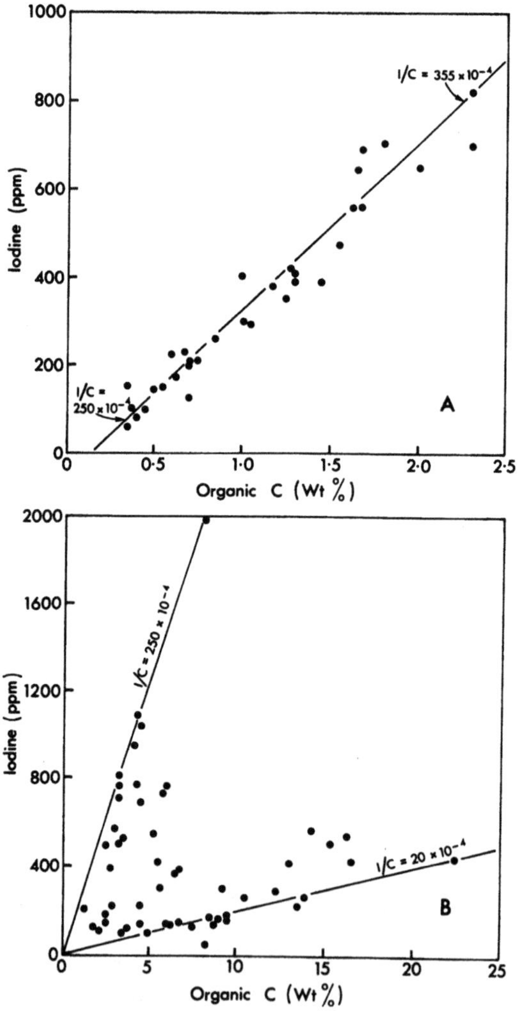

Fig. 16. Relationship between I and organic carbon in A) surface sediments of the Barents Sea (from Price et al., 1970) and B) surface sediments of the Namibian shelf.

outer shelf sediments, having the highest contents, are well oxygenated and the I levels are correlated with the carbon contents. The inner shelf sediments, and especially the diatom oozes, are highly reducing and they have the lowest I contents. The concentration of dissolved oxygen in the bottom water on the shelf also decreases fairly regularly from the shelf edge to the nearshore zone (Calvert and Price, 1971a) and parallels the isopleths of I concentration.

Unlike the other minor elements discussed so far, I is associated entirely with the organic fraction in marine sediments (Harvey, 1980). Price et al. (1970) argued that the concentrations of I in marine organisms were not sufficiently high to explain the observed concentrations in oxidized shallow marine sediments by direct supplies of organic material, and that sorption of the element by degraded organic material, either on the bottom or while in suspension, is required. From the data presented here and by Price et al. (1970), it is clear that the sorbed I is released from oxic sediments upon burial. I/C ratios of such sediments generally show a decrease from variable surface values to relatively constant values at depth. In contrast, the anoxic diatom ooze off Walvis Bay has similar I values at the surface and at depth.

These observations can be explained by control of the speciation of dissolved I in seawater by the redox potential. Under oxygenated conditions I exists mainly as the iodate ion (Sillén, 1961), but under anoxic conditions it is present entirely as iodide (Wong and Brewer, 1977). This contrasts with the behavior of Br which is present in seawater overwhelmingly as the bromide ion. In the Cariaco Trench and the Black Sea, Wong and Brewer (1977) showed that the change from iodate to iodide takes place over a very narrow depth range as one passes from the oxic to the underlying anoxic water. Consequently, the uptake of I by organic matter evidently only takes place under oxic conditions where the I is present mainly as iodate ion. No uptake takes place under anoxic conditions since the observed concentrations of I in the diatom ooze could be contributed by the I incorporated in planktonic seston. Alternatively, the difference in I contents of the oxic and anoxic sediments could be due to the uptake of both iodate and iodide by different classes of organic compounds in these two environments.

GEOCHEMICAL FACIES

The interrelationships between the major and minor elements in the surface sediments have been examined thus far in terms of the areal distributions of individual elements and pair-wise correlations. From this approach it seems clear that several elemental associations can be recognized in the surface sediments and that these can be mapped.

The complex interrelationships between such variables can be conveniently expressed by means of factor analysis (Harmen, 1960). Such an analysis was carried out on the major and minor element data in Table 2 using the Q-mode method (Imbrie and Purdy, 1962) where relationships in a sample by variable matrix were sought. Since both major and minor element data contribute more or less equally to the delineation of the sediment facies, the data were expressed in terms of the proportion of the concentration range of each individual variable thereby giving equal weight to all elements. The program of Klovan and Imbrie (1971) was used.

The analysis produced six factors, accounting for 93.8% of the total variance in the scaled data set, after varimax rotation. They are all readily identified with the main features of the geochemical variability already discussed and provide a convenient summary of the inter-element relationships. Fig. 17 shows the distribution of the scaled factor scores in each identified factor.

Factor 1 is a carbonate factor, containing most of the variance of Ca and CO_2 and about half that of Sr. Factor 2 delineates the nearshore organic-, silica- and metal-rich ooze of the inner shelf region; Mo, Cu and Ni, in addition to C and S are particularly diagnostic. Factor 3 represents terrigenous aluminosilicates and factor loadings are high for this factor in the nearshore sediments north and south of Walvis Bay. Factor 4 is a phosphorite factor with high scores for Sr, U and Y as well as P. Factor 5 consists of Ba and I, two minor elements which are enriched in outer shelf and upper slope sediments, but for different reasons --Ba because of its enrichment in slowly accumulating sediments in areas of high biological production and I because of its enrichment in oxic, organic-rich surface sediments. Finally, Factor 6, with high scores for Fe, K and Rb, represents a glauconite factor. High loadings are found in only two samples and these are also shown in Fig. 8.

DISCUSSION AND IMPLICATIONS

The modern sediments accumulating on the Namibian shelf have a predominantly biogenous origin, consisting of planktonic and benthic tests and shells, fragments of nektonic skeletons and dispersed organic material. The skeletal material comprises calcium carbonate, opaline silica and phosphatic material.

The organic-rich diatom ooze, lying in a broad belt in the nearshore zone, represents the modern equilibrium sediment facies which is a consequence of the high primary production in the Benguela Current. Apart from two smaller areas of diatom ooze, off Sylvia Hill and Dune Point, the main site of accumulation of this facies occurs where there is a change in the orientation of the coastline. Speculatively, this may be where the surface circulation entrains recently upwelled, nutrient-rich water which thereby has a longer residence time in the nearshore zone.

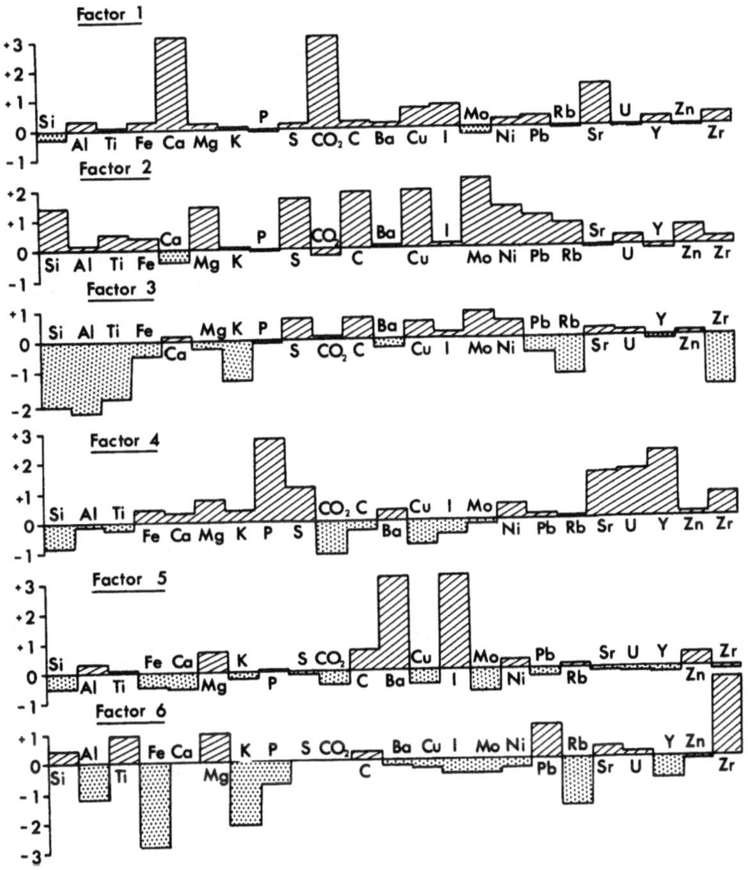

Fig. 17. Plot of scaled varimax factor cores for the six factors identified by the Q-mode analysis. Scales on left show values of scaled "correlations" between individual elements and the six factors. Values > +1 or -1 indicate significant contributions.

The location of the diatom ooze off Namibia contrasts markedly with that of a somewhat similar facies on the Peru continental margin. Here, an organic-rich mud lens occurs on the upper part of the slope adjoining a very narrow shelf (Krissek et al., 1980). The physiography of the Peruvian and Namibian margins, and the relationships between the topography and the surface and subsurface circulation in the two areas, are probably responsible for the different distributions of a similar sediment type with respect to the respective shorelines.

The presence of significant quantities of diatom silica in marine sediments is a direct reflection of the presence of upwelling in the overlying water body. Not all areas of upwelling are characterized by diatom-rich sediments, of course, because dilution by other sedimentary components or winnowing by bottom currents may lead to a diminished opal content. But wherever sediments are known to contain measureable quantities of opal (see summary in Calvert, 1974) the environment is typically an upwelling one, either in the open ocean (subarctic Pacific and the circum-Antarctic regions) or in nearshore areas (Sea of Okhotsk, Gulf of California, Peru-Chile margin, Namibian margin). Diatoms are apparently particularly well suited to the upwelling environment, their high specific growth rate providing them with a competitive advantage where nutrient supplies are variable or pulsed (Turpin and Harrison, 1979). It is therefore suggested that diatom ooze is the characteristic sediment facies of upwelling regimes.

The diatom oozes on the Peru-Chile and Namibian margins are also exceptionally rich in organic matter. It is entirely planktonic in origin, a direct reflection of upwelling and the presence of extremely arid zones on the adjacent land masses leading to minimal dilution of marine sediment components by terrigenous materials. The organic matter evidently accumulates at a high rate, suffering relatively little degradation during its brief transit time through the short water column. The combination of high diatom silica and organic matter serve to distinguish relatively shallow water upwelling sediments (hence a coastal upwelling facies) from deep ocean sediments forming beneath upwelling zones.

In addition to the remarkably high organic matter and opal contents, the Namibian shelf diatom ooze is also enriched in minor metals compared with the adjacent facies and is the site of precipitation of phosphorite. The metal enrichments are unusual by virtue of the overwhelmingly biogenous nature of the sediments, many of the oozes off Walvis Bay containing only 10% by weight lithogenous aluminosilicate material. Calvert (1976) has discussed these enrichments, pointing out that the concentrations of some minor metals are higher than in many nearshore terrigenous sediments containing little organic material. On the basis of the bulk compositions, using Al and CO_2 as measures of the amount of terrigenous aluminosilicate and biogenous carbonate, respectively, and assuming that these fractions have metal compositions given by average nearshore sediments (Wedepohl, 1971) and foraminiferal ooze (Krinsley, 1960; Arrhenius, 1963), it was shown that, on average, 67% of the Pb, 75% of the Ni, 78% of the Zn, 81% of the Cu, and 96% of the Mo were present in excess of the contributions from the terrigenous and carbonate fractions. The sediment phases which could be hosts for the excess metals are, of course, phosphorite, diatom opal, pyrite and organic matter. The first phase is unlikely to be important, in view of the available data of Price and Calvert (1978). No information is available on the metal content of diatom shells, although it is likely to be very small. The relative importance of the remaining two phases cannot be judged on the basis of the evidence available and some unequivocal methods for differentiating between these two hosts are required.

Brongersma-Sanders (1965, 1969) has argued that the Namibian shelf sediments may provide a modern analogue of bituminous, metal-rich black shales in the geological record. She has suggested that the metal contents of such rocks are readily explained by the rapid accumulation of planktonic organic material containing a suite of minor metals extracted from seawater by the plankton in areas of coastal upwelling. A direct comparison of some modern analyses of the metal contents of bulk marine plankton, where problems of contamination have been recognized (Martin, 1970; Leatherland et al., 1973; Martin and Knauer, 1973), with the excess metals in the Namibian diatom ooze shows that only in the case of Zn can a direct contribution from such plankton represent a significant fraction of the metal in the sediment. The other metals (Cu, Mo, Ni, and Pb) are present in fresh plankton in very low concentrations. The available data, therefore, cast doubt on the mechanism of metal enrichment in black shales envisioned by Brongersma-Sanders (1976, 1969).

A limited amount of evidence suggests that some organic fractions of modern sediments contain high concentrations of minor metals. Nissenbaum and Swaine (1976) have reviewed the available data showing that the humic acid fraction, in particular, is an important host for metals, and Calvert and Morris (1977) have shown that this observation also applies to the diatom oozes on the Namibian Shelf. The mechanism of enrichment of the metals remains unresolved, however. It is possible that settling organic material, albeit having low absolute metal concentrations, is the metal carrier, the metals being recycled diagenetically and complexed by organic condensation and polymerization products during the formation of humic substances. Confirmation of the importance of the organic fraction in this process awaits unequivocal methods for isolating this phase and for determining its original metal content.

The sulfide fraction of the sediments could also play an important role in the minor metal geochemistry of anoxic sediments. The mechanisms of formation of iron sulfides are known quite well (Berner, 1964) but the precipitation of other metal sulfides in such sediment has not been examined in detail. The concentration levels of minor metals in the pore waters of anoxic sediments are thought to be too high for control by sulfide precipitation (Presley et al., 1972) and other types of equilibria have been discussed (Gardiner, 1973). The coprecipitation of minor metals by iron sulfides also warrants further examination. Korolev (1958) and Bertine (1972) have shown that Mo, for example, is very effectively removed from anoxic seawater by coprecipitation with FeS, and Calvert (1976) has suggested that such a mechanism could explain the remarkably high levels of this metal in many modern anoxic sediments. In addition, Volkov and Fomina (1971) contend that much of the Mo in some Black Sea sapropels is present in finely-divided MoS_3 on the surfaces of pyrite. The question of whether the metals of interest here are presently associated with the phase or phases initially responsible for their incorporation into the sediments, or whether they have been diagenetically recycled, as also alluded to by Volkov and Fomina (1971), cannot be answered with the available information.

The origin of phosphorite has been a recurring problem in sedimentary geochemistry, and until Baturin (1969) described the occurrence of putative modern phosphorite on the Namibian shelf and Baturin, Merkulova and Chalov (1972) proved its contemporary age using uranium-series dating, no modern environment of phosphorite formation was known. The diatom oozes off Namibia and Peru-Chile (Veeh et al., 1973; Burnett and Veeh, 1977; Burnett, Beers and Roe, 1982) are now recognized as modern environments of carbonate fluorapatite precipitation. Baturin (1971a; 1971c) has eloquently described the various stages of phosphorite formation from its initial diagenetic precipitation in diatom ooze as impure, friable concretions to the concentration of lithified concretionary and pelletal forms in coarser-grain facies by winnowing. The phosphorite evidently forms in the upwelling environment off coastal Namibia and Peru-Chile by virtue of the high rate of accumulation of organic material having suffered little degradation and consequent phosphorus loss (Price and Calvert, 1978), and because the abundant phosphatic fish debris contributes substantially to the dissolved phosphate levels in the sediment pore waters (Suess, 1981).

Modern phosphorites on the eastern Australian continental slope (Kress and Veeh, 1980; O'Brien and Veeh, 1980; Cook and Marshall, 1981) however, have formed in a poorly-productive environment; upwelling occurs only sporadically along this margin. The associated sediments are accumulating slowly and have low concentrations of organic material. O'Brien et al. (1981) have suggested that the deposits have a bacterial origin and that the phosphorites could have only formed here as a consequence of bacterial activity in an environment of low primary production. Whatever the mechanisms of formation in this case, this particular occurrence appears to cast doubt on the identification of an upwelling environment by the presence of phosphorite. Thus, it would be more prudent to conclude that phosphorite can form by more than one mechanism, that it is unwise to use a single criterion for identifying any particular environment, and that a combination of criteria provide a more unequivocal basis for such an identification. In this respect, the combined presence of diatom silica, planktonic organic material and contemporary phosphorite probably provide a sound basis for identifying coastal upwelling conditions in the geological record.

ACKNOWLEDGEMENTS

This paper has benefited from the constructive reviews of C. Reimers and E. Suess. Special thanks go to E. Suess for informative discussion on the geochemistry of Ba in marine sediments. Expedition

CIRCE was funded by the U.S. Office of Naval Research and the National Science Foundation. We thank Tj. H. Van Andel for ship time on RV "Argo" and for ancillary information on the bathymetry and shallow structure of the Namibian shelf.

REFERENCES

Arrhenius, G., 1963, Pelagic sediments. in: THE SEA, Vol. 3, M.N. Hill, ed., John Wiley, New York, 655-727.

Avilov, I.K. and Gershanovich, D.Y., 1970, Investigation of the relief and bottom deposits of the southwest Africa shelf, Oceanology, 10:229-323.

Baturin, G.N., 1969, Authigenic phosphate concretions in recent sediments of the southwest Africa shelf, Doklady Akademii Nauk S.S.S.R., 189:227-230.

Baturin, G.N., 1971a, Stages of phosphorite formation on the ocean floor, Nature, 232:61-62.

Baturin, G.N., 1971b, Uranium in oceanic ooze solutions of the southeastern Atlantic, Doklady Akademii Nauk S.S.S.R., 198:224-226.

Baturin, G.N., 1971c, Formation of phosphate sediments and water dynamics, Oceanology, 11:372-376.

Baturin, G.N., Kochenov, A.V. and Shimkus, K.M, 1967, Uranium and rare metals in the sediments of the Black and Mediterranean Seas, Geokhimiya, 1:41-50.

Baturin, G.N., Merkulova, K.I. and Chalov, P.I., 1972, Radiometric evidence for recent formation of phosphatic nodules in marine shelf sediments, Marine Geology, 13:M37-M41.

Belova, I.V., 1970, Zinc in Holocene Black Sea sediments, Doklady Akademii Nauk S.S.S.R., 193:433-436.

Bennett, J.H. and Manuel, O.K., 1969, On iodine abundance in deep-sea sediment, Journal of Geophysical Research, 73:2302-2303.

Berner, R.A., 1964, Iron sulphides formed from aqueous solution at low temperatures and atmospheric pressure, Journal of Geology, 72:293-306.

Bertine, K.K., 1972, The deposition of molybdenum in anoxic waters, Marine Chemistry, 1:43-53.

Bertine, K.K. and Turekian, K.K., 1973, Molybdenum in marine deposits, Geochimica et Cosmochimica Acta, 37:1415-1434.

Bojanowski, R. and Paslawska, S., 1970, On the occurrence of iodine in bottom sediments and interstitial waters of the southern Baltic Sea, Acta Geophysica Polonica, 18:277-286.

Bremner, J.M., 1980a, Physical parameters of the diatomaceous mud belt off South West Africa, Marine Geology, 34:M67-M76.

Bremner, J.M., 1980b, Concretional phosphorite from S.W. Africa, Journal of the Geological Society of London, 137:773-786.

Brongersma-Sanders, M., 1948, The importance of upwelling water to vertebrate paleontology and oil geology, Koninklijke Nederlandsche Akademic van Wetenschappen, Section 2, 45:1-112.

Brongersma-Sanders, M., 1951, On conditions favoring the preservation of chlorophyll in marine sediments, Proceedings of the Third World Petroleum Congress, Section 1, 401-413.

Brongersma-Sanders, M., 1965, Metals of Kupferschiefer supplied by normal seawater, Geologische Rundschau, 55:365-375.

Brongersma-Sanders, M., 1966, The fertility of the sea and its bearing on the origin of oil, Advancement of Science, 23:41-46.

Brongersma-Sanders, M., 1967, Barium in pelagic sediments and in diatoms, Koninklijke Nederlandsche Akademie van Wetenschappen, Series B, 70:93-99.

Brongersma-Sanders, M., 1969, Permian wind and the occurrence of fish and metals in the Kupferschiefer and marl slate, in: "Sedimentary Ores," C.H. James, ed., University of Leicester Press, 61-68.

Brongersma-Sanders, M., Stephan, K.M., Kwee, T.G., and Debrun, M., 1980, Distribution of minor elements in cores from the southwest Africa shelf with notes on plankton and fish mortality, Marine Geology, 37:91-132.

Burnett, W.C. and Veeh, H.H., 1977, Uranium-series disequilibrium studies in phosphorite nodules from the west coast of South America, Geochimica et Cosmochimica Acta, 41:755-764.

Burnett, W.C., Beers, M.J. and Roe, K.K., 1982, Growth rates of phosphate nodules from the continental margin off Peru, Science, 215:1616-1618.

Busch, W.H. and Keller, G.H., 1981, The physical properties of Peru-Chile continental margin sediments - the influence of coastal upwelling on sediment properties, Journal of Sedimentary Petrology, 51:705-719.

Calvert, S.E., 1974, Deposition and diagenesis of silica in marine sediment, in: "Pelagic Sediments: On Land and Under the Sea," K.J. Hsu and H.C. Jenkyns, eds., Special Publication International Association of Sedimentology, 1:273-299.

Calvert, S.E., 1976, The mineralogy and geochemistry of nearshore sediments, in: "Chemical Oceanography," 2nd edition, J.P. Riley and R. Chester, eds., Academic Press, London, 6:187-280.

Calvert, S.E. and Price, N.B., 1970, Minor metal contents of recent organic-rich sediments off south west Africa, Nature, 227: 593-595.

Calvert, S.E. and Price, N.B., 1971a, Upwelling and nutrient regeneration in the Benguela Current, October 1968, Deep-Sea Research, 18:505-523.

Calvert, S.E. and Price, N.B., 1971b, Recent sediments of the South West African shelf, in: "Geology of the East Atlantic Continental Margin," F.M. Delaney, ed., Institute of Geological Science Report 70/16, 171-185.

Calvert, S.E. and Morris, A.J., 1977, Geochemical studies of organic-rich sediments from the Namibian shelf. II. Metal-organic associations, in: "A Voyage of Discovery," M. Angel, ed., Pergamon Press, London, 580-667.

Chester, R. and Elderfield, H., 1968, The infra-red determination of opal in siliceous deep-sea sediments, Geochimica et Cosmochimica Acta, 32:1128-1140.

Cook, P.J. and Marshall, J.F., 1981, Geochemistry of iron and phosphorus-rich nodules from the east Australian continental shelf, Marine Geology, 41:205-221.

Copenhagen, W.J., 1953, The periodic mortality of fish in the Walvis Region, Investigational Report 14, Division of Fisheries Union of South Africa, 34 pp.

Curray, J.R., 1965, Late Quaternary history, continental shelves of the United States, in: "The Quaternary of the United States," H.E. Wright and D.G. Frey, eds., Princeton University Press, 723-735.

Curtis, C.D., 1966, The incorporation of soluble organic matter into sediments and its effect on trace element assemblage, in: "Advances in Organic Geochemistry," G.D. Hobson and M.C. Louis, eds., Pergamon Press, 1-13.

Dehairs, F., Chesselet, R., and Jedwab, J., 1980, Discrete suspended particles of barite and the barium cycle in the open ocean, Earth and Planetary Science Letters, 49:528-550.

Emery, K.O., 1960, "The Sea off Southern California," John Wiley, New York, 366 pp.

Fresnel, J., Galle, P. and Gayral, P., 1979, Résultats de la microanalyse des cristaux vacuolaires chez deux chromophytes unicellulaires marines: *Exanthemachrysis gayraliae*, *Pavlova* sp. (Prymnésiophycées, Pavlovacées), Comptes Rendus de l'Académie des Sciences de Paris, 288:823-825.

Gardiner, L.R., 1973, Chemical models for sulphate reduction in closed anaerobic marine environments, Geochimica et Cosmochimia Acta, 37:53-68.

Glagoleva, M.A., 1961, Zirconium in recent Black Sea sediment, Doklady Akademii Nauk S.S.S.R., 193:184-187.

Goldberg, E.D., 1958, Determination of opal in marine sediments, Journal of Marine Research, 17:178-182.

Goldberg, E.D. and Arrhenius, G.O.S., 1958, Chemistry of Pacific pelagic sediments, Geochimica et Cosmochimica Acta, 13:153-212.

Gross, M.G., 1967, Concentrations of minor elements in diatomaceous sediments of a stagnant fjord, in: "Estuaries," G.H. Lauff, American Association for the Advancement of Science, Washington, 273-282.

Gucluer, S.M. and Gross, M.G., 1964, Recent marine sediments in Saanich Inlet, a stagnant marine basin, Limnology and Oceanography, 9:359-376.

Gulbrandsen, R.A., 1966, Chemical composition of the phosphorites in the Phosphoria formation, Geochimica et Cosmochimica Acta, 30:-769-778.

Hart, J.J. and Currie, R.I., 1960, The Benguela Current, Discovery Reports, 31:123-298.

Harman, H.H., 1960, "Modern Factor Analysis," University of Chicago Press, Chicago, 469 pp.

Harvey, G.R., 1980, A study of the chemistry of iodine and bromine in marine sediments, Marine Chemistry, 8:327-332.

Haughton, S.H., 1969, "Geological History of Southern Africa," Geological Society of South Africa, Johannesburg, 535 pp.

Hirst, D.M., 1962, The geochemistry of modern sediments from the Gulf of Pari. II. The location and distribution of trace elements, Geochimica et Cosmochimica Acta, 26:1147-1187.

Horstmann, R.L., 1957, The distribution of lithium, rubidium and caesium in igneous and sedimentary rocks, Geochimica et Cosmochimica Acta, 12:1-28.

Imbrie, J. and Purdy, E.G., 1962, Classification of modern Bahamian carbonate sediments, in: "Classification of Carbonate Rocks," W.E. Ham, ed., American Association of Petroleum Geologists, Memoir, 1:253-272.

Klovan, J.E. and Imbrie, J., 1971, An algorithm and FORTRAN-IV program for large-scale Q-mode factor analysis and calculation of factor scores, Journal of Mathematical Geology, 3:61-77.

Kochenov, A.V., Baturin, G.N., Kovaleva, S.A., Emel'Yanov, E.M., and Shimkus, K.M., 1965, Uranium and organic matter in the sediments of the Black and Mediterranean Seas, Geokhimiya, 3:302-313.

Kolodny, Y. and Kaplan, I.R., 1973, Deposition of uranium in the sediment and interstitial water of an anoxic fjord, in: "International Symposium on Hydrogeochemistry and Biogeochemistry," Vol. I, E. Ingerson, ed., The Clarke Co., Washington, 418-442.

Kolpack, R.L. and Bell, S.A., 1968, Gasometric determination of carbon in sediments by hydroxide absorption, Journal of Sedimentary Petrology, 38:617-620.

Korolev, D.F., 1958, The role of iron sulphides in the accumulation of molybdenum in sedimentary rocks of the reduced zone, Geokhimiya, 4:452-463.

Kress, A.G. and Veeh, H.H., 1980, Geochemistry and radiometric ages of phosphatic nodules from the continental margin of northern New South Wales, Australia, Marine Geology, 36:143-157.

Krinsley, D., 1960, Trace elements in the tests of planktonic foraminifera, Micropaleontology, 6:297-300.

Krissek, L.A., Scheidegger, K.R. and Kulm, L.D., 1980, Surface sediments of the Peru-Chile continental margin and the Nazca plate, Geological Society of America, Bulletin, 91:321-331.

Leatherland, T.M., Burton, J.D., Culkin, F., McCartney, K.J., and Morris, R.J., 1973, Concentrations of some trace metals in pelagic organisms and of mercury in north-east Atlantic Ocean Water, Deep-Sea Research, 20:679-685.

Logan, R.F., 1960, The Central Namib Desert, South West Africa, National Academy of Science, National Research Council, Washington, Publication No. 758.

Lubchenko, I.Y., 1970, Lead in Holocene Black Sea sediments, Doklady Akademii Nauk S.S.S.R., 193:445-448.

Martin, J.H., 1970, The possible transport of trace metals via moulted copepod exoskeletons, Limnology and Oceanography, 15:-756-761.

Martin, J.H. and Knauer, G.A., 1973, The elemental composition of plankton, Geochimica et Cosmochimica Acta, 37:1639-1653.

Morris, R.J. and Calvert, S.E., 1977, Geochemical studies of organic-rich sediments from the Namibian shelf. I. The organic fractions, in: "A Voyage of Discovery," M. Angel, ed., Pergamon Press, Oxford, 647-665.

Moum, J., 1965, Falling drop used for grain-size analysis of fine-grained materials, Sedimentology, 5:343-347.

Nissenbaum, A. and Swaine, D.J., 1976, Organic matter-metal interactions in recent sediments: The role of humic substances, Geochimica et Cosmochimica Acta, 40:809-816.

Norrish, K. and Hutton, J.T., 1969, An accurate x-ray spectrographic method for the analysis of a wide range of geological samples, Geochimica et Cosmochimica Acta, 33:431-454.

O'Brien, G.W. and Veeh, H.H., 1980, Holocene phosphorite in the east Australian continental margin, Nature, 288:690-692.

O'Brien, G.W., Harris, J.R., Milnes, A.R., and Veeh, H.H., 1981, Bacterial origin of east Australian continental margin phosphorites, Nature, 294:442-444.

Pilipchuk, M.F., 1972, Some problems of the geochemistry of molybdenum in the Mediterranean Sea, Litologiya i Poleznye Iskopaenye, 2:25-31.

Pilipchuk, M.F. and Volkov, I.I., 1968, The geochemistry of molybdenum in the Black Sea, Litologiya i Poleznye Iskopaenye, 4:5-27.

Presley, B.J., Kolodny, Y., Nissenbaum, A., and Kaplan, I.R., 1972, Early diagenesis in a reducing fjord, Saanich Inlet, British Columbia - II. Trace element distribution in interstital water and sediment, Geochimica et Cosmochimica Acta, 36:1073-1090.

Price, N.B. and Calvert, S.E., 1973, The geochemistry of iodine in oxidised and reduced recent marine sediments, Geochimica et Cosmochimica Acta, 37:2149-2158.

Price, N.B. and Calvert, 1977, The contrasting geochemical behaviors of iodine and bromine in recent sediments of the Namibian shelf, Geochimica et Cosmochimica Acta, 41:1769-1775.

Price, N.B. and Calvert, S.E., 1978, The geochemistry of phosphorites from the Namibian shelf, Chemical Geology, 23:151-170.

Price, N.B., Calvert, S.E. and Jones, P.G.W., 1970, The distribution of iodine and bromine in the sediments of the southwest Barents Sea, Journal of Marine Research, 28:22-34.

Reynolds, R.C., 1963, Matrix corrections in trace element analysis by X-ray fluorescence: estimation of the mass absorption coefficient by Compton scattering, American Mineralogist, 48:1133-1143.

Schuette, G., and Schrader, H., 1981, Diatom taphocoenoses in the coastal upwelling area off South West Africa, Marine Micropaleontology, 6:131-155.

Senin, Y.M., 1968, Characteristics of sedimentation on the shelf of southwestern Africa, Litologiya i Poleznye Iskopaeme, 4:108-111.

Shepard, F.P., and Moore, D.G., 1954, Sedimentary environments differentiated by coarse-fraction studies, American Association of Petroleum Geologists, Bulletin, 38:1792-1802.

Shishkina, O.V., and Pavlova, G.A., 1965, Iodine distribution in marine and oceanic bottom muds and their pore fluids, Geochemistry International, 2:559-565.

Sillén, L.G., 1961, The physical chemistry of sea water, in: "Oceanography," M. Sears, ed., American Association for the Advancement of Science, Washington, 549-581.

Simpson, E.S.W., 1971, The geology of the south-west African continental margin, in: "The Geology of the East Atlantic Continental Margin," F.M. Delany, ed., Institute of Geological Sciences, Report 70/16, 153-170.

Suess, E., 1981, Phosphate regeneration from sediments of the Peru continental margin by dissolution of fish debris, Geochimica et Cosmochimica Acta, 45:577-588.

Summerhayes, C.P., Birch, G.F., Rogers, J., and Dingle, R.V., 1973, Phosphate in sediments off south west Africa, Nature, 243: 509-511.

Swaine, D.J., 1962, The trace element content of fertilisers, Technical Communication 52, Commonwealth Agricultural Bureau, Slough, England, 306 pp.

Thompson, G., and Bowen, V.T., 1969, Analyses of coccolith ooze from the deep tropical Atlantic, Journal of Marine Research, 27:32-38.

Thurber, D.L., 1962, Anomalous $^{234}U/^{238}U$ in nature, Journal of Geophysical Research, 67:4518-4520.

Trask, P.D., 1953, Chemical studies of sediments of the western Gulf of Mexico, Papers in Physical Oceanography and Meteorology, Massachusetts Institute of Technology, 12:49-120.

Turekian, K.K., 1964, The geochemistry of the Atlantic Ocean, Transactions of the New York Academy of Science, 26:312-330.

Turekian, K.K., and Tausch, E.H., 1964, Barium in deep-sea sediments of the Atlantic Ocean, Nature, 201:696-697.

Turpin, D.H., and Harrison, P.J., 1979, Limiting nutrient patchiness and its role in phytoplankton ecology, Journal of Experimental Marine Biology and Ecology, 39:151-166.

van Andel, Tj.H., 1964, Recent marine sediments of the Gulf of California, in: "Marine Geology of the Gulf of California," Tj.H. van Andel and G.D. Shor, eds., American Association of Petroleum Geologists, Memoir, 3:216-310.

van Andel, Tj.H. and Calvert, S.E., 1971, Evolution of sedimentary wedge, Walvis Shelf, Southwest Africa, Journal of Geology, 79:-585-602.

Veeh, H.H., 1967, Deposition of uranium from the ocean, Earth and Planetary Science Letters, 3:145-150.

Veeh, H.H., Burnett, W.C., and Soutar, A., 1973, Contemporary phosphorites on the continental margin of Peru, Science, 181: 844-845.

Veeh, H.H., Calvert, S.E., and Price, N.B., 1974, Accumulation of uranium in sediments and phosphorites on the southwest African Shelf, Marine Chemistry, 2:189-202.

Vinogradov, A.P., 1939, Iodine in marine muds, Trudy Biogeokhimiya Akademii Nauk S.S.S.R., 5:19-32.

Volkov, I.I., and Fomina, L.S., 1971, Dispersed elements in sapropel of the Black Sea and their interrelationship with organic matter, Litologiya i Poleznye Iskopaemye, 6:3-15.

Wedepohl, K.H., 1971, Environmental influences on the chemical composition of shales and clays, in: "Physics and Chemistry of the Earth," L.H. Ahrens, F. Press, S.K. Runconn, and H.C. Urey, eds., Pergamon Press, New York, 8:307-333.

Wellington, J.H., 1955, "Southern Africa," Vol. 1, Physical Geography, Cambridge University Press, Cambridge, 528 pp.

Wong, G.T.F., and Brewer, P.G., 1977, The marine chemistry of iodine in anoxic basins, Geochimica et Cosmochimica Acta, 41:151-159.

Wright P.L., 1972, "The Geochemistry of Recent Sediments of the Barents Sea," Ph.D. thesis, University of Edinburgh, 226 pp.

Zeigler, J.M., Whitney, G.G., and Hayes, C.R., 1960, Woods Hole rapid sediment analyzer, Journal of Sedimentary Petrology, 30:490-495.

UPWELLING AND PHOSPHORITE FORMATION IN THE OCEAN

William C. Burnett and Kevin K. Roe

Department of Oceanography, The Florida State University
Tallahassee, Florida 32306, U.S.A.

David Z. Piper

Pacific-Arctic Branch of Marine Geology
U.S. Geological Survey
Menlo Park, California 94025, U.S.A.

ABSTRACT

Phosphorites on the current sea floor are known to occur in both upwelling and non-upwelling environments. Uranium-series and rare earth element (REE) studies display certain characteristics which may be a consequence of the type of phosphorite formation prevalent in a coastal upwelling environment. Quaternary phosphate nodules exposed at the sediment-seawater interface in the upwelling zone off Peru/Chile are shown to: (a) display unidirectional growth downward into the underlying sediments at rates which are slow compared to rates of associated sediment accumulation; (b) contain small but measurable amounts of excess ^{231}Pa relative to ^{230}Th; and (c) have very low concentrations of REE with a shale-like pattern rather than one resembling seawater. Results from phosphorites sampled from low-productivity areas (e.g., Pacific seamounts, the Agulhas Bank) indicate extreme differences in these characteristics when compared to those from Peru/Chile. Radiochemical and geochemical approaches thus offer promise for distinguishing "upwelling" from "non-upwelling" suites of phosphorites.

INTRODUCTION

Ocean floor phosphorites occur along many of the continental margins of the world as nodules, irregular masses, sands, pellets, and oolites. The worldwide distribution of ancient sedimentary phosphorites on land (Fig. 1) shows that these deposits tend to occur at

Fig. 1. World distribution of sedimentary phosphorite occurrences. Deposits shown on land (crosses) include all ages. Submarine deposits are divided into Quaternary and all others based on uranium-series disequilibrium dating.

low latitudes. This is even more evident when the deposits are plotted on paleomagnetic reconstructions (Sheldon, 1964; Cook and McElhinny, 1979). The observation that marine phosphorites often occur along the western margins of continents in areas of upwelling currents and related phenomena, has contributed to the concept that phosphorites are a sedimentary response to upwelling. Phosphorite formation based on Kazakov's (1937) model was thought to result from direct inorganic precipitation of marine apatite from nutrient-enriched seawater. Although more recent studies support precipitation from anoxic pore solutions either interstitially (Atlas, 1975; Burnett, 1977; Suess, 1981) or at the sediment-water interface (Burnett, Beers and Roe, 1982), the link to upwelling currents as a primary source for phosphorus still holds.

There are several submarine phosphate deposits, however, that are not associated with modern upwelling. Extensive deposits occur, for example, off the southeast coast of the United States (Manheim, Pratt and McFarlin, 1980; S.R. Riggs, pers. comm.). The recent discovery that phosphorites are currently forming off New South Wales, Australia, an area of little or no upwelling and little organic productivity, adds fuel to the argument that all phosphorites are not directly linked to upwelling (Kress and Veeh, 1980; O'Brien and Veeh, 1980; this volume; O'Brien et al., 1982). This deposit, as a modern analog of an "east coast" phosphogenic province (McKelvey, 1967), eliminates the necessity of explaining non-upwelling associated phosphorites as a result of formation during an earlier geologic period when upwelling may have been prevalent in the area.

Another non-upwelling environment where phosphorites are known to occur is the summits of seamounts. Although occurrences along continental margins have been known since the days of the Challenger Expedition (Murray and Renard, 1891), it wasn't until the Mid-Pacific Mountains were sampled in 1950 that seamount phosphorites were discovered (Hamilton, 1956). Subsequently, further expeditions have established the presence of phosphorites as a fairly regular feature of seamounts in many areas throughout the world's oceans (Hamilton and Rex, 1959; Heezen et al., 1973; Slater and Goodwin, 1973; Baturin, 1978; Jones and Goddard, 1979). The origin of this type of phosphorite is unclear. Suggestions include: (a) submergence of an oceanic island which contained surficial guano-derived phosphate (Bezrukov, 1973); (b) replacement of previously existing materials in the submarine environments, i.e., phosphate derived from seawater (Marlow, 1971; Heezen et al., 1973); and (c) volcanogenic inputs of phosphorus (Kharin, 1974, cited in Baturin, 1978). No matter what their exact mode of formation, it seems unlikely that the environment of deposition was characterized by intense upwelling and high organic productivity.

Thus, although many phosphorites in the geologic record may indeed be of an "upwelling facies" as suggested by Sheldon (1964), it is clear from the study of deposits on the modern sea floor that phosphorites may form in non-upwelling regions as well. Thus, the question arises: can we differentiate between phosphorites formed in upwelling environments from those which were not? This paper will address this question by presenting some recent uranium-series and rare earth element (REE) results from phosphorites sampled from different environments as an example of how a geochemical approach may assist in this characterization. Clearly, if phosphorites are to be used as indicators of past centers of coastal upwelling, it is necessary to develop the tools for resolving the "upwelling" from the "non-upwelling" suite of phosphorites.

GROWTH RATES OF MARINE PHOSPHORITES

Presumably, if one can discern the growth mechanisms and histories of phosphorites from different environments, it would assist in establishing criteria for recognition of "upwelling" phosphorites. Any theory concerning the origin of marine phosphorite depends critically on establishing a rate and type of growth. Do phosphate nodules grow rapidly? Do they accrete in all directions in a fashion somewhat like manganese nodules? These questions and others are important not only for an understanding of the origin of phosphorite but for an evaluation of their significance toward the marine geochemical balance of phosphorus (Froelich et al., 1982).

Samples were selected for growth rate studies principally from previously analyzed samples (Burnett and Veeh, 1977) from the continental shelf of Peru. The phosphate deposits in this area of intense upwelling consist of both small (millimeter scale) pelletal forms

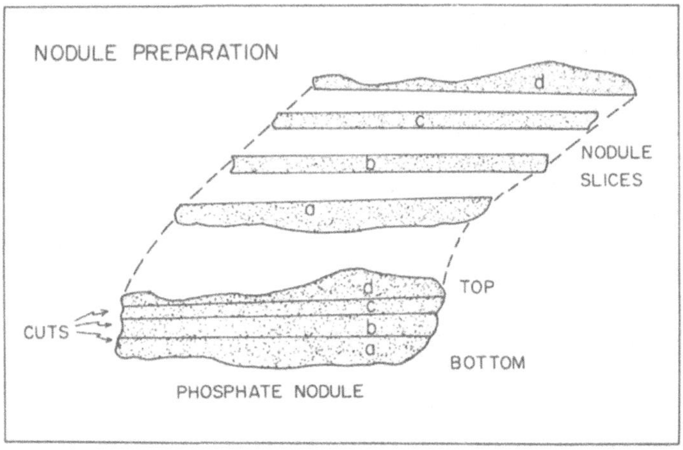

Fig. 2. Schematic cross-sectional view of typical phosphate nodule from the Peru shelf showing our sampling technique for single-nodule analysis. These nodules are characteristically 2-3 cm thick by about 10 cm long - a "slab" morphology.

dispersed in the organic-rich sediment and large (several centimeters long by 2-3 cm thick) nodules occurring at the sediment-water interface. The large nodules off Peru tend to be concentrated at the upper and lower boundaries of the intersection of the oxygen minimum layer with the continental margin. A more complete analysis of the phosphate nodule distribution and information concerning the mineralogy and chemistry of these deposits has been presented elsewhere (Burnett, 1977; Burnett, Veeh and Soutar, 1980).

Our approach to determining growth rates and histories of individual nodules was to analyze uranium-series isotopes from discrete layers oriented perpendicular to the direction of assumed growth. Determination of radiometric ages (Veeh and Burnett, 1982) within each layer will thus allow an evaluation of the rate and direction of growth. We prepared nodules by cutting as many slices as convenient through a nodule by a standard thin section saw (Fig. 2). Nodules from off Peru were selected for analysis if they were: (a) young, providing a higher probability of measuring a gradient in radiogenic ^{230}Th; (b) relatively flat, enabling a series of contiguous slices a few millimeters thick to be cut through each nodule; and (c) able to be oriented with respect to top and bottom by noting an oxidized layer on the exposed (top) surface (Fig. 3). After each nodule was sectioned, the individual slices were powdered, dried, radiochemically separated and purified, and analyzed for uranium-series isotopes (^{238}U, ^{234}U, ^{232}Th, and ^{230}Th) by isotope dilution alpha spectrometry (Burnett and Veeh, 1977).

If radiometric ages within these nodule slices are plotted against their thickness, several different two-dimensional growth

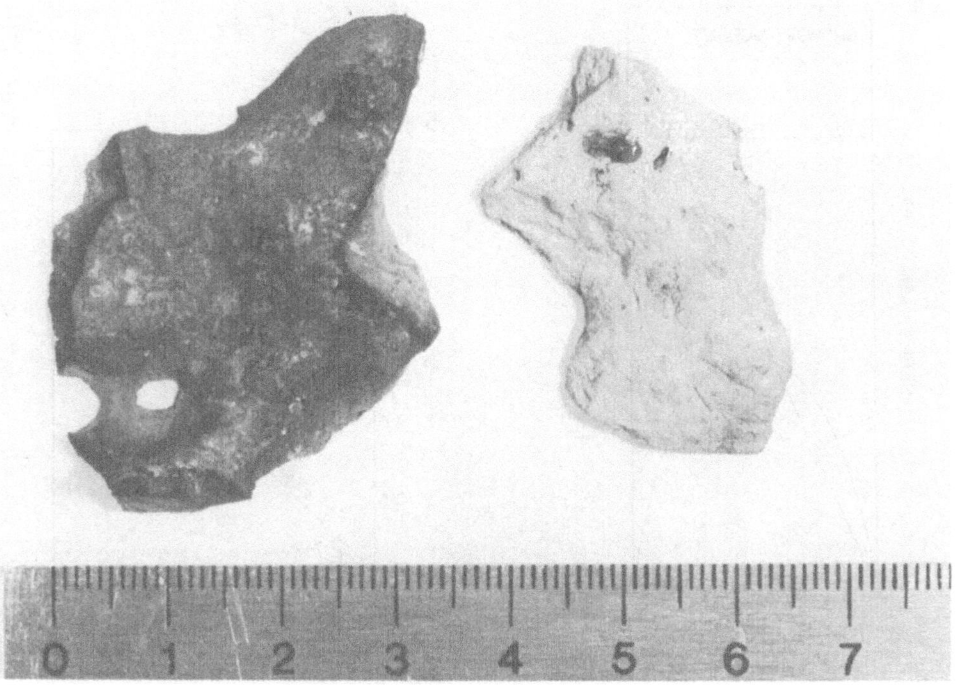

Fig. 3. Photograph of the "top" (left) and "bottom" (right) surfaces of Peru nodule PD18-30. The striking difference in color is due to an oxidized coating which is often present on one side of these nodules. We have assumed that this surface represents the portion exposed to the overlying water, the "top" surface.

models are possible (Fig. 4). If the phosphate nodules grow very rapidly, or if all slices are very old, it is unlikely that a gradient could be discerned in the $^{230}Th/^{234}U$ ages versus nodule thickness plot (model #1 in Fig. 4). If growth is constant and unidirectional from top to bottom throughout a nodule's history, a straight line with a zero intercept at the bottom surface should result (model #2). If nodules grow upward, on the other hand, a negative slope would result with the youngest age at the topmost layer. Bi-directional growth would result in the oldest layer somewhere in the interior with progressively younger layers toward both bottom and top (model #3). Variations of these three possibilities are possible but these models seem to represent the most likely situations.

Radiochemical results from nodules off Peru have been presented in tabulated form elsewhere (Burnett et al., 1982). We present here age-nodule thickness plots for nodules PD15-17 (~15°S latitude, 370 m water depth) and PD18-30 (~18°S, 385 m) in Figs. 5 and 6, respectively. Ages are shown both corrected and uncorrected for common thori-

4

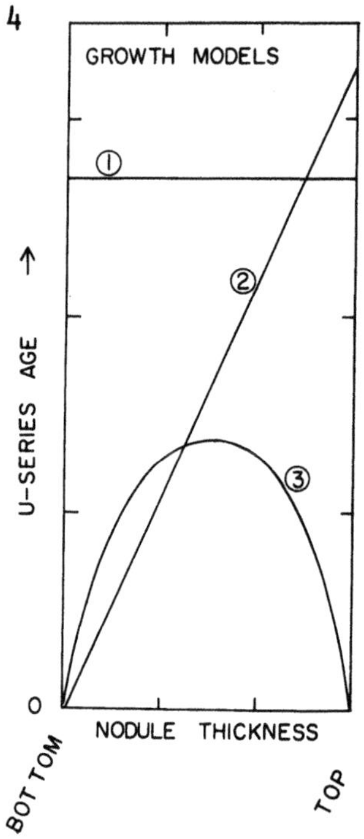

5

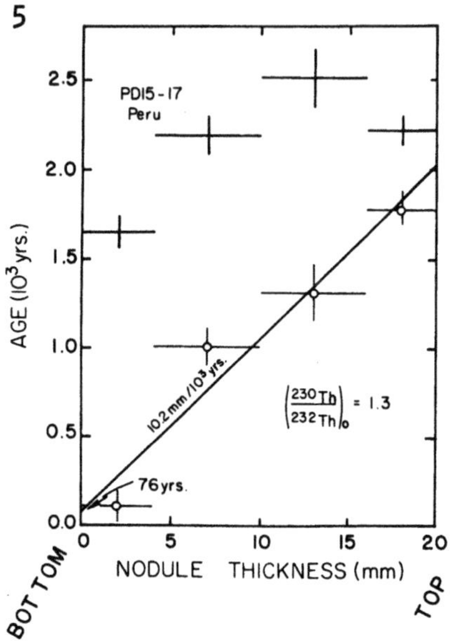

Fig. 4. Schematic plot of age versus thickness for authigenic phosphate nodules. Possibilities will depend on rates and growth histories of the nodules as revealed by systematic uranium-series analysis of discrete layers within single nodules. Three possible models (see text) are illustrated.

Fig. 5. Common-thorium corrected (open circles) and uncorrected (cross-bars) uranium-series ages versus thickness for Peru nodule PD15-17. Initial $^{230}Th/^{232}Th$ activity ratio of 1.3 used for the common thorium correction.

um. The correction was applied following the method of Kaufman and Broecker (1965). Assuming an initial $^{230}Th/^{232}Th$ activity ratio of the contaminating fraction and measuring the amount of ^{232}Th in each sample, an amount of ^{230}Th may be subtracted from the total measured after a decay correction. We used an initial $^{230}Th/^{232}Th$ value of 1.3 for PD15-17 and 4.0 for PD18-30 to make these corrections. These are the maximum reasonable values in both cases (higher values result in negative age intercepts) so the "best" values must lie between the corrected and uncorrected ages. We feel that the better linear fits of the data and near zero age intercepts argue favorably for our corrected ages. For determination of growth rate and direction, it actually makes little difference which set of data is used, as the slopes are not markedly affected.

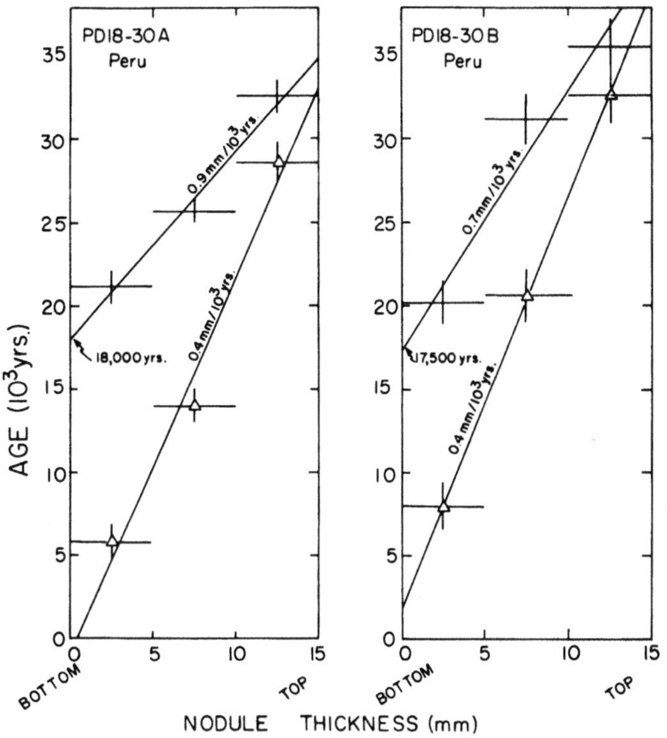

Fig. 6. Corrected (open triangles) and uncorrected (cross-bars) ages versus nodule thickness for two different sides of Peru nodule PD18-30. Both left and right sides of this nodule were sampled and prepared independently to evaluate how the sample's inherent inhomogeneity would influence our growth rate measurements. Although some distinct differences between comparable layers did result, the calculated growth rates for the two sequences are in excellent agreement. An initial value of 4 was used for the $^{230}Th/^{232}Th$ activity ratio for the common ^{230}Th correction.

The age-nodule thickness plot of our youngest and most densely analyzed sample, PD15-17, shows a distinct age gradient with apparent constant growth at a rate of about 10 mm/1000 yrs over the last 2000 years. The results of PD18-30 show a similar trend but with a much slower growth rate (0.4-0.9 mm/1000 yrs). Thus, our measurements in these nodules (and others which have been analyzed but not shown) from off Peru all conform to growth model #2, unidirectional constant growth toward the bottom surface. Thus, phosphate nodules from the upwelling zone off Peru apparently grow down into the soft underlying mud rather than concentrically as a manganese nodule.

The observation that phosphate nodules grow slowly is surprising in view of the rapid sediment accumulation rates (mm to cm per year) reported for the continental shelf of Peru (DeMaster, 1979; Henrichs, 1980). These nodules apparently remain at the sediment-water interface for periods which are long compared to rates of associated sediment accumulation. Even the youngest of the nodules we have examined in this way, PD15-17, must have remained at the sediment-water interface for at least two thousand years - a period long enough for 2-20 meters of sediment to accumulate. Why aren't these nodules buried?

This situation is analogous to the manganese nodule paradox, i.e., million year old nodules resting on sediment surfaces no older than a few thousand years. The mechanism responsible for this has yet to be resolved. The feeding activity of epifaunal organisms may remove sediment from nodule surfaces (Paul, 1976) while bioturbation may provide the downward sediment flux which leaves nodules exposed and undisturbed (Piper and Fowler, 1980). Off Peru, both sediment accumulation rates and nodule growth rates are approximately three orders of magnitude higher compared to the deep sea where manganese nodules are found. Even though the environments and the rate factors are quite different, we see no reason why the physical mechanism(s) responsible should be different. As a working hypothesis, we propose that biological activity is one likely agent which may prevent nodules (whether of manganese or phosphate) from becoming buried.

Although we have no direct documentation of the role organisms may have in phosphate nodule maintenance, the observation that these nodules form preferentially near the 0.25 mℓ/ℓ isopleth of dissolved

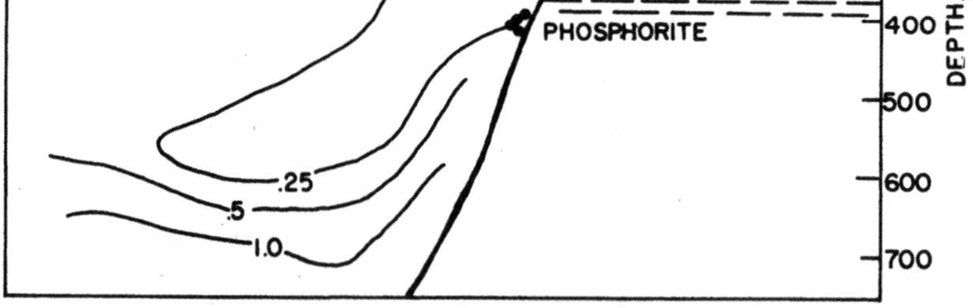

Fig. 7. Schematic diagram of the continental shelf and upper slope off Peru. Phosphorite tends to be concentrated at the upper and lower boundaries of the O_2-minimum with a zone of laminated (non-bioturbated) sediments in between.

oxygen concentration off Peru (Fig. 7) provides circumstantial evidence that organisms may actively participate in preventing nodule burial. Phosphate nodules may not form within the boundaries of the oxygen minimum zone (where bioturbation is absent) because the agent responsible for their maintenance at the sediment surface is not available. Additionally, in bioturbated areas, organisms may increase the net upward flux of dissolved phosphate from sediment pore waters through their burrowing activity.

A scenario recently suggested by Soutar and his colleagues (Soutar, Johnson and Baumgartner, 1981) to explain the destruction of sediment varves off Peru, postulates a mobile population of a galatheid crab, *Pleuroncodes*, which may occupy the sea floor just above and below the zone of minimum dissolved oxygen. Occasional excursions of these "mud sweepers" during times of increased oxygen may be responsible for disruption of some of the fine laminae within the low oxygen zone. Perhaps they are also responsible for sweeping mud off phosphate nodules. During times of decreased oxygen (expansion of the oxygen minimum zone), phosphorites that were actively forming at the sediment surface near the prior boundaries would then be in a situation where bioturbation is absent or minimal. If nodules are maintained at the sediment surface by benthic activity, then an interruption of this activity should result in nodule burial. In fact, buried nodules with overlying laminated sediments have been observed in box cores sampled off Peru (Soutar et al., 1981). Recent sediments near the low oxygen boundaries, thus may represent a unique combination of well-preserved organic matter with effective bioturbation to maintain nodules at the sediment-seawater interface. Since sediment accumulation rates on the Peru shelf are approximately 2-3 orders of magnitude faster than in the deep sea, resolution of the relationship between manganese nodules and biological activity may best be approached by further study of sediment-organism-nodule dynamics on the Peru shelf.

As an alternative to a "faunal intervention" hypothesis, it is also possible that nodules are accreting in localized areas of non-deposition. The sediments of the Peru shelf and slope are characterized by many disconformities with 500- to 8000-year hiatuses a common feature (DeVries and Schrader, 1981; Reimers and Suess, this volume). It seems plausible, therefore, that a phosphate nodule could grow at the sediment-seawater interface while a weak bottom current prevents significant sediment accumulation. Since our nodule samples are limited at present to dredge haul collections, we cannot demonstrate whether or not nodules actually rest on "young" sediment. Combined radiochemical study of <u>both</u> oriented nodules and the underlying sediment will be necessary to resolve this problem.

The observation that these phosphate nodules grow down into the sediment together with our growth rate data suggests that the source of phosphorus and other elements contained within these nodules is sediment interstitial waters. Accumulation rates of phosphorus in these nodules (reported in Burnett et al., 1982) based on our measurements of growth rates, phosphorus concentrations and bulk densities range from about 10 to 200 x 10^{-9} μmols/cm^2•sec reflecting differences in measured growth rates. Suess (1981) calculated an average diffusive phosphate flux from the sediments into bottom waters of 65 x 10^{-9} μmols/cm^2•sec based on pore water profiles of dissolved phosphate in five cores from the Peru upwelling zone. This flux is more than sufficient to supply phosphate to some nodules, while low but still of the same magnitude required to supply the fastest growing samples. We feel these estimates are sufficiently close to suggest that the upward diffusive flux of phosphate to the sediment-water interface on the Peru shelf is sufficient to supply the phosphorus contained in these nodules.

In an attempt to assess whether or not there are any differences in growth rates or directions for an authigenic phosphorite sampled from a calcareous facies, "east-coast" phosphorite province, we analyzed slices from phosphate rock GB-1 sampled from Agulhas Bank, South Africa. This sample is composed of approximately 50% sand-

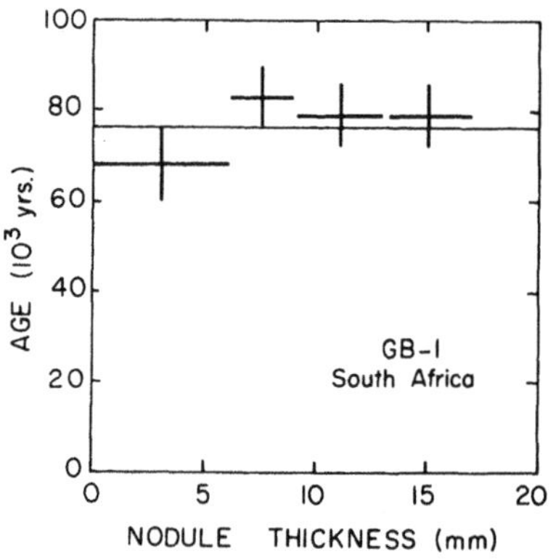

Fig. 8. Uncorrected uranium-series ages versus thickness for authigenic phosphorite GB-1 sampled from the Agulhas Bank. A correction for common thorium has no significant effect on this data set.

sized quartz set in a carbonate fluorapatite cement (Birch, 1979). Its overall size and appearance as well as chemical and mineralogical composition is much like the nodules we have worked on from the Peru/Chile shelf. Unfortunately, the results from these analyses (Fig. 8) are somewhat ambiguous. Although the age-thickness plot shows no apparent slope (resembling growth model #1, Fig. 4), it may be that we are just unable to resolve a few thousand year age difference in a sample as old as GB-1 (approximately 75,000 yrs B.P.). Future work of this type will be required in order to conclusively evaluate whether differences in rates and directions of phosphate nodule growth exist between upwelling and non-upwelling environments.

THORIUM AND PROTACTINIUM IN MARINE PHOSPHORITES

Even though thorium and protactinium have similar marine chemistries, ^{231}Pa and ^{230}Th are known to be fractionated in the deep open ocean. ^{230}Th is preferentially scavenged by particulates in low sediment flux environments whereas horizontal advection redistributes dissolved ^{231}Pa to a greater extent than ^{230}Th (Anderson, 1981). The major sinks for advected ^{231}Pa are high sediment flux regimes where dissolved ^{230}Th and ^{231}Pa are apparently scavenged without further fractionation. The resulting $^{230}Th/^{231}Pa$ activity ratio in oligotrophic deep-sea sediment is thus greater than the instantaneous production ratio (11.2) from uranium isotopes (Turekian and Chan, 1971). Conversely, $^{230}Th/^{231}Pa$ activity ratios of less than 11.2 have been measured in productive and coastal regions such as the Central American margin (Anderson, 1981), the western African margin (Mangini and Diester-Haass, this volume), and the Antarctic convergence (DeMaster, 1979). Slowly growing authigenic minerals such as manganese nodules (Ku and Broecker, 1967; 1969) and phillipsite (Bernat and Goldberg, 1969) also display $^{230}Th/^{231}Pa$ activity ratios below the production ratio. Since MnO_2 is known to adsorb dissolved ^{231}Pa and ^{230}Th without fractionation (Scott, 1977; Anderson, 1981) and phillipsite may do the same, it is of interest to determine if preferential adsorption of non-radiogenic ^{231}Pa is characteristic of accretionary phosphorites which formed in upwelling regimes.

We measured $^{231}Pa/^{235}U$ and $^{230}Th/^{234}U$ activity ratios in a suite of Quaternary phosphorite nodules from the Peru/Chile continental shelf to determine if these ratios are concordant as has been shown for aragonitic coralline material (Ku, 1968) and phosphorites from non-upwelling environments (Veeh, 1982; Birch et al., in press). The shorter half-life of ^{231}Pa (34,300 years versus 75,200 years for ^{230}Th) allows the $^{231}Pa/^{235}U$ activity ratio to be a more sensitive geochronometer within younger systems as well as a useful check on $^{230}Th/^{234}U$ and the closed system assumption for marine phosphorites.

We determined ^{231}Pa by measuring its alpha-emitting granddaughter ^{227}Th by a modification of a previously described technique (Mangini and Sonntag, 1977). Thorium was separated, purified and electroplated as previously described (Burnett and Veeh, 1977) but without addition of a $^{232}U/^{228}Th$ yield tracer. The ^{230}Th present in the unspiked sample split acts as an internal tracer with which the specific count rate of ^{227}Th can be determined after corrections for background, decay of ^{227}Th, and interfering isotopes. Excellent agreement between ^{231}Pa determinations via ^{227}Th and direct ^{231}Pa measurements of phosphorites validate this assumption (Roe et al., 1982).

Results of these analyses are presented in detail in our earlier publication (Roe et al., 1982). When plotted on a "concordia" plot of $^{231}Pa/^{235}U$ versus $^{230}Th/^{234}U$ (Fig. 9) these samples appear to have more ^{231}Pa than would be predicted on the basis of their ^{230}Th age. Secondary loss of ^{230}Th or total uranium as well as ^{231}Pa addition are possible explanations for these observations. Loss of ^{230}Th is unlikely due to its known immobility in marine environments. Slow leakage of recoil ^{234}U is known to occur in sea floor phosphorites but this loss is thought to be small and has little effect on $^{230}Th/^{234}U$ activity ratios over the first 200,000 years or so (Veeh and Burnett, 1982). There is little evidence to suggest loss of total uranium from marine phosphorites although this explanation remains a possibility. We prefer the explanation that protactinium and thorium have been added to these nodules during their exposure to seawater and pore waters at the sediment-seawater interface. If both thorium and protactinium are being adsorbed simultaneously, the $^{230}Th/^{231}Pa$ activity ratio of these unsupported constituents will presumably be less than the production ratio of 11.2 because of their prior fractionation in the ocean; thus, the apparent "excess" of ^{231}Pa. Since both the particulate and dissolved $^{230}Th/^{231}Pa$ activity ratios in coastal zones are approximately the same at 3-11 (Anderson, 1981), further fractionation during adsorption is unnecessary. Presumably, excess protactinium and thorium may be added to the bottom surfaces

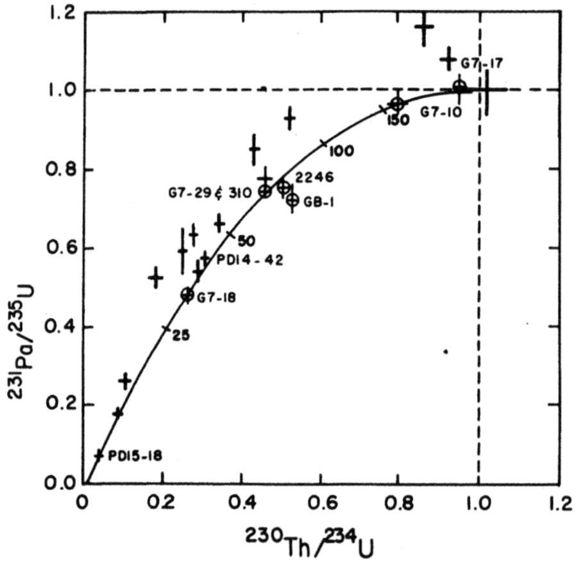

Fig. 9. $^{231}Pa/^{235}U$ versus $^{230}Th/^{234}U$ in marine phosphorites. Curved line represents theoretical development of these activity ratios in marine apatites assuming a closed system. Numbers on this "concordia" refer to age in thousands of years. Peru/Chile nodules shown as cross-bars (PD15-18 and PD14-42 from Veeh, 1982). Samples from "east coast"-type environments shown as open circles with corresponding error bars. These samples are from off New South Wales (G7 designations, Veeh, 1982) and the Agulhas Bank (GB-1, 2246, 310, Birch et al., in press).

of nodules from regenerative sources in the uppermost sediments as well as by exposure to the overlying water column. It would be useful in the future to analyze successive layers in phosphate nodules for their ^{231}Pa and ^{230}Th activities to test this concept.

Other studies to date have found concordant $^{231}Pa/^{235}U-^{230}Th/^{234}U$ activity ratios from indurated ocean-floor phosphorites (Fig. 9). Almost all of the samples previously analyzed, however, have come from non-upwelling, east-coast-type environments. Veeh (1982) reported concordant protactinium-thorium ages for samples from off the coast of New South Wales, Australia, and for two samples from the Peru shelf. Although both of the samples that Veeh analyzed from off South America were concordant, they were both in a relatively young age range (less than 40,000 yrs B.P.) where not much excess ^{231}Pa is expected based on our results. Additionally, even some of our older samples do display concordant $^{231}Pa/^{235}U-^{230}Th/^{234}U$ results which we interpret as being due to burial below the sediment-water interface. Since all our samples have thus far been from dredge hauls, it is possible that some nodules were buried below the soft organic-rich mud. In addition to Veeh's results, a set of three samples from the Agulhas Bank (Birch et al., in press) also show concordant $^{231}Pa/^{235}U-^{230}Th/^{234}U$ activity ratios. The sediments of this region are characterized by carbonate lithologies and phosphatized limestones. Oceanographically, this area more closely resembles the area off New South Wales than Peru/Chile. Thus, results to date indicate concordant ^{230}Th and ^{231}Pa ages for Quaternary phosphorites except those from active coastal upwelling zones.

Although ^{231}Pa in phosphorites may be a useful indicator of a high particle flux, upwelling environment, its applicability will be limited to late Quaternary age deposits because of its short half-life. However, a similar geochemical indicator for phosphorites in the geologic record may be possible if stable trace element analogs for $^{230}Th/^{231}Pa$ can be identified.

RARE EARTH ELEMENT GEOCHEMISTRY

Differences in rare earth element patterns and abundances between marine phosphorites from different localities have recently been noted by Altschuler (1980) and Kolodny (1981). These authors did not, however, attempt to characterize these differences directly in terms of their oceanographic setting. In order to evaluate if any differences in REE geochemistry may be systematically related to depositional environment, we analyzed two suites of phosphorite samples, one from the Peru/Chile shelf and another from North Pacific seamounts (the Musicians Seamounts). The seamount samples are phosphatized limestones and volcanics whose principal phosphatic component is a carbonate fluorapatite similar in its major element chemistry to the South American shelf samples.

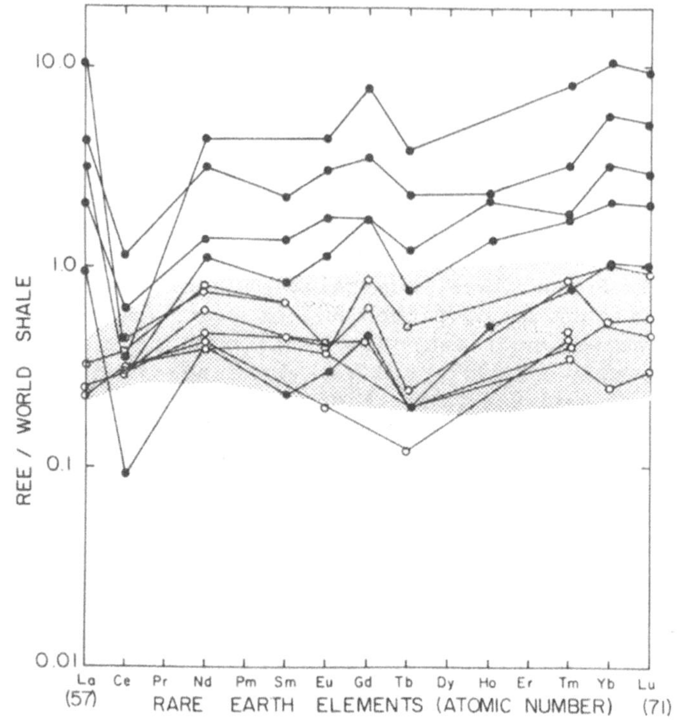

Fig. 10. Rare earth element (REE) data normalized to world shale (Piper, 1974) for Peru/Chile samples (open circles) and Pacific seamounts (closed circles). The stippled region represents the normal limits of the Peru/Chile sample set.

All samples were analyzed by instrumental neutron activation analysis (INAA) by standard techniques employed by the U.S. Geological Survey. A correction for fission product interference was made where appropriate. For the shelf samples, this correction exceeded 20% for La, Ce, and Nd because of the high uranium content of these samples (50-460 ppm). None of the seamount samples had high enough concentrations of uranium (3.7-5.0 ppm) to require a significant fission product correction. Although the large corrections for the light REE's in the Peru/Chile samples may cause some degree of uncertainty (perhaps ± 20%), this should have little bearing on overall trends and differences between the two sample sets. We show all the REE data (Fig. 10) as well as averages of each sample set and a pattern for open seawater for comparison (Fig. 11). All our results have been normalized to an average world shale (Piper, 1974) in order to smooth the odd-even variation among REE and emphasize geochemical characteristics.

At first glance the REE patterns of the seamount phosphorites differ from those of the shelf phosphorites in two ways: (a) the seamount phosphorites have higher concentrations of REE, and (b) they have a strong negative Ce anomaly, which is not present at all in the shelf samples. However, other subtle differences are also present. In the seamount samples, La is enriched relative to all but the heavy end members (Yb and Lu) and Eu is greater than Sm, similar to their relation in seawater. In the shelf samples, La is essentially equal to the other REE, and Eu is less than Sm. The Ce/La ratio in the seamount samples (0.25) is close to that ratio for seawater (0.35) but unlike that in the Peru/Chile samples (3.1). The low Tb values in both sets of samples suggest that its anomalous behavior may merely be an artifact of normalization, i.e., the normalizing value of 1.23 may be too large.

The similarity of the seamount patterns to that of seawater, in essentially every detail, suggests that these elements (and perhaps the phosphorus as well) have been derived directly from seawater. Supporting this interpretation is the distribution of Sc, often considered with the REE because of similarities in chemical properties.

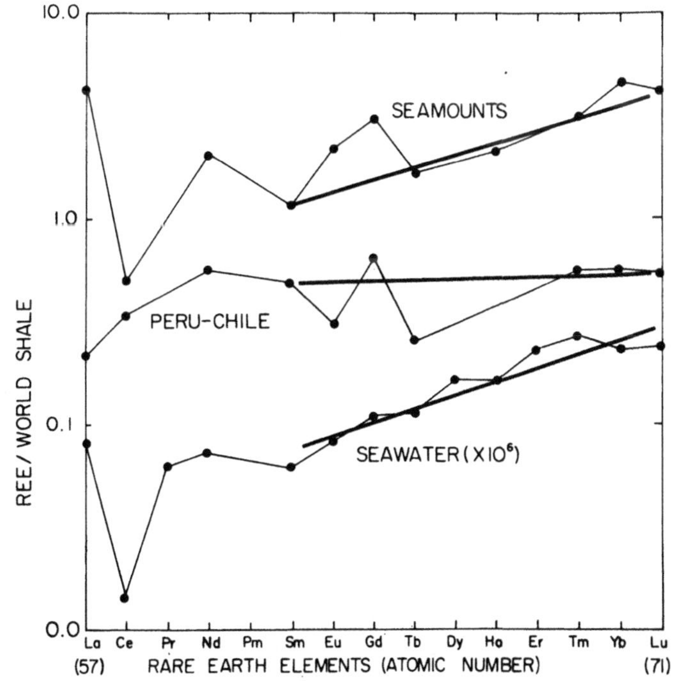

Fig. 11. Average REE analyses normalized to world shale for Pacific seamounts, Peru/Chile, and seawater. The heavy lines are intended to show the monotonic trends in the heavy REE. The seawater data is from Martin, Hogdahl and Philippot (1976).

In the seamount phosphorites, the La/Sc ratio is approximately 50 and in Pacific seawater samples it averages 25 (Hogdahl, 1968).

The REE in these seamount samples are unusual, however, because of their high concentrations. Their P/La ratio, for example, is approximately 10^3 whereas this ratio in seawater has a minimum value of approximately 3×10^4 and in plankton of about 10^5. It would seem that the REE may play a significant role in maintaining charge balance during the PO_4^{3-} substitution for CO_3^{2-}. The alternative explanation, that the phosphorite inherits its REE from the pre-existing $CaCO_3$, can be discounted. Although the REE pattern for the phosphate is similar to that of marine carbonate (Elderfield et al., 1981), the phosphate contains approximately 100-times higher concentrations.

The REE in the shelf samples may reflect their source. These samples need contain only about 25% lithogenic debris to account for the REE pattern and concentrations. Most of these samples do, in fact, contain about 20-30% detrital mineral grains and the sample with the highest REE concentration contains the highest acid insoluble reside. Furthermore, the Sc/La ratio for shale is 0.3 and for the shelf phosphorites may vary between 0.1 to 1.0 with most values close to 0.7. Thus, the actual phosphatic material may be essentially devoid of REE. The dissimilarity of the two types of deposits is therefore even more striking than their bulk compositions suggest.

Comparisons of these two types of deposits with ancient marine phosphorite (Altschuler, 1980; Table 2) gives rather intriguing results. REE in these latter deposits resemble the seamount deposits in their pattern and absolute abundances, rather than the REE in the nodules from the Peru/Chile shelf. Obviously, the phosphorites of the Phosphoria, Bone Valley, and other formations, each covering thousands of km^2 and often several meters thick, did not accumulate on seamounts. The current popularly held hypothesis concerning their origin, however, is that they accumulated in an upwelling environment, similar to that along the Peru/Chile shelf of today (Sheldon, 1981; Bentor, 1980). Yet, their chemical compositions suggest otherwise.

One possible explanation for this apparent discrepancy is that we are looking at two entirely different phenomena, large tabular nodules from the Peru/Chile shelf and pelletal- to clay-size material in the case of the ancient deposits. In the case of the nodules, we now know that they grow at the surface by slow accretion of phosphate which has diffused from some depth below the surface via sediment pore water. Pelletal phosphorite, however, also is present in the underlying siliceous sediment (Burnett, 1977). We suggest that this pelletal debris forms at depth in the sediment, in intimate contact with the decomposing organic matter which provides the phosphorus and other components such as REE, and it is this precipitate that is analogous to ancient phosphatic deposits.

SUMMARY

Our measurements of uranium decay-series isotopes from successive layers in single phosphate nodules have shown that these nodules grow down into the soft underlying mud at rates that may be controlled by the upward diffusive flux of dissolved phosphate regenerated in the sediment. Furthermore, the slow (millimeters per thousand year) growth rates require that either some mechanisms maintain these nodules at the sediment-seawater interface while the associated sediment is deposited 2-3 orders of magnitude faster or that nodules are formed in local areas of non-deposition. The observation that these nodules grow actively near but not within the boundaries of the zone of minimum dissolved oxygen suggests that the feeding and burrowing activity of benthic organisms may play a decisive role in nodule maintenance.

Results of our protactinium-thorium measurements indicate that phosphorites exposed on the sea floor in high productivity upwelling environments may scavenge ^{231}Pa and ^{230}Th with an activity ratio of $^{230}Th/^{231}Pa$ significantly below the "production" ratio because of prior fractionation between these two nuclides in the ocean. In contrast, there is no evidence for excess ^{231}Pa in Quaternary phosphorites from east-coast environments. If a stable analog for protactinium-thorium can be identified, it may serve as a useful indicator of phosphorites formed in upwelling regions.

Rare earth element geochemistry also shows promise as an indicator of different depositional environments of marine phosphorites. There is a clear difference both in the concentrations and shale normalized patterns between a Peru/Chile upwelling suite of samples and a Pacific seamount open ocean sample set. Further studies of REE and other trace elements will be useful to verify if these conclusions are generally applicable. A summary of our observations as well as an extension of the division of sedimentary phosphorites into "east coast" and "west coast" provinces first proposed by McKelvey (1967) is given in Table 1. We have added an "open ocean" division for those deposits formed on topographic highs far from land. These divisions and characteristics are based on few measurements and in many cases are somewhat tentative and arbitrary--our hope is that they can be refined by further analysis in light of our observations.

ACKNOWLEDGEMENTS

Samples reported on in this study were kindly provided by A. Soutar (Scripps Institution of Oceanography), R. Moberly and S. Sinton (University of Hawaii), and G. Birch (University of Cape Town). We thank T. DeVries and R. Berner for their very helpful reviews of this paper. The uranium-series studies reported were supported by grants from the National Science Foundation (OCE-8007047 and OCE-8117583). This paper is a contribution to the International Geological Correlation Program (IGCP) Project 156 (Phosphorites).

Table 1. Characteristics and examples of ocean floor phosphorites from east coast, west coast, and open ocean phosphogenic provinces

Characteristic	East Coast Province	West Coast Province	Open Ocean
Oceanographic environment:	Western Boundary Current: warm, low to moderate productivity, little coastal upwelling	Eastern Boundary Current: cold, high to very high productivity, well-developed oxygen minimum, intense coastal upwelling	Mid-gyre regions, little productivity
Commonly associated sediment:	Foraminiferal ooze, hardgrounds, C-org content moderate	Diatomaceous muds, C-org content high	Carbonate oozes, shallow and deep water limestones, volcano-clastics
Commonly associated minerals:	Calcite, dolomite, glauconite, goethite, Mn oxides	Quartz, glauconite, pyrite	Calcite, volcanic minerals, glass, pala-gonite, Fe-Mn oxides
Geochemistry:			
CaO/P_2O_5-	High (2-3)	Near francolite value (1.5)	High (1.5-2)
F/P_2O_5-	High (∿0.12)	Low to moderate (0.09-0.11)	High (∿0.12)
U-	Moderate to high (50-200 ppm)	Moderate (∿100 ppm)	Very low (∿5 ppm)
%U(IV)-	High (70-80%)	Moderate (40-70%)	Very low
REE-	Seawater pattern concentrations high	Resembles shale pattern, concentrations low	Seawater pattern, concentrations high
$^{231}Pa/^{235}U$-	Concordant with $^{230}Th/^{234}U$	Discordant with $^{230}Th/^{234}U$	---
Dominant mode of apatite crystallization:	Replacement of carbonate	Precipitation from aqueous phase	Replacement of carbo-nate, volcanic materials
Origin of phosphorus:	Uncertain: from land? regeneration in estuaries?	Surface productivity: organic matter and fish debris regenerated on sea floor	Uncertain: from seawater? guano? volcanism?
Examples:	Chatham Rise, New South Wales, Agulhas Bank, Blake Plateau?	Peru/Chile, Namibia, California, Portugal	Musician Seamounts, Mid-Pacific Mountains, Annan Seamount, Blake Plateau?

REFERENCES

Altschuler, Z.S., 1980, The geochemistry of trace elements in marine phosphorites - Part I. Characteristic abundances and enrichment, in: "Marine Phosphorites," Y.K. Bentor, ed., Society of Economic Paleontologists and Mineralogists Special Publication No. 29, 9-30.

Anderson, R.F., 1981, "The Marine Geochemistry of Thorium and Protactinium," Ph.D. Dissertation, M.I.T./Woods Hole Oceanographic Institution, Woods Hole, 287 pp.

Atlas, E.L., 1975, "Phosphate Equilibria in Seawater and Interstitial Waters," Ph.D. Dissertation, Oregon State University, Corvallis, 154 pp.

Baturin, G.N., 1978, "Phosphorites," Nauka Press, Moscow, 232 pp. (in Russian).

Bentor, Y.K., 1980, Phosphorites - the unsolved problems, in: "Marine Phosphorites," Y.K. Bentor, ed., Society of Economic Paleontologists and Mineralogists Special Publication No. 29, 3-18.

Bernat, M. and Goldberg, E.D., 1969, Thorium isotopes in the marine environment, Earth and Planetary Science Letters, 5:308-312.

Bezrukov, P.L., 1973, Main scientific results of the 54th voyage of the RV "Vityaz" in the Indian and Pacific Oceans, February-May, 1973, Okeanologiya 5 (in Russian).

Birch, G.F., 1979, Phosphorite pellets and rock from the western continental margin and adjacent coastal terrace of South Africa, Marine Geology, 33:91-116.

Birch, G.F., Thomson, J., Burnett, W.C., and McArthur, J., in press, Pleistocene phosphorites on the Agulhas Bank, South Africa, Nature.

Burnett, W.C., 1977, Geochemistry and origin of phosphorite deposits from off Peru and Chile, Geological Society of America, Bulletin, 88:813-823.

Burnett, W.C. and Veeh, H.H., 1977, Uranium-series disequilibrium studies in phosphorite nodules from the west coast of South America, Geochimica et Cosmochimica Acta, 41:755-764.

Burnett, W.C., Veeh, H.H. and Soutar, A., 1980, U-series oceanographic and sedimentary evidence in support of recent formation of phosphate nodules off Peru, in: "Marine Phosphorites," Y.K. Bentor, ed., Society of Economic Paleontologists and Mineralogists Special Publication No. 29, 61-72.

Burnett, W.C., Beers, M.J. and Roe, K.K., 1982, Growth rates of phosphate nodules from the continental margin off Peru, Science, 215:1616-1618.

Cook, P.J. and McElhinny, M.W., 1979, A reevaluation of the spatial and temporal distribution of sedimentary phosphate deposits in the light of plate tectonics, Economic Geology, 74:315-330.

DeMaster, D.J., 1979, "The Marine Budgets of Silica and ^{32}Si," Ph.D. Dissertation, Yale University, New Haven, 308 pp.

DeVries, T.J. and Schrader, H., 1981, Variation of upwelling/oceanic conditions during the latest Pleistocene through Holocene off the central Peruvian coast: a diatom record, Marine Micropaleontology, 6:157-167.

Elderfield, H., Hawkesworth, C.J., Greaves, M.J., and Calvert, S.E., 1981, Rare earth element zonation in Pacific ferromanganese nodules, Geochimica et Cosmochimica Acta, 45:1231-2134.

Froelich, P.N., Bender, M.L., Luedtke, N.A., Heath, G.R., and DeVries, T.J., 1982, The marine phosphorus cycle, American Journal of Science, 282:474-511.

Hamilton, E.L., 1956, Sunken islands of the Mid-Pacific Mountains, Geological Society of America, Memoir, 64, 97 pp.

Hamilton, E.L. and Rex, R.W., 1959, Lower Eocene phosphatized ooze from Sylvania Guyot, U.S. Geological Survey Professional Paper, 260-W.

Heezen, B.C., Mathews, J.L., Catalano, R., Natland, J., Coogan, A., Tharp, M., and Rawson, M., 1973, Western Pacific Guyots, in: "Initial Reports DSDP," 20, B.C. Heezen, I.D. McGregor et al., U.S. Government Printing Office, Washington, 653-702.

Henrichs, S.M., 1980, "Biogeochemistry of Dissolved Free Amino Acids in Marine Sediments," Ph.D. Dissertation, Woods Hole Oceanographic Institution, Woods Hole, 253 pp.

Hogdahl, O., 1968, Distribution of the rare earth elements in seawater, NATO Research Grant #203 Semi-annual Progress Report No. 6, 23 pp.

Jones, E.J.W. and Goddard, D.A., 1979, Deep-sea phosphorite of Tertiary age from Annan Seamount, eastern equatorial Atlantic, Deep-Sea Research, 26A:1363-1379.

Kaufman, A. and Broecker, W., 1965, Comparison of ^{230}Th and ^{14}C ages for carbonate materials from Lakes Lahonton and Bonneville, Journal of Geophysical Research, 70:4039-4054.

Kazakov, A.V., 1937, The phosphorite facies and the genesis of phosphorites, Scientific Institute of Fertilizers and Insecto-Fungicides Transactions (in Russian), 142:95-113.

Kolodny, Y., 1981, Phosphorites, in: "The Oceanic Lithosphere," C. Emiliani, ed., THE SEA, Vol. 7, Wiley and Sons, New York, 981-1023.

Kress, A.G. and Veeh, H.H., 1980, Geochemistry and radiometric ages of phosphatic nodules from the continental margin of northern New South Wales, Australia, Marine Geology, 36:143-157.

Ku, T.-L., 1968, Protactinium-231 method of dating coral from Barbados Island, Journal of Geophysical Research, 73:2271-2276.

Ku, T.-L. and Broecker, W.S., 1967, Uranium, thorium and protactinium in a manganese nodule, Earth and Planetary Science Letters, 2: 317-320.

Ku, T.-L. and Broecker, W.S., 1969, Radiochemical studies on manganese nodules of deep-sea origin, Deep-Sea Research, 16:625-637.

Mangini, A. and Sonntag, C., 1977, ^{231}Pa dating of deep-sea cores via ^{227}Th counting, Earth and Planetary Science Letters, 37:251-256.

Manheim, F.T., Pratt, R.M. and McFarlin, P.F., 1980, Composition and origin of phosphorite deposits of the Blake Plateau, in: "Marine Phosphorites," Y.K. Bentor, ed., Society of Economic Paleontologists and Mineralogists Special Publication No. 29, 117-138.

Marlowe, J.I., 1971, Dolomite, phosphorite, and carbonate diagenesis on a Caribbean seamount, Journal of Sedimentary Petrology, 41: 809-827.

Martin, J.M., Hogdahl, O. and Philippot, J.C., 1976, Rare earth element supply to the ocean, Journal of Geophysical Research, 81: 3119-3124.

McKelvey, V.E., 1967, Phosphate deposits (a summary of salient features of the geology of phosphate deposits, their origin and distribution), U.S. Geological Survey Bulletin, 1252-D:1-21.

Murray, J. and Renard, A.F., 1891, "Deep Sea Deposits: Scientific Results of the Exploration Voyage of H.M.S. 'Challenger', 1872-1876," Longmans, London, 525 pp.

O'Brien, G.W. and Veeh, H.H., 1980, Holocene phosphorite on the East Australian continental margin, Nature, 288:690-692.

O'Brien, G.W., Harris, J.R., Milnes, A.R., and Veeh, H.H., 1982, East Australian continental margin phosphorites: a bacterial origin, Nature, 294:442-444.

Paul, A.Z., 1976, Deep-sea bottom photographs show that benthic organisms remove sediment cover from manganese nodules, Nature, 263:50-51.

Piper, D.Z., 1974, Rare earth elements in the sedimentary cycle: a summary, Chemical Geology, 14:285-304.

Piper, D.Z. and Fowler, B., 1980, New constraint on the maintenance of Mn nodules at the sediment surface, Nature, 286:880-883.

Roe, K.K., Burnett, W.C., Kim, K.H., and Beers, M.J., 1982, Excess protactinium in phosphate nodules from a coastal upwelling zone, Earth and Planetary Science Letters, 60:39-46.

Scott, M.R., 1977, Abstract, Separation of Th, Pa and Ra in the marine environment, Transactions American Geophysical Union, 58: 1153.

Sheldon, R.P., 1964, Paleolatitudinal and paleogeographic distribution of phosphorite, U.S. Geological Survey Professional Paper, 501-C:106-113.

Sheldon, R.P., 1981, Ancient marine phosphorites, Annual Review of Earth and Planetary Sciences, 9:251-284.

Slater, R.A. and Goodwin, R.H., 1973, Tasman Sea Guyots, Marine Geology, 14:81-99.

Soutar, A., Johnson, S.R. and Baumgartner, T.R., 1981, Modern depositional analogs to the Monterey, in: "The Monterey Formation and Related Siliceous Rocks of California," R.E. Garrison and R.G. Douglas, eds., Society of Economic Paleontologists and Mineralogists, Pacific Section, Publication 16:123-148.

Suess, E., 1981, Phosphate regeneration from sediments of the Peru continental margin by dissolution of fish debris, Geochimica et Cosmochimica Acta, 45:577-588.

Turekian, K.K. and Chan, L.H., 1971, The marine geochemistry of uranium isotopes ^{230}Th and ^{231}Pa, in: "Activation Analysis in Geochemistry and Cosmochemistry," A.O. Brunfelt and E. Steinnes, eds., Universitetsforlaget, Oslo, 311-320.

Veeh, H.H., 1982, Concordant ^{230}Th and ^{231}Pa ages of marine phosphorites, Earth and Planetary Science Letters, 57:278-284.

ARE PHOSPHORITES RELIABLE INDICATORS OF UPWELLING ?

Geoffrey W. O'Brien and H. Herbert Veeh

School of Earth Sciences
Flinders University of South Australia
Bedford Park, South Australia 5042, Australia

ABSTRACT

Most models of phosphorite genesis involve upwelling as an essential element, if only to provide a mechanism for continuous nutrient supply to ocean surface water and hence sustain a high flux of phosphorus to the sediment via organic matter. Although recent studies have confirmed a close link between upwelling and phosphorite formation on the continental margins of Namibia and Peru-Chile, the simplified assumption that phosphorites in the sedimentary record invariably are indicative of upwelling can be challenged on several grounds: (a) some major phosphorite deposits in the geologic record cannot be adequately explained in terms of reasonable upwelling models; (b) formation of phosphorites in modern times, as demonstrated by uranium-series age determinations and other methods, is not confined to major upwelling centers, but also occurs on the continental margin of Eastern Australia, an area of only moderate seasonal upwelling, and quite limited organic productivity. Apatite in phosphorites off Eastern Australia is most likely formed during post-mortem alteration of organic phosphorus originally present in bacterial cells. Populations of transiently motile, facultatively chemolithotropic bacteria may proliferate within the upper continental slope sediments during times when the environment became quite restrictive, such as when nutrient supply was drastically limited. Additional clues for differentiating phosphorites might be found in the rare earth element pattern of the "Eastern Boundary Current"-type; i.e., absence of the Ce-anomaly. Phosphorites or phosphatic sediments lacking these features do not reliably indicate upwelling and should not be used indiscriminately in the reconstruction of former upwelling centers.

INTRODUCTION

With direct reference to the theme of the present volume, the purpose of this chapter is to examine the reliability of phosphorites as indicators of past upwelling conditions by drawing attention to a modern phosphorite deposit situated in an area of normal oceanic productivity, and which thus appears to be unrelated to upwelling. This phosphorite deposit, which is located on the upper continental slope of the East Australian continental margin, has been confirmed as being in part of Holocene age by the same dating techniques as those applied to Peru-Chile phosphorites and the Namibian deposits. We believe that the general geographical, geological and oceanographic characteristics of this deposit qualify it as a modern analogue of a "Western Boundary Current"-type phosphorite (see below), and propose that its origin may be explained without recourse to oceanic or coastal upwelling. We will further propose a mechanism for the origin of these particular phosphorites and offer criteria by which phosphorites not related to upwelling might be recognized in the sediment record and distinguished from those that are.

HISTORIC CONCEPTS OF UPWELLING AND PHOSPHORITES

Kazakov (1937) proposed that a direct link existed between oceanic upwelling and the genesis of marine sedimentary phosphorites. His model suggested that as cold upwelling waters ascended to the surface, changing chemical conditions in the water column favored the inorganic precipitation of carbonate fluorapatite, which then slowly sank to the sea floor to form phosphorite. This model was subsequently adopted by many workers and was used to explain the genesis of numerous phosphorite deposits in the geologic record. While more recent experimental (Atlas, 1975) and sedimentological studies (Baturin, 1971a; 1971b; Burnett, 1977) have shown Kazakov's model to be incorrect as far as the chemical precipitation of carbonate fluorapatite is concerned, a close and possibly genetic association between coastal upwelling and marine phosphorites has nonetheless been demonstrated by many workers (Sheldon, 1964a; McKelvey, 1967) and in fact has been used with some success in the guidance of exploration efforts (Sheldon, 1964b).

While phosphorites are found in the geologic record of virtually all ages since lower Proterozoic times, some workers believe that there is a non-random distribution of phosphorite deposits on a global scale, both spatially and temporally (Cook and McElhinny, 1979). In light of the previous discussion of the association between upwelling and phosphorites, it can be seen that the study of phosphorites becomes an important aspect of paleoceanography (Fisher and Arthur, 1977; Sheldon, 1980; Arthur and Jenkyns, 1981).

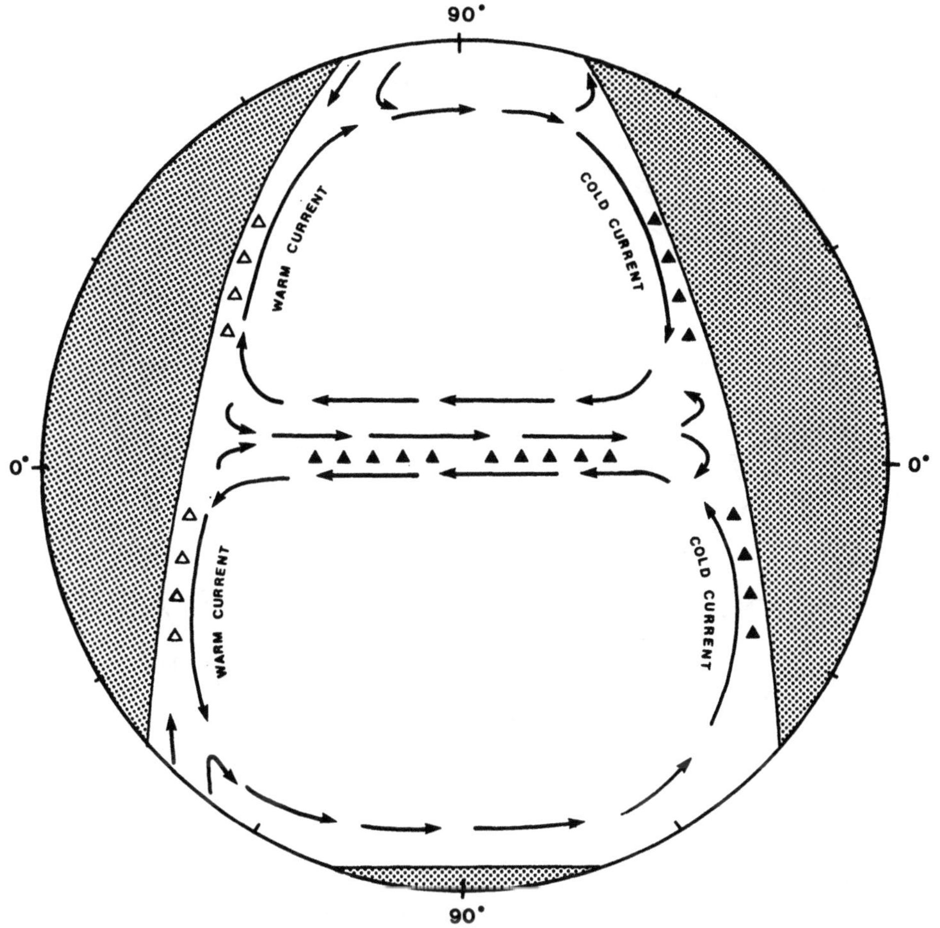

Fig. 1. Surface currents and upwelling in a model ocean; after McKelvey (1967); (Δ) upwelling caused by divergence, (Δ) dynamically caused upwelling.

Attempts to relate the distribution of marine phosphorites to ocean circulation and, in particular to coastal and oceanic upwelling, have been made by Brongersma-Sanders (1957), Sheldon (1964a), McKelvey (1967), and Baturin (1971a; 1971b). McKelvey (1967) for example, first drew attention to the fundamental difference between phosphorite deposits forming in different oceanographic environments. Using an idealized ocean model (Fig. 1) not unlike the present Pacific (Fig. 2), he argued that phosphorites formed in an area of coastal upwelling associated with divergence in the eastern boundary currents and are typically associated with chert and black shale, whereas phosphorites that formed in the warmer waters of the current western boundary currents consist of phosphatic limestones and sand-

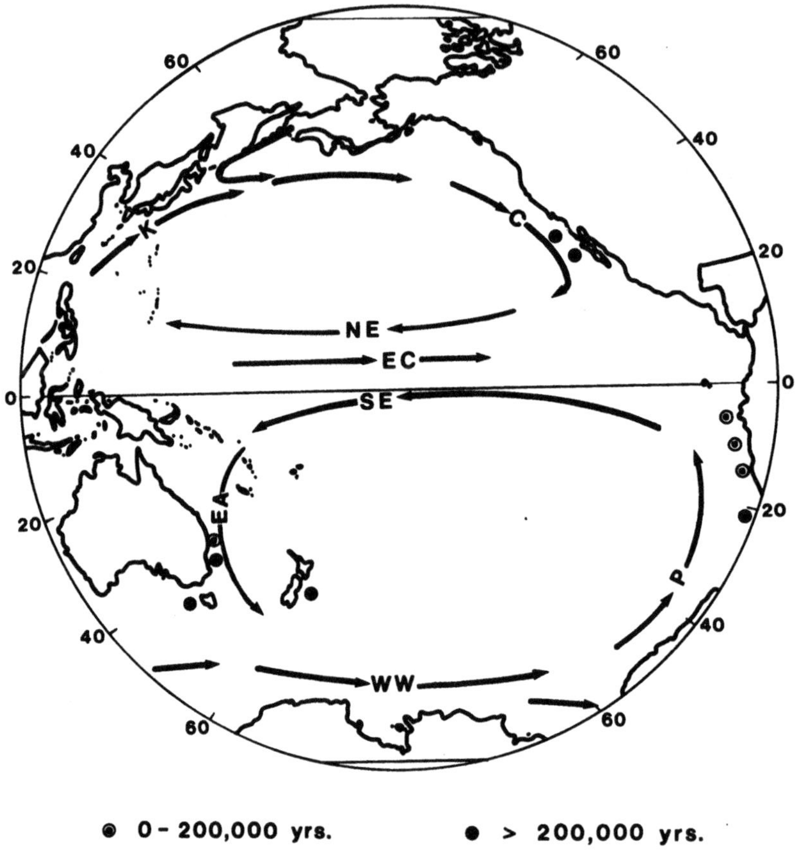

Fig. 2. Distribution of sea floor phosphorites with known radiometric ages in relation to major ocean currents in the Pacific. K = Kuroshio Current, C = California Current, NE = North Equatorial Current, EC = Equatorial Counter Current, SE = South Equatorial Current, P = Peru Current, EA = East Australian Current, WW = West Wind Drift.

stones. The latter normally are less extensive and of lower grade, unless subsequently enriched by reworking and/or chemical weathering. In the following, we shall refer to these two different types of phosphorite deposits as "Eastern Boundary" (or diatomaceous) and "Western Boundary" (or calcareous) types, respectively.

While McKelvey's study, and others, provided circumstantial evidence for Kazakov's upwelling model in the sense that they supported upwelling as an essential factor in phosphorite genesis, they were limited by inadequate age control on the phosphorite deposits. The application of uranium-series disequilibrium age-dating techniques to marine phosphorites, first by Kolodny and Kaplan (1970), and subsequently by other workers (Baturin, Merkulova and Chalov, 1972; Veeh, Burnett and Soutar, 1973; Veeh, Calvert and Price, 1974; Burnett and Veeh, 1977; Burnett, Veeh and Soutar, 1980; O'Brien and Veeh, 1980; O'Brien et al., 1981; Veeh, 1982; Burnett, this volume) marked the beginning of a new development in phosphorite research, i.e., the introduction of an absolute time frame into the distribution of marine phosphorites and hence the ability to more accurately relate the formation of phosphorites to known upwelling regions. Thus, Kolodny (1969) and Kolodny and Kaplan (1970) showed that many phosphorites occurring on the ocean floor in areas of present day upwelling are in fact relict deposits, their true ages being in excess of 800,000 years. Deposits off the coast of California are a case in point. On the other hand, more recent studies by Baturin et al. (1972), Veeh et al. (1973; 1974), Burnett and Veeh (1977), Burnett et al. (1980) confirmed the occurrence of modern phosphorite nodules in organic-rich diatomaceous sediments on the continental margins of Peru-Chile and Namibia, long known to be classic examples of wind-induced coastal upwelling regions associated with eastern boundary currents, and characterized by intense biological productivity in surface water overlying the phosphorite deposits. Both of these areas could be considered as modern analogues of "Eastern Boundary Current"-type phosphorites.

In contrast to Kazakov's (1937) model for phosphorite genesis, these more recent studies emphasize a biological pathway for the phosphorus from seawater to sediment. The role of upwelling lies mainly in a continued supply of nutrients to the surface water, leading to a high biological productivity and hence a high flux of organic matter to the sediments. Apatite is thought to form diagenetically within the sediment from phosphate released by bacterial decomposition of organic matter (Baturin, 1971a; 1971b; Atlas, 1975; Burnett, 1977). However, several phosphorite deposits in the geologic record do not easily fit this "Upwelling Model" (Bushinski, 1964; Cook, 1976; Bentor, 1980) when paleogeographic reconstructions are applied. Furthermore, the sediments associated with these deposits often have low contents of both organic carbon and diatomaceous silica, in contrast to the sediments of modern day "Eastern Boundary Current" upwelling areas. Examples of possible "non-upwelling" phosphorites are the Tertiary phosphorite deposits of the Atlantic coastal plain of the southeastern United States (Pevear, 1966) believed to be of estuarine origin, and the Precambrian phosphorites of Udaipur and Jhubua in India, which appear to be directly associated with stromatolites (Banerjee, Basu and Srivastava, 1980). Thus, caution must be exercised when using phosphorites indiscriminately as indicators of past upwelling centers. In the following discussion, aspects of the depositional environments of Late Pleistocene to Holocene phosphatic nodules from the East Australian continental margin will be considered. An appraisal will then be made of whether a direct cause and effect relationship exists between coastal upwelling and phosphogenesis in this environment.

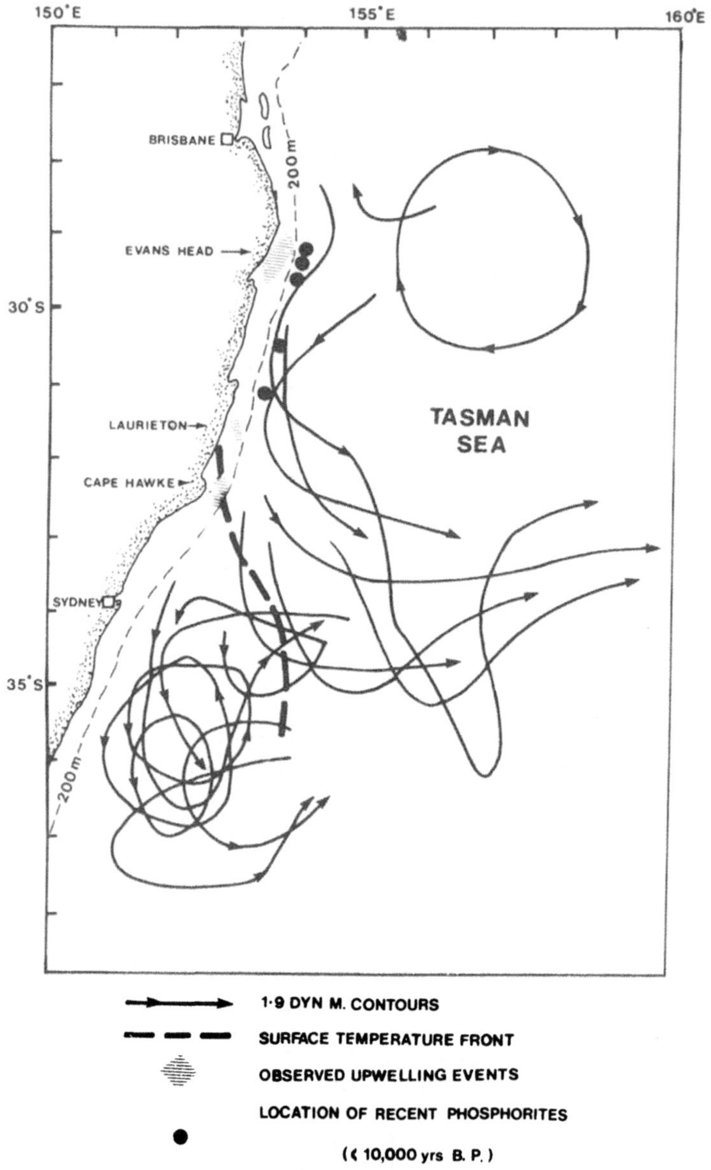

Fig. 3. The East Australian Continental Margin, showing the East Australian Current (indicated by a composite picture of the 1.9 dyn.m contour of surface dynamic height relative to 1300 db, obtained during several summer cruises, and by a distinctive surface temperature front based on satellite infrared images), and the location of Holocene phosphorites in relation to known upwelling events; oceanographic data from Rochford (1972; 1975); Godfrey et al. (1980b).

THE EAST AUSTRALIAN CONTINENTAL MARGIN

Geology and Oceanography

The continental shelf of the central East Australian coast is 25-40 km wide, with an unusually steep slope and shelf break between 210 and 450 m (Jones, Davies and Marshall, 1975; Marshall, 1979). The oceanography in this area is dominated by the East Australian Current which has characteristics of a western boundary current, at least north of Cape Hawke (Fig. 3). South of Cape Hawke, the current turns eastward away from the coast, and breaks up into a series of anticyclonic eddies. There have been reports of seasonal upwelling along the coast, in particular off Evans Head (Rochford, 1972), Laurieton (Rochford, 1975) and Cape Hawke (Pearce, pers. comm.), recognized by the temperature structure and moderate nutrient enrichment in surface water (Fig. 4). These upwelling events are of short duration and occur over the inner portion of the shelf, usually within 15 km of the coast.

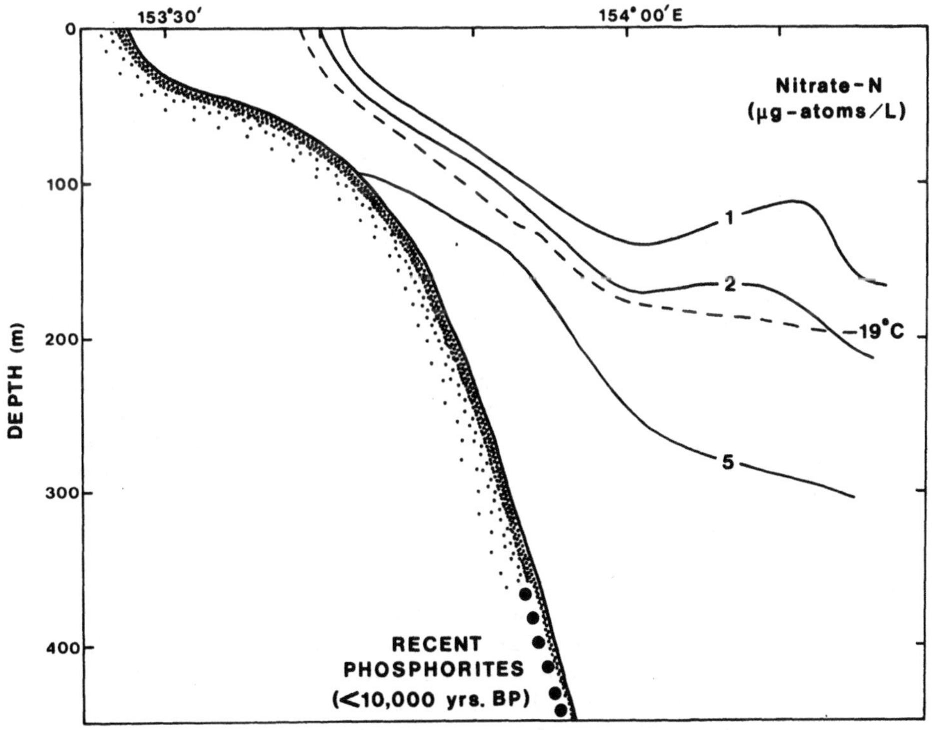

Fig. 4. Typical depth distribution of nitrate, and 19°C isotherm off Evans Head, NSW, during upwelling period (after Rochford, 1972), in relation to Holocene phosphorite occurrences in the same area.

Table 1. Organic carbon budget in the East Australian phosphogenic province compared to that of Peru

		East Australia	Peru*
Productivity	($gC/m^2/yr$)	36	400
Predicted** Flux at 370 m	(")	4	44
Measured Burial Rate	(")	0.04	17
Net Loss	(")	3.96	27

* Suess, 1980.

** $$C_{flux(z)} = \frac{C_{prod}}{0.0238\,z + 0.212}$$

where C_{prod} = Primary production in surface water

C_{flux} = Predicted flux of C_{org} at depth z(m)

Biological productivities as high as 1 $gC/m^2/day$ have been observed seasonally over the inner shelf and especially in several of the adjacent estuaries (B.D. Scott, pers. comm.). However, the surface waters over the slope show only normal oceanic productivities of the order of 0.1 $gC/m^2/day$ (Jitts, 1965) and in no way come close to the continuous high productivities encountered in upwelling areas associated with eastern boundary currents, as off Peru and Namibia (Wooster and Reid, 1963). Using the empirical relationship between surface productivity and measured C-org flux developed by Suess (1980), the expected C-org flux at a depth of 370 m is considerably lower than that expected in major upwelling centers (Table 1). As a consequence, the East Australian slope sediments lack the characteristics commonly associated with sediments beneath areas of upwelling, such as high contents of organic matter, opaline silica and fish bones. Rather, the sediments are typically calcareous sands, composed of quartz, glauconite pellets and planktonic foraminiferal tests. The organic carbon contents of the sediments are low, usually less than 0.5 wt-% (O'Brien et al., 1981). O'Brien et al. (1981) reported that the C-org and P-flux into the sediments on the East Australian continental margin was several orders of magnitude lower than those into the sediments off Peru. The low C-org flux is obviously the combined result of the low organic productivity in the surface water off Eastern Australia and the very low sedimentation rate in the area (less than 1.0 $cm/10^3$ years) (O'Brien et al., 1981). This low sedimentation rate would further reduce the organic matter contents by prolonging its exposure on the sea floor.

Phosphorite Ages

O'Brien and Veeh (1980) reported Holocene (less than 10,000 years) uranium-series disequilibrium ages for three phosphatic nodules from the East Australian upper continental slope. Two of these nodules, from a water depth of 365 m were zero and 2,000 years old, while a nodule from a depth of 376 m was 5,000 years old. In view of the low productivity and virtual absence of upwelling over the slope in this region, two explanations for these "young" phosphorites are possible; either: The age determinations of phosphorite nodules from the East Australian continental margin are invalid; or, upwelling is not an essential factor in the genesis of these phosphorites.

We consider the age determinations to be valid on the following grounds: In a recent evaluation of uranium-series dating as applied to carbonates and phosphorites, it was shown that the reliability of the uranium-series ages of marine phosphorites is at least comparable, if not better than that of reef corals (Veeh and Burnett, 1982). If we accept uranium-series ages of phosphorites from the Peru-Chile and Namibian upwelling regions, there is no reason to question the ages of phosphorites from Eastern Australia. In fact, uranium-series isotopic data of East Australian phosphorites show a degree of internal consistency, including excellent concordancy between ^{230}Th and ^{231}Pa ages, which is unsurpassed by any other previous studies (Veeh, 1982).

In some areas on the continental slope off Eastern Australia, fossil solitary corals have the interstices between their septa filled with sediment containing apatite, suggesting phosphate deposition subsequent to the death of the coral. On paleontologic evidence these corals are of a Recent species - *Caryophyllia planilamellata* Dennat 1906 (J. Wells, pers. comm.). The phosphatic material within the coral always yields radiometric ages younger than the coral within which it is enclosed (O'Brien and Veeh, 1980), supporting the contention that modern phosphate deposition is a feature of the East Australian upper continental slope sediments.

MECHANISM OF PHOSPHORITE FORMATION

It thus appears that the second of the previously mentioned alternatives is the correct one, i.e., the genesis of phosphorites from the East Australian continental margin is not related to coastal upwelling. In view of the fact that upwelling appears as an essential factor in the genesis of phosphorites from off Peru-Chile and Namibia, we may assume that a different phosphogenetic mechanism is responsible for the East Australian deposits.

In a recent paper, O'Brien et al. (1981) reported that all of the phosphate in Late Pleistocene to Holocene phosphatic nodules from the East Australian continental margin was located within bacterial cell structures. These authors proposed that a population of transiently motile, facultatively chemolithotrophic bacteria proliferated within the upper continental slope sediments during times when the environment became quite restrictive, such as when the nutrient supply was drastically limited. The bacterial biomass became a major component of the sediments, in spite of the probable slow bacterial growth rates, because the bacteria were well adapted to occupy the restricted ecological conditions. The slow sedimentation rate in the region, while obviously having a controlling influence on nutrient input to the sediments, also ensured that the slowly growing bacterial population was not significantly "diluted" with sediment.

The bacterial cells are now composed mostly of apatite. This apatite most likely formed during the *post-mortem* alteration of organic phosphorus originally present in bacterial cells. Thus, bacteria initially "fix" the phosphate in the sediment, which is then "post-depositionally" altered to apatite. O'Brien et al. (1981) reported evidence that bottom current velocities during times of phosphogenesis were significantly lower than present day current velocities. Calculated bottom current velocities on the continental slope in the region of the phosphorites are as high as 50 cm/sec (Godfrey, Cresswell and Boland, 1980a) and surface sediments are usually winnowed (O'Brien et al., 1981). In this regard, new observations reported in the present paper are of some relevance.

Figure 5 is a scanning electron micrograph of a section of the bacterial matrix within a Pleistocene East Australian phosphorite nodule. Two coccolith tests are shown lying in a bacterial matrix, and appear to have been deposited there and subsequently engulfed by bacterial growth. It is considered unlikely that deposition of such small tests as coccoliths would occur in a regime of high current velocities. This is taken as evidence for low bottom current velocities during times of phosphogenesis. Note also in Plate I that several of the bacterial cells show evidence of cell division.

As previously mentioned, phosphate, -and hence bacterial cells-, is often concentrated in fossil solitary corals. In addition, mollusc shells also contain phosphatic fillings. The consistent presence of mineralized bacteria within solitary corals, and mollusc shells would seem rather too common to be simply a chance happening. We believe that the concentrations of bacteria within these sites indicates their preference for growth in a sheltered site. It is likely that transiently motile bacteria could establish permanent phospholipid anchorages (Norkrans, 1980) only in interstitial sites where they were protected from the influence of a strong current. Thus, on a small scale at least, we have an indication that the bacteria present within the East Australian phosphorites prefer a sheltered "quiet" growth environment. The bacteria which proliferated to form the phosphatic nodules (as opposed to phosphatic fillings) com-

Fig. 5. Scanning electron micrograph of bacterial matrix in Late Pleistocene phosphatic nodule from the East Australian continental margin. Note the presence of dividing bacteria in the matrix, indicating active bacterial growth prior to mineralization.

prising most of the phosphorite found on the East Australian continental margin most likely did so at the sediment-water interface. By analogy with the bacteria found in the solitary corals and mollusc shells, the bacteria in the phosphorite nodules would require a "quiet" growth site. The presence of fine biogenic carbonate in the bacterial matrix (Fig. 5) indicates that this requirement was indeed met.

We have thus demonstrated that phosphogenesis on the East Australian continental margin seems to be favored by lower current velocities than those currently present over much of the upper slope. The exact mechanism of mineralization of the bacterial population is uncertain. However, several workers (Ennever, 1963; Rizzo, Scott and Bladen, 1963) have reported mineralization of bacterial cells with apatite under laboratory conditions. In addition, the enzyme carbonic anhydrase, a common constituent of bacteria (Veitch and Blakenship, 1963) has been reported as a catalyst in the formation of carbonate-hydroxyapatite (McConnell et al., 1961). Possibly if the bacterial population has been free of predation and grazing, then enzymes such as carbonic anhydrase present in the cells will facilitate the rapid conversion of cellular phosphates to apatite, following the death of the organism. It is also possible, though purely in the realm of speculation, that an inhibitory mechanism is active in the organism *in vivo* which prevents mineralization and that upon cell death, this mechanism ceases to operate allowing mineralization to proceed.

In contrast to the Late Pleistocene to Holocene nodules from the East Australian continental margin, the apatite in Middle Miocene ferruginous nodules from the same region is mostly present as large, euhedral crystals which have formed during the diagenesis of initially bacterially "fixed" phosphate (O'Brien et al., 1981). This observation is significant, for it suggests that the original depositional textures of marine phosphorites can be lost, even without exposure to sub-aerial weathering. If we can consider the East Australian continental margin phosphorites to be a modern analogue of a "Western Boundary Current" phosphogenic province, then it is quite likely that many other phosphorites in fact had a bacterial origin though the evidence has now been obscured by recrystallization.

Riggs (1979a; 1979b) described bacterial populations in the Tertiary phosphorites of Florida, though he was uncertain of the role they played in phosphogenesis. He also questioned the direct cause-effect relationship between phosphorite formation and upwelling, considering upwelling to be only one of several factors involved in phosphogenesis.

With specific reference to the East Australian continental margin, we see no evidence that coastal upwelling plays any role in the genesis of the phosphorites in the region. The active phosphogenic region is some 35 to 40 km from the coast, far removed from the sporadic upwelling region which is located 10 to 15 km from the coast. In addition, it is clear from Fig. 3 that some of the Holocene phosphorite occurs in regions from which no coastal upwelling has been reported. The low sedimentation rate in the area is of vital importance for it allows the normally slow bacterial activity to leave an imprint in the sediment. The slow sedimentation rate probably results from the strength of the East Australian current (O'Brien and Veeh, 1980). One can thus consider the East Australian Current to be a vital factor in phosphorite genesis. As most western boundary currents have some upwelling associated with them, it is not surprising that upwelling has been reported on the East Australian continental margin. The "association" of the upwelling there and the phosphorites is however, altogether fortuitous. Both the phosphorites and upwelling owe their existence to the East Australian Current, however no relationship exists between the upwelling and the phosphorites. It is therefore conceivable that some other phosphorites in the geologic record owe their existence to the various effects of ocean currents as well, with upwelling simply fortuitously associated with the currents. It appears that since Kazakov's (1937) paper, workers have become mesmerized by the "Upwelling Model" and have tried to fit most phosphorites into this scheme, an approach which has been recently criticized (Bentor, 1980).

RARE EARTH ELEMENT GEOCHEMISTRY OF PHOSPHORITES

We have demonstrated that no clear relationship exists between upwelling and phosphorite genesis on the East Australian continental margin. In dealing with a modern phosphogenic system, we are indeed fortunate because the various interrelationships (or lack of them) between the biological, geological, geochemical and oceanographic aspects of the region can be seen reasonably clearly. However, we are not nearly so fortunate when dealing with ancient phosphorite deposits --all that we have to study are suites of rocks that represent the "time-integrated" result of perhaps millions of years of development. From these rocks we have to reconstruct the depositional environment and paleoceanography of the phosphogenic system. The results and interpretations of such reconstructions are often ambiguous, hence it would be useful if some reliable criteria were available to distinguish, for example, ancient phosphorites that are obviously genetically related to upwelling ("Eastern Boundary Current") from those that are not ("Western Boundary Current"). We propose that rare earth element (REE) data may serve this purpose.

The different REE patterns of phosphorites from different areas and their possible significance as environmental indicators was first recognized by Altschuler, Berman and Cuttitta (1967), who pointed out that phosphorites associated with organic-rich sediments do not show the cerium deficiency observed in marine apatite from the Bone Valley formation of Florida, suggesting differences in depositional environment, such as proximity of land or depth of water, as a likely cause. Altschuler (1980) also stressed the importance of distinguishing between REE patterns obtained for pure apatite and those obtained for whole rock phosphorite samples, where the REE pattern of the apatite may be masked by non-phosphatic sediment components.

Similarly, Kolodny (1982) considered the REE pattern of different types of sea floor phosphorites as resulting from the mixing between a seawater component and a particulate matter component. The latter could be terrigenous debris and/or suspended organic matter (plankton and fecal material). The possibility of a biogenic carrier phase for the tranport of REE from seawater to sediment has also been considered by Piper (1974). Although the absolute concentration of rare earth elements in plankton is much smaller than in riverborne

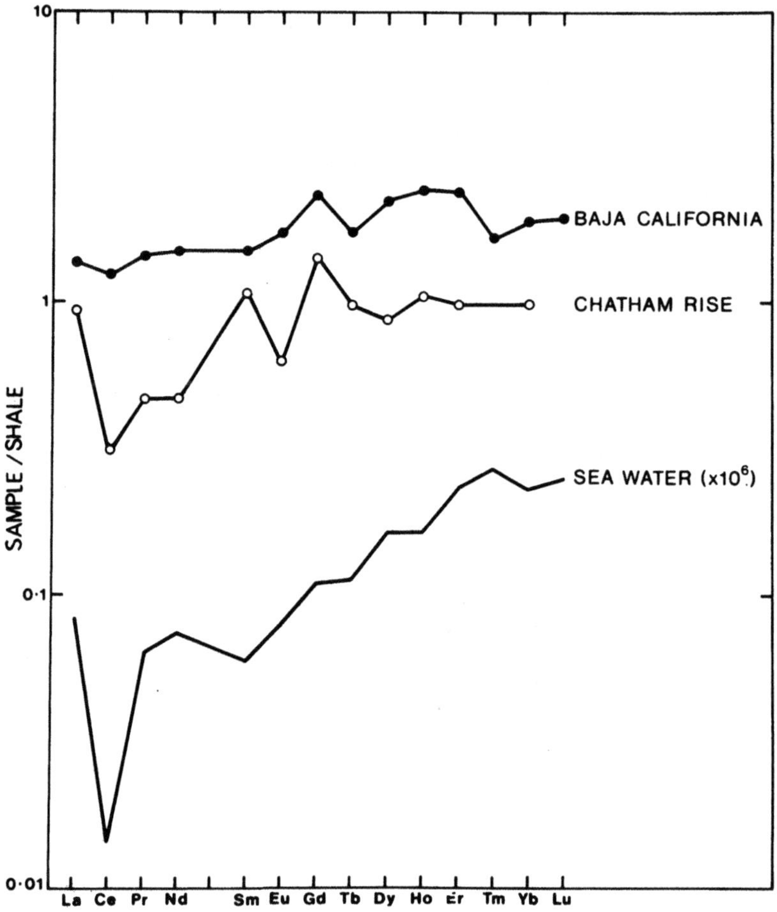

Fig. 6. Shale-normalized REE pattern in seawater (North Atlantic Deep Water) and phosphorites from different areas. Source of data: Average shale (Piper, 1974), seawater (Hogdahl, Melson and Bowen, 1968). Phosphorite from sea floor off Baja California (Goldberg et al., 1963). Phosphorite from the Chatham Rise (Cullen, 1980).

terrigenous debris or in shale (Elderfield et al., 1981), it represents considerable enrichment over the REE concentrations in seawater with a factor as high as 10^5 for Ce in phytoplankton (Lowman, Rice and Richards, 1971).

In Fig. 6, the shale-normalized pattern of rare earth elements in two phosphorites from contrasting depositional environments are shown together with the REE pattern of seawater. The phosphorite from Baja California was collected on the continental shelf in an area of enhanced coastal upwelling and hence should be representative

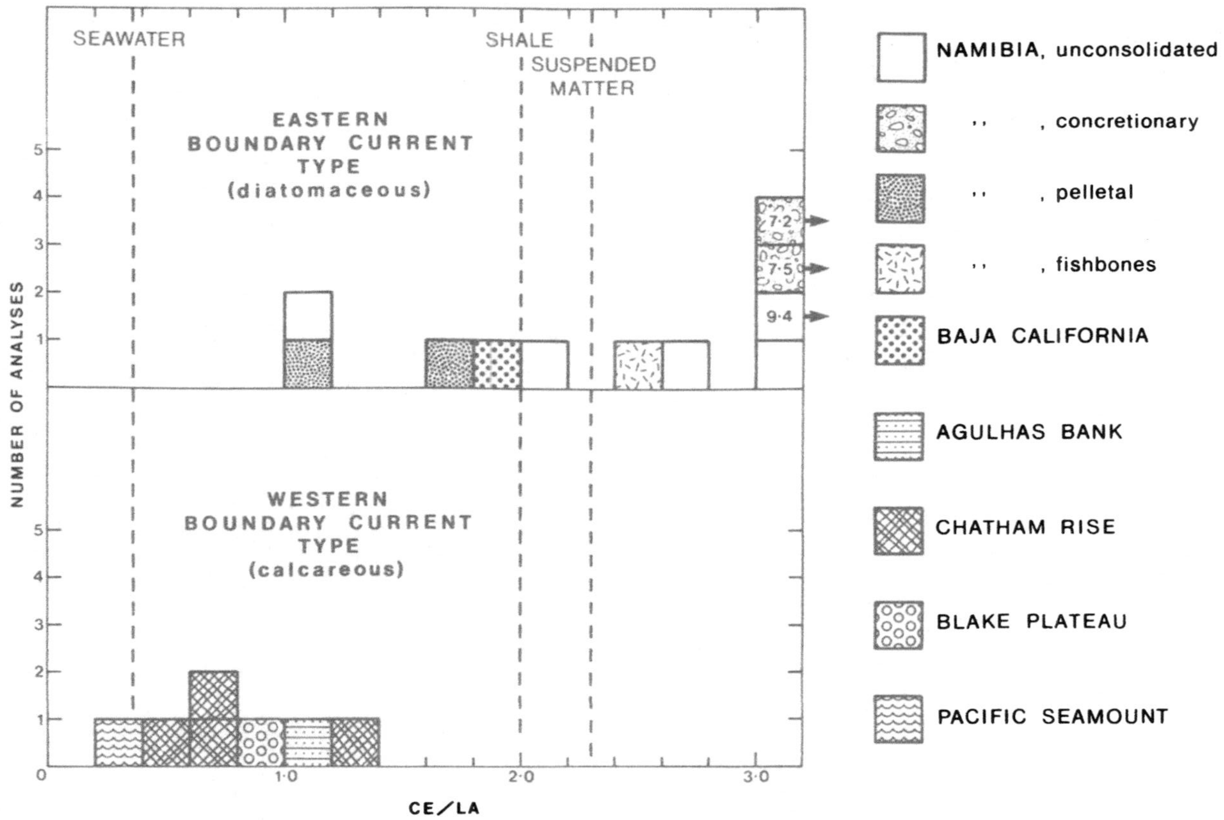

Fig. 7. Histogram of Ce/La ratios in sea floor phosphorites from different oceanographic environments, together with Ce/La ratios of seawater, shale and marine suspended matter. Source of data: Namibia (Price and Calvert, 1978), Baja California (Goldberg et al., 1963), Chatham Rise (Cullen, 1980), Agulhas Bank, Blake Plateau, Pacific Seamount (Kolodny, 1982), Seawater (Hogdahl et al., 1968), Shale (Piper, 1974), Suspended matter (Martin, Hogdahl and Philippot, 1976).

of an "Eastern Boundary Current"-type phosphorite. By contrast, the phosphorite deposit from the Chatham Rise southeast of New Zealand occurs in an area of normal oceanic productivity. The seawater pattern is typical of deep ocean water removed from continental influences (Hogdahl, Melson and Bowen, 1968). It should be noted however, that the REE pattern of shallow water (less than 100 m), in particular near continental margins, tends to resemble the REE pattern of shale and hence does not display a negative Ce anomaly (Hogdahl et al., 1968; Elderfield and Greaves, 1982).

The phosphorite from the Chatham Rise shows a pronounced Ce deficiency with respect to shale, not unlike the pattern of REE in deep ocean water. On the other hand, the phosphorite from the sea floor off Baja California shows no such anomaly. A similar REE pattern, although with lower concentration levels, has been reported by Burnett et al. (this volume) for modern phosphorites from the upper continental slope off Peru. Inasmuch as the most conspicuous difference in the REE pattern of different phosphorites appears to stem from the Ce-anomaly (Kolodny, 1982), the Ce/La ratios, for which more published data are available, may be used in lieu of the complete REE patterns (Fig. 7). Although there is some overlap between the Ce/La ratios of "Eastern" and "Western Boundary Current"-type phosphorites, the gradation from a "pure end-member", in this case a Pacific seamount phosphorite, with a Ce/La ratio identical to that of deep ocean water, towards "Eastern Boundary Current"-type phosphorites showing increasing enrichment in Ce with respect to La is readily apparent.

These data can be interpreted as follows: Phosphorite forming on the sea floor in areas of normal oceanic productivity and receiving only negligible input of terrigenous debris closely reflect the Ce anomaly of deep ocean water. On the other hand, phosphorites forming in areas of high biological productivity, such as the upwelling centers off Baja California, Namibia or Peru receive additional contributions of REE from decomposing organic matter (i.e., plankton, fecal pellets) which is relatively enriched in Ce. The scatter of Ce/La ratios in Fig. 7 then merely reflects the different proportional inputs from these two end member sources, i.e., seawater and organic matter, combined with various contributions of REE associated with terrigenous debris. The relative magnitude of these inputs and hence the resulting REE patterns, would depend on a number of factors including the flux of organic matter to the sea floor, depth of water, and the proximity of land. The very high Ce/La ratios shown for several phosphorite samples from Namibia, however, are difficult to explain by this model. Unless the Ce/La ratio in organic matter *per se* is significantly higher than that reported for "suspended matter" (probably a mixture of terrigenous debris and plankton), the possibility of post-depositional alteration of the chemical composition of phosphorites as suggested by Price and Calvert (1978) should also be considered.

Alternatively the range of Ce/La ratios of different types of phosphorite could also be explained in terms of the anomalous chemical behavior of Ce which, as the only REE susceptible to oxidation from the trivalent to tetravalent state, is subject to separation from the other REE in the marine environment. In fact, such a mechanism was proposed by Goldberg et al. (1963) to explain the marked enrichment of Ce in Mn nodules which can accommodate Ce^{+4} (as CeO_2 in the MnO_2 lattice) more readily than the trivalent members of the REE group. According to R.A. Schmitt (pers. comm.), the negative Ce anomaly in open ocean water is caused by the oxidation of Ce^{3+} to Ce^{4+} which, as insoluble $Ce(OH)_4$, is rapidly removed from seawater. The different Ce/La ratios shown in Fig. 7 would then merely reflect the degree of Ce depletion in response to different redox conditions in the water to which a given phosphorite has been exposed, from the most oxic (deep ocean water), to more reducing (oxygen minimum in the water column overlying organic-rich sediments).

It is the organic matter imprint on the REE pattern of phosphorites which, if confirmed by future studies, would make the study of rare earth elements in phosphorites a potential tool for the recognition and identification of past upwelling centers in the sediment record. In order to develop this concept further, a more systematic survey of REE in phosphorites from different areas as well as in particulate organic matter of various kinds, would be required. The presently available data are too limited to be conclusive and have been obtained by different investigators using different techniques, making a direct comparison of results unreliable.

CONCLUSIONS

The uncritical acceptance of phosphorites in the sediment record as reliable indicators of upwelling is certainly not warrented. Although upwelling plays an important, and perhaps even essential role in the formation of many major phosphorite deposits, such deposits would be recognized by other indicators of enhanced biological productivity, such as high organic carbon and opaline silica contents in the associated sediments. The occurrence of the three important nutrient elements C, P and Si and their preservation in the sediments as C-org, apatite and chert has been cited by Kolodny (1982) as an essential criterion of upwelling. Additional clues might be found in the REE pattern of the phosphorites of the "Eastern Boundary Current"-type, i.e., absence of the Ce-anomaly. Phosphorites or phosphatic sediments lacking these features do not reliably indicate upwelling and should not be used indiscriminately in the reconstruction of former upwelling centers.

ACKNOWLEDGEMENT

This study is based on material collected by C.C. von der Borch (Flinders University), and during cruises of RV "Sprightly" (CSIRO Division of Fisheries and Oceanography) and RV "Tangaroa" (New Zealand Oceanographic Institute, courtesy D. Cullen). SEM work was carried out at the University of Adelaide Electron Optical Center. We thank J. Harris and A.R. Milnes (CSIRO Division of Soils), and A. Pearce, J.S. Godfrey and B. Scott (CSIRO Division of Fisheries and Oceanography) for technical advice and helpful suggestions. Financial support was provided by the Australian Research Grants Committee.

REFERENCES

Altschuler, Z.S., 1980, The geochemistry of trace elements in marine phosphorites, part I. Characteristic abundances and environment, Society of Economic Paleontologists and Mineralogists, Special Publication No. 29:19-30.

Altschuler, Z.S., Berman, S. and Cuttitta, F., 1967, Rare earths in phosphorites - geochemistry and potential recovery, U.S. Geological Survey Professional Paper 575-B:45-90.

Arthur, M.A. and Jenkyns, H.C., 1981, Phosphorites and paleoceanography, Oceanologica Acta, Proceedings 26th International Geological Congress, Geology of Oceans Symposium, Paris, 83-96.

Atlas, E.L., 1975, "Phosphate Equilibria in Sea Water and Interstitial Waters," Ph.D. Dissertation, Oregon State University, Corvallis, 154 pp.

Banerjee, D.M., Basu, P.C. and Srivastava, N., 1980, Petrology, mineralogy and origin of the Precambrian Aravallian phosphorite deposits of Udaipur and Jhaba, India, Economic Geology, 75:1181-1199.

Baturin, G.N., 1971a, Stages of phosphorite formation on the ocean floor, Nature, 232:61-62.

Baturin, G.N., 1971b, Formation of phosphate sediments and water dynamics, Oceanology, 11:373-376.

Baturin, G.N., Bliskovskiy, V.Z. and Minyev, D.A., 1972, Rare-earth elements in phosphorite from the sea floor, Doklady Academii Nauk S.S.S.R., 207:954-957 (in Russian).

Baturin, G.N., Merkulova, K.I. and Chalov, P.I., 1972, Radiometric evidence for recent formation of phosphatic nodules in marine shelf sediments, Marine Geology, 13:M37-M41.

Bentor, Y.K., 1980, Phosphorites - the unsolved problems, Society of Economic Paleontologists and Mineralogists, Special Publication No. 29, 3-18.

Brongersma-Sanders, M., 1957, Mass mortality in the sea, in: "Treatise on Marine Ecology and Paleoecology, Vol. 1," J.W. Hedgepeth, ed., Geological Society of America, Memoir, 67:941-1010.

Burnett, W.C., 1977, Geochemistry and origin of phosphorite deposits from off Peru and Chile, Geological Society of America, Bulletin, 88:813-823.

Burnett, W.C. and Veeh, H.H., 1977, Uranium-series disequilibrium studies in phosphorite nodules from the west coast of South America, Geochimica et Cosmochimica Acta, 41:755-764.

Burnett, W.C., Veeh, H.H. and Soutar, A., 1980, U-series, oceanographic and sedimentary evidence in support of recent formation of phosphate nodules, Society of Economic Paleontologists and Mineralogists, Special Publication No. 29, 61-71.

Bushinski, G.I., 1964, On shallow water origin of phosphorite sediments, in: "Deltaic and Shallow Marine Deposits," L.M.J.V. van Straaten, ed., Elsevier, Amsterdam, 62-70.

Cook, P.J., 1976, Sedimentary phosphorite deposits, in: "Handbook of Stratabound and Stratiform Ore Deposits," Vol. 7, K.H. Wolf, ed., Elsevier, New York, 505-535.

Cook, P.J. and McElhinny, M.W., 1979, A reevaluation of the spatial and temporal distribution of sedimentary phosphate deposits in the light of plate tectonics, Economic Geology, 74:315-330.

Cullen, D.J., 1980, Distribution, composition and age of submarine phosphorites on Chatham Rise, east of New Zealand, Society of Economic Paleontologists and Mineralogists, Special Publication, 29:139-148.

Elderfield, H. and Greaves, M.J., 1982, The rare earth elements in seawater, Nature, 296:214-219.

Elderfield, H., Hawkesworth, C.J., Greaves, M.J., and Calvert, S.E., 1981, Rare earth element geochemistry of oceanic ferromanganese nodules, Geochimica et Cosmochimica Acta, 45:513-528.

Ennever, J., 1963, Microbiologic calcification, Annals New York Academy of Sciences, 109:4-13.

Fischer, A.G. and Arthur, M.A., 1977, Secular variations in the pelagic realm, Society of Economic Paleontologists and Mineralogists Special Publication, 25:19-50.

Godfrey, J.S., Cresswell, G.R. and Boland, F.M., 1980a, Observations of low Richardson numbers and undercurrents near a front in the East Australian Current, Journal of Applied Meteorlogy, 10:301-307.

Godfrey, J.S., Cresswell, G.R., Golding, T.J., Pearce, A.P. and Boyd, R., 1980b, The separation of the East Australian Current, Journal of Physical Oceanography, 10:429-440.

Goldberg, E.D., Koide, M., Schmitt, R.A. and Smith, R.H., 1963, Rare-earth distributions in the marine environment, Journal of Geophysical Research, 68:4209-4217.

Hogdahl, O.T., Melson, S. and Bowen, V., 1968, Neutron activation analysis of lanthanide elements in sea water, Advances in Chemistry Series, 73:308-325.

Jitts, H.R., 1965, The summer characteristics of primary productivity in the Tasman and Coral Seas, Australian Journal of Marine and Freshwater Research, 16:151-162.

Jones, H.A., Davies, P.J. and Marshall, J.F., 1975, Origin of the shelf break off southeast Australia, Journal of the Geological Society of Australia, 22:71-78.

Kazakov, A.V., 1937, The phosphorite facies and the genesis of phosphorites, Geological Investigations of Agriculture Ores, Transactions: Russian Scientific Institute of Fertilizers and Insecto-Fungicides, 142:93-113.

Kolodny, Y., 1969, Are marine phosphorites forming today? Nature, 224:1017-1019.

Kolodny, Y., 1982, Phosphorites, in: "The Oceanic Lithosphere," C. Emiliani, ed., THE SEA, Vol. 7, Wiley & Sons, New York, 981-1023.

Kolodny, Y. and Kaplan, I.R., 1970, Uranium isotopes in sea floor phosphorites, Geochimica et Cosmochimica Acta, 34:3-24.

Lowman, F.G., Rice, T.R. and Richards, F.A., 1971, Accumulation and redistribution of radionuclides by marine organisms, National Research Council, Panel on Radioactivity in the Marine Environment, National Academy of Sciences, Washington, 161-199.

Marshall, J.F., 1979, The development of the continental shelf of northern New South Wales, Bureau of Mineral Resources Journal of Australian Geology and Geophysics, 4:281-288.

Martin, J.M., Hogdahl, O. and Philippot, J.C., 1976, Rare earth element supply to the ocean, Journal of Geophysical Research, 81: 3119-3124.

McConnell, D., Frajola, W.J. and Deaner, D.W., 1961, Relation between the inorganic chemistry and biochemistry of bone mineralization, Science, 133:281-282.

McKelvey, V.E., 1967, Phosphate deposits, U.S. Geological Survey Bulletin, 1252-D:1-21.

Norkrans, B., 1980, Surface microlayers in aquatic environments, Advances in Microbial Ecology, 4:51-85.

O'Brien, G.W. and Veeh, H.H., 1980, Holocene phosphorite on the east Australian continental margin, Nature, 288:690-692.

O'Brien, G.W., Harris, J.R., Milnes, A.R., and Veeh, H.H., 1981, Bacterial origin of East Australian continental margin phosphorites, Nature, 294:442-444.

Pevear, D.R., 1966, The estuarine formation of the United States Atlantic Coastal Plain phosphorites, Economic Geology, 61:251-256.

Piper, D.Z., 1974, Rare earth elements in the sedimentary cycle: a summary, Chemical Geology, 14:285-304.

Price, N.B. and Calvert, S.E., 1978, The geochemistry of phosphorites from the Namibian shelf, Chemical Geology, 23:151-170.

Riggs, S.R., 1979a, Petrology of the Tertiary phosphorite system of Florida, Economic Geology, 74:195-220.

Riggs, S.R., 1979b, Phosphorite sedimentation in Florida - a model phosphogenic system, Economic Geology, 74:285-314.

Rizzo, A.A., Scott, D.B. and Bladen, H.A., 1963, Calcification of oral bacteria, Annals New York Academy of Science, 109:14-22.

Rochford, D.J., 1972, Nutrient enrichment of East Australian coastal waters. I. Evans Head upwelling, CSIRO Australia Division of Fisheries and Oceanography, Technical Paper No. 33, 16 pp.

Rochford, D.J., 1975, Nutrient enrichment of East Australian coastal waters. II. Laurieton upwelling, Australian Journal of Marine and Freshwater Research, 26:223-243.

Sheldon, R.P., 1964a, Paleolatitudinal and paleogeographic distribution of phosphorite, U.S. Geological Survey Professional Paper, 501-C:C106-C113.

Sheldon, R.P., 1964b, Exploration for phosphorite in Turkey - a case history, Economic Geology, 59:1159-1175.

Sheldon, R.P., 1980, Episodicity of phosphate deposition and deep ocean circulation - an hypothesis, Society of Economic Paleontologists and Mineralogists, Special Publication, 29:239-247.

Suess, E., 1980, Particulate organic carbon flux in the oceans - surface productivity and oxygen utilization, Nature, 288:260-263.

Veeh, H.H., 1982, Concordant ^{230}Th and ^{231}Pa ages of marine phosphorites, Earth and Planetary Science Letters, 57:278-284.

Veeh, H.H. and Burnett, W.C., 1982, Carbonate and phosphate sediments, in: "Uranium-Series Disequilibrium: Application to Environmental Problems in the Earth Sciences," M. Ivanovich and R.S. Harmon, eds., Oxford University Press, Cambridge, 459-480.

Veeh, H.H., Burnett, W.C. and Soutar, A., 1973, Contemporary phosphorites on the continental margin of Peru, Science, 181:844-845.

Veeh, H.H., Calvert, S.E. and Price, N.B., 1974, Accumulation of uranium in sediments and phosphorites on the South West African shelf, Marine Chemistry, 2:189-202.

Veitch, F.P. and Blakenship, L.C., 1963, Carbonic anhydrase in bacteria, Nature, 197:76-77.

Wooster, W.S. and Reid, J.L., 1963, Eastern boundary currents, in: "The Sea", vol. 2, M.N. Hill, ed., Interscience, New York, 253-280.

UNCONSOLIDATED PHOSPHORITES, HIGH BARIUM AND DIATOM ABUNDANCES IN SOME NAMIBIAN SHELF SEDIMENTS

Margaretha Brongersma-Sanders

Houtlaan 3
2334 CJ Leiden
The Netherlands

ABSTRACT

North of Walvis Bay, upwelling is only important during winter months as evidenced by thermoclines which develop in the inshore area when upwelling is at a minimum. A seasonal succession of upwelling and thermocline development favors productivity of phytoplankton. The thermocline development worsens the oxygen deficiency and zooplankton is uncommonly scarce in the inshore areas. Accordingly, downward transport of phytoplankton debris in fecal pellets is very limited. A narrow band of contemporary phosphorites occurs near the western border of this "thermocline area" where zooplankton are more abundant and settling of phosphorus-containing particulate matter is important. The narrow band occurs along the eastern border of the diatomaceous mud belt and the mean rate of sedimentation is relatively low due to occasional erosion of the non-cohesive surface mud layer. The high supply of phosphorus-containing matter combined with limited release and low rate of sedimentation favors the formation of phosphorites. The horizontal distribution of fish debris shows a striking resemblance to that of the phosphorites, but whether or not fish bones contribute importantly to the dissolved phosphate of the interstitial waters, and thus to the formation of phosphorites, has to be investigated.

High barium concentrations are found in the diatomaceous muds. In nearshore cores these concentrations occur in the topmost strata, possibly due to upward migration in the strongly reducing environment. Plankton is a likely source of the high Ba concentrations. The suggestion that *Chaetoceros* species are the main source is not supported by recent analyses of cultured diatoms.

INTRODUCTION

Phosphorites are currently forming in regions of coastal upwelling (Baturin, this volume; Burnett, Roe and Piper, this volume). As they also occur in non-upwelling environments, the presence of phosphorites alone is no sure indication of upwelling (O'Brien and Veeh, this volume). Off Namibia and off Peru contemporary phosphorites form in sediments rich in biogenous silica, bones and scales of fish, H_2S, organic carbon, and a number of minor elements. Diatomaceous muds, accumulations of fish bones, and high concentrations of these minor elements also occur in non-upwelling environments. Any single one of these peculiarities is no sure indicator of upwelling, but jointly they are useful indicators of past centers of coastal upwelling.

The combination reminds one of the chert/organic carbon-rich shale/phosphorite association frequently occurring in the ancient geological record. In the fossil deposits the phosphorites usually do not occur in, but next to, above or below, the strata with highest organic carbon, e.g., in the Mead Peak and Retort phosphatic shale members of the Permian Phosphoria Formation (Maughan, 1980).

Off Peru contemporary phosphorites occur on the shelf and the upper slope where the oxygen minimum layer impinges on the bottom. They are not found in the core of this layer, but in two narrow bands nearly coinciding with its upper and lower boundaries (Veeh, Burnett and Soutar, 1973; Burnett, Veeh and Soutar, 1980). Large nodules occur preferentially near the 0.25 mℓ/ℓ isopleth of dissolved oxygen (Burnett et al., this volume). Off Namibia contemporary phosphorites occur in a narrow band bordering the landward side of the diatomaceous mud belt (Fig. 1; Bremner, 1980a; this volume). Are the oxygen conditions in that area comparable to those in the narrow bands of Peru? In the present paper the special role of the seasonal thermoclines will be emphasized which may have an important bearing on oxygen conditions and primary production in the waters of the Namibian inner shelf and thereby on the formation of unconsolidated phosphorites.

High barium concentrations are characteristic of pelagic sediments below highly productive surface water. This paper also considers the Ba-rich sediments in the highly productive neritic area off Namibia, and their relation to the barium contents of unconsolidated phosphorites.

UNCONSOLIDATED PHOSPHORITES

Hydrography of the Namibian Shelf and the Importance of Thermoclines

Oxygen contents as low as those in the minimum layer off Peru do not occur in the waters above the continental border and the outer

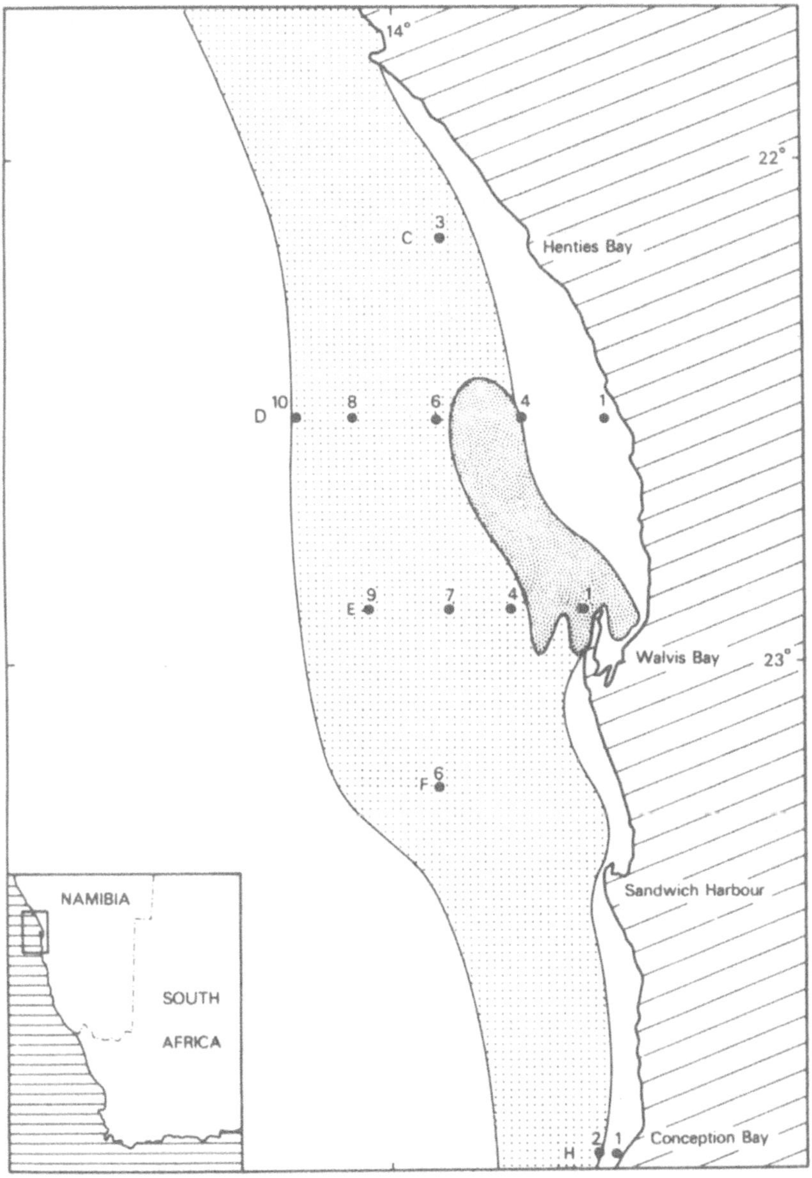

Fig. 1. Core locations and sediment distribution; adapted from Eisma (1969). White inshore area = grey sand; dark dotted area = green mud, strong H_2S smell, no remains of macrobenthos; light dotted area = green mud, weak or hardly any H_2S smell, occasional mollusc fragments. By permission from Marine Geology, Copyright (c) 1980, Elsevier Publ. Co.

shelf, but they occur periodically in the shallow coastal areas. The oxygen contents of the bottom water decrease shorewards due to local input and decomposition of organic matter (Hart and Currie, 1960). Another factor contributing to the low oxygen contents of the Namibian inshore area is the seasonal thermoclines which prevent vertical exchange and renewal of oxygen in subsurface waters. Thermoclines cannot develop where there is strong upwelling. Upwelling occurs all along the coast of Namibia, but it is most intense, irrespective of the season, between 26° and 25°S. The intensity decreases with latitude, particularly north of Walvis Bay (23°S), where upwelling is only important in winter and early spring and the intensity of the process decreases even to insignificance in the warmer time of the year (Stander, 1964). When upwelling is negligible, thermoclines may develop. They are strongest and occur most fre-

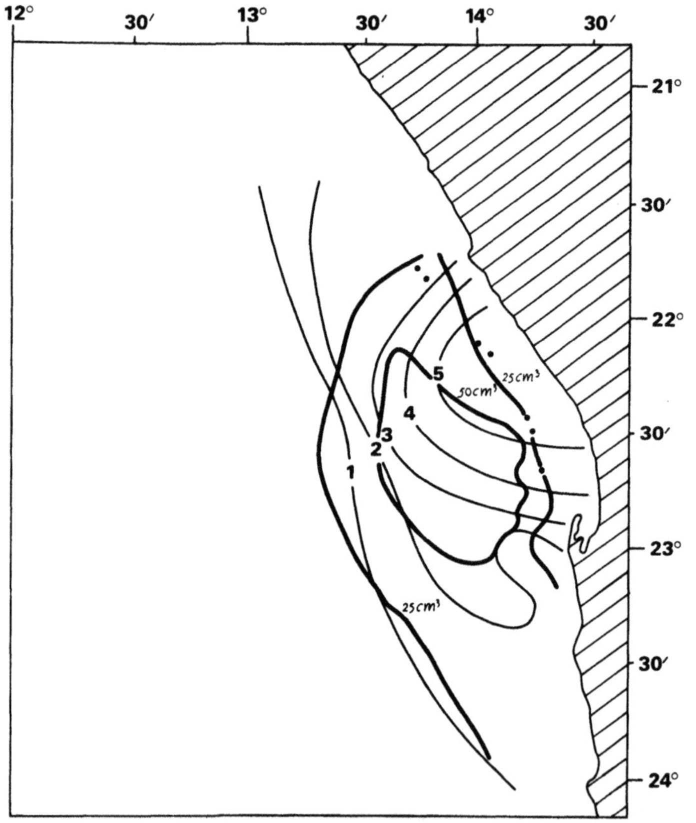

Fig. 2. Distribution of plankton and unconsolidated phosphorites off the Namibian coast. Thin lines = phytoplankton contours, numerals indicate numbers of cells in millions per liter (after DuPlessis, 1967). Thick lines = zooplankton contours in cm^3/ℓ (after Kollmer, 1963). Dots = unconsolidated phosphorites.

quently in the inshore area north of Walvis Bay, extending 30-40 miles offshore. This is related to, among other factors, the wind force being less in the inshore area than farther offshore (DuPlessis, 1967).

The thermoclines not only prevent aeration of the subsurface waters, they also favor the productivity in the surface waters. In many lakes a seasonal thermocline leads to a decrease in primary production due to exhaustion of nutrients, but in the very fertile waters off Namibia they favor increased productivity (DuPlessis, 1967). The fertility of the surface water north of Walvis Bay is the result of winter upwelling *in situ*, and of northerly transportation of surface water from the area of strong upwelling further south. In the warm and stable water above the thermocline an uncommonly big bloom of phytoplankton may develop. In the area around Henties Bay (22°00'-22°30'S), there is a marked interrelationship between thermoclines and phytoplankton productivity, the thermocline gradients and the number of phytoplankton cells per liter both decreasing in an offshore direction (Fig. 2; DuPlessis, 1967, Charts 10, 22). Highest zooplankton abundance is not found close to the shore, but 15-40 miles offshore; inshore zooplankton are uncommonly scarce, offshore they are uncommonly abundant (Fig. 2; Kollmer, 1963). The scarcity in the inshore waters has been attributed to, among other factors, the low oxygen contents of the coastal waters (Division of Sea Fisheries, South Africa, 1971). Dearth of zooplankton also occurs in the oxygen depleted waters in the core of the Peru oxygen minimum layer (Judkins, 1980).

PHOSPHORITE AND SEDIMENTARY FACIES

Above the belt of diatomaceous muds off Namibia the properties of the water and the nature and quantity of the plankton are far from uniform. On the seaward side of the belt the oxygen content of the bottom water is higher than on the landward side. Along the seaward side the muds have the highest contents of organic carbon (Calvert and Price, 1971a; Bremner, 1974a; 1974b) and maximum concentrations of a number of minor elements (Calvert, 1976; Brongersma-Sanders et al., 1980). It is a noteworthy fact that the maximal C-org contents do not occur in the area where the oxygen contents are lowest. The high C-org contents along the seaward side are probably related to the extreme abundance of both phyto- and zooplankton, the fecal pellets providing a quick downward transportation of particulate organic matter (Rogers, 1973). Unconsolidated phosphorites occur on the landward side of the diatomaceous mud belt (Bremner, 1980a).

Oxygen and temperature profiles were made in lines normal to the coast by the Dutch 1969 Expedition (Eisma, 1969). Profiles along the D-line (22°30'S) are given in Fig. 3a and 3b. Diatomaceous muds occur from D_4 to D_{10}; shorewards of D_4 the sediment consists of ter-

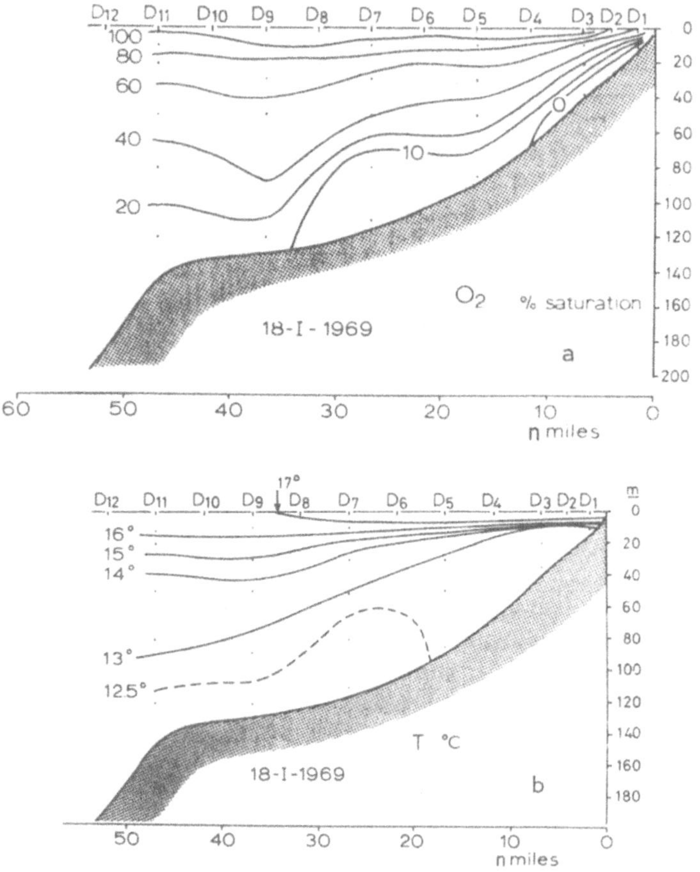

Fig. 3. (a) Oxygen profile along D-line, normal to the coast at 22°30'S; (b) temperature profile along D-line, normal to the coast at 22°30'S; both after Eisma (1969).

rigenous greenish-gray sand and gravel. The profiles were made in January, i.e., in the southern summer when strong thermoclines had developed; they occurred as far west as Sta. D_4 (Fig. 3b). The oxygen profile (Fig. 3a) shows a shoreward decrease of dissolved oxygen, the contents becoming negligible in the bottom water east of D_4. This means that D_4 is situated on the western border of the area with strong seasonal thermoclines and negligible oxygen contents of the bottom water.

A profile made by the Circe Expedition off Cape Cross (21° 51.5'S) (Calvert and Price, 1971b) shows a similar situation. The station closest to land (Sta. 188) is situated on the landward border of the diatomaceous muds. Although it was early in spring, a marked thermocline had developed at Sta. 188 (but not farther west). The oxygen content of the bottom water on Sta. 188 was extremely low (0.1 mℓ/ℓ), nitrite was high apparently due to bacterial nitrate respiration, and the phosphate concentration (3 μg-at/ℓ) was higher than anywhere farther offshore. Contemporaneous phosphorites were found in sediments of the same station.

On the landward border of the diatomaceous mud belt, unconsolidated phosphorites were found by various expeditions, i.e., off Cape Cross (Circe Expedition: Veeh, Calvert and Price, 1974; Price and Calvert, 1978; South Africa Expedition: Bremner, 1980a), and southwest of Henties Bay (Russian Expedition: Baturin, Kochenov and Petelin, 1970; Baturin, Kochenov and Senin, 1971: e.g., on Sta. 157: 22°28'S, 14°14'E). Sta. 157 is close to Sta. D_4 (22°30'S, 14°15.8'E) of the Dutch 1969 Expedition. The peculiar minor element concentrations in the top strata of a core taken on Sta. D_4 suggested phosphatization (Table 1; Brongersman-Sanders et al., 1980). As the high phosphate contents occur only in the unconsolidated upper sections of core D_4 it is likely that the phosphatization is recent. Analysis of a large number of bottom samples by Bremner (1980a) revealed that contemporaneous phosphorites occur in a narrow discontinuous band along the landward side of the diatomaceous mud belt.

Table 1. Concentration of minor elements (in ppm) and P_2O_5 (in %) in upper parts of core D_4

Depth (cm)	As	Ba	Cc	La	Sr	U	P_2O_5
0- 3	24	548	30	15	565	74	3.7
3- 8	41	291	nd	37	323	179	3.1
8- 13	32	790	nd	42	1046	264	9.2
13- 18	12	759	24	13	537	42	2.5
18- 23	14	1210	24	12	422	37	1.4
Average n = 19 23-118	16	214	33	16	241	19	0.9

nd = no data.

FACTORS FAVORING FORMATION OF PHOSPHORITES

The formation of phosphorites requires a combination of three factors:

1. low mean rate of sedimentation to prevent quick burial of the phosphorites;
2. limited release of dissolved phosphate from the sediment;
3. large supply of phosphorus-containing organic material and strong phosphate regeneration.

(1) On the Peruvian shelf the rate of sedimentation is high (mm to cm per year). Quick burial of phosphorite nodules is possibly prevented by activity of benthic invertebrates. As these animals cannot live in anoxic waters, this might explain why phosphorites are not formed in the core of the oxygen minimum layer (Burnett et al., this volume). In the Namibian mud belt the rate of sedimentation is approximately 0.3 mm per year (Calvert, 1976); the mean rate will be lower on the landward border of the belt due to periodic erosion of the non-cohesive surface layer (Bremner, 1980b). In this area removal of sediment by biological activity is probably not required for the formation of large phosphorite nodules.

(2) Under anoxic conditions phosphate is free to diffuse into the bottom water. This has been studied best in freshwater systems (Mortimer, 1941; Ruttner, 1953; Syers, Harris and Armstrong, 1973; Carignan and Flett, 1981; and many others). In highly eutrophic lakes phosphate is adsorbed during winter and spring on $Fe(OH)_3$-complexes infiltrating the topmost millimeters of the sediment. Subsequently during development of a summer thermocline, the dissolved oxygen contents approach zero, and the redox decreases such that Fe^{3+} is reduced to the more soluble Fe^{2+}, and the phosphate released in this process (Ruttner, 1953). Also, in the marine environment phosphate is released as the sediment passes through this redox boundary and the iron is reduced (Krom and Berner, 1980).

Manheim, Rowe and Jipa (1975) suggest that the absence of phosphorites in the core of the Peru oxygen minimum layer is related to release of phosphate from the sediment. In a large part of the Namibian mud belt where the bottom water is not completely anoxic, release will be limited; but it will be periodically important in the "thermocline area." Possibly part of the phosphate released from these sediments will precipitate somewhat farther west, along the western border of the "thermocline area."

(3) Due to a seasonal succession of upwelling and thermoclines, phytoplankton productivity is uncommonly high in inshore waters. Sinking of the phytoplankton debris will be hampered by the thermocline, and downward transportation of the debris in rapidly sinking fecal pellets will not be important because of the scarcity of zooplankton. Along the western border of the "thermocline area" where oxygen contents are slighly higher in subsurface waters and zooplankton become more abundant, the water will be quickly cleared of particulate organic matter. This will lead to a large supply of phosphate-containing fecal pellets along the outer border of the zooplankton-poor inshore area.

Zooplankton are also important for removing terrigenous matter from the water (Dunbar and Berger, 1981; Scheidegger and Krissek, this volume). Bremner (1980a) suggests that phosphate released from the anoxic sediments is adsorbed on mica-illite floating in the bottom water. If so, phosphate-containing illite probably is carried to the bottom in fecal pellets.

Though plankton make an important contribution to the dissolved phosphate contents of the interstitial waters, and with that probably to the phosphorite-phosphorus, the contribution by rapid dissolution of fish remains must not be overlooked. Murray and Renard (1891) suggested that phosphorites are formed in areas where catastrophic killing of marine animals occurs, calcium phosphate being an important constituent of some invertebrates and of vertebrate bones and teeth. Mass mortality of fish occurs repeatedly in all regions of upwelling and in other areas where phosphorites are found on the sea floor (Brongersma-Sanders, 1957). The hypothesis of Murray and Renard had been abandoned since "decomposable" organic matter, mainly derived from plankton, was considered to be the main source. Recent research in Peru continental margin sediments, however, has made it seem likely that a high percentage of the interstitial water phosphate may be derived from dissolution of fish debris. This is supported by abundance of fish debris and associated high dissolved PO_4 contents and by the interstitial fluoride distribution (Suess, 1981).

On the Namibian shelf the upper layer (0-25 cm) of diatom muds "often contain beds and lenses up to 10 cm thick, consisting almost entirely of relatively fresh fish bones and scales, predominantly of sardines" (Baturin, 1974). The horizontal distribution of fish debris in the muds shows a striking resemblance to that of contemporary phosphorites (Bremner, 1980a). Bremner remarks:

> "A high degree of correlation exists between areas rich in P_2O_5, areas with abundant phosphorite concretions, and areas where fish debris is common.... This led me to believe initially that concretion formation was somehow related to the influx of fish debris. However, sample 4028 shows that diatomaceous mud in these areas is less than 10 cm thick, and overlies a coarse-grained pavement of relict detritus left behind in the course of a previous regressive-transgressive episode."

Phosphatic nodules and fish remains scattered through the diatom muds become concentrated by removal of fine material by hydrodynamic processes. This indeed might explain the joint occurrence of phosphorite and fish bones. Nevertheless, the question remains to be settled (e.g., by determination of the fluoride distribution [Suess, 1981]) whether or not an important contribution to the phosphate contents of the interstitial waters of the diatom muds is made by dissolution of fish bones. If the contribution is important, this might have a bearing on the formation of phosphorites.

BARIUM AND HIGH PRIMARY PRODUCTION

Barium-Silica Association

Several authors have reported enrichment of Ba in pelagic sediments beneath regions of high biological productivity (Goldberg and Arrhenius, 1958; Goldberg, 1961; Turekian and Tausch, 1964). This suggests an association of Ba with the biological cycle. The validity of this hypothesis is supported by the Ba distribution in ocean water: depletion in surface waters, enrichment in deeper layers (Wolgemuth and Broecker, 1970). Barium is apparently concentrated by planktonic organisms in the surface water, but which of these organisms is principally responsible for the downward transport of Ba remains an open question.

Goldberg (1961) observed Ba concentrations up to 0.25% in diatom oozes from the Bering Sea. It appeared, however, that the diatom skeletons in these sediments were Ba-poor. To rationalize these findings the present author pointed out that the recognizable frustules in the sediment need not be (and usually are not) the remains of species predominating in the surface waters (Brongersma-Sanders, 1967). Vinogradova and Koval'skiy (1962) had shown that some diatoms (*Chaetoceros* and *Rhizosolenia* sp.) are very rich in Ba. These are diatoms that easily dissolve after death. In many parts of the oceans they abound in surface waters, but in the underlying sediments skeletons of *Chaetoceros* and *Rhizosolenia* are strikingly scarce (ref. in Brongersma-Sanders, 1967; reported from Namibian coastal area by Rogers, 1974). Resting spores, however, have been detected recently in Namibian shelf sediments (Boon, 1978; Schuette and Schrader, 1981).

The high Ba concentrations found by Vinogradova and Koval'skiy (1962) made it likely that diatoms, and particularly *Rhizosolenia* and *Chaetoceros* species, are an important source of Ba in pelagic sediments. The vertical profiles of Ba (and Ra) in the water column generally resemble those of silica, and this supported the hypothesis that siliceous plankton are a major carrier of these elements (Bacon and Edmond, 1972; Ku and Lin, 1976; Chung, 1980). However, the validity of the hypothesis that diatoms play the leading part is not strengthened by recent results. Of the marine planktonic organisms investigated by Martin and Knauer (1973), the Sr-concentrating group (radiolarians and dinoflagellates) were highest in barium. Of the cultured organisms analyzed by Bankston et al. (1979), highest Ba concentrations were found in dinoflagellates (range 50-328 ppm of ash), and the flagellate *Platymonas* (589 ppm). In the diatom genera *Fragilaria, Phaeodactylum, Skeletonema, Thalassiosira* barium ranged from 11-49 ppm in the ashed samples; in the coccolithophorid *Emiliania huxleyi* the concentrations were very low (<1 ppm). Barium contents of cultured *Rhizosolenia alata* and *Chaetoceros lauderi* were determined by Dehairs, Chesselet and Jedwab (1980). The Ba values found by these authors were much lower than those reported for these genera by Vinogradova and Koval'skiy (1962), i.e., cultured in natural seawater, the contents of *C. lauderi* were below the detection limit (∿1 ppm), those of *R. alata* 116 ppm of ash. Barium appeared to be entirely associated with the silica fraction. Though the Ba concentration of *Rhizosolenia* is much lower than those reported previously, *R.* species still may play an important part in the downward transportation of Ba in the oceans. This seems unlikely, however, for *Chaetoceros*.

BARIUM IN SEDIMENTS OF THE NAMIBIAN INNER SHELF

In deep-sea sediments of the southern Atlantic the highest Ba concentrations (>4000 ppm) have been found off Namibia (Turekian and Tausch, 1964). High concentrations have also been found in cores taken by the Dutch 1969 Expedition on the Namibian inner shelf. The cores were frozen on shipboard and kept frozen until opened in the laboratory for further analysis. Element concentrations were determined by instrumental neutron activation analysis (INAA) using a computer-controlled Ge(Li)-detector gamma ray spectrometer (DeBruin and Korthoven, 1972). Supplementary analyses of Ba were carried out by atomic absorption. Part of the samples were analyzed for P_2O_5 by the colorimetric method. The core locations are shown in Fig. 1, and concentrations of barium are listed in Table 2 on a total sediment basis (salt-free). The concentrations of the shelf sediments are not directly comparable to those of the deep-sea sediments mentioned above, as the latter are given on a carbonate-free basis. Core D_1 consists of muddy greenish gray sand; the other cores from the D- and E-line are olive-green diatomaceous muds.

In D_1, D_4 and E_1 high Ba concentrations (>1000 ppm) only occur in the upper parts, in E_1 also at depth of 108-113 cm, in other sections of these cores the concentrations range from 110-470 ppm. In E_4 high values occur in the upper 18 cm, but in addition the deeper layers, between 103-118 cm, are particularly enriched in barium. In D_6 and E_7 high concentrations are not limited to the uppermost section but are observed throughout D_6 and the upper 63 cm of E_4.

The high concentrations in the upper parts of the nearshore cores could conceivably be related to upward migration of Ba in the highly reducing sediments. The reducing conditions decrease in an offshore direction, a smell of H_2S being weak or hardly noticeable in cores as far offshore as D_6 and E_7 (Eisma, 1969). In these sediments the upward migration, and eventual release, are probably rather limited.

In E_4, 108-113 cm, the isolated high Ba concentration coincides with an isolated high Zn concentration (344 ppm; the range in other sections of the cores of diatomaceous muds is 12-83 ppm of Zn). The

Table 2. Ba concentrations (in ppm) of cores from the D- and E-line

Depth (cm)	D-Profile					E-Profile			
	D 1	D 4	D 6	D 8	D 10	E 1	E 4	E 7	E 9
0- 3	1860	548	-	470	110	1810	444	2920	-
3- 6	465	-	459	-	-	-	-	-	-
3- 8	-	291	-	573	200	432	1160	294	187
6- 9	966	-	392	-	-	-	-	-	-
8- 13	-	790	-	317	160	351	1210	550	87
9- 12	468	-	-	-	-	-	-	-	-
12- 15	360	-	598	-	-	-	-	-	-
13- 18	-	759	-	305	-	428	738	363	167
15- 18	326	-	425	-	-	-	-	-	-
18- 23	-	1210	-	569	-	300	534	1060	258
18- 21	281	-	-	-	-	-	-	-	-
21- 24	314	-	625	-	-	-	-	-	-
23- 28	-	456	-	405	-	165	435	1490	261
24- 27	366	-	-	-	-	-	-	-	-
27- 30	353	-	965	-	-	-	-	-	-
28- 33	-	354	-	406	-	369	175	1140	110
30- 33	218	-	916	-	-	-	-	-	-
33- 35	272	-	-	-	-	-	-	-	-
33- 38	-	114	-	371	-	265	437	371	102
35- 39	322	-	603	-	-	-	-	-	-
38- 43	-	165	-	93	-	360	392	3660	97
39 -42	404	-	-	-	-	-	-	-	-
42- 45	400	-	391	-	-	-	-	-	-
43- 48	-	165	-	106	-	261	321	2040	d
45- 48	294	-	663	-	-	-	-	-	-
48- 51	369	-	-	-	-	-	-	-	-
48- 53	-	108	-	162	-	143	615	459	93
51- 54	299	-	721	-	-	-	-	-	-
53- 58	-	162	734	171	-	178	808	829	99
58- 63	-	273	441	102	-	284	376	1230	150
63- 68	-	228	546	122	-	180	186	191	130
68- 73	-	289	955	319	-	284	241	189	-
73- 78	-	108	469	182	-	261	189	183	-
78- 83	-	222	-	d	-	150	658	469	-
83- 88	-	108	-	d	-	165	364	140	-
88- 93	-	165	-	-	-	177	411	91	-
93- 98	-	166	-	-	-	213	398	282	-
98-103	-	164	-	-	-	107	176	516	-
103-108	-	278	-	-	-	140	2330	569	-
108-113	-	322	-	-	-	1048	80000	178	-
113-118	-	218	-	-	-	175	1050	172	-
118-123	-	-	-	-	-	139	-	-	-

d = detected.

Ce/La ratio of 2.3 in 108-113 cm is above the core average (1.8); in the overlying 103-108 cm the ratio attains 3.0, the highest value of all samples investigated (for Ce and La concentrations see Brongersma-Sanders et al., 1980, Table 2). Upward migration of rare earth elements and reprecipitation of the easily oxidized Ce in the top might be responsible for the high Ce/La ratio. Relatively high oxygen contents in the bottom water possibly prevented escape of Ba, Zn and Ce. It is noteworthy that conditions were favorable for high barium concentrations when 108-113 cm was deposited in both E_1 and E_4. X-ray diffraction analyses did not reveal barite in the Ba-rich sections of E_4; Ba-bearing calcite might be present.

POSSIBLE SOURCES OF BARIUM IN SEDIMENTS OF THE NAMIBIAN SHELF

Barium is associated on the one hand with the detrital fraction, K-feldspars and micas being important Ba carriers, and on the other hand with organic matter (Puchelt, 1972). In D_1 Ba will mainly be associated with detrital matter, but this is unlikely for the high concentrations in the middle of the diatomaceous mud belt where the supply of terrigenous material is extremely low. Mica, which easily floats on the water, settles rather far offshore. Bremner (1980b) has shown that much mica settles in a narrow band bordering the landward side of the diatomaceous muds in water depths of 22-37 m. In E_4, at water depth of 91.5 m, very minor amounts of mica have been found (Boon, 1978). The low iron contents of the offshore cores also argue against an important mica supply. Plankton is the most likely source of the high Ba concentrations.

Data on plankton in water and sediment of the Namibian shelf (Hart and Currie, 1960; Kollmer, 1962; 1963; Schuette and Schrader, 1981) show the following: In the nearshore area where E_1 is located, the plankton consist mainly of dinoflagellates, diatoms, coccolithophorids, and ciliates. Dinoflagellates appear throughout the year producing tremendous blooms (red tide) in Walvis Bay and all along the coast as far north as Cape Cross (21°51.5'S). Diatoms occur in approximately equal numbers with *Thalassiosira*, *Thalassionema* and *Delphineis karsteni* predominating. Further offshore (E_4, E_7) the bulk of the phytoplankton consists of diatoms where *Chaetoceros* is a very important component. Blooms of coccolithophorids are observed occasionally and dinoflagellates are moderately abundant. Zooplankton consist mainly of copepods, euphausids, and the radiolarian *Aulosphaera*. In the warmer water farther west (E_9) calcareous organisms are abundant. The "oceanic" diatom association contains large numbers of *Rhizosolenia*, among others; this genus is scarce in the cold neritic area.

In view of the high Ba concentrations reported by Vinogradova and Koval'skiy (1962) for *Chaetoceros*, and the abundance of this genus in the area of E_4 and E_7, it has been suggested by Brongersma-

Sanders et al. (1980) that *Chaetoceros* is the principal source of Ba. The low concentrations found by Dehairs et al. (1980) in cultured *Chaetoceros* do not support this suggestion. As high Ba concentrations have been reported for dinoflagellates (Bankston et al., 1979) these organisms might be responsible for part of the Ba in sediments of the inshore area. This is unlikely for the area of E_4 and E_7 where the bulk of the phytoplankton consists of diatoms. Determination of Ba in plankton and resting spores is required to settle the question of which organisms are primarily responsible for high Ba in the sediments.

Phosphorites isolated from the diatom muds contain variable amounts of Ba, the concentrations ranging from 400-1100 ppm in unconsolidated phosphorites (P_2O_5 4-16%), and from 200-250 ppm in concretionary phosphorite ((P_2O_5 31.6-32.6%) (Price and Calvert, 1978). Bremner (1980a) reports a mean value of 79 ppm for concretionary phosphorites. The phosphatized upper sections of core D_4 have rather high Ba concentrations; however, they do not co-vary with the concentrations of P_2O_5: D_4 8-13 cm Ba 790 ppm P_2O_5 9.2%, 13-18 cm Ba 759 ppm P_2O_5 2.5%, 18-23 cm Ba 1210 ppm P_2O_5 1.4%.

With regard to the occurrence of Ba in phosphorites, substitution for Ca in the carbonate-fluorapatite lattice has been suggested by Price and Calvert (1978), and by Bremner (1980a). Bremner found a positive correlation of Ba and P_2O_5 in concretionary phosphorites, and this argues for substitution; he states that on the other hand the large ionic size of Ba --large relative to that of Ca-- argues against this conclusion. Barium contents of the unconsolidated phosphorites analyzed by Price and Calvert are much higher than those in the concretions; the unconsolidated phosphorites have higher contents of impurities such as SiO_2, and exceptionally high SiO_2/Al_2O_3 ratios might indicate the presence of diatomaceous opal. As Ba concentrations just as high or even higher than those in the phosphorites were found in non-phosphatized sections of our cores, we might suggest that most of the Ba found in phosphorites is incorporated in the impurities.

REFERENCES

Bacon, M.P. and Edmond, J.M., 1972, Barium at Geosecs III in the Southwest Pacific, Earth and Planetary Science Letters, 16:66-74.

Bankston, D.C., Fisher, N.S., Guillard, R.R.L., and Bowen, V.T., 1979, Studies of element incorporation by marine phytoplankton with special reference to barium, U.S. Department of Energy, Environmental Quarterly Reports, 1:509-531.

Baturin, G.N., 1974, On the geologic consequences of mass mortality of ichthyofauna in the Ocean, Oceanology, 14:80-84.

Baturin, G.N., Kochenov, A.V. and Petelin, V.P., 1970, Phosphorite formation on the shelf of Southwest Africa, Lithology and Mineral Resources, 3:266-276.

Baturin, G.N., Kochenov, A.V. and Senin, Yu.M., 1971, Uranium concentrations in Recent ocean sediments in zones of rising currents, Geochemistry International, 8:281-286.

Boon, J.J., 1978, "Molecular Biogeochmistry of Lipids in Four Natural Environments," Doctoral Thesis, Technical University, Delft, The Netherlands, 215 pp.

Bremner, J.M., 1974a, Texture and composition of surficial continental margin sediments between the Kunene River and Sylvia Hill, S.W.A., Sancor Marine Geology Progress, Technical Report, Cape Town, 6:34-43.

Bremner, J.M., 1974b, Further analysis of the Sardinops gravity cores collected from the shelf near Walvis Bay, South West Africa, Sancor Marine Geology Progress, Technical Report, Cape Town, 6:55-59.

Bremner, J.M., 1980a, Concretionary phosphorite from SW Africa, Journal of the Geological Society, London, 137:773-786.

Bremner, J.M., 1980b, Physical parameters of the diatomaceous mud belt off Southwest Africa, Marine Geology, 34:M67-M76.

Brongersma-Sanders, M., 1957, Mass mortality in the Sea, in: "Treatise on Marine Ecology and Paleoecology 2," J.W. Hedgpeth, ed., Geological Society of America, Memoir 67, 941-1010.

Brongersma-Sanders, M., 1967, Barium in pelagic sediments and in diatoms, Koninklijke Nederlandse Akademie Wetenschappen, Proceedings, Series B., 70:93-99.

Brongersma-Sanders, M., Stephan, K.M., Kwee, T.G., and DeBruin, M., 1980, Distribution of minor elements in cores from the southwest African shelf with notes on plankton and fish mortality, Marine Geology, 37:91-132.

Burnett, W.C., Veeh, H.H. and Soutar, A., 1980, U-series, oceanographic and sedimentary evidence in support of contemporary formation of phosphate nodules off Peru, Society of Economic Paleontologists and Mineralogists, Special Publications 29, IAS Congress Phosphorite Symposium, Jerusalem, 61-71.

Calvert, S.E., 1976, The mineralogy and geochemistry of near-shore sediments, in: "Chemical Oceanography," J.P. Riley and R. Chester, eds., 2nd ed., Academic Press, London, 187-280.

Calvert, S.E. and Price, N.B., 1971a, Recent sediments of the South West African shelf, in: "The Geology of the East Atlantic Continental Margin," F.M. Delany, ed., Institute Geological Sciences Report 70/16, 171-185.

Calvert, S.E. and Price, N.B., 1971b, Upwelling and nutrient regeneration in the Benguela Current, October 1968, Deep-Sea Research, 18:505-523.

Carignan, R. and Flett, R.J., 1981, Postdepositional mobility of phosphorus in lake sediments, Limnology and Oceanography, 26:361-366.

Chung, Y., 1980, Radium-barium-silica correlations and a two-dimensional radium model for the world ocean, Earth and Planetary Science Letters, 49:309-318.
DeBruin, M. and Korthoven, P.J.M., 1972, Computer-oriented system for nondestructive neutron activation analysis, Analytical Chemistry, 44:2382-2385.
Dehairs, F., Chesselet, R. and Jedwab, J., 1980, Discrete suspended particles of barite and the barium cycle in the ocean, Earth and Planetary Science Letters, 49:528-550.
Division of Sea Fisheries, South Africa, 1971, Annual report of the Director of Sea Fisheries for the calendar year 1971, Cape Town, 39:1-32.
Dunbar, R.B. and Berger, W.H., 1981, Fecal pellet flux to modern sediment of Santa Barbara Basin (California) based on sediment trapping, Geological Society of America, Bulletin, Pt. 1, 92: 212-218.
DuPlessis, E., 1967, Seasonal occurrence of thermoclines off Walvis Bay, Southwest Africa, Investigational Report of the Marine Research Laboratory, Southwest Africa, 13:1-13.
Eisma, D., 1969, Sediment sampling and hydrographic observations off Walvis Bay, S.W. Africa, Dec. 1968-Jan. 1969, NIOZ Internal Publications 1969-1, Texel, The Netherlands, 8 pp.
Goldberg, E.D., 1961, Chemical and mineralogical aspects of deep-sea sediments, in: "Physics and Chemistry of the Earth," L.H. Ahrens, ed., 4:281-302.
Goldberg, E.D. and Arrhenius, G., 1958, Chemistry of Pacific pelagic sediments, Geochimica et Cosmochimica Acta, 13:153-212.
Hart, T.J. and Currie, R.I., 1960, The Benguela Current, Discovery Reports, 31:123-298.
Judkins, D.C., 1980, Vertical distribution of zooplankton in relation to oxygen minimum off Peru, Deep-Sea Research, 27A:475-487.
Kollmer, W.E., 1962, The annual cycle of phytoplankton in the waters off Walvis Bay, 1958, Research Report of the Marine Research Laboratory Southwest Africa, 4:1-44.
Kollmer, W.E., 1963, Notes on zooplankton and phytoplankton collections made off Walvis Bay, Investigational Report of the Marine Research Laboratory Southwest Africa, 8:1-78.
Krom, M.D. and Berner, R.A., 1980, Adsorption of phosphates in anoxic marine sediments, Limnology and Oceanography, 25:797-808.
Ku, T.L. and Lin, M.C., 1976, ^{226}Ra distribution in the Atlantic Ocean, Earth and Planetary Science Letters, 32:236-248.
Manheim, F., Rowe, G.T. and Jipa, D., 1975, Marine phosphate formation off Peru, Journal of Sedimentary Petrology, 45:243-251.
Martin, J.H. and Knauer, G.A., 1973, The elemental composition of plankton, Geochimica et Cosmochimica Acta, 37:1639-1653.
Maughan, E.K., 1980, Relation of phosphate, organic carbon, and hydrocarbons in the Permian Phosophoria Formation, Western United States of America, International Conference on Comparative Geology of Phosphate and Oil Deposits, Orleans, France, 6-7 Nov. 1979, Document Bureau Recherches Géologiques et Minières, 24:63-91.

Mortimer, C.H., 1941; 1942, The exchange of dissolved substances between mud and water in lakes, Journal of Ecology, 29:280-329; 30:147-201.

Murray, J. and Renard, A.F., 1891, Report on the Deep-Sea Deposits, "Report on the Scientific Results of the Voyage of H.M.S. Challenger During the Years 1873-1876;" Vol. 29, 525 pp.

Price, N.B. and Calvert, S.E., 1978, The geochemistry of phosphorites from the Namibian shelf, Chemical Geology, 23:151-170.

Puchelt, H., 1972, Barium, in: "Handbook of Geochemistry," K.H. Wedepohl, ed., Springer, Berlin, II-4, 56, B-O.

Rogers, J., 1973, Texture, composition and depositional history of unconsolidated sediments from the Orange-Lüderitz continental shelf, and their relationship with Namib desert sands, GSO/UCT Marine Geology Progress Technical Report, Cape Town, 5:67-88.

Rogers, J., 1974, Surficial sediments and tertiary limestones from the Orange-Lüderitz shelf, Sancor Marine Geology Progress, Technical Report, Cape Town, 6:24-28.

Ruttner, F., 1953, "Fundamentals of Limnology," University of Toronto Press, Toronto, 242 pp.

Schuette, G. and Schrader, H., 1981, Diatom taphocoenoses in the coastal upwelling area off Southwest Africa, Marine Micropaleontology, 6:131-155.

Stander, G.H., 1964, The Benguela Current of South West Africa, Investigational Report of the Marine Research Laboratory Southwest Africa, 12:1-43.

Suess, E., 1981, Phosphate regeneration from sediment of the Peru continental margin by dissolution of fish debris, Geochimica et Cosmochimica Acta, 45:577-588.

Syers, J.K., Harris, R.F. and Armstrong, D.E., 1973, Phosphate chemistry in lake sediments, Journal Environmental Quality, 2:1-14.

Turekian, K.K. and Tausch, E.H., 1964, Barium in deep-sea sediments of the Atlantic Ocean, Nature, 201:696-697.

Veeh, H.H., Burnett, W.C. and Soutar, A., 1973, Contemporary phosphorite on the continental margin of Peru, Science, 181:845-847.

Veeh, H.H., Calvert, S.E. and Price, N.B., 1974, Accumulation of uranium in sediments and phosphorites on the Southwest African shelf, Marine Chemistry, 2:189-202.

Vinogradova, Z.A. and Koval'skiy, V.V., 1962, Elemental composition of Black Sea plankton, Doklady Academii Nauk S.S.S.R., Earth Sciences Section, 147:217-219.

Wolgemuth, K. and Broeker, W.S., 1970, Barium in sea water, Earth and Planetary Science Letters, 8:372-378.

SULFIDE SPECIATIONS IN UPWELLING AREAS

Jacques Boulègue

Laboratoire de Géologie Appliquée, Université de Paris VI
4 place Jussieu, 75230 Paris Cedex 05, France

Jérôme Denis

Laboratoire de Géochimie des Eaux, Université de Paris 7
2 place Jussieu, 75251 Paris Cedex 05, France

ABSTRACT

In anoxic waters of upwelling areas the main sulfur species are hydrogen sulfide, polysulfide ions and organo-sulfur compounds. Generally, their composition can be explained by equilibrium considerations. The redox state of the sulfur system reflects the redox state of the environment. Hydrogen sulfide and volatile organo-sulfur compounds are transported into oxygenated waters either by diffusion or by bubbling out of the sediment. The resulting redox boundary is generally enriched in elemental sulfur and thiosulfate. Chemical models of sulfide speciation can be compared to field data from the upwelling area near Walvis Bay (Namibia). The data acquired in the laboratory and in the field allow a description of some characteristics of the sulfur speciations in upwelling areas and their possible influence on sedimentary processes.

INTRODUCTION

The sulfur system in aqueous reducing conditions has been recently studied in the laboratory (Cloke, 1963; Chen and Morris, 1972; Giggenbach, 1972; Boulègue, 1976; 1978b; Gourmelon, Michard and Boulègue, 1977; Boulègue and Michard, 1978; 1979) and in the field (Boulègue, 1977; 1978a; Boulègue et al., 1979; Boulègue, Lord and Church, 1982). Among other objectives, the investigation of the formation of polysulfide ions (HS_n^-, S_n^{2-}, n> 1) was emphasized. These ions are rapidly formed upon reaction of hydrogen sulfide with elemental sulfur or with incomplete oxidation of hydrogen sulfide.

Equilibria in the H_2S-S_8-H_2O systems are rapid and reversible and the ratio of the concentrations of polysulfide to hydrogen sulfide imposes the redox potential. Since these species are electroactive, they generally give a measurable potential, and thus enable an assessment of environmental conditions.

Polysulfide ions can play an important role in the transfer of sulfur from the sediment to the supernatant waters. Owing to easy dismutation, they may also yield elemental sulfur which is relatively inert compared to other sulfur species in the geochemical cycle.

We present here a brief survey of the chemistry and geochemistry of reduced sulfur species of importance in upwelling areas. The laboratory data are then compared with field data from Walvis Bay. These results allow a characterization of sulfur speciations in upwelling areas and their possible impact on the sediment record.

ANALYSIS OF SULFUR SPECIES

Sulfur species, within the sediment, are involved in reactions between pore waters and minerals. Much information concerning the sulfur species and the thermodynamic state of the sediment can be deduced from the composition of the interstitial waters. This requires specific handling of the sediment and analytical procedures.

Numerous sulfur species are very reactive upon contact with air. Thus, the pore waters must be extracted under inert atmosphere with gas-tight instrumentation. Environmental parameters: must be measured immediately after extraction of the pore water: pH, rest potential of the platinum electrode (expressed as pe in the following) and potential of a sulfide specific electrode ($E_{S^{2-}}$). Titrations of the sulfur species must also be done immediately to prevent possible bacterial interference.

The following species should be expected in the pore waters and the overlying waters of anoxic sediments: hydrogen sulfide (H_2S, HS^-), organo sulfur compounds (RSH, RS_mR', m>1), polysulfide ions (HS_n^-, S_n^{2-}, n≥1), elemental sulfur (S_8), thiosulfate ($S_2O_3^{2-}$), sulfite (HSO_3^-, SO_3^{2-}) and sulfate (SO_4^{2-}). Classical methods, such as iodometric titrations, sulfide specific electrode potentiometry and colorimetric measurements cannot detect the difference between most of the above species due to their non-specificity, or they are time consuming or not precise enough. Most of them also involve large volumes of water (more than 10 ml).

The use of the Ag/Ag_2S membrane electrode in combination with mercuric ion titration under a nitrogen atmosphere enables a rapid and accurate determination of the above species except for sulfate (Boulégue and Popoff, 1979; Boulégue et al., 1979). It can be employed very easily in the field study of anoxic sediments (Boulègue et al., 1982). The identification of the sulfur species and the equilibrium existing between them must be completed by rest potential measurements combined with UV spectroscopic measurements (Boulègue, 1978a; 1978b; Boulègue and Michard, 1979; Boulègue et al., 1982).

The main inorganic sulfides, elemental sulfur, organically bound sulfur as well as carbonate minerals and organic matter can be quantitatively determined in one run on the solid fraction of the sediment by differential thermal analysis combined with gas analyses (Chantret, 1971; 1973; Chantret and Boulègue, 1980).

CHEMISTRY AND GEOCHEMISTRY OF THE SULFUR SYSTEM IN ANOXIC WATERS

The H_2S-S_8-H_2O System

In aqueous solutions, hydrogen sulfide, elemental sulfur and sulfate are the stable species of sulfur (Valensi, 1950; Garrels and Naeser, 1958). However, these species only reach equilibrium in extreme conditions rarely found in the sediments. Hydrogen sulfide and sulfate are chemically non-reactive with each other in sedimentary conditions. Only bacterial processes can catalyze their transformation. In the absence of dissolved oxygen one can consider that metastable equilibrium is established between hydrogen sulfide, polysulfide and thiosulfate, and be maintained for periods of time much larger than the time of residence of water in a given natural system. Thus, the H_2S-S_8-H_2O system constitutes a good chemical model of sulfur in anoxic waters.

Table 1. Thermodynamic equilibrium data for the H_2S-$S_8(\alpha)$-H_2O system at 298°K

Reaction	Boulègue (1978a)	Cloke (1963)	Teder (1969)
$(1/4)\ S_8 + HS^- \rightleftarrows S_3^{2-} + H^+$	pK_3 = 12.5	13.19	11.75
$(3/8)\ S_8 + HS^- \rightleftarrows S_4^{2-} + H^+$	pK_4 = 9.52	9.74	10.07
$(1/2)\ S_8 + HS^- \rightleftarrows S_5^{2-} + H^+$	pK_5 = 9.41	9.50	9.41
$(5/8)\ S_8 + HS^- \rightleftarrows S_6^{2-} + H^+$	pK_6 = 9.62	9.79	9.43
$HS^-_4 \rightleftarrows S_4^{2-} + H^+$ $pK_{H/4}$ = 7.0	Schwarzenbach and Fischer (1960)		
$HS^-_5 \rightleftarrows S_5^{2-} + H^+$ $pK_{H/5}$ = 6.1	Schwarzenbach and Fischer (1960)		

Thermodynamic and electrochemical study of the corresponding systems H_2S-H_2O and H_2S-S_8-H_2O has been done by Cloke (1963), Teder (1969), Giggenbach (1972), Boulègue (1976; 1978b; 1978c), Boulègue and Michard (1978; 1979). The main thermodynamic data are given in Table 1. Polysulfide ions are produced according to the general reaction

$$(n-1)/8 \quad S_8 + HS^- \rightleftarrows S_n^{2-} + H^+ \qquad (1).$$

As shown in Table 2 and in Fig. 1, they can constitute a large fraction of the total reduced sulfur species.

The physical state of elemental sulfur in the H_2S-S_8-H_2O system is one of the key parameters of the amount of polysulfide forming in

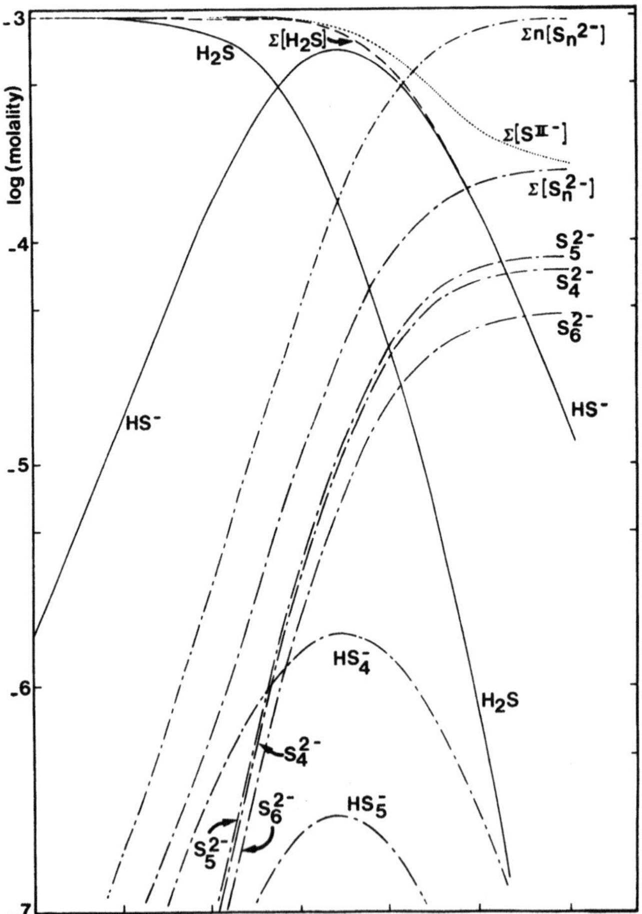

Fig. 1. Distribution of sulfur species in the system H_2S-$S_8(\alpha)$-seawater for $\Sigma[S] = 10^{-3}$ g-atom/kg • log (molality) versus pH at 298°K.

Table 2. Characteristic parameters for the H_2S-$S_8(\alpha)$-seawater system and the H_2S-S_8 (colloid)-seawater system at 298°K

$\bar{n} = \Sigma n[S_n^{2-}]/\Sigma[S_n^{2-}]$; $\chi_S = \Sigma(n-1)[S_n^{2-}]/\Sigma[S^{II-}]$;

$B = [HS^-].(OH^-)/\Sigma(n-1)[S_n^{2-}]$; where square brackets correspond to concentrations; $\Sigma[S^{II-}] = \Sigma[H_2S] + \Sigma[S_n^{2-}]$

pH	4	5	6	7	8	9	10
$\bar{n}$	4.14	4.24	4.38	4.74	4.85	4.86	4.86
χ_S	0.000014	0.00015	0.0019	0.038	0.457	2.248	3.605
10^6xB	0.011	0.107	0.714	1.597	1.820	1.848	1.851
$\bar{n}$	4.16	4.28	4.45	4.88	4.95	4.96	4.97
χ_S	0.000025	0.00026	0.0036	0.079	0.834	2.920	3.829
10^6xB	0.006	0.062	0.383	0.768	0.892	0.903	0.904

the system. Elemental sulfur is mostly found as stable rhombic sulfur (S_8[a]) and metastable colloidal sulfur (S_8[c]). The Gibbs free energy of $S_8[\alpha]$ is zero, that of S_8[c] is larger than 3.5 kJ•mole^{-1} (Boulégue, 1976; 1978c). Thus, according to reaction (1) the amount of polysulfide ions in the presence of colloidal elemental sulfur will be at least twice the amount in the presence of rhombic sulfur (see Table 2). This effect has been found to be effective in the field (Boulégue, 1977; 1978b; Boulégue et al., 1982). In natural environments, on the other hand, it is colloidal elemental sulfur which is found. The oxidation of hydrogen sulfide by oxygen, bacteria, MnO_2 and iron(III) minerals is the main source of colloidal elemental sulfur.

Polysulfide ions are important intermediaries in the geochemical cycle of sulfur in sedimentary environments. The production of pyrite involves polysulfide as the reaction ions (Sweeney and Kaplan, 1973; Rickard, 1975). Elemental sulfur acts as a reservoir of potentially reactive sulfur via its reaction with hydrogen sulfide to yield polysulfide.

Oxidation of Hydrogen Sulfide

The oxidation of hydrogen sulfide can be expected as soon as it is in contact with oxidants. In upwelling areas oxidation is most likely by ascent of bubbles of hydrogen sulfide through oxic waters or through the mixing of anoxic and oxic waters at redox boundaries. The oxidation reaction can be mediated and catalyzed by bacterial processes. Kinetic data on the oxidation of hydrogen sulfide can be found in Avrahami and Golding (1968), Cline and Richards (1969),

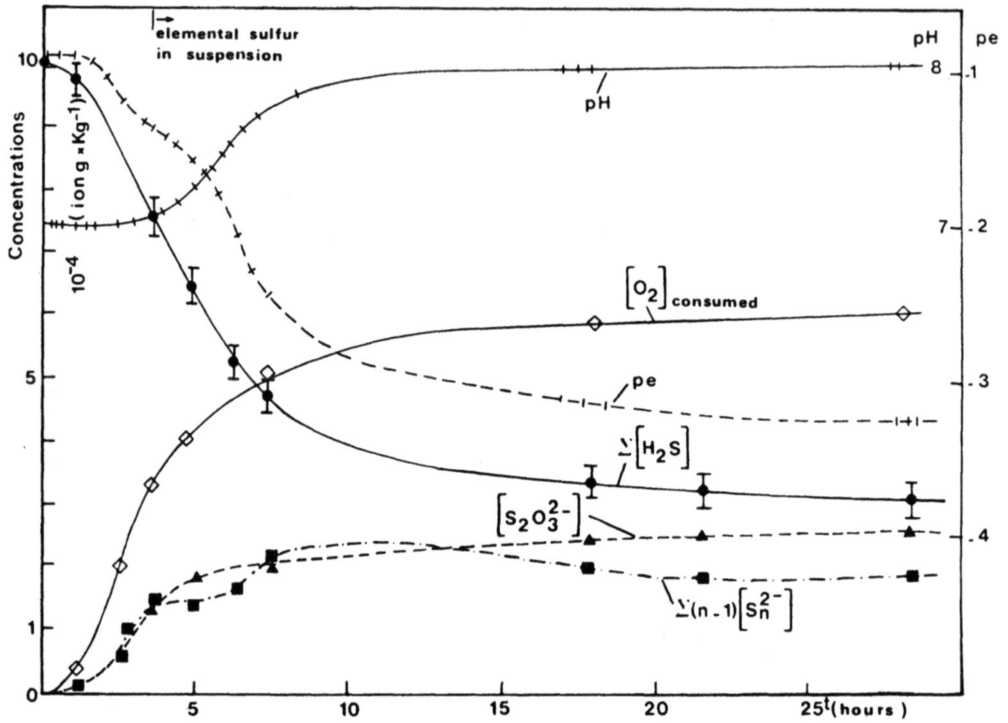

Fig. 2. Incomplete oxidation of hydrogen sulfide by oxygen in NaCl (0.7 M). Initial conditions: $\Sigma[H_2S] = 1.05 \times 10^{-3}$ mol/kg; $[O_2] = 6.9 \times 10^{-4}$ mol/kg; pH = 7.0; T = 288°K.

Koundo and Keier (1970), Chen and Morris (1972), Boulégue (1972) and Gourmelon et al. (1977). However, most of these studies are incomplete due to inadequate analyses of reaction products.

Polysulfide and elemental sulfur are important products in seawater during the incomplete oxidation of hydrogen sulfide, as can be found at redox boundaries (see Fig. 2). They are important intermediaries, along with thiosulfate, during the complete oxidation to sulfate. A diagram of the oxidation process can be given according to Boulégue (1972) and Chen and Morris (1972) (see Fig. 3). In this process, it appears that elemental sulfur is the only product which can be removed from the reacting system by sedimentation. This can be of importance in natural systems.

If the production of hydrogen sulfide is large enough to consume most of the oxygen in a given area, it can lead to the appearance of hydrogen sulfide, polysulfide ions, elemental sulfur and thiosulfate at the surface of the sea. Such a phenomenon has been described in the Walvis Bay area (Pieterse and van der Post, 1967). In this case "discoloration of seawater" can be attributed to the formation of milky, colloidal elemental sulfur.

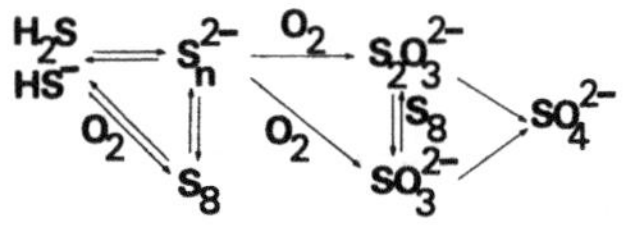

Fig. 3. Reaction scheme for the complete oxidation of hydrogen sulfide by oxygen.

Formation of Metal Sulfides

The most important sulfides forming in anoxic environments are iron sulfides: amorphous iron sulfide (FeS), greigite (Fe_3S_4) and pyrite (FeS_2). Solubility diagrams for most transition metal sulfides in the H_2S-H_2O and H_2S-S_8-H_2O systems can be found in Boulègue (in press). The kinetics of formation of FeS and FeS_2 have been studied by Rickard (1974; 1975). Polysulfide ions are necessary intermediaries between FeS and FeS_2 and during the direct formation of FeS_2. Most often the reaction of the formation of FeS_2 involves preliminary formation of FeS which is then transformed into Fe_3S_4; then the passage from Fe_3S_4 to FeS_2 is generally a solid state controlled process (Sweeney and Kaplan, 1973; Boulègue et al., 1982). The kinetics of formation of polysulfide ions are more rapid than the kinetics of formation of pyrite (Rickard, 1975).

Consideration of the kinetics is of importance since the formation of FeS and FeS_2 depends on them. In numerous sediments where FeS_2 is observed to form and where it constitutes the major sulfur and iron phase, the concentrations of iron and sulfide correspond to large oversaturation with respect to pyrite, while being close to saturation with FeS or Fe_3S_4. These last minerals are only observed as secondary minerals. However, since their kinetics of formation is rapid and since the formation of pyrite mostly involves solid state processes, they control the concentrations of iron in the pore waters. Possibilities for application of these results as well as their limitations will be discussed in the following with the help of field data from the Walvis Bay area.

GEOCHEMISTRY OF SULFUR AND SOME TRACE METALS IN THE WALVIS BAY AREA

During the expedition Walda (May-June 1971) of RV "Jean Charcot", opportunity was given to sample sediments and waters from the Walvis Bay area (Boulègue, 1974). The floor of the bay is an extension of the so-called azoic zone which extends patchily for some 300 km along the coast of southwest Africa and from the shore line to approximately 50 km west. The floor of this zone is covered with a thick layer of dark green mud mostly consisting of a sulfide diatomaceous ooze (Copenhagen, 1953; Calvert and Price, 1971). The sampling location (22°59'S; 14°28'E) is in an area of dark mud, rich in organic matter and with a strong smell of hydrogen sulfide. The top centimeters of the sediment show enrichment in elemental sulfur. In places, yellow patches of elemental sulfur can be seen on the surface of the sediment. They are sometimes associated with bacterial veils (*Beggiatoaceae*, *Thiobacteriaceae*).

Sampling and Analyses

The samples consist of (1) surface seawater: WB1; (2) water from the sediment interface with varying amounts of overlying water: WB3a: 5 cm, WB4a: 10 cm and WB5a: 25cm; (3) interstitial water from the top 5 cm of the sediment at the same sampling locations as the interface water: WB3b, WB4b and WB5b. The pore waters were extracted under nitrogen pressure with gas-tight instrumentation. All analyses were done within hours of sampling on the RV "Jean Charcot". Hydrogen sulfide, zero valence sulfur of polysulfide ($\Sigma[n-1][S_n^{2-}]$) and thiosulfate were also measured by UV spectroscopy; pH, rest potentials of a Pt electrode (pe) and of a sulfide electrode were also measured. These last measurements were also done by direct insertion of the electrodes into the sediment prior to pore water extraction. These data enabled us to compute the sum of hydrogen sulfide and polysulfide species and to check their state of equilibrium (Boulégue and Michard, 1979). Elemental sulfur and organo-sulfur compounds were extracted with chloroform and assessed by UV spectroscopy. Organo-sulfur compounds represented about 5% of the total dissolved reduced sulfur species in WB4 and WB5. Manganese and Fe were analyzed by AAS on filtered water samples (0.1 μm). The results of the analyses are given in Table 3.

Discussion of Sulfur Speciations

All interface and pore water samples contained hydrogen sulfide, polysulfides and elemental sulfur. A comparison of the measured concentrations with the pe-pH data (see Fig. 4) shows that they are in

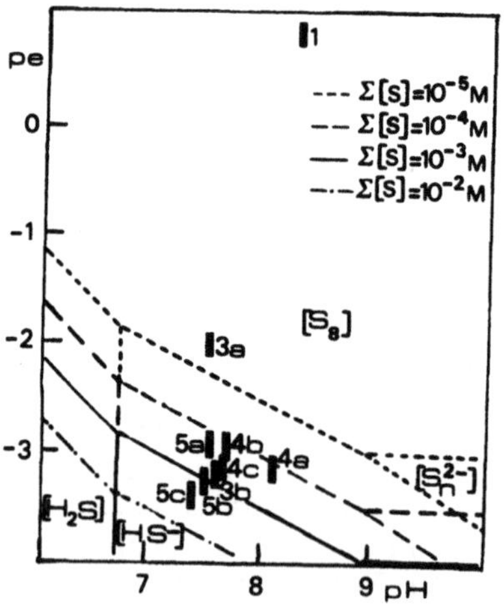

Fig. 4. pe-pH diagram for the H_2S-S_8-seawater system at different total dissolved sulfur concentrations ($\Sigma[S]$). The measured values for the Walvis Bay samples are given by black rectangles corresponding to the uncertainties of the measurements.

Table 3. Environmental parameters, concentrations of sulfur species and trace metals in samples from Walvis Bay (Namibia). All concentrations are given in 10^{-4} mole/kg seawater; (+) in S_8-column indicates the presence of elemental sulfur, (*) correspond to data computed from pe-pH and E_S^{2}-measurements.

Sample	Description	pH	pe	$\Sigma[S^{II-}]$	$\Sigma[H_2S]$	$\Sigma[S_n^{2-}]$	$\Sigma(n-1)[S_n^{2-}]$	S_8	$[S_2O_3^{2-}]$	[Fe]	[Mn]
WB1	surface w.	8.26	0.84							$<10^{-5}$	0.005
WB3a	interface w.	7.55	-2.0	0.015	0.012	0.003	0.0012	(+)	0.15	0.005	0.035
WB3b	interstit.w.	7.60	-3.15	5.51	5.24	0.27	1.05	(+)	0.15	0.007	0.050
WB4a	interface w.	8.10	-3.1	1.50	1.15	0.35	1.40	(+)	1.45	$5x10^{-4}$	0.033
WB4b	interstit.w.	7.7	-2.9	2.50	2.35	0.15	0.60	(+)	0.25	0.013	0.050
WB4c	sediment	7.67	-3.1	3 (*)				(+)			
WB5a	interface w.	7.57	-2.85	2.51	2.26	0.25	0.95	(+)	1.25	$5x10^{-4}$	0.040
WB5b	interstit.w.	7.50	-3.25	26.93	25.90	1.03	3.85	(+)	0.25	0.0026	0.087
WB5c	sediment	7.37	-3.35	30 (*)				(+)			

good agreement. Each pe-pH couple corresponds to a range of total dissolved reduced sulfur (Σ[S]) concentrations in the H_2S-$S_8(\alpha)$-seawater system (Boulégue and Michard, 1979). The measured concentrations of hydrogen sulfide and polysulfide are in a range corresponding to the pe-pH data. This is indicative of the consistency of the results. It also shows that the redox potential can be determined in such anoxic environments with the concentrations of hydrogen sulfide and polysulfides (Boulégue and Michard, 1979).

The origin of hydrogen sulfide in the sediment is due to bacterial sulfate reduction and anaerobic microbiological breakdown of organic substances (Stander, 1962). The hydrogen sulfide in the interface water can result from diffusion out of the underlying sediment and/or local production. The observation of sulfide oxidizing bacteria (*Beggiatoaceae, Thiobacteriaceae*) makes the first alternative more plausible.

A closer examination of the results of the titrations shows that in the pore waters the amount of polysulfide ions is close to that found in the H_2S-$S_8(\alpha)$-seawater system. The presence of polysulfides in the pore water may be due to an input of elemental sulfur owing to the sulfide oxidizing bacterial activity and aging of the colloidal sulfur. A limited input of oxygen could also have been possible since little burrowing activity was observed at the sampling locations. This could explain the low concentrations of thiosulfate found in the pore waters.

In the interface waters the number of polysulfide ions is close to that predicted in the presence of colloidal sulfur. The stoichiometry of the ratio of polysulfide to thiosulfate is close to the one observed during the incomplete oxidation of hydrogen sulfide (see Fig. 2). Thus, in the interface waters the repartition of sulfur species should result from a steady-state of the oxidation process of hydrogen sulfide diffusing from the sediment.

Trace Metals: Fe and Mn

The concentrations of iron and manganese observed in the pore and interface waters are in the range expected for anoxic environments. They show an enrichment as compared to the oxic surface water of Walvis Bay. The low concentrations observed in oxic waters generally correspond to control by the oxide phase. In anoxic sediments, one may expect control by the sulfide, carbonate or phosphate phase.

In the presence of the H_2S-S_8-H_2O system, one should expect control of dissolved iron by sulfide formation (Fe_3S_4 or FeS) as documented by field data from various environments (Boulégue, 1977; 1978a; in press; Boulégue et al., 1979; 1982). These possibilities have been tested as can be seen in Figs. 5 and 6. Details of calculations can be found in Boulégue (1977; 1978a; in press) and

Boulégue et al. (1982). The pore waters are very close to saturation with respect to amorphous FeS and supersaturated with respect to Fe_3S_4. Since pyrite is found in the Walvis Bay sediments investigated, this is an example of the preceeding predictions on the formation of metal sulfides.

The interface waters are all undersaturated with respect to FeS and Fe_3S_4 and supersaturated with respect to FeS_2. They also are undersaturated with respect to iron carbonate and vivianite in the conditions encountered in Walvis Bay. It is possible that the con-

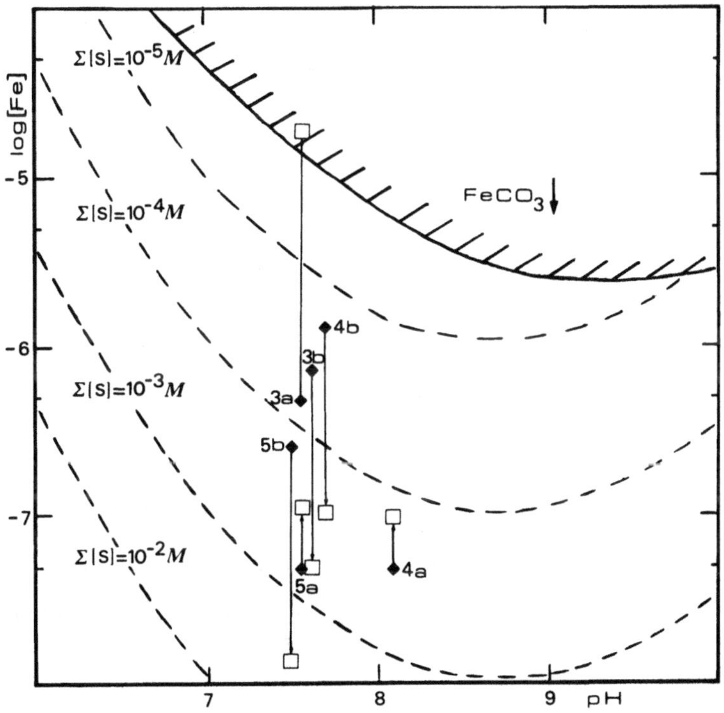

Fig. 5. Solubility diagram of greigite (Fe_3S_4) in the H_2S-S_8-seawater system. The dotted lines correspond to saturation for a given total dissolved sulfur concentration (Σ[S]) in the H_2S-S_8-seawater system (Boulégue, in press). The black diamonds correspond to measured concentrations of Fe in the Walvis Bay samples. The open squares correspond to the saturation of Fe with respect to Fe_3S_4 formation in the conditions imposed by the sulfur species and other possible ligands in the Walvis Bay samples. The arrow between the diamond and the square for a given sample indicates undersaturation when pointing up and supersaturation when pointing down. The corresponding saturation indexes are given by the length of the arrow measured on the concentration scale.

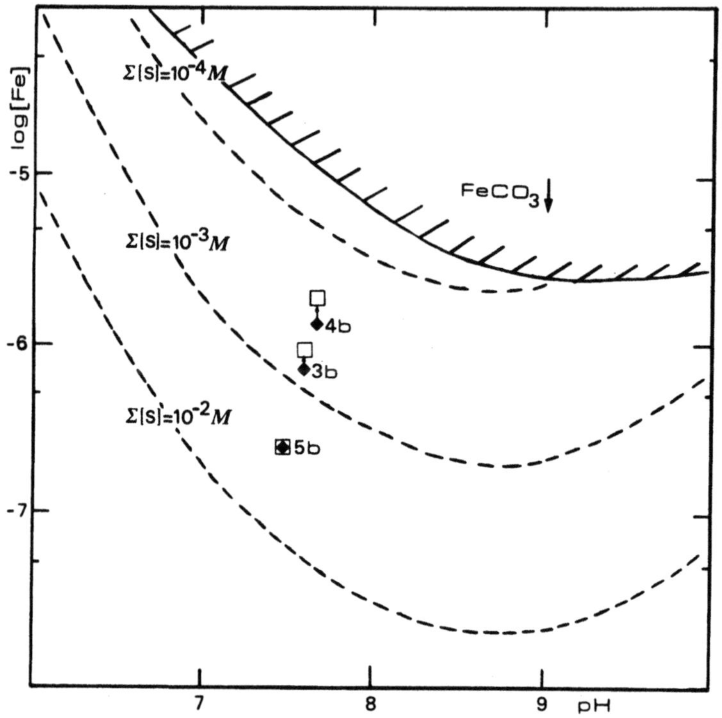

Fig. 6. Solubility diagram of amorphous iron sulfide (FeS) in the H_2S-S_8-seawater system. Other headings are the same as in Fig. 5.

centrations of iron in the interface waters are controlled by diffusion from the sediment. Control by oxidation of dissolved iron(II) is also possible but less likely. The Gibbs free energy of oxidation of iron(II) is less favorable than that of hydrogen sulfide, as is the kinetic constant (Stumm and Morgan, 1970; Gourmelon et al., 1977). Thus, the oxidation of dissolved iron should be limited in the presence of hydrogen sulfide.

Going from the pore waters to the interface waters, manganese exhibits the same behavior as iron. Manganese did not appear to be controlled by any solid phase. The similar behavior of Fe and Mn signifies that their concentrations within the interface waters are probably controlled by diffusion from the sediment and, in the case of iron, by limited oxidation.

SULFIDE SPECIATIONS AND THE SEDIMENT RECORD IN UPWELLING AREAS

One should expect that the sulfur speciations in the anoxic sediments of upwelling areas will exhibit the general characteristics of anoxic sediments. Among the sulfur species, certain ones may have relatively more importance in upwelling areas and may be traced back during diagenesis. Some of these characteristics are examined in the following.

Hydrogen sulfide, polysulfide and organo-sulfur compounds will be found in the pore waters of upwelling areas and may diffuse into the overlying waters. The organic-rich sediments of upwelling areas can be expected to be especially enriched in organo-sulfur compounds. Methylsulfide, dimethylsulfide, dimethyldisulfide, and cysteine will result from the microbial breakdown of proteins. Organo polysulfide compounds will also be found and are especially enriched due to elemental sulfur reaction with organic matter (Boulégue et al., 1982). Data on these species in upwelling areas are not available at present. The processes leading to their formation and their accumulation have been documented in the study of organic-rich sediments from salt marsh areas, however (Boulégue et al., 1982).

Rapid formation of metal sulfides can also be expected in upwelling regions. Metals that are enriched in diatoms such as Fe, Ni, Cu, Zn, Pb (Vinogradova and Koval'skiy, 1962; Calvert and Price, 1971) will be able to coprecipitate with iron sulfide or form their own metal sulfides.

The presence of elemental sulfur, even in very low amounts, is a general characteristic of reducing environments. It can be strongly enriched in upwelling areas. Elemental sulfur results from the chemical and microbiological oxidation of hydrogen sulfide in the sediment and in the overlying waters. The sediment is enriched by sedimentation of elemental sulfur from the overlying waters. There it can react with hydrogen sulfide to yield polysulfide ions. They will, in turn, react with organic matter leading to noticeable enrichment of sulfur within the humic fractions. Such processes have been observed in an anoxic basin off the coast of California (Nissenbaum and Kaplan, 1972), in swamps (Casagrande and Ng, 1979) in the sediments off Peru (Reimers and Suess, this volume), and in salt marshes (Boulégue et al., 1982). Unreacted elemental sulfur will be stocked in the sediment and be available for eventual reaction during diagenesis.

During diagenesis the role of elemental sulfur and polysulfides (organic and inorganic) will be important and characteristic in several respects.

(i) In the pore waters sulfur may be transported in large quantities in the polysulfide form (e.g., S_5^{2-} and S_6^{2-}). Inorganic polysulfides can also play a role in the migration of some trace metals owing to their good complexing ability (Cu^+, Ag^+), as can organic polysulfides with other trace metals (Cu^+, Zn^{2+}, Pb^{2+}, Hg^{2+}). Thus, some trace metals may be redissolved, transported via ion diffusion as polysulfide complexes then redeposited after disproportionation of the polysulfides in the regions where different environmental conditions are found. Since the difference in chemical potentials which brings about diffusion is due to the difference in environmental conditions, this is a natural process which does not call for any other special explanation. An irreversible thermodynamic treatment of such processes is available in Boulégue and Michard (1979). Such processes will lead to the separation of the initially deposited metal sulfides. They are not limited by the amount of metal but by the amount of complexing agent able to sustain their migration. In the presence of elemental sulfur, the production of polysulfide ions will be sustained and leads to separation and local enrichment of metal sulfides.

(ii) The presence of elemental sulfur in the sediment also brings about a large enrichment of sulfur in the organic fraction. This may be one of the reasons for the post-diagenetic enrichment of some petroleum reservoirs in hydrogen sulfide and organo-sulfur compounds.

In conclusion, one can see that the laboratory and field data available today show the importance of the organic and inorganic polysulfides in the sulfur cycle of anoxic sediments. Their role can be emphasized in areas, such as those of upwelling, where elemental sulfur is expected in significant amounts. The processes of speciation, re-distribution and solid sulfide formation have just begun to be investigated and much remains to be learned of the fate of the sulfur species during diagenesis.

ACKNOWLEDGEMENTS

M. Van Rijswijck has helped with sampling in Walvis Bay. Dr. A. du Plessis has given important information about the coast off Namibia. The field study in Walvis Bay was funded by a grant from CNEXO.

REFERENCES

Avrahami, M. and Golding, R.M., 1968, The oxidation of the sulphide ion at very low concentrations in aqueous solution, Journal of the Chemical Society (A), 647-651.

Boulégue, J., 1972, Mise en évidence et dosage des polysulfures au cours de l'oxydation de l'hydrogéne sulfuré dans l'eau de mer, Comptes Rendus de l'Académie des Sciences de Paris, 275(C): 1335-1338.

Boulégue, J., 1974, Mesures électrochimiques en milieu réducteur: la lagune de Walvis Bay (R.A.S.), Compte Rendu de l'Académie des Sciences de Paris, 278(D):2723-2726.

Boulégue, J., 1976, Equilibres dans le systéme H_2S-S_8 (colloide)-H_2O, Comptes Rendus de l'Académie des Sciences de Paris, 283(D):591-594.

Boulégue, J., 1977, Equilibria in a sulfide rich water from Enghien-les-Bains, Geochimica et Cosmochimica Acta, 41:1751-1758.

Boulégue, J., 1978a, Metastable sulfur species and trace metals (Mn, Fe, Cu, Zn, Cd, Pb) in hot brines from the french Dogger, American Journal of Sciences, 278:1348-1411.

Boulégue, J., 1978b, Electrochemistry of reduced sulfur species in natural waters. I - The H_2S-H_2O system, Geochimica et Cosmochimica Acta, 42:1439-1445.

Boulégue, J., 1978c, Solubility of elemental sulfur in water at 2980°K, Phosphorus and Sulfur, 5:127-128.

Boulégue, J., in press, Trace metals (Fe, Cu, Zn, Cd) in anoxic environments, in: "Trace Metals in Sea Water", C.S. Wong, ed., Plenum Press, New York.

Boulégue, J. and Michard, G., 1978, Constantes de formation des ions polysulfures S_6^{2-}, S_5^{2-} et S_4^{2-} en phase aqueuse, Journal Francais d'Hydrologie, 9:27-34.

Boulégue, J. and Michard, G., 1979, Sulfur speciations and redox processes in reducing environments, in: "Chemical Modeling in Aqueous Systems," E.A. Jenne, ed., A.C.S. Symposium Series 93, Washington, 25-50.

Boulégue, J. and Popoff, G., 1979, Nouvelles méthodes de détermination des principales espéces ioniques du soufre dans les eaux naturelles, Journal Francais d'Hydrologie, 10:83-90.

Boulégue, J., Ciabrini, J.P., Fouillac, C., Michard, G., and Ouzounian, G., 1979, Field titrations of dissolved sulfur species in anoxic environments, Geochemistry of Puzzichello waters (Corsica, France), Chemical Geology, 25:19-29.

Boulégue, J., Lord, C.J. and Church, T.M., 1982, Sulfur speciations and associated trace metals (Fe, Cu) in the pore waters of Great Marsh, Delaware, Geochimica et Cosmochimica Acta, 46:453-464.

Calvert, S.E. and Price, N.B., 1971, Recent sediments of the South West African shelf, in: "The Geology of the East Atlantic Continental Margin," F.M. Delaney, ed., Institute of Geological Sciences Report, No. 70/16, Natural Environmental Research Council, Cambridge, 175-195.

Casagrande, D.J. and Ng, L., 1979, Incorporation of elemental sulfur in coal as organic sulfur, Nature, 282:598-599.

Chantret, F., 1971, Analyseur thermique pour de dosage simultané du SO_2 et du CO_2, "Thermal Analysis", Burkhauser Verlag, Basel, 313-324,.

Chantret, F., 1973, Étude des minerais d'uranium par analyse thermique, Bulletin d'Information Technique, C.E.A., 181:25-31.

Chantret, F. and Boulegue, J., 1980, Mise en évidence de soufre élémentaire et de sidérite dans les gisements d'uranium de type roll, Comptes Rendus de l'Académie des Sciences de Paris, 290(D):73-76.

Chen, K.Y., and Morris, J.C, 1972, Kinetics of oxidation of aqueous sulfide by O_2, Environmental Sciences and Technology, 6:529-537.

Cline, J.D. and Richards, F.A., 1969, Oxygenation of hydrogen sulfide in sea water at constant salinity, temperature and pH, Environmental Science and Technology, 3:838-843.

Cloke, P.L., 1963, The geologic role of polysulfides. I - The distribution of ionic species in aqueous sodium polysulfide solutions, Geochimica et Cosmochimica Acta, 27:1265-1298.

Copenhagen, W., 1953, The periodic mortality of fish in the Walvis region, Investigation Report, Division of Sea Fishing, South Africa, 14, 35 pp.

Garrels, R.M. and Naeser, C.R., 1958, Equilibrium distribution of dissolved sulphur species in water at 25°C and 1 atm total pressure, Geochimica et Cosmochimica Acta, 15:113-130.

Giggenbach, W., 1972, Optical spectra and equilibrium distribution of polysulfide ions in aqueous solution at 20°, Inorganic Chemistry, 11:1201-1207.

Gourmelon, G., Michard, G. and Boulégue, J., 1977, Oxydation partielle de l'hydrogéne sulfuré en phase aqueuse, Comptes Rendus de l'Académie des Sciences de Paris, 284(C):269-272.

Koundo, N.N. and Keier, N.P., 1970, Catalytic activity of phtalocyanins during the oxidation of hydrogen sulfide in aqueous phases, Kinetic i Katalysi, 2:91-99 (in Russian).

Nissenbaum, A. and Kaplan, I.R., 1972, Chemical and isotopic evidence for the *in situ* origin of marine humic substances, Limnology and Oceanography, 17:570-582.

Pieterse, F. and van der Post, D.C., 1967, The pilchard of South Africa, Oceanographical conditions associated with red-tides and fish mortalities in the Walvis Bay region, South West Africa Marine Research Laboratory Investigation Report, 14, 125 pp.

Rickard, D.T., 1974, Kinetics and mechanism of the sulfidation of goethite, American Journal of Science, 274:941-952.

Rickard, D.T., 1975, Kinetics and mechanism of pyrite formation at low temperature, American Journal of Science, 275:636-652.

Schwarzenbach, G. and Fischer, A., 1960, Die Acidität der Sulfane und die Zusammensetzung wässeriger Polysulfidlösungen, Helvetica Chimica Acta, 43:1365-1390.

Stander, G.H., 1962, The pilchard of South West Africa. General hydrography of the waters off Walvis Bay, South West Africa, 1957-1958, South West Africa Marine Research Laboratory Investigational Report, 5, 63 pp.

Stumm, W. and Morgan, J.J., 1970, "Aquatic Chemistry," Wiley Interscience, New York, 583 pp.

Sweeney, R.E. and Kaplan, I.R., 1973, Pyrite framboid formation : laboratory systems and marine sediments, Economic Geology, 68: 618-634.

Teder, A., 1969, The spectra of polysulfide solutions, part II - The effect of alkalinity and stoichiometric composition of equilibrium, Arkiv Kemi, 31:173-198.

Valensi, G., 1950, Contribution au diagramme potentiel - pH du soufre, Comptes Rendus 2éme réunion Circulaire d'Informations Techniques, Centre Education, Milan, Polytechnica, 51-68.

Vinogradova, Z.A. and Koval'skiy, V.V., 1962, Elemental composition of the Black Sea plankton, Doklady Akademii Nauk. SSSR, 147:217-219 (in Russian).

EXCESS TH-230 IN SEDIMENTS OFF NW AFRICA TRACES UPWELLING IN THE PAST

Augusto Mangini

Institut für Umweltphysik der Universität
Im Neuenheimer Feld 366
D-69 Heidelberg, Federal Republic of Germany

Liselotte Diester-Haass

Geographisches Institut, Universität des Saarlandes
D-66 Saarbrücken, Federal Republic of Germany

ABSTRACT

Core sections deposited during periods of upwelling conditions off northwestern Africa received deposition fluxes of Th(ex)-230 and Pa(ex)-231 which exceed production in the water column up to four times. These high flux rates coincide with Th(ex)-230/Pa(ex)-231 ratios approaching three. From these findings we conclude that in upwelling areas scavenging of Th-230 and Pa-231 by particulates is more effective than in the open ocean environment such that deposition appears to be controlled by the kinetic scavenging process and not by the amount of available Th-230 in the water column. The highest nuclide fluxes occurred during the glacial O-18 stages 2 and 4 and were accompanied by increased opal content, the latter being a sensitive indicator of increased fertility. Therefore, Th-230 geochemistry might have recorded upwelling during the last 200,000 years. Our study shows that Th-230 not only may be used as a tool for sediment dating, but also as a potentially useful tracer for upwelling when applied in combination with other sedimentological parameters, such as oxygen isotopes.

INTRODUCTION

Ocean areas with increased bioproductivity are known to play a key role in the geochemistry of dissolved and particle reactive elements (Turekian, 1977; Demaster, 1979; Anderson, 1981). Higher bioproductivity leads to higher suspended particulate matter concentrations and thereby to higher than usual deposition rates of organic matter into sediments (Müller and Suess, 1979; Suess, 1980). The consequences are increased scavenging of particle reactive elements, such as Pb, Th and Pa, as well as increased incorporation of dissolved elements, such as U and Mo, into the reducing sedimentary environment (Veeh, 1967; Bertine and Turekian, 1973; Mangini, 1978). Therefore, sediments from upwelling areas offer a unique opportunity to evaluate the efficiency of these sinks for anthropogenic pollutants and to better understand the geochemical pathways of elements in the oceans.

In this context Th-230 and Pa-231 are interesting tracers since their production in the oceans, by decay of dissolved U-234 and U-235, has been more or less constant over the last 300,000 years due to the long residence time of uranium in the ocean (Mangini et al., 1979). Production may be considered homogeneous throughout the water column due to the homogeneous distribution of uranium (Ku et al., 1977). The latter assumption is not a trivial one for many other tracers, where an anthropogenic component is difficult to separate from a natural one and where, due to climatic changes, variations in supply of the natural components cannot be excluded. We have studied Th-230 and Pa-231 distribution in a water column profile and in four cores taken off northwestern Africa, two of them from areas where upwelling is known to have occurred in the past. Excess Th-230 and Pa-231 represent the amounts of Th-230 and Pa-231 which have been produced in the water column and scavenged by particles. Their fluxes into the sediments are then used as indicators for scavenging efficiency during different environmental conditions.

DATA

The locations of the sediment cores and the water profile are given in Table 1. Tables 2-5 display depth intervals of the cores

Table 1. Location of the studied cores and water profile

Core No.	Location		Water Depth (m)
135 19	5°39.5'N	19°51.0'W	2862
123 29	19°22' N	19°55.8'W	3320
123 10	23°29.9'N	18°43.0'W	3080
123 09	26°50.3'N	15°06.6'W	2760
Station 513 Th-230 profile	27°45' N	27 W	5450

Table 2. ^{230}Th and ^{231}Pa in Core 12309-3

Depth (cm)	$\delta^{18}O$ Stage	Age (kyrs)	Sedimentation Rate ($cm \cdot ky^{-1}$)	Nuclide Activities $^{230}Th_{ex}$ ($dpm \cdot g^{-1}$)	Nuclide Activities $^{231}Pa_{ex}$ ($dpm \cdot g^{-1}$)	Age Corrected $^{230}Th_{ex}$ ($dpm \cdot g^{-1}$)	Age Corrected $^{231}Pa_{ex}$ ($dpm \cdot g^{-1}$)	Ratio ($\frac{Th}{Pa}$)	Flux Rates $^{230}Th_{ex}$ ($dpm \cdot g^{-1} \cdot ky^{-1}$)	Flux Rates $^{231}Pa_{ex}$ ($dpm \cdot g^{-1} \cdot ky^{-1}$)
79-81	2	20,000	12.1*	3.13±0.3	0.91±0.30	3.75	1.37	2.73	average	10.8
99-101		23,000		3.41±0.3	0.73±0.60	4.2	1.15	3.65		9.04
Average						4			29.0	
119-121	3	25,000	9.1*	2.78±0.3		3.22				
158-160		28,000		2.56±0.2	0.60±0.50	3.31	1.02	3.2		5.9
189-191		31,600		2.22±0.2		2.97			average	
210-217		35,500		2.95±0.3		4.09				
248-250		39,400		2.60±0.2	0.29±0.20	3.74	0.63	6.2	20.2	3.68
290-292		46,700		2.78±0.3		4.28				
330-332		52,700		2.47±0.2		4.02				
350-352		55,700		2.12±0.2	0.49±0.14	3.54	1.45	2.44		8.5
392-394		62,000		2.23±0.2		3.94				
432-434		68,000		2.17±0.2		4.05				
Average						3.70				
450-452	4	69,400	11*	2.66±0.2		5.05				
470-472		71,000		2.97±0.3		5.73			average	
490-492		72,600		2.41±0.2	0.46±0.20	4.72	1.93	2.44	33.2	13.8
510-512		74,200		2.37±0.2		4.62				
Average						5.03				
550-552	5	76,000		1.53±0.2		3.09				
570-572		78,000		1.47±0.1		3.01				
558-560		80,000		1.40±0.1		2.94				
Average						3.0				

* Based on oxygen isotope stratigraphy and ^{14}C-dating.

Table 3. ^{230}Th and ^{231}Pa in Core 12310

Depth (cm)	$\delta^{18}O$ Stage	Age (kyrs)	Sedimentation Rate ($cm \cdot ky^{-1}$)	Nuclide Activities $^{230}Th_{ex}$ ($dpm \cdot g^{-1}$)	Nuclide Activities $^{231}Pa_{ex}$ ($dpm \cdot g^{-1}$)	Age Corrected $^{230}Th_{ex}$ ($dpm \cdot g^{-1}$)	Age Corrected $^{231}Pa_{ex}$ ($dpm \cdot g^{-1}$)	Ratio $(\frac{Th}{Pa})$	Flux Rates $^{230}Th_{ex}$ ($dpm \cdot g^{-1} \cdot ky^{-1}$)	Flux Rates $^{231}Pa_{ex}$ ($dpm \cdot g^{-1} \cdot ky^{-1}$)
		H								
2-4	1	O	2.5	5.59±0.05	0.54±0.07		0.57	9.80		0.85
10-15		L		5.67±0.06		no corr.			8.1	
19-21		O		4.95±0.05						
26-30		C		5.51±0.06	0.34±0.05		0.47	11.7		0.70
Average		E				5.43				
		N								
		E								
56-67	2	16	6.6	3.86±0.19	0.38±0.03	4.47	0.57	7.8	19.2	2.3
94-104		22		4.18±0.23	0.33±0.06	5.14	0.60	8.6		2.4
Average						4.81				
190-200	3	40	4.1	2.64±0.15	0.15±0.03	3.82	0.43	8.9		1.06
228-238		47		2.85±0.14		4.39			11.2	
273-283		55		3.20±0.10		5.32				
Average						4.51				
317-327	4	63	7.0	3.85±0.08		6.89			28.9	
365-375	5	72	3.3	2.15±0.05		4.17				
401-411		83		3.10±0.05		(6.65)				
425-435		92		1.92±0.06		4.49			8.9	
480-490		108		1.73±0.10		4.70				
495-505		117								
512-522		123		2.43±0.30		7.58				
540-550	6	128		2.23±0.14		7.27				
Average						4.45				

Table 4. ^{230}Th and ^{231}Pa in Core 12329

Depth (cm)	$\delta^{18}O$ Stage	Age (kyrs.)	Sedimentation Rate ($cm \cdot ky^{-1}$)	$^{230}Th_{excess}$ ($dpm \cdot g^{-1}$)	Age Corrected $^{230}Th_{excess}$ ($dpm \cdot g^{-1}$)	Flux Rates ($dpm \cdot g^{-1} \cdot ky^{-1}$)
2	1	Holocene	2.0	3.34±0.38	no corr.	4.0
10				5.48±0.30		6.6
50	2	19		2.95±0.20	3.51	6.5
70		27		2.68±0.45	3.43	6.4
100		36		2.40±0.20	3.35	6.22
150	3	52	3.1	1.71±0.49	2.76	5.1
170		59		1.84±0.50	3.17	5.4
210	4	71		2.35±0.50	4.53	8.42
260	5	80	3.3			
320		100		1.90±0.10	4.8	9.47
380		120		1.13±0.40	3.42	6.8
420	6		4.0			
480						

Table 5. Th-230 and Pa-231 data for Core 13519

Section (cm)		S ($cm \cdot ky^{-1}$)	^{230}Th excess ($dpm \cdot g^{-1}$)	Flux ^{230}Th excess ($dpm \cdot g^{-2} \cdot ky^{-1}$)
10-12			5.18±0.15	
30-32			4.94±0.10	
50-52			3.78±0.16	
70-72			3.40±0.20	
93-95			2.40±0.20	
110-112		average:	2.90±0.12	average:
130-132	I	1.3	1.85±0.25	5
	II	(Th-230	1.72±0.12	
150-152		and 0-18	2.12±0.10	
190-192		stratigraphy*)	1.53±0.12	
210-212			1.45±0.15	
230-232			1.00±0.10	
250-252			0.96±0.10	
270-272			0.91±0.12	
290-292			0.94±0.07	
310-312			0.65±0.10	
350-352			0.39±0.10	
370-372			0.38±0.10	
390-392			0.24±0.10	

* Müller et al., this volume

studied, along with their oxygen isotope stratigraphies, sedimentation rates, excess Th-230 (Th(ex)-230) and Pa-231 (Pa(ex)-231) activities, age corrected Th-230 to Pa-231 ratios and their respective fluxes into the sediment. Excess Th-230 and Pa-231 values are derived from total Th-230 and Pa-231 activities subtracting the fraction of supported Th-230 and Pa-231 concentrations assumed to be in radioactive equilibrium with uranium. In cores 12310, 12309, and 12329 corrections for ingrowth of Th-230 and Pa-231, due to authigenic uranium enrichment in several core sections, have been made as described in detail by Mangini (1978) for core 12310. The overall equilibrium component amounts to 10-40% of total Th-230, depending on the level of authigenic uranium content, which varies downcore. Ages used for corrections are listed in Tables 2-4. They were derived from correlations based on oxygen isotope stratigraphy according to Diester-Haass, Schrader and Thiede (1973), Thiede (1977) and Lutze et al. (1979). Procedures for thorium and uranium isotope determination have been described elsewhere (Mangini and Sonntag, 1977) as are procedures for the water profile studies (Mangini and Key, in press).

RESULTS

The excess activities of Th-230 and Pa-231 display the following prominent features with regard to sedimentation and flux rates. Generally, sedimentation rates (see Fig. 1) are in good agreement with the rates derived from stratigraphic correlations and absolute C-14 dating methods, ranging from 1.3 to 12 cm/1000 y, within different core sections. We find, however, similar Th-230 activities in core sections with sedimentation rates differing by more than six times. This implies that fluxes into the sediment must have varied during the sedimentation histories of these cores. These deposition fluxes of Th(ex)-230 and Pa(ex)-231 were calculated from the relationship

$$1) \qquad F = S \cdot C \cdot d,$$

where S is the partial sedimentation rate for each interval derived from the O-18 stratigraphy, C-14 dating, and paleontological correlations; C are the age corrected Th(ex)-230 and Pa(ex)-231 contents, averaged for different core units as shown in Tables 2-5; and d is the *in situ* density of the sediment. In our calculations we assume a value of $d = 0.6$ g/cm^3, being an average derived from physical property data for four cores from W. Africa (Thiede, Suess and Müller, 1982) with $CaCO_3$ contents between 40 and 75 wt-%. We assign to d a variability of 20%, based on Lyle and Dymond's (1976) relationship for d as a function of the $CaCO_3$ content. The computed excess fluxes for Th-230 and Pa-231 are given in Tables 2-5 and of Th(ex)-230 are plotted against the sedimentation rate in Fig. 2. For comparison the range of Th(ex)-230 fluxes to be expected from Th-230 production in the water column and water depths is indicated by the dashed lines. Production in the water column amounts to 2.6 dpm/cm$^2\cdot$1000 yr for

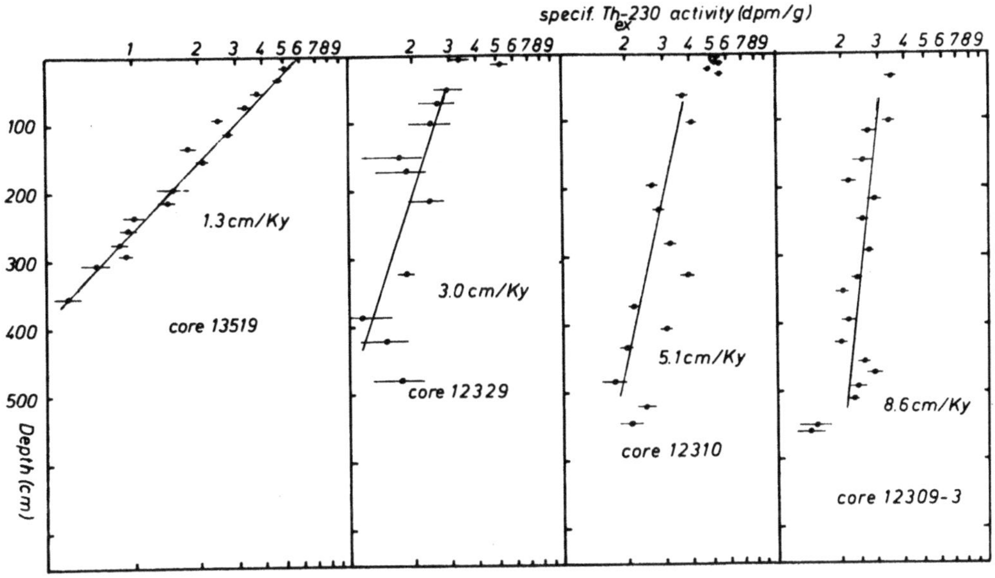

Fig. 1. Logarithmic plot of the ^{230}Th-excess concentration (N) against depth (x) in the cores. Average sedimentation rates (S) derived from the slope of the curve are listed. The sedimentation rate was derived from the expression

$$S = - \frac{\lambda \cdot x}{\ln N/N_o}$$

where λ is the decay constant of ^{230}Th.

each 1000 m of water column so that expected fluxes at the four sites are in dpm/cm^2·1000 yr:

Core No.	13519	12329	12310	12309
Flux	7.4	8.6	8.0	7.2

These estimates have the following implications. First, a comparison between fluxes into the sediments and the expected fluxes shows that in cores 12309 and 12310 observed fluxes exceed production rates up to four times. This difference is much larger than the range due to uncertainties of sedimentation rates and bulk densities used in the calculations. Second, in cores 12309 and 12310 the highest fluxes occurred during the colder O-18 stages 2 and 4. In core 12310 fluxes during warmer stages approximate those expected by production, whereas in core 12309 the flux during stage 3 also exceeds production somewhat. By comparison, in core 12329 we find only slight differences in the deposition fluxes throughout the various oxygen isotope stages, generally closely agreeing to those predicted from production rates in the water column. And third, core 13519 shows a deposition flux 37% lower than the expected one, however, this deficiency is not relevant. It results from calculation of the Th-230 flux with an

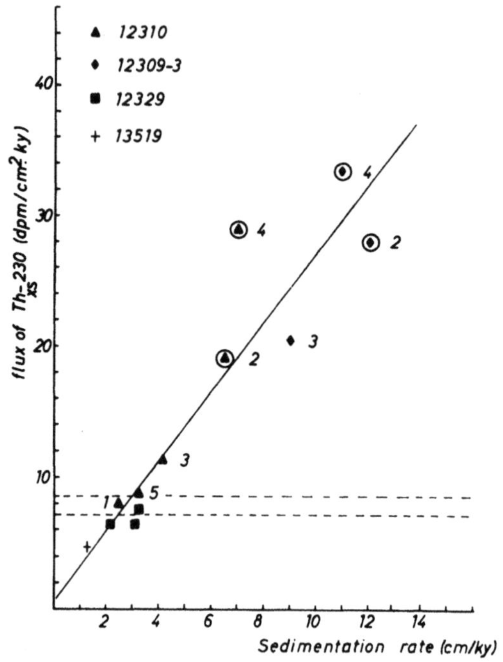

Fig. 2. Plot of ^{230}Th fluxes against sedimentation rates during different oxygen isotope stages; numbers in parentheses refer to oxygen isotope stages. Fluxes were derived from the expression $F = S \cdot c_o \cdot d$, where S is the average sedimentation rate, c_o the ^{230}Th-excess content corrected to zero age, and d the bulk density of the sediment.

average accumulation rate of 1.3 cm/ky, which does not take into account different partial rates during different climatic conditions. These partial sedimentation rates range between 1.84 and 0.96 cm/ky for the last 300,000 years (Sarnthein, pers. comm.).

The water column profile of Th-230 at station 513 displays a linear increase of Th-230 activity with depth from a value of 0.02 dpm/1000 ℓ at the surface to 0.2 dpm/1000 ℓ at 4000 m (Mangini and Key, in press). Division of the concentration of 0.1 dpm/1000 ℓ at 2500 m water depth by the production rate of Th-230 in the water column of 0.0258 dpm/1000ℓ/y yields an average residence time for Th-230 in the water column of five years. This is nearly 5 times lower than reported for the Central Pacific (Nozaki, Horibe and Tsubota, 1981). The lower Th-230 concentration is attributed to the higher particulate concentration at station 513 so that scavenging is more efficient than in the Central Pacific. The present-day concentration of Th-230 in the water column is important information because it allows a rough estimate of how much Th-230 may be brought into upwelling areas via water circulation, as well as providing a length scale on which scavenging processes in coastal areas will influence the open ocean system. To derive a length scale, we assume horizontal eddy diffusion type transport of Th-230, where the governing equation will be

$$2)\qquad dC/dt = D \cdot \partial^2 C/\partial x^2 - C/\tau + P ,$$

where C is the concentration of Th-230 at the locality x, D the eddy diffusion coefficient (cm^2/sec), τ the residence time of Th-230 in the water column, and P the Th-230 production by uranium decay. Using boundary conditions of $C(0) = 0$ and $C(\infty) = C$, the solution of the equation will be $C(x) = C \cdot \exp(- x/\sqrt{D \cdot \tau})$. The length scale derived from this equation is given by $x\pm = \sqrt{D \cdot \tau}$. This means that at a distance $x = x\pm$ from the preferential sink area (some 400 km, $D \sim 10^7$ cm^2/sec; $\tau = 5y$) with very low Th-230 concentration in the water column, the deposition flux into the sediment will still be reduced by nearly one-third in comparison to the open ocean situation. When estimating the scale length, however, one has to be aware that τ is not a constant across the ocean. For example, $\tau \sim 20$ years near Bermuda (Anderson, 1981), yielding a more than doubled scale length.

DISCUSSION

The scope of the discussion will be to determine how Th(ex)-230 and Pa(ex)-231 fluxes into the sediments might be linked to increased particle production and thereby to upwelling conditions. We first will discuss how fluxes vary during the accumulation histories of the cores and then present the inventories of Th(ex)-230 and Pa(ex)-231 and their relation to production in the water column.

Th-230 and Pa-231 Fluxes

The most interesting observation on cores 12310, 12309 and 12329 is that the increase in the sediment accumulation rate apparently causes higher radionuclide fluxes. This does not need to be the case generally because a higher supply of terrigenous material or sand, which does not scavenge Th-230 from the water column, will increase the bulk accumulation rate but will not affect the deposition flux. In such a case, one should expect lower specific activities in the sediment and a flux close to the one predicted by production in the water column. Under such circumstances, the radionuclide flux rates would lie between the dashed lines of Fig. 2 but would be shifted towards higher sedimentation rates. For cores 12310 and 12309, however, we find an almost proportional increase in radionuclide fluxes with the sedimentation rate. Fluxes higher than production in the water column must, therefore, be attributed to additional lateral supply of Th(ex)-230 from elsewhere into the area of deposition. Such a transport may either occur on the ocean floor by near bottom transport, causing sediment to be preferentially deposited at certain

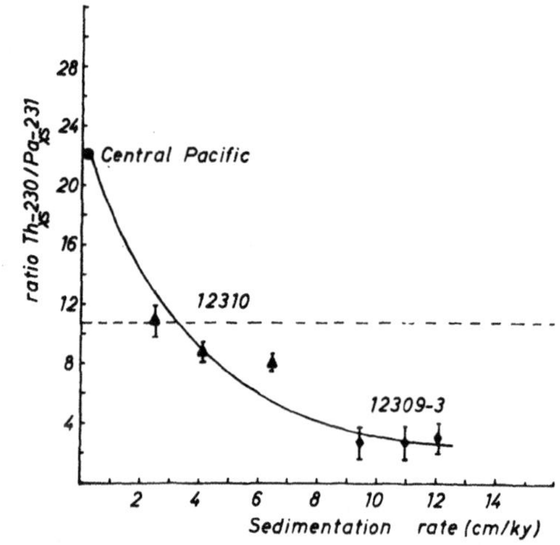

Fig. 3. Plot of the zero age corrected ^{230}Th(ex)/^{231}Pa(ex) ratio in different core sections against sedimentation rate. The dashed line represents the ^{230}Th(ex)/^{231}Pa(ex) ratio expected from production in the water column (i.e., 10.8), when no fractionation between ^{230}Th and ^{231}Pa has occurred. The value in the left upper corner is the ^{230}Th(ex)/^{231}Pa(ex) ratio derived for cores from the Central Pacific (Mangini, 1979).

places and not in others, or by horizontal mixing in the water column supplying excess Th-230. There are sedimentological and geochemical arguments against "bottom focusing" as the main cause of excess burial of Th-230 and Pa-231. Lutze et al. (1979) and Fütterer (this volume) have estimated the amount of material supplied from shallower areas to be less than 30% in core 12309, based on their studies of benthic shallow water species. The geochemical argument results from the Th(ex)-230/Pa(ex)-231 ratio. As shown on Fig. 3, the ratio in different core sections (12310 and 12309) displays a decreasing trend with increasing accumulation rate, just the opposite of that expected for downslope transported material. Because of the shorter half-life of Pa-231 in comparison to Th-230, older mobilized material will generally shift the Th(ex)-230/Pa(ex)-231 ratios towards higher values, not to lower ones as found here.

Finally, the fact that the Th(ex)/Pa(ex)-231 ratio in the sediments approaches a value of 3 at higher accumulation rates also seems relevant. This is exactly the ratio reported for sediments from areas of high productivity, such as the Antarctic Convergence (Demaster, 1979). It is lower than the ratio of 10.8 expected from production by decay of U-234 and U-235 dissolved in the ocean. A qualitative explanation of these low ratios (Demaster, 1979; Anderson, 1981) is as follows: In pelagic areas with low particle concentrations -far off coasts and regions of high bioproductivity-

Pa-231 is removed from the water column less efficiently than is Th-230. Consequently, the Th(ex)-230/Pa(ex)-231 ratio in the sediments from such areas will exceed the ratio expected from production. Inversely, in high scavenging areas, such as off northwest Africa especially during upwelling conditions, both Th-230 and Pa-231 are rapidly removed, and the Th(ex)-230/Pa(ex)-231 ratio will approach the value in seawater which is ∿3. This trend is probably best illustrated by the relationship between the Th-230/Pa-231 ratio in sediments and particle concentration in the water column in different oceans (Fig. 4) as has been compiled by Anderson (1981). One may see the same general trend in cores 12310 and 12309 (Fig. 3) where we have plotted Th(ex)-230/Pa(ex)-231 ratios against accumulation rates in core sections recording the climatic O-18 stages 1 through 5. We therefore draw the conclusion that fluxes higher than production in the water column, accompanied by Th(ex)-230/Pa(ex)-231 ratios approaching 3, must be a consequence of increased radionuclide scavenging from the water column and not due to "bottom focussing".

The Th-230 Inventory in the Sediment

In areas where the Th-230 deposition exceeds production in the water column, and where supply of Th-230 via bottom transport may be excluded, one has to ask whether water column circulation, i.e., horizontal eddy diffusion, can have supplied as much Th-230 to this area as one finds in the sediments. With a simple box model the steady state equation is given by

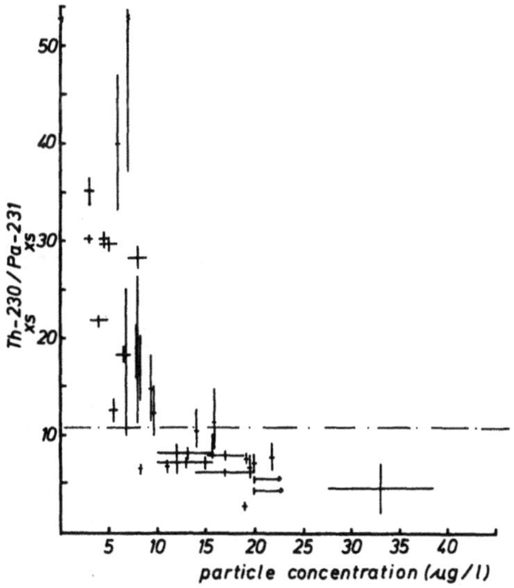

Fig. 4. Particulate $^{230}Th/^{231}Pa$ activity ratios plotted against the concentration of suspended particles as compiled by Anderson (1981) for locations in the Atlantic and Pacific oceans.

3) $$(F - P) = D \cdot dC/dx,$$

where F is the flux into sediments, and P the production of Th-230 in the water column, D the eddy diffusion coefficient and dC/dx the gradient of the Th-230 concentration. To evaluate the right-hand side of this equation we use the following numbers:

$D = 10^7$ cm^2/sec (Kaufman, Trier and Broecker, 1973),
C = 0.14 dpm/1000 ℓ, the average value from our profile in the Atlantic (Mangini and Key, in press), and
x = 400 km, comparable to the length scale of Th-230.

The conclusion to be drawn from this estimate is that the Th(ex)-230 supply to the sediments appears to be controlled by kinetic scavenging and not by the amount of Th-230 available in the water column.

Th-230, Pa-231 Fluxes, Particle Production and Upwelling

Most important with respect to scavenging of Th-230 and Pa-231 and its relation to upwelling is the fact that during the past cold O-18 stages 2 and 4, for which strong upwelling is assumed (Diester-Haass et al., 1973; Lutze et al., 1979), we find fluxes into sediments much higher than today or during past warmer stages when there was little or no upwelling activity.

A reliable sedimentological parameter for increased fertility in the northwest African upwelling area is the opal content (as the ratio of radiolarians versus planktonic foraminifers), which increases with increasing opal production in the surface waters (Diester-Haass, 1977; 1978; this volume).

A plot of this ratio versus Th(ex)-230 flux (Fig. 5) shows a rather good correlation for the cores 12310 and 12309; increasing opal content is accompanied by increasing Th-230 fluxes in glacial stages (2 and 4). Core 12329, which is situated about 450 km off the continent and thus far away from the area of coastal upwelling, has very low opal contents and here the Th(ex)-230 fluxes are much smaller than in the glacial, strongly upwelling-influenced sediments of cores 12310 and 12309. Fluxes are of the same order of magnitude as in the oxygen isotope stages 1 and 5 of core 12310 which are warm, interglacial periods with less upwelling influence (Diester-Haass, 1977; 1978). The present study thus not only gives proof that upwelling areas are more effective sinks for Th-230 and Pa-231 than the open ocean, this being an important step in understanding their geochemical behavior in the marine environment, but it also reveals that Th-230 may be a potentially useful tracer for upwelling during the past 200,000 years.

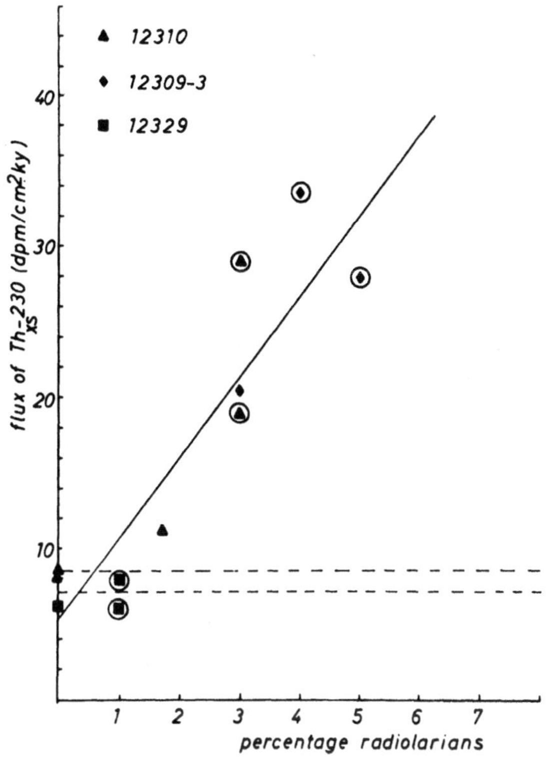

Fig. 5. Plot of the ^{230}Th-excess fluxes against opal contents (as % of radiolarians from sum radiolarians plus planktonic foraminifers). Radiolarian data from Diester-Haass (1977). Mean fluxes of individual oxygen isotope stages.

The task for future research will be to identify the Th and Pa scavenging phase. Occurrence of increased scavenging during upwelling periods, where productivity is estimated at 2-3 times higher than in "normal" periods (Müller et al., this volume), suggests a connection with productivity. One should expect that either more particles are available in the water column increasing the overall scavenging affinity, and/or the speciation of particles may be shifted towards higher "Th-230 and Pa-231 scavenger" concentrations, i.e., higher opal contents. Further sediment trap studies may provide an answer to this question.

Authigenic Uranium Incorporation and Coastal Upwelling

Higher fluxes of organic matter to the sediments, as a consequence of higher productivity during upwelling events, may lead to higher than "normal" C-org contents in the sediments (0.3-0.4%). In the cores sampled off northwestern Africa, C-org contents as high as

1% have been reported in core sections deposited under upwelling conditions. In the more reducing environment beneath a thin oxidized layer at the sediment surface, authigenic uranium may be incorporated from the water column, presumably in an amorphous phase, as reported by K. Meyer (1973) for U-rich sediments from the Walvis Bay. Sediments from upwelling areas thus become an additional sink for U dissolved in the water column. In core 12310, where both C-org and U data are available, it has been shown earlier that a 3.5 fold U-enrichment occurs (Mangini, 1978). Splitting of U into a detritic and an authigenic component becomes possible due to the different activity ratios of U-234/U-238 in the water column (1.15 ± 0.01) and in detritic material (∿1). The authigenic deposition rate of 9 μg/ cm^2 1,000 y evaluated for this core is, however, much lower than the rates reported for shallower areas nearshore, such as Walvis Bay where reducing sediments are presently accumulating (Veeh, 1967; 1974).

CONCLUSIONS

Our study of Th and Pa isotopes in sediment cores from northwestern Africa covers a time interval of 100,000 to 200,000 years B.P., where, during oxygen isotope stages 2 and 4, upwelling and a paleoproductivity higher than during "normal" Interglacial periods have been reported. This study shows that Th-230 and Pa-231 produced in the water column were scavenged more efficiently during upwelling activity periods than during the warmer oxygen stages 1, 3 and 5, where upwelling activity was less or non-existent. Fluxes into the sediments are up to 4 times higher than expected from production in the water column, which means an additional lateral supply into the area of preferential scavenging. A simple box model calculation shows that fluxes into the sediments must be controlled by the kinetic scavenging process and not by the amount of available radioisotope in the water column. The Th-230 deposition flux into the sediment shows proportionality to opal content, which is considered a reliable parameter for increased fertility in the northwest African upwelling area. This study shows that Th-230 may not only be used as a tool for sediment dating, but also as a potentially useful tracer for upwelling during the last 200,000 years when used in combination with other sedimentological data, such as O-18.

REFERENCES

Anderson, R.F., 1981, "The Marine Geochemistry of Thorium and Protactinium," Ph.D. Thesis, Woods Hole Oceanographic Institution, Woods Hole.

Bertine, K.K. and Turekian, K.K., 1973, Molybdenum in marine deposits, Geochimica et Cosmochimica Acta, 37:1415-1434.

Demaster, D.J., 1979, "The Marine Budgets of Silica and Si-32," Ph.D. Thesis, Yale University, New Haven, 308 pp.

Diester-Haass, L., 1977, Radiolarian/planktonic foraminiferal ratios in a coastal upwelling region, Journal of Foraminiferal Research, 7(1):25-33.

Diester-Haass, L., 1978, Sediments as indicators for upwelling, in: "Upwelling Ecosystems," R. Boje and M. Tomczak, eds., Springer, Heidelberg, 261-281.

Diester-Haass, L., Schrader, H.J. and Thiede, J., 1973, Sedimentological and paleoclimatological investigations of two pelagic ooze cores off Cape Barbas, N.W. Africa, "Meteor"Forschungs-Ergebnisse, C16:19-66.

Kaufman, A., Trier, R.M. and Broecker, W.S., 1973, Distribution of Ra-228 in the world ocean, Journal of Geophysical Research, 78: 8827-8848.

Ku, T.-L., Knauss, K.G. and Mathieu, G., 1977, Uranium in open ocean: concentration and isotopic composition, Deep-Sea Research, 24: 1005-1017.

Lutze, G.F., Sarnthein, M., Koopmann, B., Pflaumann, U., Erlenkeuser, H., and Thiede, J., 1979, "Meteor" cores 12309, late Pleistocene reference section for interpretation of the Neogene of site 397, in: "Initial Reports DSDP," 47, Part I, U. von Rad, W.B.F. Ryan, et al., eds., U.S. Government Printing Office, Washington, 727-739.

Lyle, M.W. and Dymond, J., 1976, Metal accumulation rates in the S.E. Pacific. Errors introduced from assumed bulk densities, Earth and Planetary Science Letters, 30:164-168.

Mangini, A., 1978, Thorium and uranium isotope analyses on "Meteor" core 12310, "Meteor"Forschungs-Ergebnisse, C29:1-5.

Mangini, A., 1979, Detailed ^{230}Th and ^{231}Pa chronology in 13 Central Pacific box cores, Transactions, American Geophysical Union, 61:084.

Mangini, A. and Sonntag, C., 1977, Pa-231 dating of deep-sea cores via Th-227 counting, Earth and Planetary Science Letters, 37: 251-256.

Mangini, A. and Key, R.M., in press, A Th-230 profile measured in the Atlantic Ocean, Earth and Planetary Science Letters.

Mangini, A., Sonntag, C., Bertsch, G., and Müller, E., 1979, Evidence for a higher, natural U-content in world rivers, Nature, 79:337-339.

Meyer, K., 1973, Uran Prospektion von Südwestafrika, Erzmetall, 26(7):313-317.

Müller, P.J. and Suess, E., 1979, Productivity, sedimentation rate, and sedimentary organic matter in the oceans - I. Organic carbon preservation, Deep-Sea Research, 26A:1347-1362.

Nozaki, J., Horibe, J. and Tsubota, H., 1981, The water column distributions of Th-isotopes in the Western N. Pacific, Earth and Planetary Science Letters, 54:203-216.

Suess, E., 1980, Particulate organic carbon flux in the oceans-surface productivity and oxygen utilization, Nature, 288:260-263.

Thiede, J., 1977, Aspects of variability of the glacial and interglacial N. Atlantic eastern boundary current (last 150,000 years), "Meteor"Forschungs- Ergebnisse, C28:1-36.

Thiede, J., Suess, E. and Müller, P., 1982, Late Quaternary fluxes of major sediment components to the sea floor at the northwest African continental slope, in: "Geology of the Northwest African Continental Margin," U. von Rad et al., eds., Springer, Heidelberg, 605-631.

Turekian, K.K., 1977, The fate of metals in the oceans, Geochimica et Cosmochimica Acta, 41:1139-1144.

Veeh, H.H., 1967, Deposition of uranium from the ocean, Earth and Planetary Science Letters, 3:145-150.

Veeh, H.H., 1974, Accumulation of uranium in sediments and phosphorites on the S.W. African shelf, Marine Chemistry, 2:189-202.

A NOTE ON CRETACEOUS BLACK SHALES AND RECENT SEDIMENTS FROM OXYGEN DEFICIENT ENVIRONMENTS: PALEOCEANOGRAPHIC IMPLICATIONS

Hans J. Brumsack

Geochemisches Institut, Goldschmidtstrasse 1
D-3400 Göttingen
Federal Republic of Germany

ABSTRACT

Published data from several Holocene organic carbon-rich continental margin sediments have been used to geochemically characterize their environment of deposition. It can be shown that a relationship exists between the accumulation rate of organic carbon and excess heavy metals (Cu, Ni, Zn). Sediments from stagnant anoxic basins such as the Black Sea, however, are geochemically different. Several Cretaceous black shales from the Atlantic Ocean show characteristics close to those of the Black Sea data. Therefore, their paleoenvironment should be much more comparable to the present Black Sea than to areas of upwelling. The trace element geochemistry of Cretaceous black shales from the Atlantic cannot be explained by a simple mixing model, using average shale and plankton data as end members. A significant contribution of metals from normal seawater to these sediments, such as from sulfide precipitation or scavenging of reduced species, is indicated.

INTRODUCTION

It is widely accepted that two different marine environments give rise to organic-carbon-rich, anaerobic sediments:

(1) stagnant, anoxic basins with slow bottom-water renewal (e.g., Black Sea) and
(2) oxygen-minimum zones underlying highly fertile upwelling water masses (e.g., Gulf of California).

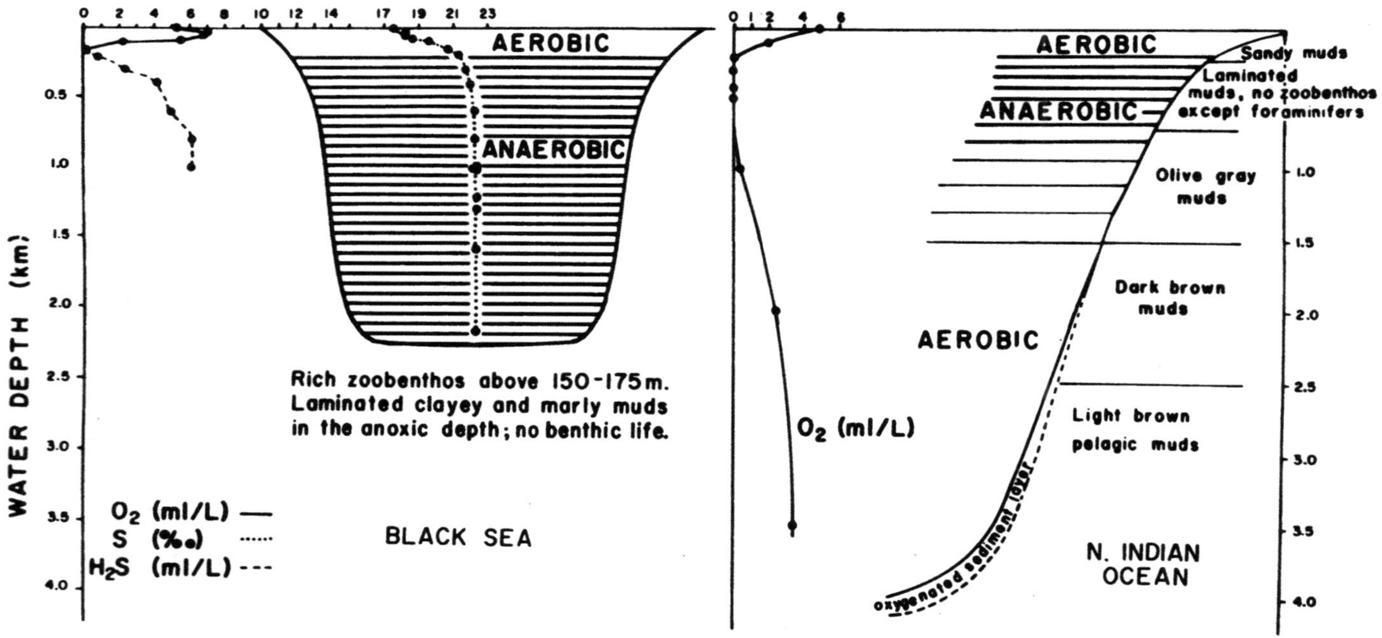

Fig. 1. Schematic representation of the two environments leading to organic-carbon-rich sediments (from: Thiede and van Andel, 1977). By permission from Earth & Planet. Sci. Letts., Copyright (c) 1977, Elsevier Publ. Co.

Table 1. Geochemistry of Holocene organic carbon-rich continental margin sediments

Element	1	2	3	4	5	6	7	8	9	10	11
% Fe	4.72	1.84	2.39	4.50	5.00	3.40	2.10	1.8	1.4	3.0	4.85
% Corg	5.9	9.35	4.58	3.8	4.0	3.0	6.0	12.2	21.2	7.8	<0.5
ppm Mn	720	115	335	465	335	268	175	200	200	316	600
ppm Zn	135	54	84	111	117	108	70	92	98	97	110
ppm Ni	51	87/108	65	57	54	46	58	87	(160)	77/66	68
ppm Cu	37	68	70	43	42	29	35	61	76	51	39
ppm Cr	81	145		100	88	119	115			108	90
ppm V	99			125	142	152	115			127	130
ppm Pb		12		9	10	11	6			10	23
ppm Cd		32?				2				2?	0.2
ppm Ba	620	240				600				480	580

1 = 7 clays, Baltic Sea (K. Bostrøm, pers. comm.)
2 = SW African shelf (Calvert, 1976; Brongersma-Sanders et al., 1980)
3 = Saanich Inlet (Presley et al., 1972)
4 = San Pedro Basin (Presley et al., 1972)
5 = Santa Monica Basin (Bruland et al., 1974, calculated from natural fluxes)
6 = Santa Barbara Basin (Dymond et al., 1981, average surface values)
7 = Soledad Basin
8 = Peru core 7706-39 (Suess, pers. comm.)
9 = Peru core 7706-36 (Suess, pers. comm.)
10 = average from 1-9, Holocene Corg-rich sediments
11 = average shale, low in Corg (Wedepohl, 1970 and pers. comm.)

It is difficult to distinguish between sediments deposited in these two environments in the geological records, as both are rich in organic carbon and/or occasionally laminated. Both environments differ with respect to the position of the redoxcline (Fig. 1). In the case of the Black Sea, it is located in the water column which contains hydrogen-sulfide and where oxygen is absent; but below oxygen-minimum zones the redoxcline is within the sediment but close to the sediment-seawater interface. The oxygen content of the water approaches zero (< 0.1 ml O_2/l), but hydrogen-sulfide usually is not present. Here I present some geochemical data which might help to characterize and distinguish both environments on the basis of their trace metal contents.

CHEMISTRY OF HOLOCENE ORGANIC-CARBON-RICH CONTINENTAL MARGIN SEDIMENTS

Table 1 lists some major and minor element concentrations in organic-carbon-rich continental margin sediments. In general, they show a very uniform composition. Due to recent developments in "clean" sampling techniques and analytical methods, more reliable data on the trace element composition of seawater (Boyle, 1976; Bruland, 1980) and marine plankton (Martin, Bruland and Broenkow, 1976; Collier and Edmond, in press) have been obtained (see Tables 2 and 3). Consistent oceanographic profiles of certain heavy metals (Cu, Ni, Cd, Zn, Cr, Ba) strongly suggest involvement of those metals in nutrient cycling. Therefore, marine plankton, besides terrigenous material, must represent an important heavy metal source in organic-carbon-rich sediments.

Table 2. Trace metal concentrations in marine plankton; all values in ppm of dry material

Fe	Mn	Zn	Ni	Cu	Cr	V	Pb	Cd	Ba
178[a]	10[a]	44[a]	6[a]	7[a]	9.6[c]	3.9[c]	5[a]	12[a]	140[a]
168[b]	8[b]	131[b]	12[b]	14[b]	1.2[d]	3.1[d]	11[f]	22[b]	55[b]
631[c]	18[c]	366[c]	13[c]	41[c]	4.9[f]	3.1[g]	2-31[g]		41[c]
		150[d]	3[d]	14[d]					
		165[e]	9[e]	13[e]					
			8[f]	39[f]					
200	11	165	9	18	5	3.5	5	12	85

a = Martin et al., 1976
b = Collier, 1981
c = Bostrøm, personal communication
d = Yamamoto and Fujita, 1966
e = Presley et al., 1972
f = Fowler, 1977
g = Martin and Knauer, 1973

In fact, the "average organic-carbon-rich continental margin sediment" (Table 1, column 10) can be explained by a very simple mixing model using "average shale" (Wedepohl, 1970) and marine plankton as end members. Fig. 2 shows the best fit for this model. The average organic-carbon-rich sediment consists roughly of about 25 ± 15% marine plankton in average shale matrix. Fig. 3 clearly shows the limitations of this simple mixing model. Some elements (e.g., Zn) have to stay close to the mixing line because their concentrations are about the same in both marine plankton and average shale.

Müller and Suess (1979) showed that the sedimentation rate correlated exponentially with the organic carbon accumulation rate. Since organic matter, mainly plankton, also represents a source for heavy metals, one should expect a similar relationship between the accumulation rate of organic carbon and the accumulation rate of heavy metals corrected for the terrigenous input. This hypothesis was tested by using published data from ten different locations (Table 4). Care was taken in selecting these data such that, besides the metals, Cu, Ni and Zn, the organic carbon --and Fe-- contents, the accumulation or sedimentation rates were determined as well. It was assumed that the Fe concentration represents the terrigenous input. The corresponding terrigenous fraction of the trace metals was subtracted from the bulk values to yield the excess heavy metals.

Table 3. Trace metal concentrations in seawater; all values in ppb, mean values represent average mid- and deep water.

Fe	Mn	Zn	Ni	Cu	Cr	V	Pb	Cd	Ba
1.8[i]	0.06[f] 0.08[i]	0.6[d] 0.6[e] 1.07[i]	0.6[a] 0.6[e] 0.23[i] 0.59[m]	0.25[a] 0.25[e] 0.39[i] 0.17[l]	0.16[c] 0.23[h]	1.2[k] 2.1[n]	0.002[o]	0.12[a] 0.11[e] 0.044[i] 0.11[n]	14[a] 22[b] 20[g]
1.8	0.08	0.6	0.6	0.3	0.2	1.5	0.002	0.1	20

[a] = Boyle, 1976
[b] = Church and Wolgemuth, 1972
[c] = Cranston and Murray, 1978
[d] = Bruland, Knauer and Martin, 1978
[e] = Bruland, 1980
[f] = Landing and Bruland, 1980
[g] = Chan et al., 1976
[h] = Campbell and Yeats, 1981
[i] = Bewers, Sundby and Yeats, 1976
[k] = Morris, 1975
[l] = Moore, 1978
[m] = Sclater, Boyle and Edmond, 1976
[n] = Riley and Taylor, 1972
[o] = Doe in Wedepohl, 1974

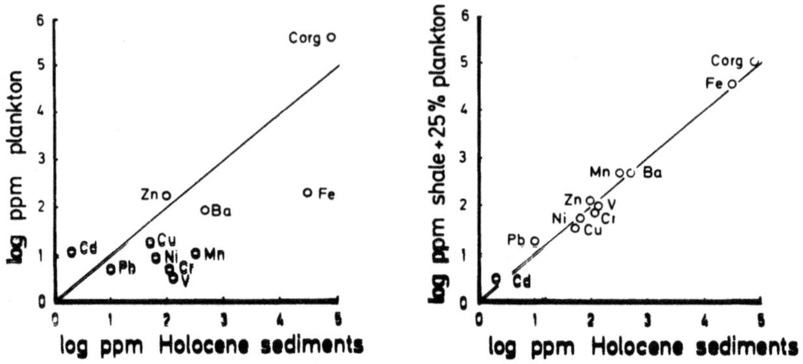

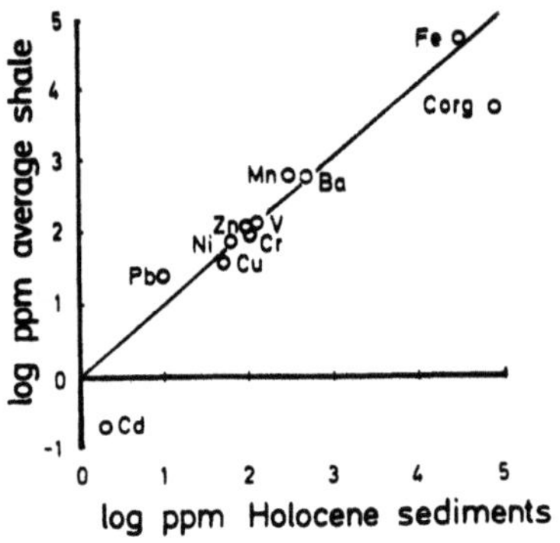

Fig. 2. Mixing model for the explanation of Holocene organic-carbon-rich continental margin sediments.

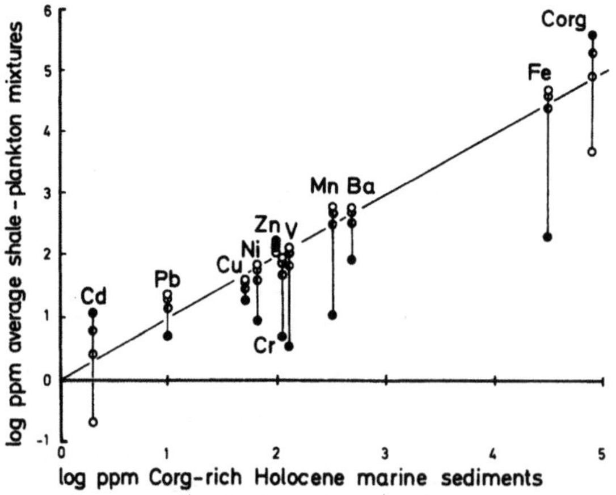

Fig. 3. Diagram showing the limitations of the mixing model shown in Fig. 2 (● = 100% plankton; ◑ = 50% plankton; ⦶ = 25% plankton; ○ = average shale).

Table 4. Data base for the calculation of the accumulation rate of organic carbon (ARCorg) and the accumulation rate of excess heavy metals (ARHMex) in organic-carbon-rich sediments.

Location	Cu	Ni	Zn	% Fe	Cu_{ex}	Ni_{ex}	Zn_{ex}	ΣHM_{ex}	%Corg	AR	$ARHM_{ex}$	ARCorg	Ref
Gulf of California	31	29	50	1.71	17	5	11	33	4.39	33	1.08	1.44	1
San Pedro Basin	43	57	111	4.50	7	0	9	16	3.8	28	0.45	1.06	2
Santa Monica Basin	42	54	117	5.00	3	0	7	10	4	24	0.24	0.96	2
Santa Barbara Basin	29	46	108	3.40	2	0	31	33	3	90	2.97	2.70	2
Soledad Basin	35	58	70	2.10	18	29	22	69	6	40	2.76	2.40	2
SW African shelf	68	88	53	1.84	53	62	11	126	9.35	55	6.93	5.14	3,4
Baltic Sea	37	71	117	3.37	10	24	41	75	5.9	55	4.13	3.25	5
Saanich Inlet	39	65	75	3.54	11	15	0	26	2.41	130	3.38	3.13	6
Peru, core 7706-39	61	87	92	1.8	47	62	51	160	12.5	32	5.12	4.00	7
Peru, core 7706-36	76	160	98	1.4	65	140	66	271	21.2	18	4.88	3.82	7
Black Sea, 11 core tops	54	121	130	4.9	15	53	20	88	2.3	2.3	0.20	0.05	8
Black Sea, 1278, 1572, 1621	142	108	130*	4.9*	103	40	20	163	13.0	2.3	0.37	0.30	9
Black Sea, core 1474 (top)	50	79	130*	4.9*	11	11	20	42	10	2.3	0.10	0.23	10
Cretaceous (14-137-7)	65	84	169	1.81	50	59	128	237	1.51	4.35	1.03	0.07	11
Cretaceous (41-367-18)	208	259	1400	2.45	188	224	1344	1756	18.60	1.84	3.23	0.34	12
Cretaceous (41-367-20)	134	154	413	4.49	98	91	311	500	6.30	0.83	0.42	0.05	13

References: 1 = Donegan, 1981; 2 = Bruland et al., 1974; 3 = Calvert, 1976; 4 = Brongersma-Sanders et al., 1980; 5 = Erlenkeuser, Suess and Willkomm, 1974; 6 = Presley et al., 1972; 7 = Dunbar, pers. comm.; 8 = Hirst, 1974; 9 = Volkov and Fomina, 1974; 10 = Cooper, Dash and Kaye, 1974; 11 = Lange et al., 1978; 12 = Brumsack, 1979; 13 = Thierstein, pers. comm..

Cu, Ni, Zn, ΣHM in ppm; AR (=accumulation rate) and ARCorg (=organic carbon accumulation rate) in mg x cm^{-2} x y^{-1}; $ARHM_{ex}$ (=accumulation rate of excess heavy metals) in µg x cm^{-2} x y^{-1}

*=see text.

Figure 4 shows that this suggested relation seems to be quite close. The Black Sea data (Degens and Ross, 1974) by contrast do not conform to this regression line. At present, reliable trace element data for Black Sea sediments are scarce. Therefore, in this compilation three different data sources have been used. The data by Hirst (1974) represent carbonate-free core tops from 11 locations, the analyses by Volkov and Fomina (1974) are from three cores located in the central Black Sea (cores 1278, 1572, 1621), and the Cooper et al. (1974) data are from the core top 1474. For this core, the sedimentation rate was known as well and therefore it is used in this report. The extremely high Zn value (3400 ppm) was excluded because such high values have never been reported elsewhere, even though they might be real. If one assumes that Zn-sulfide precipitation is taking place (Brewer and Spencer, 1974), Zn may accumulate in this area of low bulk accumulation rate as a metal sulfide rather than as part of the organic matter load. In this case, the Black Sea data point would move even further off the regression line. Therefore, truly anoxic basins should act as a sink for heavy metals either due to sulfide precipitation or redox reactions and corresponding scavenging of certain elements by particulate matter. These mechanisms are most likely responsible for the higher than expected heavy metal accumulation rate.

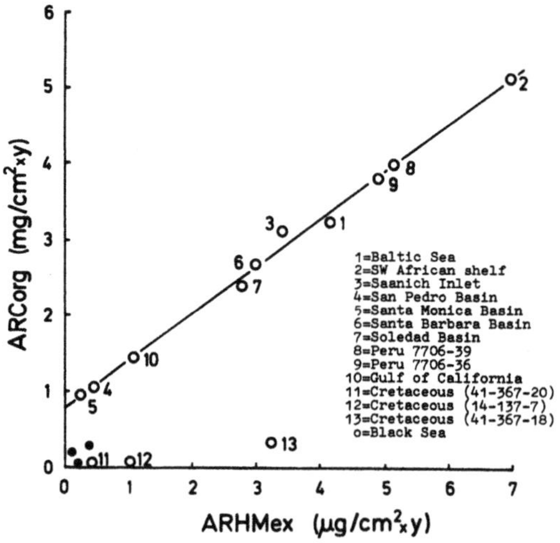

Fig. 4. Correlation of the accumulation rate of organic carbon (ARCorg) and the accumulation rate of excess heavy metals (ARHMex). According to Table 4 the concentrations of Cu, Ni and Zn were corrected for the terrigenous fraction assuming Fe to be 4.85% in terrigenous material (average shale). The term HMex (excess heavy metals) therefore refers to the sum of Cu, Ni and Zn after subtraction of the appropriate terrigenous fraction. Negative values were excluded.

PALEOENVIRONMENTAL IMPLICATIONS FOR THE GENESIS OF CRETACEOUS BLACK SHALES FROM THE ATLANTIC OCEAN

During Legs 11, 14, 36, and 41 (1970-1975) of the DSDP Project, Cretaceous black shales mostly of Cenomanian-Turonian age were sampled. It is still being debated whether these sediments were deposited under oxygen-minimum conditions or in an environment more comparable to the present day Black Sea (Schlanger and Jenkyns, 1976).

These black shales are characterized by a unique trace element composition: the elements Zn, V, Cu, Ni, and Cr are highly enriched compared to average shale (Brumsack, 1979) (Fig. 5a) or other organic-carbon-rich marine continental margin sediments (Fig. 5b). The simple mixing of marine plankton and average shale cannot account for the minor element geochemistry of these sediments (see Fig. 5a, 5b, 5c), even if a ten-fold enrichment of heavy metals in plankton is assumed (Fig. 5e). The concentrations of the elements V, Cr, Ni, Cu, and Zn would still not approach the observed concentrations.

On the other hand, these elements are very sensitive with respect to changes in redox conditions (e.g., Cr^{6+}/Cr^{3+}; Cranston and Murray, 1978) or form relatively stable sulfides of low solubilities. For this reason a seawater contribution (the heavy metal content of 500 t seawater extracted into 1 kg sediment) was considered in the mixing model. Imagine an enclosed basin of 2500 m water depth, a very low sedimentation rate of 1 m/10^6 yrs., and a residence time of about 1000 yrs. for the deep water; 1 kg of sediment would be deposited in about 10^5 yrs., during which time a water column of 10x10 cm x 50 km (500 t) would have to be stripped for its trace metal content (= 500 m/1000 yrs.). Since the actual residence time is about 1000 yrs. and the water depth 2500 m, only 20% of the trace elements need to be extracted to explain the unusual trace element composition of this sediment type. Fig. 5d shows that this hypothetical mixture fits best the "average black shale".

Estimates of the bulk accumulation rate are available for three of the analyzed black shale sequences (H. Thierstein, pers. comm.). These data points are listed in Table 4 and shown in Fig. 4. They do not conform to the regression shown by the Holocene organic-carbon-rich continental margin sediments, but plot relatively close to the Black Sea sediments. Therefore, tentatively, it is proposed that the Cenomanian-Turonian black shales of the North Atlantic have been deposited in an environment much more similar to the recent Black Sea than areas of midwater oxygen-minima along continental margins. This conclusion is supported by paleodepth reconstructions (Thierstein, 1979) which show that the highest organic carbon concentrations are found in the former basin centers and not along the Cretaceous continental margin.

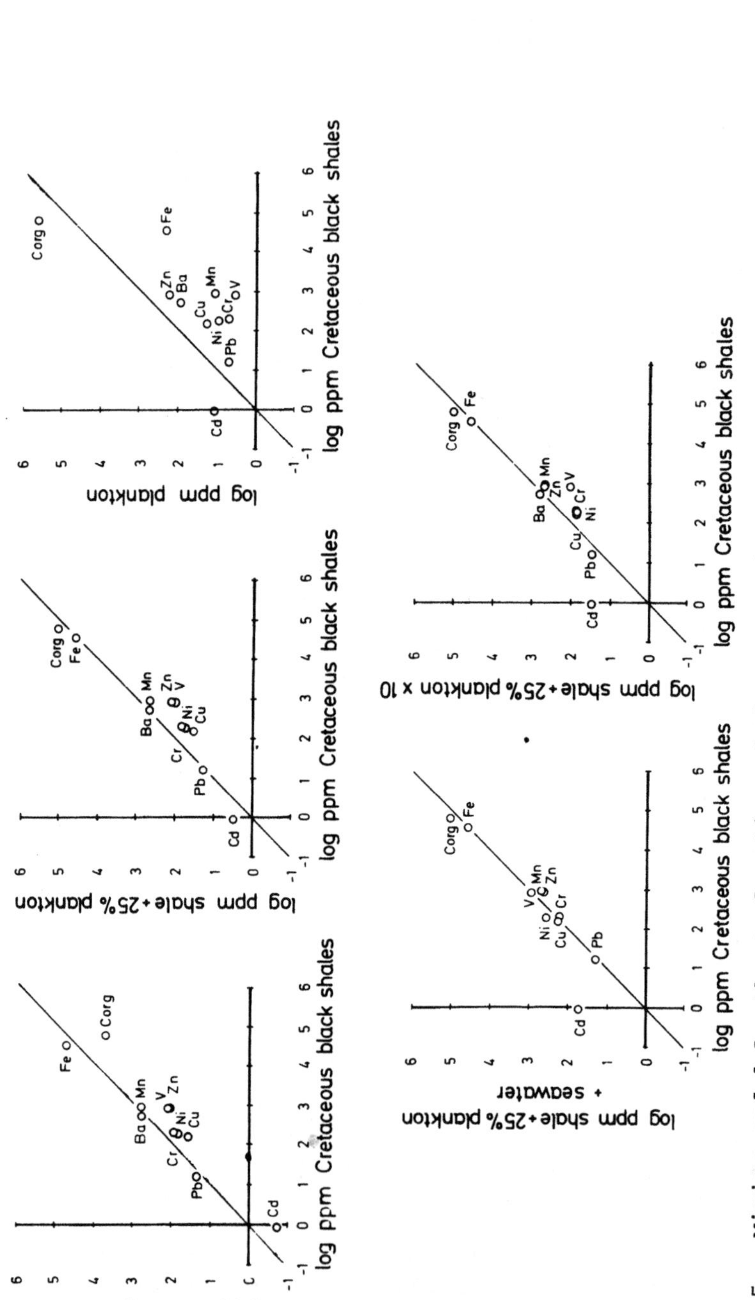

Fig. 5. Mixing model for the explanation of the trace element composition of Cretaceous black shales from the Atlantic.

CONCLUSIONS

Organic carbon-rich continental margin sediments generally show a uniform trace element composition which can best be explained by simple mixing of terrigenous and planktonic matter. For these sediments a relationship between the organic carbon accumulation rate and the excess heavy metal accumulation rate seems to exist. Truly euxinic (Black Sea-type) environments by contrast should act as sinks for heavy metals and therefore, show higher than expected excess heavy metal accumulation rates. Heavy metal analysis might for this reason be a tool to help to distinguish between upwelling and truly euxinic environments in the geological record.

ACKNOWLEDGEMENTS

I would like to thank Hans Thierstein and Joris Gieskes for helpful discussions. Financial assistance was provided by the Deutsche Forschungsgemeinschaft and the National Science Foundation.

REFERENCES

Bewers, J.M., Sundby, B. and Yeats, P.A., 1976, The distribution of trace metals in the western North Atlantic off Nova Scotia, Geochimica et Cosmochimica Acta, 40:687-696.

Boyle, E.A., 1976, "The Marine Geochemistry of Trace Metals," Ph.D. Dissertation, WHOI-MIT Joint Program in Oceanography, Woods Hole.

Brewer, P.G. and Spencer, D.W., 1974, Distribution of some trace elements in the Black Sea and their flux between dissolved and particulate phases, in: "The Black Sea," E.T. Degens and D.A. Ross, eds., American Association of Petroleum Geologists, Memoir 20, 137-143.

Brongersma-Sanders, M., Stephan, K.M., Kwee, T.G., and DeBruin, M., 1980, Distribution of minor elements in cores from the Southwest Africa shelf with notes on plankton and fish mortality, Marine Geology, 37:91-132.

Bruland, K.W., 1980, Oceanographic distributions of cadmium, zinc, nickel, and copper in the North Pacific, Earth and Planetary Science Letters, 47:176-198.

Bruland, K.W., Bertine, K., Koide, M., and Goldberg, E.D., 1974, History of metal pollution in Southern California coastal zone, Environmental Science and Technology, 8:425-432.

Bruland, K.W., Knauer, G.A. and Martin, J.H., 1978, Zinc in northeast Pacific water, Nature, 271:741-743.

Brumsack, H.-J., 1979, "Geochemische Untersuchungen an kretazischen Atlantik-Schwarzschiefern der Legs 11, 14, 36 und 41 (DSDP)," Ph.D. Dissertation, Göttingen, 57 pp.

Brumsack, H.-J., 1980, Geochemistry of Cretaceous black shales from the Atlantic Ocean (DSDP Legs 11, 14, 36 and 41), Chemical Geology, 31:1-25.

Calvert, S.E., 1976, Mineralogy and geochemistry of near-shore sediments, in: "Chemical Oceanography," J.P. Riley and R. Chester, eds., 2nd edition, Academic Press, London, 187-280.

Campbell, J.A. and Yeats, P.A., 1981, Dissolved chromium in the northwest Atlantic Ocean, Earth and Planetary Science Letters, 53:427-433.

Chan, L.H., Edmond, J.M., Stallard, R.F., Broecker, W.S., Chung, Y.C., Weiss, R.F., and Ku, T.L., 1976, Radium and barium at Geosecs stations in the Atlantic and Pacific, Earth and Planetary Science Letters, 32:258-267.

Church, T.M. and Wolgemuth, K., 1972, Marine barite saturation, Earth and Planetary Science Letters, 15:35-44.

Collier, R.W., 1981, "The Trace Element Geochemistry of Marine Biogenic Particulate Matter," Ph.D. Thesis, MIT-Woods Hole Oceanographic Institution, Woods Hole, 298 pp.

Collier, R.W. and Edmond, J.M., in press, Plankton compositions and trace element fluxes from the surface ocean, NATO Conference on Trace Elements in Seawater, Erice, Italy.

Cooper, J.A., Dash, E.J. and Kaye, M., 1974, Isotopic and elemental geochemistry of Black Sea sediments, in: "The Black Sea," E.T. Degens and D.A. Ross, eds., American Association of Petroleum Geologists, Memoir, 20, 554-565.

Cranston, R.E. and Murray, J.W., 1978, The determination of chromium species in natural waters, Analytica Chimica Acta, 99:275-282.

Degens, E.T. and Ross, D.A., 1974, "The Black Sea - Geology, Chemistry and Biology," American Association of Petroleum Geologists, Memoir, 20, Tulsa, 633 pp.

Donegan, D., 1981, "Modern and Ancient Marine Rhythmites from the Sea of Cortez and California Continental Borderland: a Sedimentological Study," M.S. Thesis, Oregon State University, Corvallis, 123 pp.

Dymond, J., Fischer, K., Clauson, M., Cobler, R., Gardner, W., Richardson, M.J., Berger, W., Soutar, A., and Dunbar, R., 1981, A sediment trap intercomparison study in the Santa Barbara Basin, Earth and Planetary Science Letters, 53:409-418.

Erlenkeuser, H.E., Suess, E. and Willkomm, H., 1974, Industrialization affects heavy metal and carbon isotope concentrations in recent Baltic Sea sediments, Geochimica et Cosmochimica Acta, 38:823-842.

Fowler, S.W., 1977, Trace elements in zooplankton particulate products, Nature, 269:51-53.

Hirst, D.M., 1974, Geochemistry of sediments from eleven Black Sea cores, in: "The Black Sea," E.T. Degens and D.A. Ross, eds., American Association of Petroleum Geologists, Memoir, 20, 430-455.

Landing, W.M. and Bruland, K.W., 1980, Manganese in the North Pacific, Earth and Planetary Science Letters, 49:45-56.

Lange, J., Wedepohl, K.H., Heinrichs, H., and Gohn, E., 1977, Notes about the specific chemical composition of "black shales" from site 367 (Leg 41), in: "Initial Reports DSDP," 41, Y. Lancelot,

E. Seibold et al., U.S. Government Printing Office, Washington, 875-877.

Martin, H.J. and Knauer, G.A., 1973, The elemental composition of plankton, Geochimica et Cosmochimica Acta, 37:1639-1653.

Martin, J.H., Bruland, K.W. and Broenkow, W.W., 1976, Cadmium transport in the California current, in: "Marine Pollutant Transfer," H.L. Windom and R.A. Duce, eds., Lexington Books, Lexington-Toronto, 159-184.

Moore, R.M., 1978, The distribution of dissolved copper in the eastern Atlantic Ocean, Earth and Planetary Science Letters, 41: 461-468.

Morris, A.W., 1975, Dissolved molybdenum and vanadium in the northeast Atlantic Ocean, Deep-Sea Research, 22:49-54.

Müller, P.J. and Suess, E., 1979, Productivity, sedimentation rate, and sedimentary organic matter in the oceans - I. Organic carbon preservation, Deep-Sea Research, 26A:1347-1362.

Presley, B.J., Kolodny, Y., Nissenbaum, A., and Kaplan, I.R., 1972, Early diagenesis in a reducing fjord, Saanich Inlet, British Columbia - II. Trace element distribution in interstitial water and sediment, Geochimica et Cosmochimica Acta, 36:1073-1090.

Riley, J.P. and Taylor, D., 1972, The concentrations of cadmium, copper, iron, manganese, molybdenum, nickel, vanadium and zinc in part of the tropical northeast Atlantic Ocean, Deep-Sea Research, 19:307-317.

Schlanger, S.O. and Jenkyns, H.C., 1976, Cretaceous oceanic anoxic events: causes and consequences, Geologie en Mijnbouw, 55: 179-184.

Sclater, F.R., Boyle, E. and Edmond, J.M., 1976, On the marine geochemistry of nickel, Earth and Planetary Science Letters, 31: 119-128.

Thiede, J. and van Andel, T.H., 1977, The paleoenvironment of anaerobic sediments in the Late Mesozoic South Atlantic Ocean, Earth and Planetary Science Letters, 33:301-309.

Thierstein, H.R., 1979, Paleoceanographic implications of organic carbon and carbonate distribution in Mesozoic deep sea sediments, in: "Deep Drilling Results in the Atlantic Ocean: Continental Margins and Paleoenvironments," M. Talwani, W. Hay and W.B.F. Ryan, eds., Maurice Ewing Series 3, American Geophysical Union, Washington, 249-274.

Volkov, I.I. and Fomina, L.S., 1974, Influence of organic material and processes of sulfide formation on distribution of some trace elements in deep water sediments of Black Sea, in: "The Black Sea," E.T. Degens and D.A. Ross, eds., American Association of Petroleum Geolosits, Memoir, 20, 456-476.

Wedepohl, K.H., 1970, Environmental influences on the chemical composition of shales and clays, in: "Physics and Chemistry of the Earth, Vol. 8," L.H. Ahrens, F. Press, S.K. Runcorn and H.C. Urey, eds., Pergamon Press, Oxford, 305-333.

Yamamoto, T. and Fujita, T., 1966, A summary of chemical abundance in seaweeds, Bulletin Kyoto University Education Series B, 29:9-12.

RAPID FORMATION OF HUMIC MATERIAL FROM DIATOM DEBRIS

Joseph R. Cronin

Dunstaffnage Marine Research Laboratory, P.O. Box 3
Oban, Argyll, PA34 4AD, United Kingdom

Robert J. Morris

Institute of Oceanographic Sciences
Wormley, Godalming, Surrey, GU8 5UB, United Kingdom

ABSTRACT

Experiments confirm the apparent rapid formation of high molecular weight humic acids in marine sediments receiving a high planktonic input. Two 'rotting' experiments are described involving a pure, cultured phytoplankton species and a field phytoplankton population from a large, enclosed experimental ecosystem. In addition, a sediment trap sample from under the ecosystem was analyzed. High molecular weight humic material was found in all samples, the relative amount increasing with the period of rotting. The results clearly indicate the rapidity with which such material is formed from natural planktonic components. The difference between lipid-extracted samples and non-extracted samples gives some clue as to the nature and strength of these early associations of phytoplankton natural products.

INTRODUCTION

Littoral marine algae exude large amounts of water soluble, extracellular organic constituents and have long been considered to be important contributors to the dissolved organic matter of the sea (e.g., Khailov and Burlakova, 1969; Sieburth, 1969). In particular, brown algae, which dominate the temperate intertidal zone, release large amounts of polysaccharides, nitrogenous compounds, carbohydrates and polyphenolic compounds (Craigie and McLachlan, 1964; Sieburth and Jensen, 1968). Craigie and McLachlan (1964) suggested that these algae are responsible for the yellow discoloration of sea water, the

Gelbstoff of Kalle (1937). Yentsch and Reichert (1962) found the formation of yellow colored substances from minced algae, and it has been shown (Fogg and Boalch, 1958; Craigie and McLachlan, 1964) that yellow-brown coloring matter is excreted by brown algae, apparently being formed from colorless precursors. These precursors were believed by Kalle (1966) to be carbohydrates and amino acids, who pointed out that the Gelbstoff of seawater is a complex mixture of compounds probably formed *in situ* from algal materials.

Sieburth and Jensen (1968) studied the properties of Gelbstoff, both from marine waters and as exudates from *Fucus* sp. They concluded that it may well be an intermediate in organic aggregate formation from dissolved algal substances. These workers later suggested that a mechanism involving the interaction of simple phenols, proteinaceous and carbohydrate matter might be responsible for the formation of marine humin substances (Gelbstoff), the larger molecular complexes being continually precipitated to form organic aggregates (Sieburth and Jensen, 1969). This suggestion was to some extent supported by the work of Rashid and Prakash (1972) who reported that the gross structural features of humic compounds isolated from decomposed thalli of *Fucus* and *Laminaria* and their exudates, was similar to those of humic compounds found in marine sediments. These workers reported significant amounts of high molecular weight (>200,000) humic acids in the decomposed residues of both *Fucus* and *Laminaria*.

In a study of the possible origin of kerogen (insoluble organic matter), Philp and Calvin (1976; 1977) found that a kerogen-like residue could be isolated from recently deposited algal mats and degraded cultures of blue-green algae and bacteria after an exhaustive extraction of the samples with organic solvents and 6N HCl hydrolysis. They suggested that the insoluble residues from algae and bacteria were the basic building blocks for kerogen.

Russian scientists first suggested that marine humics were authigenic (Bordovskiy, 1965; Kasatochkin et al., 1968) as the only way to explain their occurrence in marine sediments far removed from continental land masses. This idea was strongly supported by the data of Nissenbaum and Kaplan (1972). In particular the $^{13}C/^{12}C$ ratios found by these workers for marine humics indicated a source in the degradation products of marine plankton. The suggested pathway of marine humic formation was via water-soluble complexes and fulvic acids.

The occurrence of substantial proportions of humic and fulvic acids in organic-rich marine surface sediments lying under areas of high productivity has been reported by several workers (e.g., Romankevich and Artemyer, 1969; Ishiwatari, 1971; Brown et al., 1972; Nissenbaum and Kaplan, 1972, and references therein; Morris and Calvert, 1977; MacFarlane, 1978; Simoneit et al., 1979). The rapid deposition of planktonic material to the sediment interface often appears to give rise to sedimentary conditions which favor the preservation of organic matter, allowing its subsequent incorporation into the sediment after burial. Such early preservation of organic matter in these environments has been explained in various ways. Demaison and Moore (1980) suggest that anoxic conditions in organic-rich sediments, by reducing bioturbation, restricts bacterial activity and results in only short exposure times of the organic matter to oxidants such as sulphate. In areas of limited water exchange, such as inland seas, Degens and Paluska (1979) have proposed that the occurrence of hypersaline waters close to the sea-sediment interface may protect the labile material from early diagenesis by restricting the composition of the bacterial populations and also by a 'curing' or dehydrating action on the organic matter entering the sediment.

Some of the organic-rich marine sediments previously investigated included very young sediments which in spite of their immaturity still contained high levels of humic material (e.g., Morris and Calvert, 1977). This suggested that the relatively rapid formation of high molecular weight compounds may be a very important step, in certain sedimentary areas, during the incorporation of planktonic organic matter into sediments. More detailed work (Cronin and Morris, 1981) found the major store of organic carbon in a young organic-rich diatomaceous sediment to be in the form of high molecular weight (>300,000 nominally), labile, humic acids, the levels reducing with the age of the sediment and being replaced by more refractory humic material. There was also a significant age difference between the high and low molecular weight humic fractions from one sedimentary horizon, the latter being older. These workers concluded that the planktonic input (primarily diatoms) at this sedimentary site was rapidly complexed and converted to high molecular weight compounds, primarily humic acids. The large conglomerates thus produced by the initial complexation/association reactions were suggested as being a primary 'vehicle' for 'locking up' the natural product input to the sediment for its subsequent incorporation into the deeper sediment.

This paper presents the results of some simple experiments carried out to further examine the later hypothesis. Two "rotting" experiments, designed to simulate sedimentary conditions where there is a large input of diatoms, are described, one using a monospecific diatom culture and the other using a natural mixed diatom population from a large enclosed ecosystem. In addition a sample was taken from a sediment trap under the experimental ecosystem. The object of the work was to look for the appearance of high molecular weight humic material during the natural decomposition of diatom debris.

MATERIAL AND METHODS

Samples

(a) A pure culture of *Chaetoceros calcetans* was taken and placed in a clean perspex tube (1 x 0.1 m) sealed at one end. The tube was filled with filtered seawater, placed in the dark and nitrogen bubbled through the mixture for 1/2 hour. The tube was then sealed and stored in the dark at 15°C for 8 weeks, after which time the solid material was separated out by settling and centrifugation and divided into two portions.

(b) During a study of the initial spring bloom in an enclosed water column at Loch Ewe, Scotland, freshly sedimented particulate matter (i.e., material sedimented in 24 hrs) was collected from the bottom of the experimental bag. A microscopic examination showed the particulate matter consisted of phytoplankton remains (>90%) and amorphous organic matter. The phytoplankton debris was dominated by the diatoms *Skeletonema costatum*, *Thalassiosira* spp and *Chaetoceros* spp.

The sedimented material was quickly transferred to a tube, as described above, and filled with filtered seawater from the experimental bag, then placed in the dark and nitrogen was bubbled through the mixture for 1/2 hour. The tube was then sealed, stored in the dark for 4 weeks at 15°C and sampled as described. Living phytoplankton were collected from the bag by filtering a 10 liter water sample through a pre-cleaned glass fiber filter (GFF). The sample was taken at a depth of 3 meters in the experimental bag during the height of the spring bloom. The filter plus phytoplankton was then immediately stored frozen under nitrogen.

(c) A sample was taken from a sediment trap under the experimental bag (20 m depth) at Loch Ewe. The trap sample was anoxic and consisted of material which had settled out during a 19-day period covering the spring bloom. Skeletal remains in the sample were overwhelmingly those of diatoms (*Thalassiosira* spp, *Skeletonema costatum*, *Chaetoceros* spp, *Navicula* spp, *Nitzschia delicatissime*). The water temperature during the period of the spring bloom was around 7°C.

Humic Extraction and Analysis

The extraction of humic substances was carried out on undried samples, however a dry weight/wet weight ratio was obtained by freeze drying a small portion of each sample in a covered weighing bottle. One portion of the material from (a) and (b) [see samples] was first extracted with chloroform/methanol solvent (2:1) with sonication and allowed to stand overnight at 5°C under nitrogen. After centrifuging, the samples were washed several times with solvent and the chloroform and methanol/water phases separated according to the method of

Folch, Lees and Sloane-Stanley (1957). The extracted lipid was weighed after rotary evaporation of the chloroform and the residue kept [samples a(i) and b(i)].

The samples [a(i)(lipid extracted), a(ii); b(i)(lipid extracted), b(ii); c] were extracted for humic compounds. The samples (weighing between 0.1-0.3 g of dry weight) were first suspended in 50 mℓ of distilled water under nitrogen for two hours. The suspension was centrifuged and the water decanted off and kept (extract I). The residue was extracted under nitrogen once with a 50 mℓ solution 0.1 m in NaOH and $Na_4P_2O_7$ and repeatedly with 0.1 m NaOH till the extract was colorless. For each extraction the mixture was agitated in an ultrasonic bath for one hour then allowed to stand for several hours in the dark at room temperature. The extracts were clarified by centrifuging at 17,000 g for 1/2 hour, decanted, combined and kept (extract II). The residue was refluxed with 50 mℓ 0.5 m NaOH under nitrogen for 20 hours. The 0.5 m NaOH extract was clarified by centrifuging, decanted off and kept. The residue was washed several times with distilled water, the supernatent being added to the 0.5 m NaOH extract (extract III) and the residue being freeze-dried and weighed.

A separate extraction on the frozen phytoplankton sample was performed by sonicating the GFF filter in a solution 0.1M in NaOH and $Na_4P_2O_7$ for 1 hour then allowing it to stand for several hours in the dark at room temperature. The resulting extract was then examined directly for the presence of humic material.

Extracts I from a(ii) and b(ii) [i.e., not lipid extracted] and from c were found to have color and a visible spectrum similar to typical humic material. No further work was performed on these extracts during the present study but they will be the subject of future study.

From their color, extracts II and III clearly contain humic material and were further separated into molecular size fractions by use of a series of Amicon ultrafiltration membranes in an Amicon Thin Channel UF cell. To ensure comparability of results, a standard procedure was adopted for the fractionations. The extracts were made up to a standard volume (250 mℓ) with distilled water and passed through membranes of decreasing nominal molecular weight cut-off (300,000; 100,000; 30,000; 10,000). The sample solution was pushed through the first membrane under nitrogen pressure until 30-40 mℓ remained in the cell. This was washed with distilled water as the filtration proceeded. A total of 250 mℓ of filtrate was collected and a further 200 mℓ of wash water was passed through the sample. This was usually sufficient to reduce the pH of the filtrate to 9 or less. The retentate was freeze-dried and weighed (>300,000 fraction). The filtrate was passed through the other membranes in the series and the other fractions (>100,000; >30,000; >10,000) collected. The filtrate from the final membrane (<10,000) was treated in one of two ways. It was either dialysed in a counter current dialysis cell using Technicon cuprophane dialysis membrane or it was acidified to pH 2.2 and passed through a 15 x 2.5 cm column of Amberlite XAD2 resin. The column was washed with distilled water and eluted with 30 mℓ of molar NH_3 in ethanol. The ethanol was removed on a rotary evaporator and the residual solution freeze-dried and weighed (<10,000 fraction). Elemental C, H, N analyses were carried out on some of the molecular weight fractions from the *Chaetoceros* experiment using a Perkin Elmer 240 analyzer.

Table 1. Distribution of humic molecular weight fractions in the residue from the *Chaetoceros* 'rotting' experiment

(i) Weight of fractions (mg) from pre-lipid extracted sample (approx. 0.2 g dry wt)

Nominal Molecular* Mass Cut Off	0.1 m NaOH Extract	0.5 m NaOH Extract
>300,000	29.0	4.5
>100,000	25.4	5.7
>30,000	15.3	1.3
>10,000	19.5	T
<10,000	17.7	T

Weight of lipid extracted 50.9 mg
Weight of residue 40.0 mg

(ii) Weight of fractions (mg) from non-lipid extracted sample (approx. 0.3 g dry wt)

Nominal Molecular* Mass Cut Off	0.1 m NaOH Extract	0.5 m NaOH Extract
>300,000	102.0	3.9
>100,000	27.9	ND
>30,000	20.3	ND
>10,000	37.6	ND
<10,000	45.3	9.0

Weight of residue 52.4 mg

* Values of molecular mass cut off refer to protein standards as described by the suppliers of the membrane, Amicon Ltd.

ND = Not determined T = Trace

RESULTS AND DISCUSSION

No evidence of any humic material was detected in the visible spectrum of the phytoplankton extract. Significant amounts of high molecular weight humics were extracted from all the other samples. Our assumption is that the majority of this material must have been formed *in situ* during the rotting of the diatom debris. The distribution of the various humic molecular weight fractions in the samples are given in Tables 1, 2 and 3. A considerable proportion of the humics were found to have a nominal molecular weight in excess of 300,000, particularly in the non-lipid extracted samples.

Table 2. Distribution of humic molecular weight fractions in the residue from the mixed diatom 'rotting' experiment

(i) Weight of fractions (mg) from pre-lipid extracted sample (approx. 0.2 g dry wt)

Nominal Molecular Mass Cut Off	0.1 m NaOH Extract	0.5 m NaOH Extract
>300,000	15.7	0.3
>100,000	7.3	T
>30,000	5.6	T
>10,000	15.4	T
<10,000	1.3	T

Weight of lipid extracted 30.0 mg
Weight of residue 21.9 mg

(ii) Weight of fractions (mg) from non-lipid extracted sample (approx. 0.2 g dry wt)

Nominal Molecular Mass Cut Off	0.1 m NaOH Extract	0.5 m NaOH Extract
>300,000	36.7	8.2
>100,000	10.1	2.6
>30,000	13.5	6.4
>10,000	32.0	8.6
<10,000	3.1	12.4

Weight of residue 27.2 mg

The appearance of these humic fractions in the samples gives some confirmation to the suggestion (Cronin and Morris, 1982) that under the right sedimentary conditions high molecular weight material is rapidly formed from decomposing phytoplankton material. The consistencies of the three sets of samples (experiments a, b and c) were very different, making the determination of comparable dry- wt./wet-wt. ratios difficult (see methods). Thus, weights of humic fractions on a 'mg/g dry weight of sample' basis cannot be reliably compared from one experiment to another.

For the *Chaetoceros* samples (experiment a, Table 1), it appears that the chloroform-methanol breaks up some of the higher molecular weight complexes (particularly those nominally >300,000). The weight of lipid extracted accounts for approximately half of the decrease in the higher molecular weight humics, suggesting that the complexes are loose associations of the phytoplankton natural products, including the lipids. The mixed field phytoplankton samples (experiment b, Table 2), while probably containing some background refractory humic material present in Loch Ewe waters from land runoff, also show a considerable loss of higher molecular weight material with the chloroform-methanol extraction.

From the results, although the starting products are different, the amount of high molecular weight humic material per unit of final residue clearly increases as the duration and temperature of the 'rotting' period is increased, suggesting that the rate of formation of these complexes may be controlled by a number of parameters. However, even in the sediment trap sample with a short duration (19 days) and low temperatures (7°C), significant levels of humics were

Table 3. Distribution of humic molecular weight fractions in the sediment trap sample

Weight of fractions (mg) from the trap sample (approx. 0.07 g dry wt)

Nominal Molecular* Mass Cut Off	0.1 m NaOH Extract	0.5 m NaOH Extract
>300,000	12.9	3.5
>100,000	1.7	3.0
>30,000	1.4	1.5
>10,000	1.4	1.8
	Weight of residue 37.6 mg	

* See Table 1.

found (>25% of the total organic matter) indicating that very rapid rates of formation, even at low temperatures, are involved. At the end of the longest experiment (experiment a, duration 8 weeks at 15°C) the vast majority (>80%) of the organic matter in the sample appeared to be present as a high molecular weight humic complex.

Table 4 gives the CHN ratios for the various humic molecular weight fractions in the 0.1 m NaOH extract of the residue from the *Chaetoceros* experiment (experiment a, sample ii). The CHN data shows very little change in the major element composition between the different molecular weight fractions, suggesting that they are all inter-related. A comparison of this data with CHN ratios obtained from the work of Rashid and Prakash (1972) --humic acids present in the decomposed residues of *Fucus* and *Laminaria*--; Sieburth and Jensen (1969) --exudate from *Fucus*--; and Stuermer, Peters and Kaplan (1978) --marine humics present in diatomaceous, organic-rich marine sediment--, does allow some conclusions (see Table 4). Carbon is present in the brown algal humics in a significantly higher proportion compared to nitrogen and hydrogen, than in the humics coming from diatoms or the diatomaceous marine sediment. Presumably this is due to the different natural product assemblages of the diatoms and brown algae giving rise to different humic acid compositions.

Table 4. Elemental composition (C, H, N ratios) of the humic molecular weight fractions in the 0.1 m NaOH extract of the residue from the *Chaetoceros* 'rotting' experiment

Sample	Carbon	Hydrogen	Nitrogen
Chaetoceros Humic Fractions (0.1 m NaOH Extract):			
>300,000	76	11	14
>100,000	74	11	15
>30,000	74	12	14
>10,000	70	16	15
Fucus Exudate (Sieburth and Jensen, 1969)	82	8	10
Fucus 'Humic' Extract (Rashid and Prakash, 1972)	83	8	9
Humics from Diatomaceous Organic-Rich Marine Sediment (Stuermer et al., 1978)	75	10	14

A comparison of the C/N ratio obtained from the data in Table 4 (average 5:1) with the general figure of 7:1 reported for net and cultured phytoplankton (Redfield, 1934; Fleming, 1940; Redfield, Ketchum and Richards, 1963; Holm-Hansen, 1972) indicates that more carbon-than-nitrogen-compounds are lost during the early stages of breakdown of the diatom cells. This is to be expected as the component diatom lipids will tend to be more labile than the proteins.

CONCLUSION

We believe the following conclusions can be drawn from the results of this work:

(1) Sedimentary conditions involving the rapid deposition of diatom cells provide the necessary conditions to allow the rapid association and polymerization of natural products from the decomposing cells (i.e., carbohydrate, protein and lipid).

(2) The association/polymerization reactions rapidly (within a matter of days) give rise to high molecular weight (>300,000) humic complexes.

(3) The complexes appear to be very labile and can easily be broken down by chloroform-methanol treatment, suggesting extremely weak bonds/associations between the various components.

(4) The humics formed from the decomposing diatoms are different in composition from those reported from the exudates of decomposing residues of littoral brown algae, but similar to those previously reported for diatomaceous marine sediments.

(5) We believe this work supports the suggestion (Cronin and Morris, 1982) that the rapid formation of high molecular weight humic complexes may be an important process in certain marine sedimentary areas for the initial incorporation of planktonic organic matter into sediments, particularly upwelling areas with high productivity. The large conglomerates produced could therefore be seen as the primary 'vehicle' responsible for the early preservation of the natural product input to such sediments, allowing the subsequent incorporation of the organic matter into the deeper sediments.

REFERENCES

Bordovskiy, O.K., 1965, Accumulation of organic matter in bottom sediments, Marine Geology, 3:33-82.

Brown, F.S., Baedecker, M.J., Nissenbaum, A. and Kaplan, I.R., 1972, Early diagenesis in a reducing fjord, Saanich Inlet, British Columbia. III. Changes in organic constituents of sediment, Geochimica et Cosmochimica Acta, 35:1185-1203.

Craigie, J.S. and McLachlan, J., 1964, Excretion of colored ultra-violet-absorbing substances by marine algae, Canadian Journal of Botany, 42:23-33.
Cronin, J.R. and Morris, R.J., 1982, The occurrence of high molecular weight humic material in recent organic-rich sediment from the Namibian Shelf, Estuarine, Coastal and Shelf Science, 15:17-27.
Degens, E.T. and Paluska, A., 1979, Hypersaline solutions interact with organic detritus to produce oil, Nature, 281:666-668.
Demaison, G.J. and Moore, G.T., 1980, Anoxic environments and oil source bed genesis, Organic Geochemistry, 2:9-31.
Fleming, R.H., 1940, The composition of plankton and units for reporting population and production, Proceedings Sixth Pacific Science Congress California, 3:535-540.
Fogg, G.E. and Boalch, G.T., 1958, Extracellular products in pure cultures of a brown alga, Nature, 181:789-790.
Folch, J., Lees, M. and Sloane-Stanley, G.H., 1957, A simple method for the isolation and purification of total lipids from animal tissues, Journal of Biological Chemistry, 226:497-509.
Holm-Hansen, O., 1972, The distribution and chemical composition of particulate material in marine and fresh waters, Memorie dell' Instituto Italiano di Idrobiologia, Supplemento 29:37-51.
Ishiwatara, R., 1971, Molecular weight distribution of humic acids from lake and marine sediments, Geochemical Journal, 5:121-132.
Kalle, K., 1937, Meereskundliche chemische Untersuchungen mit Hilfe des Zeisschen Pulfrich Photometers, Annalen der Hydrologie, 65: 276-282.
Kalle, K., 1966, The problem of the Gelbstoff in the sea, Oceanography and Marine Biology, Annual Review, 4:91-104.
Kasatochkin, V.I., Bordovskiy, I.K., Larina, N.K., and Cherkinskaya, K., 1968, Chemical nature of humic acids in Indian Ocean floor sediments, Doklady Akademii Nauk, S.S.S.R., 179:690-693.
Khailov, K.M. and Burlakova, Z.P., 1969, Release of dissolved organic matter by marine seaweeds and distribution of their local organic production to inshore communities, Limnology and Oceanography, 14:521-527.
MacFarlane, R.B., 1978, Molecular weight distribution of humic and fulvic acids of sediments from a north Florida estuary, Geochimica et Cosmochimica Acta, 42:1579-1582.
Morris, R.J. and Calvert, S.E., 1977, Geochemical studies of organic-rich sediments from the Namibian Shelf. I. The organic fractions, in: "A Voyage of Discovery," M. Angel, ed., George Deacon 70th Anniversary Volume, 647-665.
Nissenbaum, A. and Kaplan, I.R., 1972, Chemical and isotopic evidence for the *in situ* origin of marine humic substances, Limnology and Oceanography, 570-582.
Philp, R.P. and Calvin, M., 1976, Possible origin for insoluble organic (kerogen) debris in sediments from insoluble cell-wall materials of algae and bacteria, Nature, 262:134-136.
Philp, R.P. and Calvin, M., 1977, Kerogenous material in recent algal mats at Laguna Mormona, Baja, California, "Advances in Organic

Geochemistry," Actas del 7° Congresso Internacional de Geoquimica Organica, Madrid, 735-752.

Rashid, M.A. and Prakash, A., 1972, Chemical characteristics of humic compounds isolated from some decomposed marine algae, Journal of the Fishery Resources Board, Canada, 29:55-60.

Redfield, A.C., 1934, On the proportions of organic derivatives in sea water and their relation to the composition of plankton, "James Johnstone Memorial Volume," Liverpool, 177-192.

Redfield, A.C., Ketchum, B.H. and Richards, F.A., 1963, The influence of organisms on the composition of sea water, in: "The Sea", Vol. 2, M.N. Hill, ed., Interscience, London, 26-77.

Romankevich, Ye.A. and Artemyer, V.E., 1969, Composition of the organic matter of sediments from the Kuril-Kamchatka Trench, Oceanology, 9:644-653.

Sieburth, J.M., 1969, Studies on the algal substances in the sea. III. The production of extracellular organic matter by littoral marine algae, Journal of Experimental Marine Biology and Ecology, 3:290-309.

Sieburth, J.M. and Jensen, A., 1968, Studies on algal substances in the sea. I. Gelbstoff (humic material) in terrestrial and marine waters, Journal of Experimental Marine Biology and Ecology, 2:174-189.

Sieburth, J.M. and Jensen, A., 1969, Studies on the algal substances in the sea. II. Exudates of phaeophyta, Journal of Experimental Marine Biology and Ecology, 3:275-289.

Simoneit, B.R.T., Mazurek, M.A., Brenner, S., Crisp, P.T. and Kaplan, I.R., 1979, Organic geochemistry of recent sediments from Guaymas Basin, Gulf of California, Deep-Sea Research, 26:879-982.

Stuermer, D.H., Peters, K.E. and Kaplan, I.R., 1978, Source indicators of humic substances and proto-kerogen. Stable isotope ratios, elemental compositions and electron spin resonance spectra, Geochimica et Cosmochimica Acta, 42:989-997.

Yentsch, C.S. and Reichert, C.A., 1962, The interrelationship between water-soluble yellow substances and chloroplastic pigments in marine algae, Botanica Maritima, 3:65-74.

LATE QUATERNARY FLUCTUATIONS IN THE CYCLING OF ORGANIC MATTER OFF CENTRAL PERU: A PROTO-KEROGEN RECORD

Clare E. Reimers* and Erwin Suess

School of Oceanography, Oregon State University
Corvallis, Oregon 97331, U.S.A.
*Present address:
Marine Biology Research Division, Scripps Institution of Oceanography, University of California, San Diego
La Jolla, California 92093, U.S.A.

ABSTRACT

A downcore study of organic matter from three radiocarbon-dated Kasten cores collected between 11°15'S and 11°30'S at water depths of 186-580 m revealed that the main organic fraction consists of insoluble and non-hydrolyzable proto-kerogen. Twenty-six proto-kerogen samples were analyzed for stable carbon and nitrogen isotope ratios and element compositions. $\delta^{13}C$ values range from -21.3 to -23.2°/₀₀, and H/C and O/C atomic ratios range from 1.2 to 1.6 and 0.22 to 0.49, respectively. These results indicate that the sedimentary organic matter on the Peru margin is mainly of marine origin. $\delta^{15}N$ values were anomalously low (-0.5 to +6.2°/₀₀).

With increasing depth of burial, evidence for early stages of diagenetic alteration affecting proto-kerogens includes: decreasing hydrolyzable fraction and N/C ratios, and increasing S/C ratio. Progressive transformations in H/C and O/C ratios as described by a van Krevelen diagram, however, do not occur. It is proposed that the atomic composition of sedimentary organic matter influenced by upwelling is dependent on (1) the time detrital particles are exposed to oxygenated bottom waters, and (2) burial rate. Owing to episodic variations in late Quaternary oxygen conditions in Peru waters, the proto-kerogen H/C and O/C record reflects the degree of aerobic biodegradation prior to burial rather than post-depositional kerogen evolution. This agrees well with the bulk accumulation record and with climatic trends inferred from glacial, palynological and marine micropaleontological studies. Low $\delta^{15}N$ values probably reflect the admixture of ^{14}N-enriched mesopelagic or benthic biomass with detrital organic matter. One such mechanism of ^{14}N enrichment may be N_2-fixation by mat-forming sulfide biota.

INTRODUCTION

Rates and mechanisms of organic matter production and cycling between the euphotic and benthic regions of the ocean ultimately depend upon the availability of local sources of energy and biomass. These sources, in turn, are interrelated in a complex manner with physiographic and dynamic features of the ocean such as proximity to land masses, morphology, water depth, circulation, and the distribution of oxygen.

To infer an oceanographic history of these sources and their past magnitude of input, geologists must rely somewhat paradoxically on a very minute fraction of the total organic production that is preserved in the sediment record. This fraction generally consists of a heterogenous group of insoluble geopolymers that form from the decay products of biopolymers during microbially mediated processes of organic matter diagenesis.

As a special aspect of earlier studies on organic matter sedimentation (Suess and Müller, 1980; Reimers, 1982; Reimers and Suess, in press) we chose here to examine the stabilization of organic matter with time in three cores from the Peru continental margin. Sediments from this area are rich in organic matter deposited as the result of coastal upwelling and the related high productivity of marine algae and zooplankton in surface waters. However, benthic communities of chemotrophic sulfide and associated heterotrophic bacteria, and other macro- and meiobenthos also constitute significant organic matter sources in this region (Phleger and Soutar, 1973; Morita, Iturriaga and Gallardo, 1981; Rowe, 1981). Among these, benthic foraminifera include a number of genera with variable tolerances for low oxygen concentrations. The sulfide micro-organisms develop as patchy mats at the sediment surface when a thin oxidized surface layer lies above a reduced sulfide-containing zone (Gallardo, 1977; Henrichs, 1980). Thus, periodic depletion of oxygen from bottom waters or shifts in the location of upwelling centers could produce pronounced fluctuations in the relative contributions of benthic and planktonic detrital biomass to localized sediment records (Walsh, 1981) as well as impart different chemical signatures onto the organic matter.

In the following sections we present elemental compositions and stable carbon and nitrogen isotope characteristics of insoluble organic matter, "proto-kerogens," as evidence for these fluctuations in the Peru upwelling region. These results extend and refine an earlier partitioning model for surface sediments which interprets the magnitude and origins of organic matter reservoirs from total organic carbon and nitrogen contents, and pertains in general to the evolution of insoluble organic matter in very young sediments.

MATERIALS AND METHODS

Samples and Accumulation Rates

The location of the cores, their lithologies, stratigraphies and relationship to regional sedimentation patterns are discussed in de-

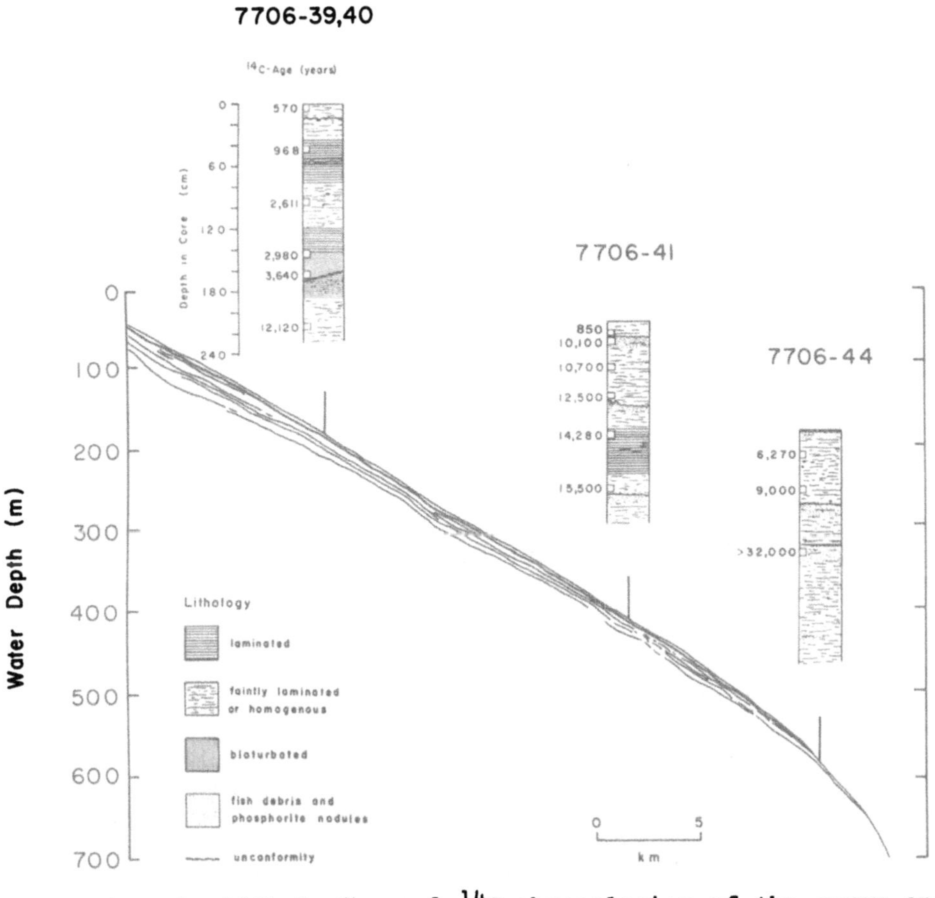

Fig. 1. The lithologies and ^{14}C-chronologies of the cores analyzed in this study in relationship to a 3.5 kHz record of the Peru upper continental slope at approximately 11°S. Very young sediments (<1,000 y B.P.) are accumulating rapidly at sites 7706-39, 7706-40, and 7706-41, but are missing from core 7706-44 which contains different time horizons not preserved in the other two sites. These sediment sequences reflect interactions between slope morphology, a fluctuating system of currents, and coastal upwelling.

Table 1. Sedimentation rates for the Peru slope cores analyzed in this study

Core-Depth Interval (cm)	Sedimentation Rate (cm/1000 yr)	Method
7706-39, 40		
0-10	160	^{210}Pb
10-55	110	^{14}C
55-145	140	^{14}C
145-165	30	^{14}C
165-200	21	^{14}C and litho-stratigraphic correlation
7706-41		
0-20	160	litho-stratigraphic correlation
20-80	33	^{14}C
80-160	41	^{14}C
7706-44		
5-80	13	^{14}C

tail by Reimers and Suess (this volume) and are illustrated in Fig. 1. In general, the cores contain a typical cross-slope sediment record between 11°S and 14°S at depths which have been swept intermittently by bottom currents creating long hiatuses. Accumulation rates for individual core sections between hiatuses were estimated from the sediment bulk density and sedimentation rates as summarized in Tables 1 and 2. Magnitudes of accumulation rates are comparable for coincident intervals in cores 7706-40 and 7706-41, but vary by a factor of four with depth. Core 7706-44 has the lowest overall accumulation rates and contains a record from different time intervals not preserved in either of the other cores. This is apparently due to the more seaward position of this core at a depth of intense reworking by a middle-slope extension of the poleward flowing undercurrent system (Reimers and Suess; Smith, both this volume).

Extractions of Soluble Organic Matter

In order to determine quantitatively the distribution of organic carbon and nitrogen in separate phases of soluble and insoluble organic compounds, we applied standard solvent extraction techniques (Deroo, Herbin and Roucache, 1978; Huc, Durand and Monin, 1978).

Table 2. Sample densities and accumulation rates

Core-Depth Interval (cm)	Dry Bulk Density (g/cm^3)	Accumulation Rate ($g/cm^2/1000$ y)
7706-39, 40		
3-6	.145	23
6-9	.172	28
15-20	.281	31
50-55	.330	36
75-80	.408	57
110-115	.331	46
135-140	.435	61
160-165	.443	13
180-185	.534	11
205-210	.564	12
7706-41		
0-4	.191	31
4-8	.213	34
15-20	.290	46
45-50	.379	13
75-80	.453	15
90-95	.457	19
110-115	.476	20
130-135	.517	21
165-170	.541	--
185-191	.813	--
7706-44		
5-10	.488	6
30-35	.519	7
42-47	.514	7
110-115	.808	--
180-185	.809	--
220-225	.896	--

Five to ten grams of sediment which had previously been desalted, freeze-dried and ground were weighed into pre-ignited porcelain thimbles and extracted for 36 hours in a Soxhlet apparatus with chloroform:methanol (5:1) mixture. The organic solvent-lipid mixture was then evaporated to a volume of less than 25 ml on a rotary evaporator and stored at 5°C for later analysis. The extracted sediment was dried at 60°C and reweighed. Separation of humic substances was then performed by multiple extraction with 0.1N NaOH solution under nitrogen at 35°C. Extractions were repeated (usually 8-10 times) until extracts were nearly colorless. Suspended solids were separated from the sodium hydroxide extract by pressure filtration through a 0.45 μm membrane filter under nitrogen. Humic and fulvic acids were separated from each other by acidifying the filtered extract with 6N HCl to precipitate humic acid, a standard procedure in soil organic matter analyses (Schnitzer and Khan, 1972). The humic acid precipitate was separated by centrifugation, washed with 0.01N HCl, freeze-dried and weighed. Fulvic acid was concentrated by vacuum evaporation, dialyzed against distilled water, freeze-dried and also weighed. The sediment remaining after these extractions was washed twice with distilled water to remove solvent salts and then freeze-dried. The residual organic matter in this fraction, termed humin, contains both hydrolyzable material and residual proto-kerogen. Standard procedures for purifying lipid and humic materials of inorganic contaminants such as silica and sulfur, were not carried out because our interest was in quantifying relative C- and N-contents rather than in individual organic compounds or molecular weight distributions.

Carbon and Nitrogen Analyses of Lipid and Humic Matter Extracts

Each of the condensed lipid extracts was transferred quantitatively to a cold 25 ml graduated cylinder, and the exact volume measured (±0.1 mℓ). Keeping the cylinder covered as much as possible to prevent solvent evaporation, three 1-mℓ aliquots were transferred to pre-weighed 7-mℓ weighing bottles. Three more 1-mℓ aliquots were transferred to pre-ignited glass cups (made from 5-mℓ ampoules), and three 2-mℓ aliquots were transferred to Kjeldahl distillation flasks. The weighing bottles and cups were dried under nitrogen overnight to remove the organic solvent mixture. When dry, the weighing bottles and cups were reweighed to determine the weight per ml of lipid extract. The glass cups plus residue were placed in crucibles and analyzed for total carbon by oxidative combustion. Standard micro-Kjeldahl distillations described by Bremner (1960) and Müller (1977) were carried out on the other 2 mℓ aliquots immediately after sampling. The precisions of these procedures are ±0.1 mg/mℓ for total mass, ±0.05 mgC/mℓ for organic carbon, and ±0.001 mgN/mℓ for organic nitrogen determinations.

The carbon and nitrogen contained in the total humic and fulvic acids, and the humin fraction were determined again by standard combustion and micro-Kjeldahl techniques on preweighed portions of the freeze-dried extracts and the residual sediment, respectively. Reproducibility of these measurements was ±2%.

Isolation of Proto-Kerogens and Hydrolysis

Proto-kerogens were separated from a portion of the residual sediment after humic substances were extracted by treatment with 6N HCl. A washing step followed and then multiple treatment with 40% HF to remove the mineral fraction. The insoluble residue (proto-kerogen) was then dialyzed against distilled water, and freeze-dried for elemental and isotopic analyses. Hydrolyzable humin carbon was quantified by shaking a split of the sediment for 16 hours in 6N HCl, measuring the organic carbon content in the remaining insoluble residue (corrected for sediment weight loss), and proportioning these values to the humin organic carbon content.

Elemental and Isotopic Analyses

All elemental analyses of the proto-kerogens were performed by Canadian Microanalytical Service (Vancouver, B.C.). The carbon, hydrogen and nitrogen contents were determined by flash combustion over Cr_2O_3 at 1050°C on a Carlo Erba Elemental Analyzer which measured CO_2, H_2O, and N_2 quantitatively by a thermal conductivity detector. Oxygen was determined on a separate sample split as CO on the same instrument modified for pyrolysis. Sulfur determinations were done by barium titration following combustion under oxygen and the formation of sulfate. The precisions of these measurements are ±0.3% for C, H, and N, and approximately ±0.5% for O and S determinations.

Stable carbon and nitrogen analyses were carried out by Global Geochemical Corporation (Canoga Park, California) and reported in standard 'δ' notation with analytical precisions of ±0.1‰ for C and ±.5‰ for N. The instrumentation and methods utilized are described by Kaplan, Smith and Ruth (1970) and Liu (1979).

Bulk Sediment and Other Analyses

Portions of the desalted, dried and ground sediment samples used for separating organic fractions, additional surface sediment samples, and two plankton samples were analyzed for bulk chemical characteristics. The plankton samples were also analyzed for stable carbon and nitrogen isotopic ratios. These samples were collected as part of a study of the distribution of floral and faunal components in surface waters in the study area (Thiede, this volume) by continuous pumping of surface water through a 75 µm screen while underway between Callao and 12°54.0'S, 76°51.0'W (PS1) and while on station at 10°11.8'S, 84°44.8'W (ES4). Total carbon and organic carbon were determined by the same LECO-induction techniques used to quantify the carbon in the organic fractions. Micro-Kjeldahl digestion was used to determine total nitrogen, and fixed nitrogen was measured as NH_3 by the methods of Silva and Bremner (1966). Total silica and alumina were measured spectrophotometrically and by atomic absorption, respectively, after digestion with HF/HNO_3. All analyses were reproducible within 2%, except fixed ammonium, which was slightly less precise.

RESULTS

Partitioning of Organic Matter from Surface Samples

The results of bulk analyses of surface samples from the three sites are listed in Table 3. From these data the concentrations of carbon and nitrogen in three forms of organic matter -detrital, biomass, and sorbed- were calculated according to a partitioning model developed by Suess and Müller (1980). The basis of this model is that the chemical characteristics of organic matter derived from marine or terrestrial detritus, *in situ* biomass, and clay-sorbed material, are each unique. Hence, the bulk elemental concentrations of surface sediments allow estimates to be made of the contents of each of the defined end members; for details see Appendix.

One limitation of this model is that it can not be applied to subsurface sediments undergoing diagenesis which leads to progressive preferential elimination of nitrogen (and phosphorus) from the solid organic phase, thus obliterating the elemental signature of original constituents. Accordingly, the model is applied here only to surface sediments from the three cores under investigation in order to:

1. characterize the starting material for the interpretation of proto-kerogen source indicators,

2. demonstrate that the contribution of biomass to the sedimentary org-N reservoir is significantly more important than to the sedimentary org-C reservoir, and

3. illustrate the effect of hydrography and sedimentation rate on the accumulation of organic matter off Peru.

Water properties off the Peru coast have been described by Wooster and Gilmartin (1961), Wyrtki (1962), Smith et al. (1971), Pak, Codispoti and Zaneveld (1980), Friederich and Codispoti (1981) and many others. Subsurface waters associated with the southward flowing Peru undercurrent are the source waters for upwelling along the coast, and lie above or in the upper part of an oxygen deficient layer (O_2 < 0.2 mℓ/ℓ) (Brockmann et al., 1980). The core of this layer also displays a nitrate deficit and a secondary nitrite maximum due to high rates of microbial denitrification (Codispoti and Packard, 1980). Under normal upwelling conditions these waters intersect the upper Peru continental slope and shelf between approximately 100 and 600 m water depth. Fluctuations in the strength of upwelling associated with the El Niño phenomenon (Wyrtki, 1975), however, are well documented and may lead to a broader and more intense oxygen minimum layer.

The sites in this study, 7706-40, 41 and 44, lie within the present day depth range of the oxygen minimum as observed under normal

Table 3. Surface sample bulk analyses and partitioning of sedimentary organic-C and organic-N into detrital, biomass and sorbed sources according to Suess and Müller (1980).

Core Depth Interval (cm)	Water Depth (cm)	Al_2O_3	N_{Fixed}	Org-C_{total}	Org-N_{total}	Detr.-C	Biom.-C	Sorb.-C	Detr.-N	Biom.-N	Sorb.-N
		(weight %)				(% of Org-C_{total})			(% of Org-N_{total})		
7706-39 0-3 cm	186	7.70	0.020	13.2	1.58	86	13	< 1	71	28	1
7706-41 0-4 cm	411	6.44	0.018	19.6	2.26	90	9	< 1	76	23	1
7706-44 0-5 cm	580	3.96	0.013	7.83	1.00	81	17	< 1	62	37	1

upwelling conditions. The surface sediments of the centermost core, 7706-41, show the highest absolute amounts of organic matter and the highest relative amounts of detritus (Table 3). Core 7706-44, which is located near the bottom boundary of the oxygen-deficient zone, has the highest relative amounts of biomass. The proportion of sorbed organic matter is negligibly small in all three surface sediments.

These observations illustrate that the preservation and sources of organic matter found in anoxic sediments are highly dependent on the period of time sedimentary particles are exposed to oxygenated bottom waters and thereby to burial rate. In the water column such exposure time can be diminished by decreasing the settling time (larger particles or shorter water column), or by decreasing the oxygen content of the water column. At the sea floor, raising the oxic-anoxic discontinuity in the sediment column with respect to the sediment-water interface, or increasing the burial rate which decreases the residence time of particles at the sea bottom, will diminish oxygen exposure. Each of these effects limits organic production by either mesopelagic or benthic organisms and reduces aerobic detrital organic matter remineralization. Hence, the preservation of detrital organic matter is favored when the period of oxygen exposure is shortened.

In this manner, then, the minimum in total benthic biomass predicted by elemental partitioning for core 7706-41 reflects first the high sedimentation rate at this site, and secondly the depth position within the center of the oxygen minimum layer. Core 7706-40 has a nearly equal sedimentation rate to core 7706-41 (Table 1), but is located closer to the fringes of the O_2-minimum zone, and so shows slightly higher amounts of biomass. The maximum in biomass predicted for core 7706-44 reflects its low sedimentation rate and its depth position near the bottom boundary of the oxygen-deficient zone. It is probable that a proportion of this biomass is due to living benthic foraminifers whose tests and biomass are preserved and concentrated in these relatively reworked anaerobic sediments (Phleger and Soutar, 1973). More importantly, however, bacterial biomass must be relatively concentrated at this site. This is suggested by total dissolved free amino acids (DFAA), an indicator of bacterial activity, which were found by Henrichs (1980) to have highest concentrations in surface sediments from the seaward edge of the oxygen-minimum zone in a transect between 15°S and 15°30'S. Henrichs (1980) also found that surface sediment DFAA concentrations were much lower at 268 m water depth but that they increased again at the landward edge of the oxygen minimum. Thus, it seems that the quantitative distribution of bacterial biomass in slope sediments parallels the distribution of total biomass and each is a reflection of both near-bottom redox conditions and dilution by detrital organic debris. These patterns are stressed here because they are important for the interpretation of downcore trends in proto-kerogen properties as indicators for source and environment.

Table 4. Carbon partitioning of extracted samples. (A)=gC of chloroform/methanol extract per gram of organic carbon; (B)=gC of fulvic acid per gram of organic carbon; (C)=gC of humic acid per gram organic carbon; (D)=gC of humin per gram organic carbon, (E)=(A+B+C+D)-1.00.

Core	Depth (cm)	Total-C (wt.%)	Org-C (wt.%)	A	B	C	D	E
7706-39	3-6	13.0	12.6	.13	.03	.10	.75	+.01
7706-39	6-9	12.0	11.7	.14	.03	.10	.73	.00
7706-40	15-20	6.9	6.8	.16	.04	.09	.72	+.01
7706-40	50-55	10.0	9.8	.09	.02	.06	.84	+.01
7706-40	75-80	9.9	9.7	.11	.03	.05	.81	.00
7706-40	110-115	7.9	7.8	.09	.02	.06	.78	-.05
7706-40	135-140	8.7	8.3	.08	.02	.07	.84	+.01
7706-40	160-165	5.2	5.1	.08	.02	.06	.85	+.01
7706-40	180-185	3.7	3.4	.06	.03	.06	.83	-.02
7706-40	205-210	3.8	3.7	.08	.03	.10	.76	-.03
7706-41	0-4	20.8	19.6	.13	.04	.19	.70	+.06
7706-41	4-8	20.0	18.6	.13	.04	.20	.70	+.07
7706-41	15-20	12.7	12.6	.09	.03	.07	.83	+.02
7706-41	45-50	11.7	11.5	.08	.03	.07	.88	+.06
7706-41	75-80	9.9	9.7	.08	.03	.08	.84	+.03
7706-41	90-95	6.3	6.2	.08	.03	.11	.82	+.04
7706-41	110-115	7.3	6.4	.07	.02	.07	.86	+.02
7706-41	130-135	6.6	5.5	.08	.03	.07	.81	-.01
7706-41	165-170	4.4	4.2	.08	.03	.09	.81	+.01
7706-41	185-191	2.6	2.5	.06	.03	.10	.85	+.04
7706-44	5-10	12.9	9.3	.05	.04	.11	.77	-.03
7706-44	30-35	12.0	8.1	.05	.03	.11	.81	.00
7706-44	42-47	12.2	8.3	.08	.04	.11	.82	+.05
7706-44	110-115	9.8	3.8	.04	.03	.06	.91	+.04
7706-44	180-185	8.9	3.7	.03	.06	.18	.80	+.07
7706-44	220-225	6.2	2.9	.04	.02	.08	.89	+.03

Fractionation of Organic Matter in Downcore Samples

The fractionation of sedimentary organic matter into lipid, fulvic and humic acids, and humin (which includes proto-kerogen) depends on the original composition and degree of alteration of the parent organic matter, and on the depth of burial. Lipids, as a whole, constitute between 5 and 25% of the biomass of marine plankton, but will generally have a more variable distribution in bacteria and land-derived organic matter (Tissot and Welte, 1978). Fulvic and humic acids which form as the result of microbial action and chemical condensation amount to between 5 and 70% of recent sedimentary organic matter (Brown et al., 1972; Nissenbaum and Kaplan, 1972), the highest values being recorded in terrigenous muds (Huc et al., 1978). The quantity of insoluble high-molecular weight components generally increases with depth due to the effects of diagenesis, but marine organic matter will usually be subject to a higher degree of remineralization (Aizenshtat, Baedecker and Kaplan, 1973). Thus, diagenetic processes may concentrate terrestrial organic matter in the residual fraction of sediments with mixed organic matter sources.

Table 5. Nitrogen partitioning of extracted samples. (A)=gN of chloroform/methanol extract per gram total nitrogen; (B)=gN of fulvic acid per gram total nitrogen; (C)=gN of humic acid per gram total nitrogen; (D)=gN of humin per gram total nitrogen; (E)=(A+B+C+D)-1.00.

Core	Depth (cm)	Total N (wt %)	A	B	C	D	E
7706-39	3-6	1.56	.03	.04	.12	.81	.00
7706-39	6-9	1.33	.03	.04	.11	.77	-.05
7706-40	15-20	0.84	.04	.04	.10	.82	.00
7706-40	50-55	1.11	.02	.03	.07	.86	-.02
7706-40	75-80	1.07	.02	.03	.07	.87	-.01
7706-40	110-115	0.87	.02	.03	.08	.83	-.04
7706-40	135-140	0.88	.02	.03	.08	.85	-.02
7706-40	160-165	0.53	.02	.03	.08	.85	-.02
7706-40	180-185	0.35	.01	.04	.07	.93	+.05
7706-40	205-210	0.32	.03	.03	.12	.81	-.01
7706-41	0-4	2.29	.04	.04	.20	.72	.00
7706-41	4-8	2.26	.04	.04	.21	.68	-.03
7706-41	15-20	1.24	.03	.03	.09	.87	+.02
7706-41	45-50	1.13	.03	.03	.08	.88	+.02
7706-41	75-80	0.95	.03	.03	.11	.88	+.05
7706-41	90-95	0.63	.02	.03	.13	.80	-.02
7706-41	110-115	0.62	.02	.03	.08	.82	-.05
7706-41	130-135	0.54	.02	.04	.09	.83	-.02
7706-41	165-170	0.39	.02	.05	.08	.89	+.04
7706-41	185-191	0.25	.02	.04	.12	.84	+.02
7706-44	5-10	1.00	.01	.04	.13	.78	-.04
7706-44	30-35	0.84	.01	.04	.12	.80	-.03
7706-44	42-47	0.89	.01	.04	.13	.79	-.03
7706-44	110-115	0.35	.02	.03	.07	.86	-.02
7706-44	180-185	0.30	.01	.07	.21	.78	+.07
7706-44	220-225	0.26	.01	.04	.09	.80	-.06

The Peru sediments, whose surface elemental compositions indicate a dominance of detrital organic matter, display changes in partitioning of org-C and -N contents as a function of depth and extractability which are listed in Tables 4 and 5. Overall the amount of extractable material is low with a slight decrease below a hiatus which separates an organic rich surface layer (∿500 years in age) from older sediments in cores 7706-40 and 7706-41. This surface layer of recent age is missing from core 7706-44 (Reimers and Suess, this volume).

The high concentrations of insoluble organic matter support the fact that these degradation products are derived from autochthonous sources deposited in a low oxygen environment. Marine kerogens may evolve early by condensation of soluble material onto insoluble polymeric debris derived from cell wall materials of bacteria and algae (Philp and Calvin, 1976). Terrestrial kerogens appear to develop more slowly and may depend on clay surfaces as catalysts for condensation-type reactions producing first simple highly oxidized melanoi-

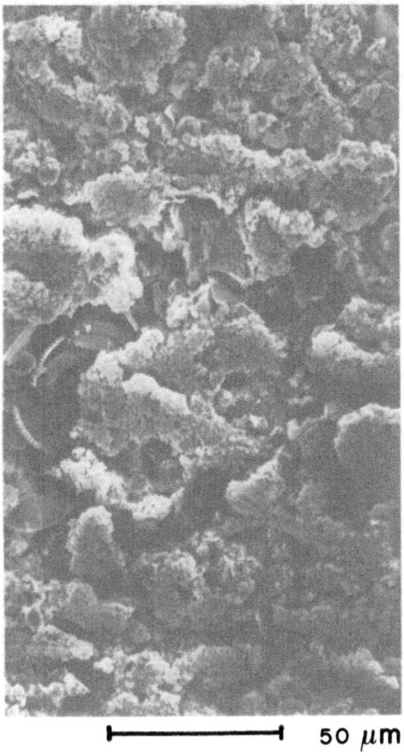

Fig. 2. Mosaic of scanning electron micrographs of sediment from core 7706-41 (2 cm). Most of the sedimentary particles are aggregated suggesting deposition as fecal pellets. During ingestion and defecation by zooplankton, easily degradable compounds are digested and structural fractions are excreted. These processes may promote organic matter stabilization. Copyright (c) 1981 C.E. Reimers.

dins and then increasingly complex fulvic acids, humic acids and humin substances (Nissenbaum and Kaplan, 1972; Hedges, 1978).

The site of insolubilization of organic matter deposited on the Peru margin could be the interior of zooplankton or anchoveta fecal pellets (Staresinic et al, this volume). There, because ingestion removes low molecular weight nutritive compounds such as amino acids and carbohydrates, micro-environments are established which concentrate non-nutritive high molecular weight fractions such as structural polysaccharides and isolate them from extensive oxygenation (Jørgensen, 1977). Fig. 2 illustrates the aggregated microstructure created by the remnants of fecal pellets in core 7706-41. Individual aggregates range in diameter from approximately 10 to 100 μm at the sediment surface but increase in size and density with increasing depth and consolidation of macro-porespace (Reimers, 1982).

Table 6. Hydrolyzable proportion of humin carbon

Core-Depth Interval (cm)	Hydrolyzable Humin Carbon (%)	Core-Depth Interval (cm)	Hydrolyzable Humin Carbon (%)
7706-39, 40		7706-41	
3-6	24	0-4	15
6-9	17	4-8	13
15-20	4	15-20	4
50-55	14	45-50	6
75-80	9	75-80	6
110-115	2	90-95	4
135-140	10	110-115	8
160-165	5	130-135	6
180-185	0	165-170	1
205-210	7	185-191	7
7706-44			
5-10	4	110-115	4
30-35	8	180-185	1
42-47	4	220-225	5

Properties of Proto-Kerogen and Hydrolysis

Generally, progressive down-core changes in the elemental composition of immature kerogen in Recent sediments and in the hydrolyzable fraction of humin organic matter are good indicators of diagenetic changes leading towards a more resistant form of sedimentary organic matter. More random variations, on the other hand, may signify changes in the source material (Deroo et al., 1978; Huc et al., 1978; Tissot and Welte, 1978).

From cores 7706-40, 41 and 44 we can piece together a record of organic matter sedimentation during approximately the last 16,000 years (Fig. 1). In cores 7706-40 and -41 a significant decrease in the hydrolyzable fraction of humin organic carbon occurs between the very young surface sediments containing relatively fresh detrital organic matter and the older sediments below (Table 6). These observations correlate with the general decrease in extractability already noted, except that hydrolyzable proportions seem to be a more sensitive indicator of intervals that have undergone extensive reworking. For example, in core 7706-40 intervals from 15 to 20 cm and from 180 to 185 cm mark hiatuses of considerable duration and have low concentrations of hydrolyzable humin carbon. The near-surface of core 7706-44, as well as the entire sediment record down to 115 cm in se-

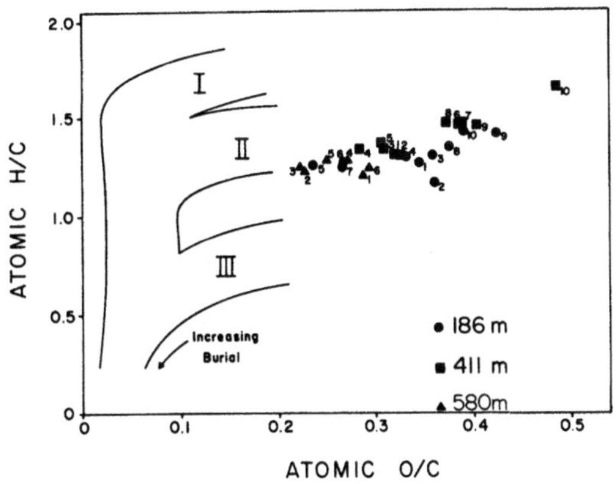

Fig. 3. Van Krevelen diagram showing atomic H/C and O/C ratios of proto-kerogens. Numbers 1-10 indicate increasing depth of burial as listed in Table 7.

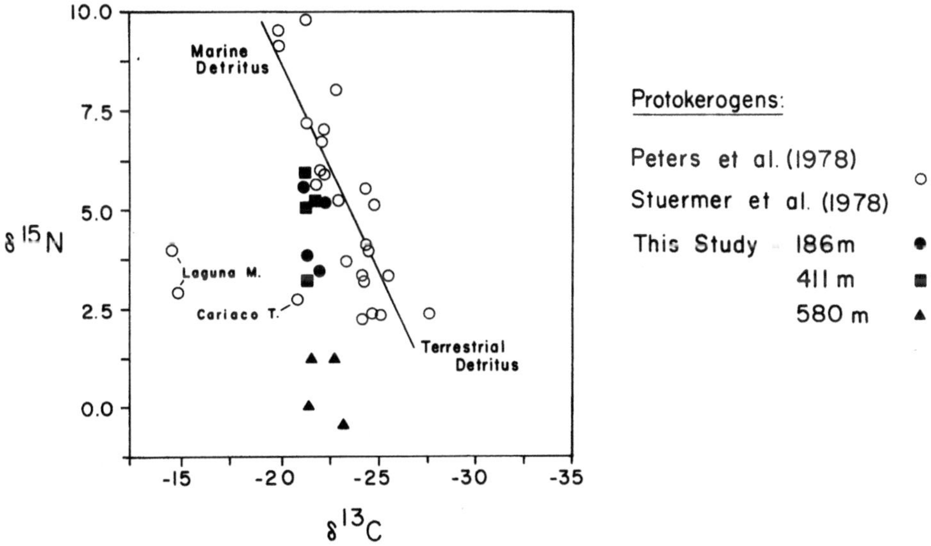

Fig. 4. $\delta^{15}N$ vs. $\delta^{13}C$ of proto-kerogens. Samples from low oxygen environments off Peru (this study); the Cariaco Trench, and Laguna Mormona, Mexico (Peters et al., 1978; Stuermer, et al., 1978) do not correspond to a typical mixing line between marine and terrestrially derived organic matter.

Table 7. Proto-kerogen data

Sample No.	Proto-Kerogen Sample		Elemental Composition					Atomic Ratios				Isotopic Ratios	
	Core	Depth (cm)	% C	% H	% N	% O	% S	H/C	N/C	O/C	S/C	$\delta^{13}C$	$\delta^{15}N$*
1	7706-39	3-6	52.27	5.59	5.53	24.10	4.35	1.27	.0907	.346	.0312	-21.24	+5.63 (+4.05)
2	7706-39	6-9	46.44	4.58	4.87	22.41	4.75	1.17	.0899	.362	.0383		+4.05
3	7706-40	15-20	42.36	4.67	4.25	20.29	4.93	1.31	.0860	.360	.0436		+2.30
4	7706-40	50-55	46.67	5.11	4.45	20.61	5.81	1.30	.0817	.333	.0466	-21.43	+3.85
5	7706-40	75-80	55.08	5.85	4.98	17.39	7.22	1.26	.0775	.237	.0491		+3.25
6	7706-40	110-115	43.94	4.70	3.82	15.69	7.66	1.27	.0745	.268	.0653		
7	7706-40	135-140	55.12	5.79	4.75	19.57	9.65	1.25	.0862	.267	.0656	-21.35	+3.36
8	7706-40	160-165	32.39	3.69	2.85	16.25	5.98	1.35	.0754	.377	.0692		+3.95
9	7706-40	180-185	21.73	2.59	1.84	12.29	5.79	1.42	.0738	.425	.0998	-22.06	+3.46
10	7706-40	205-210	22.10	2.66	1.86	11.52	5.85	1.43	.0721	.391	.0992	-22.36	+5.21
1	7706-41	0-4	52.60	5.78	5.52	22.50	2.95	1.31	.0900	.321	.0210	-21.36	+5.08
2	7706-41	4-8	54.28	5.94	5.76	23.80	2.80	1.31	.0910	.329	.0193		
3	7706-41	15-20	49.06	5.48	4.43	20.24	7.83	1.34	.0774	.310	.0598		
4	7706-41	45-50	48.50	5.42	4.31	18.39	5.64	1.34	.0762	.285	.0436	-21.84	+5.25
5	7706-41	75-80	44.62	5.09	3.94	18.27	6.24	1.37	.0757	.307	.0524	-21.25	+6.24
6	7706-41	90-95	34.06	4.14	2.88	17.49	7.15	1.46	.0725	.386	.0786		+2.71
7	7706-41	110-115	30.50	3.73	2.54	15.80	7.54	1.47	.0714	.389	.0926	-21.30±.04	+5.96
8	7706-41	130-135	31.80	3.88	2.57	15.82	9.10	1.47	.0693	.373	.107		+3.30
9	7706-41	165-170	24.67	2.99	1.99	13.32	8.74	1.46	.0691	.405	.133	-21.46±.04	+3.23 (+3.16)
10	7706-41	185-191	15.15	2.09	1.15	9.80	6.72	1.66	.0651	.486	.166		+2.78
1	7706-44	5-10	51.08	5.21	4.93	19.51	3.97	1.21	.0827	.287	.0291	-21.55	+0.03 (+2.70)
2	7706-44	30-35	55.33	5.77	5.08	16.80	4.33	1.24	.0787	.228	.0293		
3	7706-44	42-47	58.41	6.17	5.42	17.36	4.23	1.26	.0795	.223	.0271	-21.63	+1.24
4	7706-44	110-115	42.38	4.55	3.48	15.36	5.22	1.28	.0704	.272	.0461	-22.39	+1.23
5	7706-44	180-185	44.02	4.79	3.48	14.74	6.42	1.29	.0678	.251	.0546	-23.23	-0.46
6	7706-44	220-225	31.75	3.34	2.41	12.45	6.41	1.25	.0651	.294	.0756		

*( ) enclose $\delta^{15}N$ values for total-N run for comparison

diment depth, was also reworked and this is reflected in its low hydrolyzability of humin carbon.

The elemental and isotopic compositions of the proto-kerogens from each core are given in Table 7 and plotted in van Krevelen and $\delta^{13}C$ versus $\delta^{15}N$ diagrams (Figs. 3 and 4). The progressive decreases in N/C ratios and the increases in S/C as a function of depth are interpreted as further evidence for early diagenesis. Elimination of functional groups, particularly peptide linkages, from shallow samples may be responsible for these trends. This is supported by infrared spectra measured in other studies of recent and ancient sedimentary organic matter (Durand and Espitalie, 1976; Huc et al., 1978). The sulfur contents include both organic and inorganic forms of sulfur. Since the bulk of this sulfur is added to the sediment at the sediment water interface (Goldhaber and Kaplan, 1980), the increases in S/C reflect primarily post-burial microbial degradation and elimination of carbon as CO_2.

In Fig. 3 the H/C and O/C ratios of proto-kerogens are designated for each core by a different symbol and the numbers 1, 2, 3...10 are rankings to indicate increasing depth of burial (Table 7). Except within coherent stratigraphic units, successive samples do not show displacement towards the origin of the van Krevelen diagram as expected if all these proto-kerogens had evolved by diagenesis from a common source material. The data as a whole, however, support a marine autochthonous (Type II) rather than a terrestrial detrital (Type III) parent material. Optical properties of surface proto-kerogens,

Table 8. Characterization of surface proto-kerogen samples based on optical properties, Exxon Production Research Company

Location	Water Depth (m)	Kerogen Type* (%)							
		ST	PS	C	BT	AM	GA	SM	RB
7706-39, 40	186	--	--	--	--	75	--	5	20
7706-41	411	--	--	5	--	70	--	--	25
7706-44	580	--	--	5	--	70	--	--	25

* ST = Structured terrestrial organic matter
PS = Pollen and spores
C = Charcoal
BT = Biodegraded terrestrial
AM = Amorphous
GA = Gray amorphous
SM = Structured marine
RB = Round bodies
-- = none detected

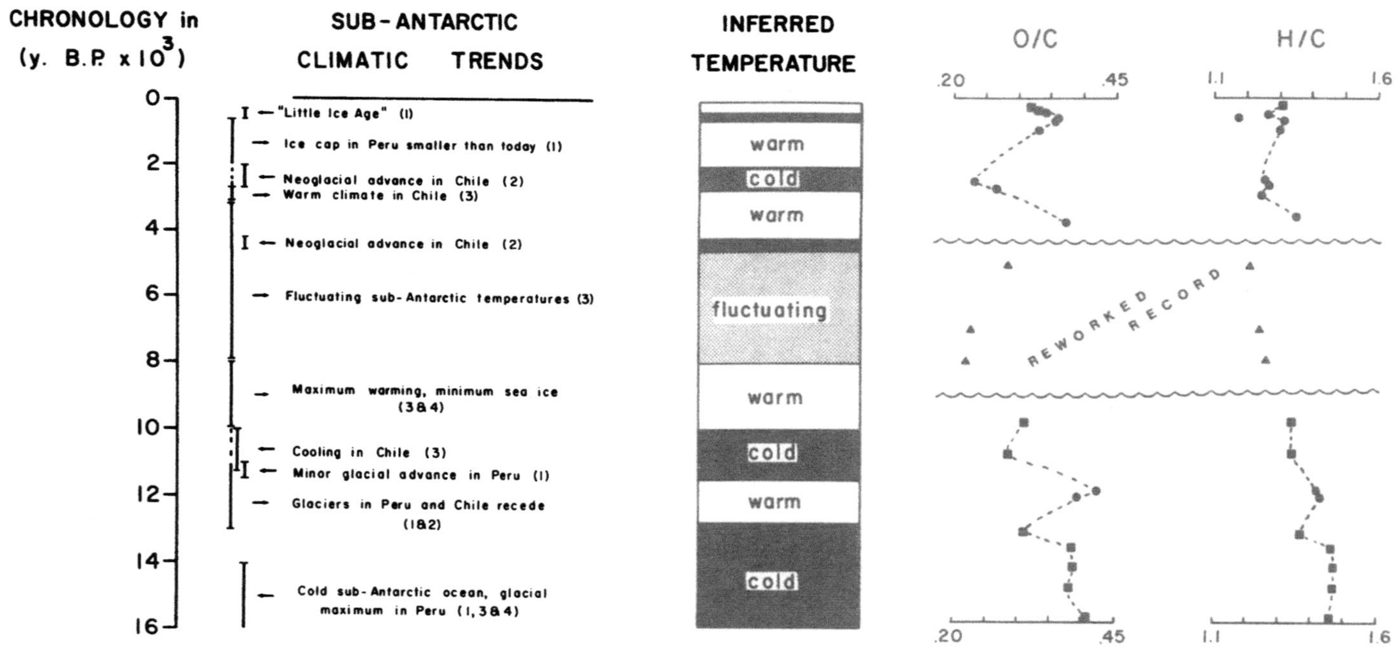

Fig. 5. O/C and H/C ratios of proto-kerogens in relationship to a sub-Antarctic temperature record inferred from studies by (1) Mercer and Palacios, 1977, (2) Mercer, 1976, (3) Heusser and Streeter, 1980, and (4) Salinger, 1981. Low O/C and H/C ratios in cores 7706-39, -40 (●) and 7706-41 (■) appear to correspond to cold climatic conditions during which the frequency of El Niño events may have been high. Low ratios in core 7706-44 (▲) probably reflect intensive reworking.

kindly made available by Exxon Production Research Company (Table 8), and stable carbon isotopic ratios are also typically marine (Degens, 1969). However, when $\delta^{13}C$ values are plotted versus $\delta^{15}N$ values (Fig. 4) the results do not correspond to a normal mixing line between marine and terrestrial organic matter sources as described by Peters, Sweeney and Kaplan (1978). Other anomalous isotopic combinations included in Fig. 4 occur in proto-kerogens from Laguna Mormona (Mexico) and in a total organic matter sample from the Cariaco Trench analyzed by Peters et al. (1978) and Stuermer, Peters and Kaplan (1978). Possible explanations for these isotopic variations and the conflicting patterns in the proto-kerogen elemental compositions are presented in the following discussion.

DISCUSSION

Significance of O/C and H/C Ratios

Peru cores 7706-40, -41 and -44 are characteristic of sediments from subtropical coastal upwelling regimes. In general, low oxygen and high sedimentation rates have permitted preservation of high amounts of marine-derived organic matter, but because the influences of climate and oceanic circulation on the depositional environment have varied throughout time, the exact partitioning and nature of organic matter sources have also varied.

The O/C and H/C ratios of the proto-kerogens are interpreted here to be indicative of the influence of a changing water column on the oxidation of detrital organic matter prior to burial. The O/C and H/C ratios of plankton are typically 0.2 to 0.3 and 1.3 to 1.5, respectively. However, O/C and H/C ratios of marine detrital organic matter trapped at approximately 400 meters depth in oxygenated waters by Honjo (1980), ranged from 0.44 to 0.56 and 1.4 to 1.8, respectively, and these values increased further in deeper traps. Such changes appear to be due to aerobic microbial degradation of detrital organic compounds and biosynthesis of more oxygenated compounds such as polysaccharides.

In oxygen deficient waters, these transformations apparently occur at a reduced rate. Thus, low O/C and H/C ratios of organic debris are preserved in rapidly formed kerogens from shallow anaerobic sediments and may be assumed to reflect the degree of aerobic biodegradation --or the lack thereof-- prior to burial. The amount of aerobic biodegradation may likewise be assumed to be dependent on the strength of the oxygen minimum layer and on the rate of burial. Walsh (1981) estimates a 90% reduction in aerobic bacterioplankton production occurred off Peru between 1966-69 (which were years following normal upwelling) and 1976-79 (which were years following El Niño events and enhancement of oxygen depleted waters). As already discussed, and further demonstrated by Walsh (1981), the consequences of El Niños also influence the partitioning of sedimentary organic matter between detrital, biomass, and sorbed sources in surface sediments. Proto-kerogens from primarily non-degraded anaerobic sedimentary bacterial biomass could have low O/C and H/C ratios.

During the last 16,000 years, the climate of western Peru and Chile has been controlled by the Antarctic ice sheet and its effects on ocean temperatures and the Peru current system. At the end of the last Pleistocene glacial, the southern climate was extremely cold and Andean precipitation may have been much less than it is today (van der Hammen, 1974). In another contribution to this volume (Reimers and Suess, this volume), we show evidence that north of about 13°S primary production and sediment accumulation along the Peru coast were also low at this time. These conditions seem to have changed,

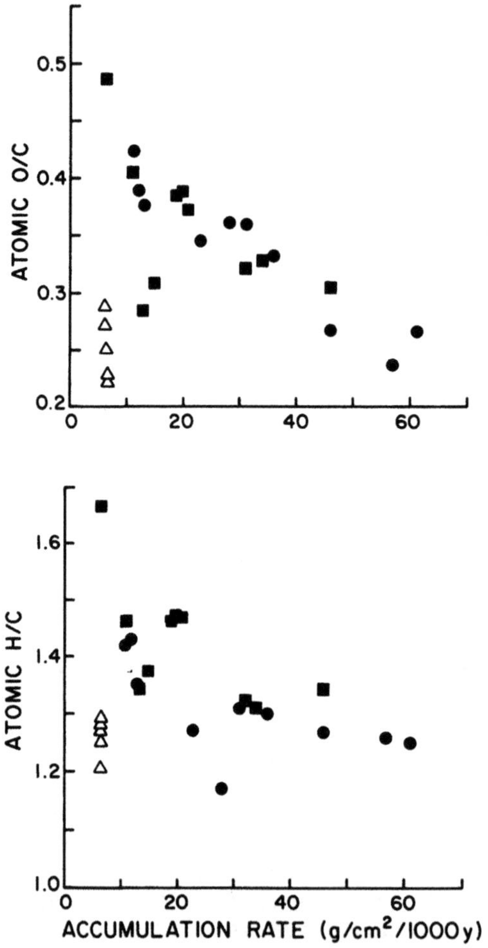

Fig. 6. Atomic O/C and H/C ratios vs. bulk sedimentation accumulation rate.

however, when about 14,000 years ago rapid climatic warming and retreat of the Antarctic ice sheet initiated the present day patterns of nearshore currents and primary production.

Holocene climatic variations, inferred from glacial geomorphology, palynology and faunal evidence from deep-sea cores (Mercer and Palacios, 1977; Heusser and Streeter, 1980; Salinger, 1981), were small in amplitude relative to the change at the end of the Pleistocene, but periods of sub-Antarctic temperatures significantly cooler than present day mean occurred between approximately 11,500 and 10,000 years B.P., and between about 600 and 300 years B.P. Less pronounced or perhaps more localized temperature minima may have also occurred in South America between approximately 4,600 and 4,200 years B.P. and between 2,700 and 2,000 years B.P. (Mercer, 1976). If at these times wind and circulation patterns were favorable for warm water to spread southward more frequently as suggested by a microfaunal record reported by DeVries and Schrader (1981), El Niño conditions may have led to prolonged periods of deoxygenated waters and high sediment accumulation as well as unaltered organic detrital input.

The Holocene record of cores 7706-40, -41 and -44 indicates that org-C and total accumulations were highest, and O/C and H/C ratios of proto-kerogens were lowest during the periods of sub-Antarctic cooling. In Fig. 5 these data are compared with a composite temperature record compiled from the works of Mercer (1976), Mercer and Palacios (1977), Heusser and Streeter (1980), and Salinger (1981). In Fig. 6, O/C and H/C ratios of the proto-kerogens are plotted versus bulk accumulation rates. Good correlations are seen for cores 7706-40 and 7706-41 which indicates that higher amounts of undegraded planktonic debris and continental mineral detritus reached the upper slope during Holocene cool periods. Further seaward, 7706-44 appears less affected, has lower sedimentation rates, and low O/C and H/C ratios. This is probably due to a history of high concentrations of anaerobic bacterial biomass at this site and generally more intense reworking.

Significance of $\delta^{15}N$ Values

The relatively low $\delta^{15}N$ values of the proto-kerogens indicate that nitrogen lighter than mean oceanic nitrate is being buried as organic nitrogen at sites 7706-40, -41, and especially, -44. Over the Peru margin the concentrations and $\delta^{15}N$ values of nitrate in the water column show large variations at different times, apparently due to fluctuations in primary production related to upwelling (Dugdale et al., 1977; Liu, 1979; Codispoti and Packard, 1980). The euphotic zone usually has the lowest concentrations of nitrate and $\delta^{15}N$ values of NO_3 ranging from +5 to +10°/₀₀ ($\delta^{15}N$ = relative to air nitrogen). Extensive denitrification and ^{15}N-enrichment to values of ca. +16°/₀₀ are typical for the remaining nitrate in oxygen-depleted zones (Cline and Kaplan, 1975; Liu, 1979).

If nitrate was the major combined species being assimilated off Peru, then under normal conditions the ^{15}N content of detrital organic matter should yield a similar range of relatively high values. This has been shown to be true for particulate matter reaching traps over the Santa Barbara Basin by Crisp et al. (1979) and employed by Sweeney and Kaplan (1980) to calculate the proportions of marine versus terrestrial organic nitrogen in Santa Barbara Basin sediments.

In the North Pacific Ocean, however, Wada and Hattori (1976) found that $\delta^{15}N$ values of plankton samples could be lowered by large isotopic fractionations during nitrate or ammonium assimilation in nitrate-rich surface-waters, or by fixation of molecular nitrogen by *Trichodesmium* and other autotrophs. In the first case, $\delta^{15}N$ values of plankton samples ranged from +3.2 to +5.1°/₀₀, and in the second case values of -1.7 to +0.5°/₀₀ were reported.

The isotopic and chemical compositions of two mixed phytoplankton and zooplankton samples taken off Callao (PS1) and just seaward of the Peru-Chile Trench (ES4) are reported in Table 9. The $\delta^{13}C$ values are typical for marine plankton in warm subtropical waters (Degens, 1969) and enriched in ^{13}C relative to the proto-kerogens from the sediments nearby. This has been shown by Degens et al. (1968) to be the natural result of diagenetic elimination of easily metabolizable ^{13}C-rich planktonic carbohydrates and proteins. The $\delta^{15}N$ values of these samples are similar to nitrate values in the euphotic zone but also isotopically heavier than many of the proto-kerogen ratios. We believe three possibilities exist for the light proto-kerogen $^{15}N/^{14}N$ ratios:

1. Processes similar to those described by Wada and Hattori (1976) are occurring off Peru, leading to highly variable and patchy distribution of $\delta^{15}N$ in particulate matter from the euphotic zone.

Table 9. Properties of plankton samples collected off central Peru

Sample	Total-C (wt. %)	Total-N (wt. %)	SiO_2 (wt. %)	Ash (wt. %)	$\delta^{13}C$ (°/₀₀)	$\delta^{15}N$ (°/₀₀)
PS 1	34.2	8.1	7.6	46	-18.43	+7.35
ES 4	--	--	3.2	14	-18.28	+4.43

2. ^{14}N is being enriched as a result of remineralization and preferential elimination of ^{15}N-rich compounds, much as is ^{12}C.

3. A source of light nitrogen is being admixed with detrital organic matter after it has left the euphotic zone but prior to final burial.

The first hypothesis must be considered in light of the significant difference in the $\delta^{15}N$ values of PS1 collected near Callao, as compared to ES4 from further offshore (Table 9). We know of no reports of *Trichodesmium* in the southern tropical Pacific Ocean, but since nitrate concentrations may be relatively high in surface waters off Peru during upwelling (Liu, 1979: Codispoti and Packard, 1980), biochemical fractionation in the assimilation of nitrate could lead to $\delta^{15}N$ values in plankton samples ranging from about +3 to +8°/oo (Wada and Hattori, 1976). Hence, patchiness in surface NO_3 concentrations related to upwelling could indirectly explain the difference between the $\delta^{15}N$ values of the plankton samples as well as most of the variability in proto-kerogen $\delta^{15}N$ values from cores 7706-40 and 7706-41 (but not 7706-44). Alternatively, if primary produced upwelling organic matter had uniform $\delta^{15}N$ values of approximately +4°/oo, higher values could reflect the relative admixture of zooplankton biomass, since biomass $\delta^{15}N$ values tend to increase with the animal's trophic level (Rau, 1981).

The isotope effect during nitrogen regeneration is not well documented (Liu, 1979). Miyake and Wada (1971) have reported up to 3°/oo variaton of $\delta^{15}N$ in residual particulate organic matter due to uneven rates of aerobic decomposition. The $\delta^{15}N$ of their partially decomposed matter, however, was higher than that of the original matter. Kerogen $\delta^{15}N$ values have been compared with total organic matter $\delta^{15}N$ values by Peters et al. (1978), and their results suggest the enrichment of ^{14}N in the more refractory kerogen fraction. We made a similar comparison of total nitrogen versus proto-kerogen nitrogen for three random samples (Table 7) but found no systematic variability. Part of this inconsistency may be due to exchangeable and fixed ammonium which can constitute up to 10% of the total nitrogen and can have a variety of sources with different isotopic compositions. More likely, fractionation of nitrogen during organic matter remineralization is not the principal cause of low proto-kerogen $\delta^{15}N$ values. This is further supported by the poor correlation between N/C ratios, which decrease downcore due to selective diagenesis and $^{15}N/^{14}N$ ratios (Table 7).

Our third hypothesis suggests that one important source of ^{14}N-enrichment is mesopelagic or benthic biomass. As already noted, the contribution of biomass to the sedimentary organic-N reservoir is significantly more important than that to the sedimentary organic-C reservoir, and increases in relative importance as detrital organic matter is degraded. In this sense hypotheses 2 and 3 are not strictly exclusive, i.e., the elimination of detritus implies that there is some production of biomass. However, certain mesopelagic or benthic organisms may also proliferate in depositional environments where aerobic nutrient regeneration has been insignificant simply because of low sedimentation rates, little predation and a large food supply. This has been shown to be the case for the sediment of 7706-44 (Table 3) which also has the lowest $\delta^{15}N$ values. Hence, we conclude that the best explanation for the low and variable proto-kerogen $\delta^{15}N$ values is that they represent a mixed organic nitrogen source that is partially synthesized within the benthic boundary layer by mesopelagic or benthic organisms.

It has been demonstrated that massive blooms of chemoheterotrophic and chemolithotrophic bacteria result from seasonal deposition of large quantities of biogenic sediment in low oxygen environments (Cohen, Krumbein and Shilo, 1977; Soutar and Crill, 1977; Klump and Martens, 1981). Some of these organisms such as sulfide oxidizing *Beggiatoa* and *Thioploca* form filamentous mats in the flocculant layer at the sediment-water interface and have been positively identified in the surface sediments of 7706-40 by R. Morita (pers. comm.). Gallardo (1977), Henrichs (1980), and Morita et al. (1981) also have reported their presence in surface samples of sediments from depths corresponding to the oxygen minimum layer off Peru and Chile. As such, these organisms could be the source of ^{14}N-enriched organic matter (Fig. 4), particularly if, like blue-green algae, cyanobacteria that are present in Laguna Mormona and in Solar Lake (Sinai) (Cohen et al., 1977), they are capable of N_2-fixation. Alternatively, they could supply large amounts of dissolved organic carbon and nitrogen to the water column to be assimilated and perhaps isotopically fractionated by mesopelagic microorganisms living in the denitrifying zone.

Obviously, a much better understanding of nitrogen cycling in the benthic boundary layer of low oxygen environments is needed to confirm the validity of our third hypothesis. However, if the natural abundance of ^{15}N in detrital organic matter is known and is not too variable, we propose that $\delta^{15}N$ values of sedimentary organic matter may be good indicators of paleo-benthic productivity. As such, $\delta^{15}N$ values of proto-kerogen suggest that off Peru benthic production has been most significantly linked with the frequency of sediment reworking as indicated by core lithologies.

CONCLUSIONS

Proto-kerogen chemical and isotopic properties reflect circulation, upwelling and oxygen conditions over the central Peru slope during the past 16,000 years. Correlation of O/C and H/C ratios with climatic and bulk accumulation records suggest that during Holocene periods of widespread climatic cooling an intensified oxygen deficient layer existed and enhanced the preservation of planktonic detrital organic matter. A marine rather than terrestrial higher plant source is evidenced by the highly insoluble nature of this organic matter, its van Krevelen classification, and its stable carbon isotopic composition.

The interpretation of the proto-kerogen O/C and H/C ratios is that higher values indicate a greater degree of aerobic biodegradation affecting detrital organic matter prior to burial. In areas with high sedimentation rates, greater aerobic biodegradation may occur if the oxygen content of the water column increases, and these effects will increase the relative contribution of oxygenated components in the organic matter reservoir. Mesopelagic and benthic biomass that is less oxygenated, however, may be concentrated under low oxygen conditions if bulk sedimentation is sufficiently reduced.

The concept of elemental fractionation illustrates that remineralization and mesopelagic and benthic biomass production has a more significant effect on organic-N sedimentary reservoirs than organic-C sedimentary reservoirs. $\delta^{15}N$ values of proto-kerogens are found to be anomalously low in sediments from the Peru slope as compared to typical planktonic material and could indicate the presence of N_2-based organisms and/or isotopic fractionation. From this initial study it is not possible to positively identify the source of this light nitrogen. However, the linkage of very low $\delta^{15}N$ values with other indicators of biomass concentration suggests that ^{14}N is most enriched by biosynthesis within the benthic boundary layer where mat-forming sulfide biota are widespread, rather than in the euphotic zone.

ACKNOWLEDGEMENTS

^{14}C ages were determined by Radiocarbon Ltd. (Lampasas, Texas). Support was provided by the Office of Naval Research, Grant N00014-76C-0067 to Oregon State University.

REFERENCES

Aizenshtat, Z., Baedecker, M.J. and Kaplan, I.R., 1973, Distribution and diagenesis of organic compounds in JOIDES sediment from the Gulf of Mexico and western Atlantic, Geochimica et Cosmochimica Acta, 37:1881-1898.

Bremner, J.M., 1960, Determination of nitrogen in soil by the Kjeldahl method, Journal of Agricultural Science, 55:11-33.

Brockmann, C., Fahrbach, E., Huyer, A., and Smith, R.L., 1980, The poleward undercurrent along the Peru coast: 5-15°S, Deep-Sea Research, 27:847-856.

Brown, F.S., Baedecker, M.J., Nissenbaum, A., and Kaplan, I.R., 1972, Early diagenesis in a reducing Fjord, Saanich Inlet, British Columbia - III. Changes in organic constituents of sediment, Geochimica et Cosmochimica Acta, 36:1185-1203.

Cline, J.D. and Kaplan, I.R., 1975, Isotopic fractionation of dissolved nitrate during denitrification in the Eastern Tropical North Pacific Ocean, Marine Chemistry, 3:271-299.

Codispoti, L.A. and Packard, T.T., 1980, Denitrification rates in the eastern tropical South Pacific, Journal of Marine Research, 38: 453-477.

Cohen, Y., Krumbein, W.E. and Shilo, M., 1977, Solar Lake (Sinai). 3. Bacterial distribution and production, Limnology and Oceanography, 22:621-634.

Crisp, P.T., Brenner, S., Venkatesan, M.I., Ruth, E., and Kaplan, I.R., 1979, Organic chemical characterization of sediment-trap particulates from San Nicolas, Santa Barbara, Santa Monica, and San Pedro Basins, California, Geochimica et Cosmochimica Acta, 43:1791-1802.

Degens, E.T., 1969, Biogeochemistry of stable carbon isotopes, in: "Organic Geochemistry," E. Eglinton and M.T.J. Murphy, eds., Springer, Berlin, 304-329.

Degens, E.T., Behrendt, M., Gotthardt, B., and Reppman, E., 1968, Metabolic fractionation of carbon isotopes in marine plankton - II. Data on samples off the coasts of Peru and Ecuador, Deep-Sea Research, 15:11-20.

Deroo, G., Herbin, J.P. and Roucache, J., 1978, Organic geochemistry of some Neogene cores from sites 374, 375, 377 and 378: Leg 42A, Eastern Mediterranean Sea, in: "Initial Reports DSDP" 42, D.A. Ross, Y.P. Neprochnov et al., U.S. Government Printing Office, Washington, 465-472.

DeVries, T.J. and Schrader, H., 1981, Variation of upwelling/oceanic conditions during the latest Pleistocene through Holocene off the central Peruvian coast: A diatom record, Marine Micropaleontology, 6:157-167.

Dugdale, R.C., Goering, J.J., Barber, R.T., Smith, R.L., and Packard, T.T., 1977, Denitrification and hydrogen sulfide in the Peru upwelling region during 1976, Deep-Sea Research, 24:601-608.

Durand, B. and Espitalie, J., 1976, Geochemical studies on the organic matter from the Douala Basin (Cameroon) - II. Evolution of Kerogen, Geochimica et Cosmochimica Acta, 40:801-808.

Friederich, G.E. and Codispoti, L.A., 1981, The effects of mixing and regeneration on the nutrient content of the upwelling waters at 15°S off Peru, in: "Coastal Upwelling," F.A. Richards, ed., American Geophysical Union Publication, Coastal and Estuarine Sciences 1, Washington, 221-227.

Gallardo, V.A., 1977, Large benthic microbial communities in sulphide biota under Peru-Chile subsurface countercurrent, Nature, 268: 331-332.

Goldhaber, M.B. and Kaplan, I.R., 1980, Mechanisms of sulfur incorporation and isotope fractionation during early diagenesis in sediments of the Gulf of California, Marine Chemistry, 9:95-144.

Hedges, J.I., 1978, The formation and clay mineral reactions of melanoidins, Geochimica et Cosmochimica Acta, 42:69-76.
Henrichs, S.M., 1980, "Biogeochemistry of Dissolved Free Amino Acids in Marine Sediments," Ph.D. Dissertation, Woods Hole Oceanographic Institute, WHOI-80-39, 253 pp.
Heusser, C.J. and Streeter, S.S., 1980, A temperature and precipitation record of the past 16,000 years in southern Chile, Science, 210:1345-1347.
Honjo, S., 1980, Material fluxes and modes of sedimentation in the mesopelagic and bathypelagic zones, Journal of Marine Research, 38:53-97.
Huc, A.Y., Durand, B. and Monin, J.C., 1978, Humic compounds and kerogens in cores from Black Sea sediments, Leg 42B - Holes 379A, B, and 380A, in: "Initial Reports DSDP," 42, D.A. Ross, Y.P. Neprochnov et al., U.S. Government Printing Office, Washington, 737-748.
Jørgensen, B.B., 1977, Bacterial sulfate reduction within reduced microniches of oxidized marine sediments, Marine Biology, 41: 7-17.
Kaplan, I.R., Smith, J.W. and Ruth, E., 1970, Carbon and sulfur concentration and isotopic composition in Apollo 11 lunar samples, Proceedings Apollo 11 Lunar Science Conference, Geochimica et Cosmochimica Acta Supplement 2:1317-1329.
Klump, J.V. and Martens, C.S., 1981, Biogeochemical cycling in an organic-rich coastal marine basin - II. Nutrient sediment-water exchange processes, Geochimica et Cosmochimica Acta, 45:101-122.
Liu, K.K., 1979, "Geochemistry of Inorganic Nitrogen Compounds in Two Marine Environments: The Santa Barbara Basin and the Ocean off Peru," Ph.D. Thesis, University of California, Los Angeles, 354 pp.
Mercer, J.H., 1976, Glacial history of southernmost South America, Quaternary Research, 6:125-166.
Mercer, J.H. and Palocios, M., 1977, Radiocarbon dating of the last glaciation in Peru, Geology, 5:600-604.
Miyake, T. and Wada, E., 1971, The isotope effect on the nitrogen in biochemical oxidation-reduction reactions, Recent Oceanographic Works of Japan, 11:1-6.
Morita, R.Y., Iturriaga, R. and Gallardo, V.A., 1981, *Thioploca*: Methylotroph and significance in the food chain, Kieler Meeresforschungen, Kiel, 5:384-389.
Müller, P.J., 1977, C/N ratios in Pacific deep-sea sediments: Effect of inorganic ammonium and organic nitrogen compounds sorbed by clays, Geochimica et Cosmochimica Acta, 41:765-776.
Nissenbaum, A. and Kaplan, I.R., 1972, Chemical and isotopic evidence for the *in situ* origin of marine humic substances, Limnology and Oceanography, 17:570-581.
Pak, H., Codispoti, L.A. and Zaneveld, J.R.V., 1980, On the intermediate particle minima associated with oxygen-poor water off western South America, Deep-Sea Research, 27:783-798.

Peters, K.E., Sweeney, R.E. and Kaplan, I.R., 1978, Correlation of carbon and nitrogen stable isotope ratios in sedimentary organic matter, Limnology and Oceanography, 23:598-604.

Philp, R.P. and Calvin, M., 1976, Possible origin for insoluble organic (kerogen) debris in sediments from insoluble cell-wall materials of algae and bacteria, Nature, 262:134-136.

Phleger, F.B. and Soutar, A., 1973, Production of benthic foraminifera in three east Pacific oxygen minima, Micropaleontology, 19:110-115.

Rau, G.H., 1981, Low $^{15}N/^{14}N$ in hydrothermal vent animals: ecological implications, Nature, 289:484-485.

Reimers, C.E., 1982, Organic matter in anoxic sediments off central Peru: Relations of porosity, microbial decomposition and deformation properties, Marine Geology, 46:175-197.

Reimers, C.E. and Suess, E., in press, The partitioning of organic carbon fluxes and sedimentary organic matter decomposition rates in the ocean, Marine Chemistry.

Rowe, G.T., 1981, The benthic processes of coastal upwelling ecosystems, in: "Coastal Upwelling," F.A. Richards, ed., Coastal and Estuarine Sciences, American Geophysical Union, Washington, 464-471.

Salinger, M.J., 1981, Paleoclimates north and south, Nature, 291: 106-107.

Schnitzer, M. and Khan, S.U., 1972, "Humic Substances in the Environment," Marcel Dekker, Inc., New York, 327 pp.

Silva, J.A. and Bremner, J.M., 1966, Determination and isotope-ratio analysis of different forms of nitrogen in soils: 5. Fixed ammonium, Soil Science Society of America Proceedings, 30:587-594.

Smith, R.L., Enfield, D.B., Hopkins, T.S., and Pillsbury, R.D., 1971, The circulation in an upwelling ecosystem: The Pisco Cruise, Investigacion Pesquera, 35:9-24.

Soutar, A. and Crill, P.A., 1977, Sedimentation and climatic patterns in the Santa Barbara Basin during the 19th and 20th centuries, Geological Society of America, Bulletin, 88:1161-1172.

Stuermer, D.H., Peters, K.E. and Kaplan, I.R., 1978, Source indicators of humic substances and proto-kerogen stable isotope ratios, elemental compositions and electron spin resonance spectra, Geochimica et Cosmochimica Acta, 42:989-998.

Suess, E. and Müller, P.J., 1980, Productivity, sedimentation rate, and sedimentary organic matter in the oceans - II. Elemental fractionation, in: "Biogeochimie de la Matiére Organique a L'interface Eau-sediment Marin," Collogues Internationaux du C.N.R.S., No. 293, Centre National de La Recherche Scientifique, Paris, 17-26.

Sweeney, R.E. and Kaplan, I.R., 1980, Natural abundances of ^{15}N as a source indicator for near-shore marine sedimentary and dissolved nitrogen, Marine Chemistry, 9:81-94.

Tissot, B.P. and Welte, D.H., 1978, "Petroleum Formation and Occurrence," Springer-Verlag, Berlin, 538 pp.

Van der Hammen, T., 1974, The Pleistocene changes of vegetation and climate in tropical South America, Journal of Biogeography, 1: 3-26.

Wada, E. and Hattori, A., 1976, Natural abundance of ^{15}N in particulate organic matter in the North Pacific Ocean, Geochimica et Cosmochimica Acta, 40:249-251.

Walsh, J.J., 1981, A carbon budget for overfishing off Peru, Nature, 290:300-304.

Wooster, W.S. and Gilmartin, J., 1961, The Peru-Chile undercurrent, Journal of Marine Research, 19:97-122.

Wyrtki, K., 1962, The oxygen minima in relation to ocean circulation, Deep-Sea Research, 9:11-23.

Wyrtki, K., 1975, El Niño - The dynamic response of the Equatorial Pacific Ocean to atmospheric forcing, Journal of Physical Oceanography, 5:572-584.

APPENDIX

The vertical flux of biogenous organic matter to the sediment surface is a function of water depth and primary productivity. Upon descent through the water column and prior to burial, biogenous detrital organic matter undergoes strong elemental fractionation by preferential removal of nitrogenous and P-containing organic compounds. At the water-sediment interface, a portion of the detritus is converted into biomass by benthic organisms, which concentrate nitrogen and phosphorus relative to carbon. These two processes are reflected in the elemental composition of sedimentary organic matter. A third process concentrates organic matter in sediments by sorption onto clays. This sorbed material is high in organic-N and devoid of organic-P; its relative abundance is high only in pelagic clays. The concentrations of each of the three forms of organic matter --detrital, biomass and sorbed-- can be calculated from the following expressions:

(1) $\text{org-C}_{\text{sorb.}} = 0.005\ Al_2O_3$

(2) $\text{org-C}_{\text{detr.}} = \dfrac{\text{org-C}_{\text{total}} - \text{org-C}_{\text{sorb}} - e(N_{\text{total}} - N_{\text{fix}})}{(1 - e/f)}$

(3) $\text{org-C}_{\text{biom.}} = \text{org-C}_{\text{total}} - \text{org-C}_{\text{detr.}} - \text{org-C}_{\text{sorb.}}$,

(4) $\text{org-N}_{\text{sorb.}} = 1/b \cdot \text{org-C}_{\text{sorb.}}$,

(5) $\text{org-N}_{\text{detr.}} = 1/f \cdot \text{org-C}_{\text{detr.}}$,

(6) $\text{org-N}_{\text{biom.}} = 1/e \cdot \text{org-C}_{\text{biom.}} = N_{\text{total}} - N_{\text{fix}} - \text{org-N}_{\text{detr.}} - \text{org-N}_{\text{sorb.}}$

where b, e and f are the characteristic C/N elemental weight ratios of sorbed organic matter, biomass, and detritus which are assumed here to have the numerical values: 1.8, 4.0, and 10, respectively. Suess and Müller (1980) also present alternative expressions based on the preferential removal of org-P relative to org-C in detrital organic matter and enrichment in benthic biomass. However, the C/P weight ratios of each of these forms of organic matter in the Peru upwelling region are uncertain, and so we chose here to utilize the C/N-model only.

ORGANIC GEOCHEMISTRY OF LAMINATED SEDIMENTS FROM THE GULF OF CALIFORNIA

Bernd R.T. Simoneit

School of Oceanography
Oregon State University
Corvallis, Oregon 97331, U.S.A.

ABSTRACT

Both varved (laminated) and homogenous sediments from DSDP/IPOD Sites 479 and 480 have been analyzed to examine any potential differences in the organic matter composition. The interstitial gas is composed mainly of methane and carbon dioxide and it is derived from biogenic sources in both types of sequences. The lipid composition is very similiar for both sequences and indicates a primarily autochthonous microbial origin, with some influx of allochthonous higher plant detritus. The non-laminated sections contain relatively more terrigenous plant wax detritus versus autochthonous residues and a much greater amount of perylene than the laminated sections. The laminated zones consist of rhythmic couplets of diatom ooze and terrigenous clay deposited in an oxygen minimum environment and the homogeneous zones consist of diatomaceous mud to muddy ooze with evidence of extensive burrowing deposited from a less pronounced oxygen minimum environment. The factors which resulted in these different lithologies also determined the distribution of organic facies as reflected in the kerogen, which has also been analyzed. The data indicate the presence of different types of kerogen in the oxic, homogeneous zones (low hydrogen index, high oxygen index, i.e., more oxidized organic matter) than in the anoxic, laminated zones (high hydrogen index, low oxygen index, i.e., more aliphatic and unaltered organic matter).

INTRODUCTION

Sites 479 and 480 of the Deep Sea Drilling Project (DSDP/IPOD) were cored 6 km apart on the Guaymas slope of the Gulf of California as indicated on the location map in Fig. 1. The sediments consisted primarily of diatomaceous ooze interbedded with dolomites and laminated sequences (Curray et al., 1979; 1982). These laminated sediments are believed to be annual varves resulting from the interaction of two seasonal events, namely, the summer rains that introduce terrigenous clays to the region and upwelling with northwesterly winter winds which result in diatom blooms (Curray et al., 1979; 1982; Schrader et al., 1980). The preservation of laminated sediment sections indicates oxygen depletion in the bottom waters.

Samples for this study were chosen from Sites 479 and 480 to examine any potential differences in the composition of the organic matter in laminated versus non-laminated sections. Primarily lipids and kerogens were analyzed and some interstitial gas data for Site 479 were also determined.

The preservation of organic matter in sedimentary organic facies is based on: (1) the types of organisms that act as a source, (2) the depositional environment, and (3) conditions during early diagenesis. In ancient sediments, organic facies are evident as subdivisions of a stratigraphic unit which can be distinguished on the basis of the chemical composition of the organic matter. Examples of the use of various geochemical parameters in differentiating the organic facies characteristic of oxic and anoxic sedimentary environments include works by Pelet and Debyser (1977), Didyk et al. (1978), Byers and Larson (1979) and Jones and Demaison (1980).

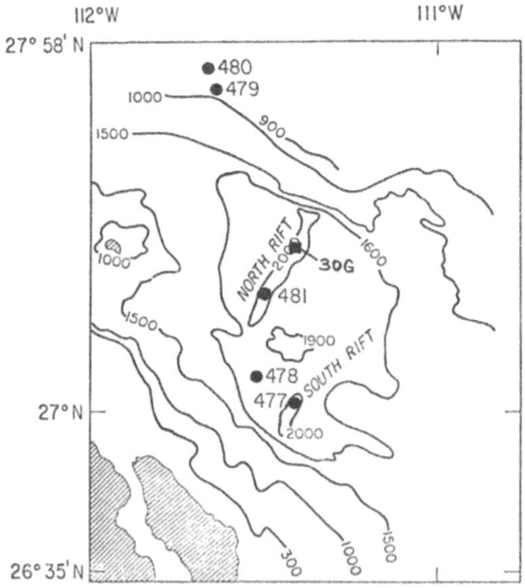

Fig. 1. Location map of the DSDP/IPOD drill sites of Leg 64 in the Guaymas Basin area.

EXPERIMENTAL

Interstitial gas samples were collected onboard DV "Glomar Challenger" in vacutainers and analyzed there for composition (Simoneit, 1982a) and onshore for stable isotope content (Galimov and Simoneit, 1982). The samples for lipid analyses from Site 480 were taken by scraping a fresh surface of split core where, based on varve counts, the intervals of 10 cm represent a maximum of 100 years of sedimentation. The small samples (0.5-2 g) were freeze-dried and then extracted repeatedly with chloroform and methanol mixture (1:1) using ultrasonication. The extracts were concentrated on a rotary evaporator and then subjected to column chromatography (microscale) on silica gel using hexane as eluent. The total hydrocarbon fractions (hexane eluates) were collected, concentrated and analyzed by gas chromatography (GC) and GC-mass spectrometry (GC-MS). The extract from sample 479-29-5, 114-116 cm was separated after esterification by thin layer chromatography (Simoneit et al., 1979) and both hydrocarbons and fatty acid esters were analyzed. The GC and GC-MS analyses were carried out as described (e.g., Simoneit, 1982b; Simoneit and Philp, 1982).

Protokerogen was isolated by demineralization with hydrochloric and hydrofluoric acids and this carbonaceous fraction also contains some of the humic substances (Peters and Simoneit, 1982). Carbon, hydrogen and nitrogen contents on the protokerogen (HCl/HF insoluble residue) were measured using a Carlo Erba Model 1106 Elemental Analyzer and pyrolysis data were determined by a Girdel Rock-Eval instrument (Peters and Simoneit, 1982) and by Curie Point pyrolysis-gas chromatography (van de Meent et al., 1980; Simoneit and Philp, 1982).

RESULTS AND DISCUSSION

Interstitial Gas

The interstitial gas at these sites is composed primarily of methane (CH_4) and carbon dioxide (CO_2) with minor amounts of heavier hydrocarbons (C_2-C_5) (Galimov and Simoneit, 1982; Simoneit, 1982a). At site 479 the CH_4 concentration decreases to about 60% of the total gas with depth and then fluctuates between 45-70% to the bottom of the hole (Fig. 2). The ethane (C_2H_6) content remains low and essentially constant, not exceeding 0.02% to a depth of about 260 m (Fig. 2). At greater depth, however, the C_2H_6 concentration increases to >0.1%, which accompanies the simultaneous increases of the C_3-C_5 hydrocarbons (Simoneit, 1982a). It should be emphasized that Hole 479 was stopped at a depth of 400 m due to the pollution hazard of a continuing increase in the concentrations of the heavy hydrocarbons with depth.

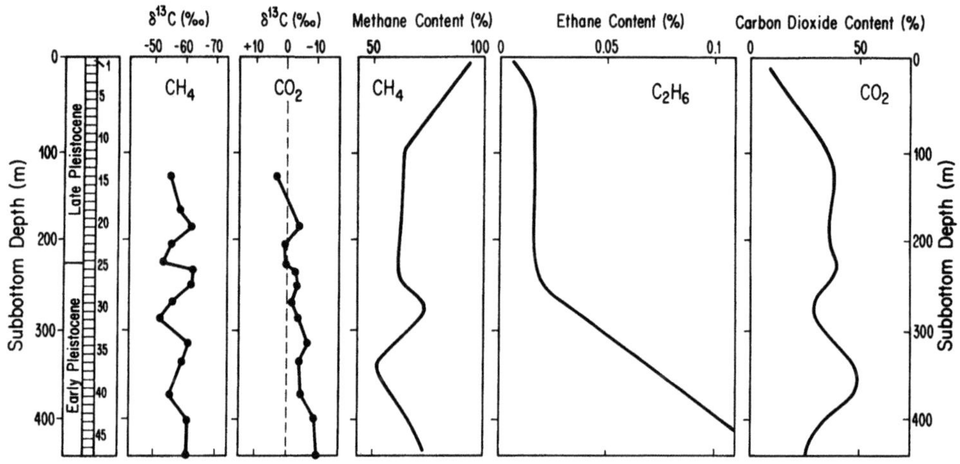

Fig. 2. Plots of the $\delta^{13}C$ and concentrations of methane and carbon dioxide and the concentration of ethane versus depth for Site 479 (Galimov and Simoneit, 1982).

The stable carbon isotope composition ($\delta^{13}C$) of the CH_4 varies within the range of -50 to -70°/₀₀ (Fig. 2). It is noteworthy that the isotopic composition of the CH_4 remains practically unaltered and does not reflect a thermogenic ^{13}C enrichment which is usually associated with an increase of the C_2-C_5 hydrocarbon concentration. The range throughout the sequence does not exceed 10°/₀₀ and is within values typical for biogenic CH_4 derived from immature organic matter (Claypool and Kaplan, 1974; Galimov, 1974; Stahl, 1974; Galimov and Simoneit, 1982).

The CO_2 of Hole 479 varies in concentration from 10-50% and has the highest values of $\delta^{13}C$ (Fig. 2). The CO_2 isotopic composition shows a tendency to depletion of the heavy carbon isotope down in the sequence. Isotopically heavy CO_2 may result from the decomposition of oxygen-bearing organic compounds (e.g., decarboxylation) in reducing environments (Galimov, 1974), as well as from the microbiological reduction of some carbon dioxide to methane (Claypool and Kaplan, 1974). [At depth of Hole 479 the high alkalinity causes precipitation, not dissolution of carbonates (Gieskes et al., 1982)]. The high biological productivity (Curray et al., 1982) and high rates of organic matter burial imply that sedimentation at Site 479 took place in environments with an oxygen deficiency. This is confirmed by the relatively high values of $\delta^{13}C$ for the CO_2, particularly in the upper part of the sequence. Due to the high carbonate concentration at greater depths, the CO_2 isotopic composition may possibly become a function of fractionation in the isotope-exchange system $CO_2 \leftrightarrow CO_3^{=}$ (Galimov and Simoneit, 1982).

Lipids

The total lipid and also hydrocarbon yields (Table 1) were high with larger concentrations in the laminated sections of Site 480. The n-alkane distributions are all quite similar with only subtle differences between the laminated (Fig. 3a, 3c) and the non-laminated (Fig. 3b, 3d) samples. They range from C_{12} to C_{33} with usually three maxima at C_{15}, C_{21}-C_{23} and C_{29} and a strong odd-to-even carbon number predominance $>C_{23}$. The homologs $<C_{23}$ are characteristic of autochthonous marine lipids where undegraded microbial residues are found as homologs to C_{19} and degraded algal material as hydrocarbons in the region of C_{19}-C_{23} (Simoneit, 1975; 1978a; 1981; in press). The homologs $>C_{23}$ with the odd/even carbon number predominance represent an allochthonous component derived from terrestrial higher plant waxes (Eglinton and Hamilton, 1963; Simoneit, 1975; 1978a; 1978b). The terrigenous alkanes are present in about equal absolute amounts in both laminated and non-laminated samples, but their relative proportions in relation to the microbial residues (to C_{19}) are greater in non-laminated samples. This indicates that during a depositional period resulting in the non-laminated sequences, the transport of terrigenous mineral detritus is high but the terrestrial organic matter associated with it is lower, or about the same as in the laminated sequences.

Isoprenoid hydrocarbons are present in significant amounts (Fig. 3), but the pristane to phytane (Pr/Ph) ratios of the Site 480 samples are slightly greater than one (Table 1). All samples contain elemental sulfur, and the amounts do not correlate with the laminated versus non-laminated sections. These data are inconclusive regarding partially euxinic conditions caused by the rapid sedimentation (Didyk et al., 1978) and must be coupled with evidence for bioturbation. However, the Pr/Ph > 1 could be due to the immature nature of the organic detritus and could invert to <1 on further maturation and diagenesis (e.g., sample 479-29-5, 114-116 cm, Fig. 3f, from greater subbottom depth).

Perylene ($C_{20}H_{12}$) is present in these samples at higher concentrations in the non-laminated sequences (Table 1). This compound has been found in many euxinic sediments and appears to be another indicator of oxygen depletion in depositional environments (e.g., Simoneit, 1978a; 1978b; Louda and Baker, 1981; Simoneit and Mazurek, 1981). Whether it is a terrigenous marker versus an autochthonous product is still under debate and investigation. Molecular markers such as steroid and triterpenoid hydrocarbon residues from autochthonous sources are not detectable; however, diterpenoid residues from terrestrial resinous plants are present in minor amounts and at higher concentrations and diversity in the non-laminated sections. The diterpenoid residues are comprised of primarily retene, dehydroabietic acid and dehydroabietane with minor amounts of dehydroabietin and methylethylphenanthrene (Simoneit, 1977). These compounds were probably transported to this sedimentary environment as resin microparticles.

Table 1. Sample descriptions and analytical results for core material from Sites 479 and 480.

Sample[a]	Subbottom depth (m)	Lithology	Total Organic Carbon (%)[b]	Total Lipids (µg/g)[b]	Total HC (µg/g)[b]	n-alkanes CPI[c]	n-alkanes Max[d]	Pr/Ph	Perylene (µg/g)[b]
480-2-3 (21-30 cm)	8.0	laminated diatom. ooze	2.7	1327	435	1.3	<u>15</u>, 21, 29	1.4	0.24
480-3-3 (107-119 cm)	13.7	non-laminated diatom. ooze	2.6	976	370	1.5	<u>15</u>, 22, 29	1.7	1.2
480-19-1 (58-59 cm)	90.9	laminated diatom. ooze	2.8	1316	490	1.4	<u>15</u>, 21, 27	1.1	0.41
480-19-1 (130-150 cm)	91.7	non-laminated diatom. ooze	2.7	748	250	1.7	<u>15</u>, 17, 22, 29	1.3	3.2
479-9-2 (28 cm)	71.3	wood		800	56	2.2	19, 22, <u>29</u>	0.83	n.d.
479-29-5 (114-116 cm)	266.7	laminated diatom. ooze	2.6	490	64	1.6	17, 29, <u>31</u>	0.9	2.0

[a] All samples are of Pleistocene age.
[b] Weights based on dry sediment.
[c] Carbon preference index, summed from C_{10} to C_{35}, odd/even.
[d] The dominant homolog is underscored.
n.d. = not detected.

A sample of wood (479-9-2-lignite) has significant lipid and hydrocarbon contents (Table 1). The n-alkanes (Fig. 3e) consist of primarily plant wax (>C_{24}) with a lesser component of autochthonous material. The autochthonous lipids were probably absorbed from the surroundings by the lignite once deposited, and the plant wax residues were transported in with the fragment.

Sample 479-29-5, 114-116 cm is located at about 267 m subbottom and is very immature and unaltered, probably in a laminated sequence. The lipid yield is high and the n-alkanes (Fig. 3f) exhibit a bimodal distribution ranging from C_{10} to C_{35} and with maxima at C_{17} and C_{31}. The homologs >C_{27} with the strong odd-to-even carbon number predominance are derived from allochthonous plant wax and the homologs <C_{24} are from autochthonous microbial sources. The Pr/Ph is less than one and both perylene and elemental sulfur are present indicating partial euxinic paleo-environmental conditions of sedimentation (Didyk et al., 1978). The n-fatty acids exhibit a bimodal distribution, with maxima at C_{16} and C_{26} and a strong even-to-odd carbon number predominance, typical of biogenic sources (Fig. 3g). The homologs >C_{21} are derived from terrigenous plant wax and those <C_{20} from autochthonous microbial sources (Simoneit, 1975; 1978a).

The molecular markers of this sample confirm its immaturity and consist of primarily triterpenoid, steroid and diterpenoid residues. The triterpenoids (Fig. 3h) are comprised of the 17β(H)-hopane series (C_{27}, C_{29} to C_{32}) with lesser amounts of 17α(H)-hopanes, hop-17(21)-ene and iso-hop-13(18)-ene. These compounds are of an autochthonous marine origin and are characteristic of geologically immature organic matter (Dastillung and Albrecht, 1976; Ourisson, Albrecht and Rohmer, 1979; Simoneit and Kaplan, 1980). The steroid residues in the hydrocarbon fractions consist of ster-2-enes and ster-4-enes in about equal proportions, diasterenes and a minor amount of steranes. They range from C_{26} to C_{30}, with C_{27} as the dominant homolog in all cases. The inferred source of these compounds is from autochthonous microbial residues in the early stages of diagenesis (Simoneit, 1978a; Huang and Meinschein, 1979; Simoneit and Philp, 1982). Diterpenoid residues are found primarily as dehydroabietic acid (Fig. 3g), retene and dehydroabietin, all at low concentrations and thus indicating a minor influx from terrigenous resinous plants (Simoneit, 1977).

The lipid data for both laminated and non-laminated sequences indicate a major origin from microbially degraded algal detritus with a minor influx of allochthonous detritus from terrigenous sources. This terrestrial component fits with an origin from sparse desert vegetation based on the low amount of resin residues and the relatively high concentrations of the odd carbon numbered n-alkanes > C_{31}.

Kerogens

The kerogen of sample 479-29-5, 114-116 cm has been analyzed by Curie Point pyrolysis-gas chromatography (Simoneit and Philp, 1982)

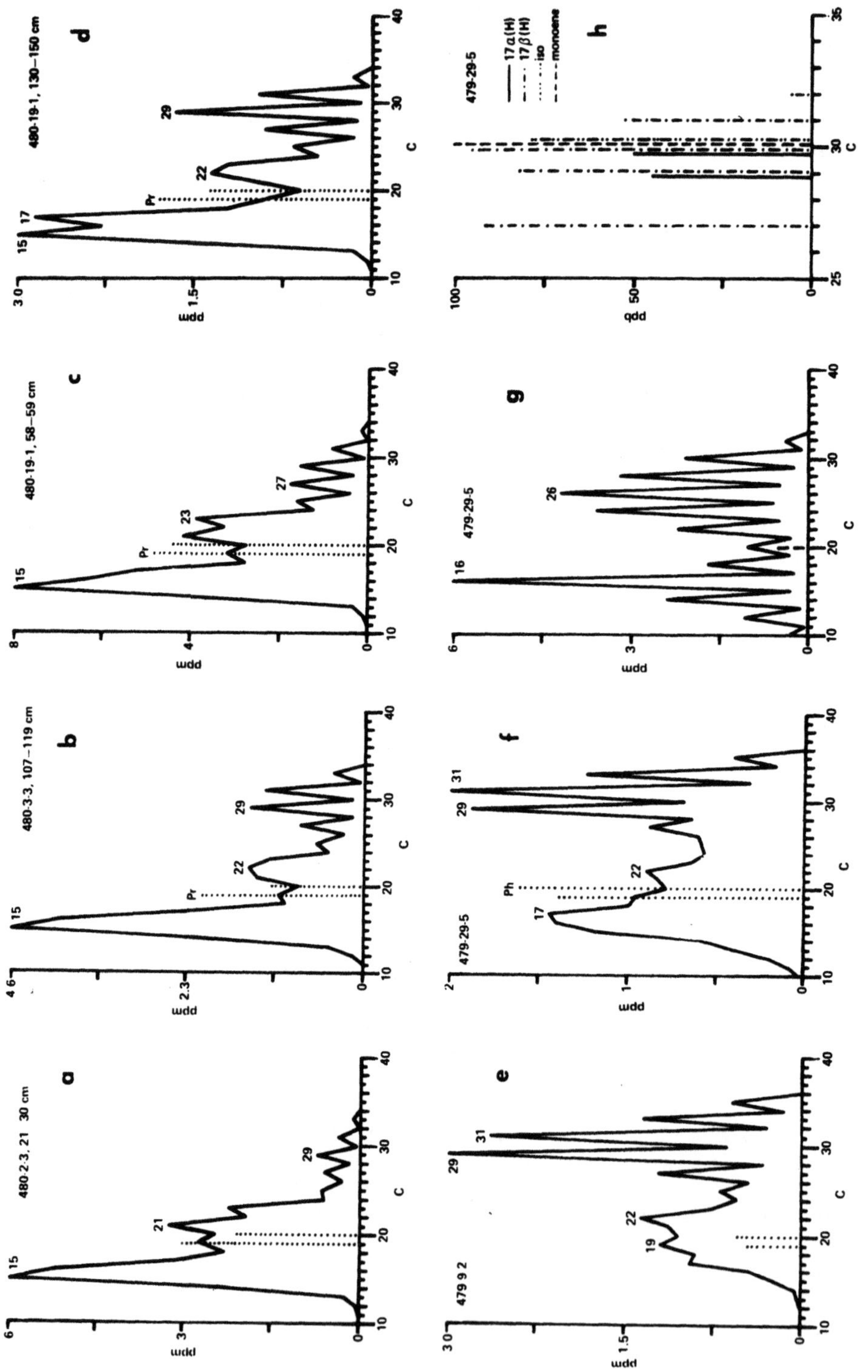

a
480-2-3, 21 30 cm
b
480-3-3, 107–119 cm
c
480-19-1, 58–59 cm
d
480-19-1, 130–150 cm
e
479 9 2
f
479-29-5
g
479-29-5
h
479-29-5
17α(H)
17β(H)
iso
monoene
ppm
ppb
C
Pr
Ph

Fig. 3. Distribution diagrams for various compound series in samples from Sites 479 and 480 (Simoneit, 1982b). (Subbottom depths are given in meters in parentheses).

n-alkanes (..... indicates isoprenoids): a) Sample 480-2-3, 21-30 cm (laminated) (8.0 m); b) Sample 480-3-3, 107-119 cm (non-laminated) (13.7 m); c) Sample 480-19-1, 58-59 cm (laminated) (90.9 m); d) Sample 480-19-1, 130-150 cm (non-laminated) (91.7 m); e) Sample 479-9-2, 28 cm (driftwood) (71.3 m); f) Sample 479-29-5, 114-116 cm (probably laminated) (266.7 m).

n-fatty acids (--- indicates dehydroabietic acid): g) Sample 479-29-5, 114-116 cm.

triterpenoids: h) Sample 479-29-5, 114-116 cm.

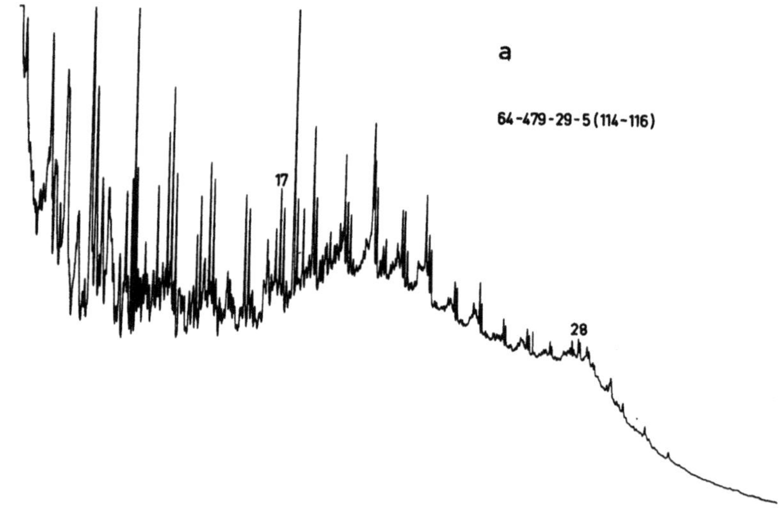
a
64-479-29-5 (114-116)
17
28

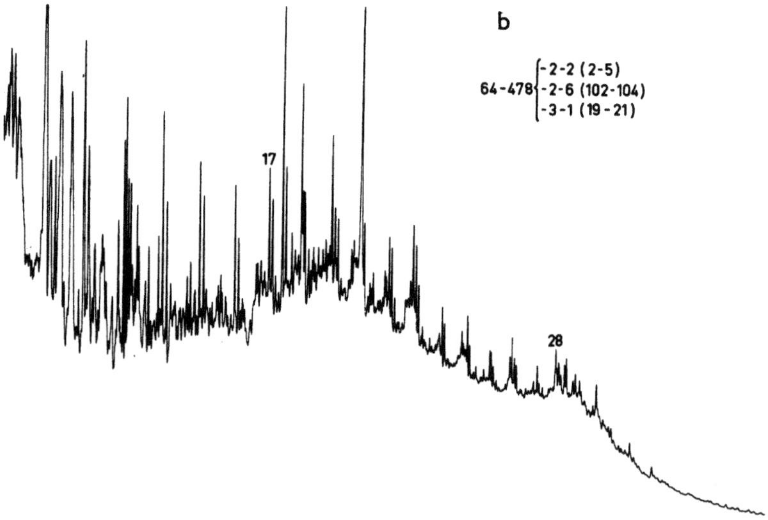
b
64-478 -2-2 (2-5)
-2-6 (102-104)
-3-1 (19-21)
17
28

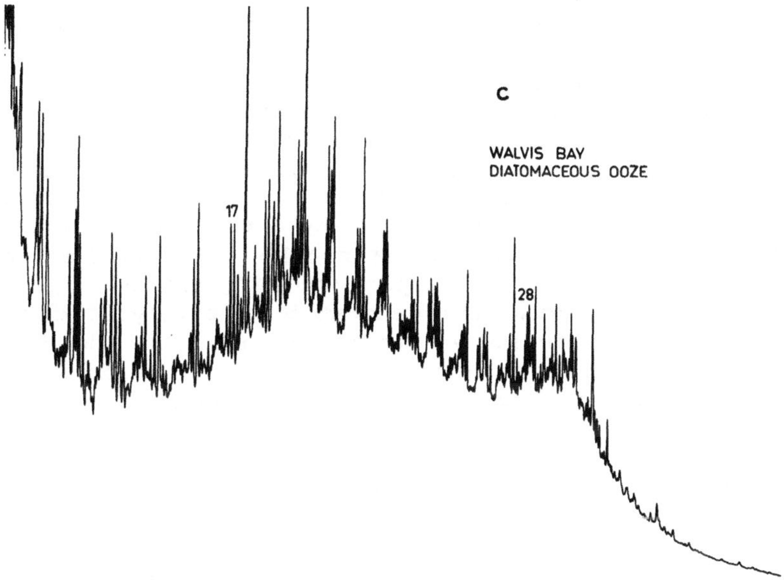

Fig. 4. Examples of Curie Point pyrolysis gas chromatograms (Cupy-GC) of kerogen concentrates from shallow and unaltered samples (Simoneit and Philp, 1982).

a) 479-29-5, 115-116 cm (226 m subbottom)
b) 478-2/3-composite (8 m subbottom)
c) Walvis Bay, southwestern Africa (0.5 m subbottom)

and the data indicate that it is of an unaltered marine origin (Fig. 4). The pyrolysis product pattern is similar to those observed for kerogens from the Cariaco Trench (van de Meent et al., 1980), other samples from the Guaymas Basin (Fig. 4b) and Walvis Bay (Namibia, Fig. 4c) (Simoneit and Philp, 1982).

Kerogens (with humic substances) from Sites 479 and 480 were analyzed by Rock-Eval pyrolysis (Peters and Simoneit, 1982). The data indicate that these protokerogens are immature and of primarily a marine origin based on the aliphatic nature of the pyrolysate. This is further substantiated by the atomic H/C and O/C data (Peters and Simoneit, 1982) and similar data for other samples from Guaymas Basin (Simoneit et al., 1979; Jenden, Simoneit and Philp, 1982).

The Rock-Eval pyrolysis method is designed to rapidly determine the hydrocarbon potential and the generative history of geological samples (Espitalié et al., 1977). It is based on the selective determination of hydrocarbons by a flame ionization detector and of the carbon dioxide generated from the sample organic matter during programmed heating in an inert atmosphere by a thermal conductivity detector. The first peak (S1) represents the free hydrocarbons that thermally volatilize from the sample (mg HC/g sample). The second peak (S2) comprises hydrocarbons generated by pyrolytic degradation of macromolecular organic matter (e.g., kerogen) in the sample (mg HC/g sample). The carbon dioxide generated from the organic matter is trapped and constitutes the third peak (S3) which is equivalent to the amount of organic carbon dioxide (mg CO_2/g sample). The Hydrogen Index (HI) corresponds to the quantity of hydrocarbon compounds from S2 normalized to the total organic carbon of the sample (mg HC/g TOC) and the Oxygen Index (OI) corresponds to the amount of CO_2 normalized to the total organic carbon (mg CO_2/g TOC). Both HI and OI exhibit good correlation with atomic H/C and O/C data, respectively (Espitalié et al., 1977).

Some Rock-Eval parameters and the total organic carbon (TOC) contents are illustrated versus depth for Site 480 in Fig. 5. An inverse relationship is apparent between the hydrogen index (HI) and the oxygen index (OI) as a function of depth. The results suggest different types of organic matter in the oxic, homogeneous zones (low HI, high OI) than in the anoxic, laminated zones (high HI, low OI). The striking change in HI, OI, and S2/S3 between 104 and 152 m subbottom depth should be noted. The deeper interval is laminated and is of sufficient thickness to be unaffected by mixing with homogeneous layers from above during coring. It should be noted that the core lithology is almost equally divided between alternating laminated and non-laminated sediment types (Schrader et al., 1980). Laminated zones consisting of rythmic couplets of diatom ooze and terrigenous clays were inferred to be deposited in oxygen depleted waters. The homogeneous zones consist of diatomaceous mud to ooze with evidence of extensive bioturbation by burrowing. The latter was interpreted to reflect tires with a less pronounced oxygen-minimum zone, allowing the existence of both epi- and in-fauna. These factors resulting in the different lithologies also appear to have determined the organic facies distribution (Peters and Simoneit, 1982).

The S2/S3 ratios and HI values for the deeper samples are significantly larger than those of the shallower sediments and the OI values of the deeper samples are smaller than those of the shallower sediments. The TOC values reflect approximately the OI values, that is, TOC is high where OI is high and the converse. This is the case for both Sites 479 and 480 (Peters and Simoneit, 1982; Simoneit, Summerhayes and Meyers, 1982). The HI and the S2/S3 values are high in the laminated zones, especially below 122 m subbottom at Site 480, and in the non-laminated sections they are relatively low. This indicates that the preservation of lipid-rich marine organic matter controls the TOC and the abundance of the marine organic matter is governed by the paleoenvironmental conditions resulting in laminated or homogeneous sediments (Simoneit et al., 1982). These results can thus be utilized to successfully differentiate organic facies in these cores.

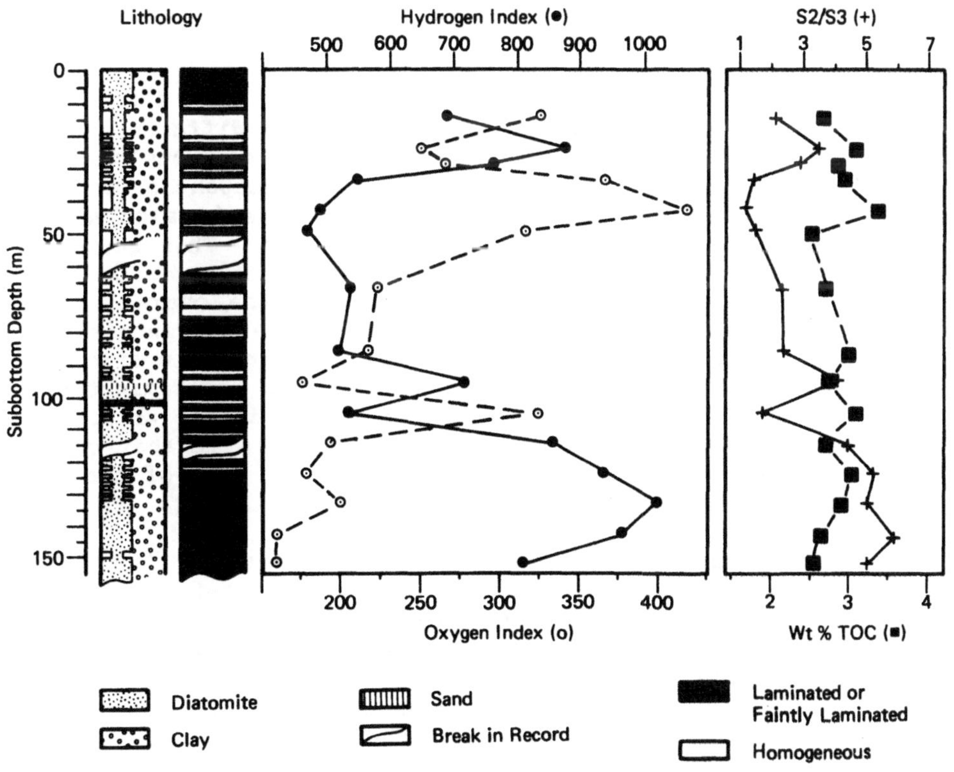

Fig. 5. Rock-Eval parameters and total organic carbon (TOC) versus depth at Site 480 (Peters and Simoneit, 1982).

The kerogen of the driftwood sample (479-9-2, 28 cm) is distinctly different and the bulk compositional data (extracted, demineralized residue) are: H/C = 0.70, N/C = 0.013 and $\delta^{13}C = -23.4°/_{oo}$. These values are outside a direct fit with an origin from peat or vascular plant detritus (C_3 or Calvin cycle), where typical data are in the ranges: H/C = 1.0, N/C = 0.03-0.06 and $\delta^{13}C = -27$ to $-28°/_{oo}$ (Stuermer, Peters and Kaplan, 1978). The H/C fits for wood that has been partially coalified (van Krevelen, 1961). The $\delta^{13}C$ range for C_4-plants (Hatch-Slack cycle) is from -13 to $-23°/_{oo}$ (Galimov, 1980). This sample may represent a primary fragment of a C_4-plant, a highly oxidized coal fragment (C_3-plant), where the light carbon isotope has been preferentially removed, and/or terrigenous organic matter which was diluted with autochthonous marine lipid detritus during the preservation and coalification processes.

CONCLUSIONS

The interstitial gas at both sites and in both types of sequences is composed primarily of methane and carbon dioxide and is derived from biogenic sources. However, at greater subbottom depths of Site 479 the gas becomes admixed with possible thermogenic hydrocarbons diffusing upward from greater depths.

The laminated and non-laminated sections of Site 480 are very similar in their lipid compositions, but there are differences in the absolute concentrations of compound groups. Terrigenous plant wax residues are more concentrated relative to the autochthonous marine components in the non-laminated sections. Perylene is also a factor of about five more concentrated in those sections. This indicates a greater transport of terrigenous mineral detritus lower in lipids during the homogeneous sedimentation, however, the overall terrestrial lipid influx is more or less constant. This terrestrial lipid detritus is probably derived primarily from vegetation in arid areas. The marine lipid residues are more concentrated in the laminated sections indicating earlier cessation of degradation in the anoxic depositional environment. The terrigenous influence is further substantiated by the kerogen and the organic composition data of the driftwood fragment (479-9-2, 28 cm).

The distributions of organic facies at Sites 479 and 480 are reflected in the kerogen data. High values of HI with low OI are observed for the laminated, anoxic zones indicating more aliphatic and unaltered organic matter; whereas, low values of HI coupled with a higher OI are found in the oxic and homogenous zones. The data for these different lithologies support the interpretation of the sedimentary processes in this region of the Gulf.

ACKNOWLEDGEMENTS

I thank Dana Blumfield and Dara Blumfield for technical assistance, E. Ruth for GC/MS data acquisition, R.P. Philp, K.E. Peters, E.M. Galimov, and P.D. Jenden for some of the analytical data, and B.P. Tissot for useful comments on this manuscript.

REFERENCES

Byers, C.W. and Larson, D.W., 1979, Paleoenvironments of Mowry shale (Lower Cretaceous), Western and Central Wyoming, American Association of Petroleum Geologists Bulletin, 63:354-361.

Claypool, G.E. and Kaplan, I.R., 1974, Origin and distribution of methane in marine sediments, in: "Natural Gas in Marine Sediments," I.R. Kaplan, ed., Plenum Press, New York, 99-139.

Curray, J.R., Moore, D.G., Aguayo, J.E. et al., 1979, Leg 64 seeks evidence on development of basin in the Gulf of California, Geotimes, 24(7):18-20.

Curray, J.R., Moore, D.G., Aguayo, J.E. et al., 1982, "Initial Reports of the Deep Sea Drilling Project," 64, U.S. Government Printing Office, Washington, 1314 pp.

Dastillung, M. and Albrecht, P., 1976, Molecular test for oil pollution in surface sediments, Marine Pollution Bulletin, 7:13-15.

Didyk, B.M., Simoneit, B.R.T., Brassell, S.C. and Eglinton, G., 1978, Geochemical indicators of paleoenvironmental conditions of sedimentation, Nature, 272:216-222.

Eglinton, G. and Hamilton, R.J., 1963, The distribution of alkanes, in: "Chemical Plant Taxonomy," T. Swain, ed., Academic Press, New York, 187-217.

Espitalié, J., Laporte, J.L., Madec, M., Marquis, F., Leplat, P., Paulet, J., and Boutefeu, A., 1977, Méthode rapide de caractérisation des roches méres de leur potentiel pétrolier et de leur degré d'évolution, Revue Institut Francais du Pétrole, 32:23-42.

Galimov, E.M., 1974, Organic geochemistry of carbon isotopes, in: "Advances in Organic Geochemistry 1973," B. Tissot and F. Bienner, eds., Éditions Technip, Paris, 439-452.

Galimov, E.M., 1980, C^{13}/C^{12} in kerogen, in: "Kerogen, Insoluble Organic Matter from Sedimentary Rocks," B. Durand, ed., Éditions Technip, Paris, 271-299.

Galimov, E.M. and Simoneit, B.R.T., 1982, Geochemistry of interstitial gases in sedimentary deposits of the Gulf of California, Leg 64, in: "Initial Reports DSDP," 64, J.R. Curray, D.G. Moore et al., U.S. Government Printing Office, Washington, 781-788.

Gieskes, J.M., Elderfield, H., Lawrence, J.R., Johnson, J., Meyers, B., and Campbell, A., 1982, Geochemistry of interstitial waters and sediments, Leg 64, Gulf of California, in: "Initial Reports DSDP," 64, J.R. Curray, D.G. Moore et al., U.S. Government Printing Office, Washington, 675-694.

Huang, W.-Y. and Meinschein, W.G., 1979, Sterols as ecological indicators, Geochimica et Cosmochimica Acta, 43:739-745.

Jenden, P.D., Simoneit, B.R.T. and Philp, R.P., 1982, Hydrothermal effects on protokerogen of unconsolidated sediments from Guaymas Basin, Gulf of California; elemental compositions, stable carbon isotope ratios and electron-spin resonance spectra, in: "Initial Reports DSDP," 64, J.R. Curray, D.G. Moore et al., U.S. Government Printing Office, Washington, 905-912.

Jones, R.W. and Demaison, G.J., 1980, (Abstract) Organic facies--stratigraphic concept and exploration tool, AAPG-SEPM-EMD Annual Convention, Denver, June 8-11, 1980.

Louda, J.W. and Baker, E.W., 1981, Geochemistry of tetrapyrrole, carotenoid and perylene pigments in sediments from the San Miguel Gap (Site 467) and Baja California Borderlands (Site 471): DSDP/IPOD Leg 63, in: "Initital Reports DSDP," 63, R.S. Yeats, B. Haq et al., U.S. Government Printing Office, Washington, 785-818.

Ourisson, G., Albrecht, P. and Rohmer, M., 1979, The hopanoids. Paleochemistry and biochemistry of a group of natural products, Pure and Applied Chemistry, 51:709-729.

Pelet, R. and Debyser, Y., 1977, Organic geochemistry of Black Sea cores, Geochimica et Cosmochimica Acta, 41:1575-1586.

Peters, K.E. and Simoneit, B.R.T., 1982, Rock-Eval pyrolysis of Quaternary sediments from DSDP/IPOD Leg 64, Sites 479 and 480, Gulf of California, in: "Initial Reports DSDP," 64, J.R. Curray, D.G. Moore et al., U.S. Government Printing Office, Washington, 925-932.

Schrader, H., Kelts, K., Curray, J., et al., 1980, Laminated diatomaceous sediments from the Guaymas Basin Slope (Central Gulf of California): 250,000-year climate record, Science, 207:1207-1209.

Simoneit, B.R.T., 1975, "Sources of Organic Matter in Oceanic Sediments," Ph.D. Thesis, University of Bristol, England, 300 pp.

Simoneit, B.R.T., 1977, Diterpenoid compounds and other lipids in deep-sea sediments and their geochemical significance, Geochimica et Cosmochimica Acta, 41:463-476.

Simoneit, B.R.T., 1978a, The organic chemistry of marine sediments, in: "Chemical Oceanography," 2nd ed., vol. 7, J.P. Riley and R. Chester, eds., Academic Press, New York, 233-311.

Simoneit, B.R.T., 1978b, Organic geochemistry of terrigenous muds and various shales from the Black Sea, DSDP Leg 42B, in: "Initial Reports DSDP," 42, Part 2, D. Ross, Y. Neprochnov et al., U.S. Government Printing Office, Washington, 749-753.

Simoneit, B.R.T., 1981, Utility of molecular markers and stable isotope compositions in the evaluation of sources and diagenesis of organic matter in the geosphere, in: "The Impact of the Treibs' Porphyrin Concept on the Modern Organic Geochemistry," A. Prashnowsky, ed., Bayerische Julius Maximilian Universität, Würzburg, 133-158.

Simoneit, B.R.T., 1982a, Shipboard organic geochemistry and safety monitoring, Leg 64, Gulf of California, in: "Initial Reports DSDP," 64, J.R. Curray, D.G. Moore et al., U.S. Government Printing Office, Washington, 723-728.

Simoneit, B.R.T., 1982b, Preliminary organic geochemistry of laminated versus non-laminated sediments from Sites 479 and 480, DSDP/IPOD Leg 64, in: "Initial Reports DSDP," 64, J.R. Curray, D.G. Moore et al., U.S. Government Printing Office, Washington, 921-924.

Simoneit, B.R.T., in press, The composition, sources and transport of organic matter to marine sediments -- The organic geochemical approach, in: "Proceedings, Symposium Marine Chemistry into the Eighties," J.A.J. Thomspon, ed., National Research Board, Canada.

Simoneit, B.R.T. and Kaplan, I.R., 1980, Triterpenoids as molecular markers of paleoseepage in recent sediments of the Southern California Bight, Marine Environmental Research, 3:113-128.

Simoneit, B.R.T. and Mazurek, M.A., 1981, Organic geochemistry of sediments from the Southern California Borderland, DSDP/IPOD Leg 63, in: "Initial Reports DSDP," 63, R.S. Yeats, B. Haq et al., U.S. Government Printing Office, Washington, 837-853.

Simoneit, B.R.T. and Philp, R.P., 1982, Organic geochemistry of lipids and kerogen and effects of basalt intrusions on unconsolidated oceanic sediments, Sites 477, 478 and 481 in Guaymas Basin, Gulf of California, in: "Initial Reports DSDP," 64, J.R. Curray, D.G. Moore et al., U.S. Government Printing Office, Washington, 883-904.

Simoneit, B.R.T., Mazurek, M.A., Brenner, S., Crisp, P.T., and Kaplan, I.R., 1979, Organic geochemistry of recent sediments from Guaymas Basin, Gulf of California, Deep-Sea Research, 26A: 879-891.

Simoneit, B.R.T., Summerhayes, C.P. and Meyers, P.A., 1982, Sources, preservation and maturation of organic matter in Pliocene and Quaternary sediments of the Gulf of California: A synthesis of organic geochemical studies from DSDP/IPOD Leg 64, in: "Initial Reports DSDP," 64, J.R. Curray, D.G. Moore et al., U.S. Government Printing Office, Washington, 939-952.

Stahl, W.J., 1974, Carbon isotope fractionations in natural gases, Nature, 251:134-135.

Stuermer, D.H., Peters, K.E. and Kaplan, I.R., 1978, Source indicators of humic substances and protokerogen: Stable isotope ratios, elemental compositions and electron-spin resonance spectra, Geochimica et Cosmochimica Acta, 42:989-997.

van de Meent, D., Brown, S.C., Philp, R.P., and Simoneit, B.R.T., 1980, Pyrolysis high resolution gas chromatography and pyrolysis gas chromatography-mass spectrometry of kerogens and kerogen precursors, Geochimica et Cosmochimica Acta, 44:999-1013.

van Krevelen, D.W., 1961, "Coal, Typology-Chemistry-Physics-Constitution," Elsevier, Amsterdam, 514 pp.

THE POTENTIAL OF ORGANIC GEOCHEMICAL COMPOUNDS AS SEDIMENTARY INDICATORS OF UPWELLING

Simon C. Brassell and Geoffrey Eglinton

Organic Geochemistry Unit, University of Bristol
School of Chemistry, Cantock's Close
Bristol BS8 1TS, United Kingdom

ABSTRACT

Many lipid structures and individual compounds are synthesized by a limited range of organisms and therefore can serve as biological markers for inputs from these biota to sediments. Such compounds include nonspecific markers that originate from either algal or bacterial or higher plant sources, whereas others signify contributions from specific classes of organisms, e.g., methanogens and coccolithophores. The recognition of such biological marker lipids in DSDP cores from the Japan and Middle America Trenches and from the Walvis Ridge, illustrates that sediments underlying areas of high productivity tend to possess a prominent signature of marine marker compounds of both algal and bacterial origin. In the Japan Trench sediments, intact marker lipids occur at depths of *ca.* 800 m, illustrating the survival of features of the molecular signature during diagenesis. At present no biological markers that are restricted to upwelling environments, rather than simply high productivity, have been identified. A fuller understanding of the inputs of, and processes affecting, organic matter in upwelling systems should, however, lead to the recognition of lipid characteristics diagnostic of such environments which may, in turn, be used to assess upwelling conditions in the geological record.

INTRODUCTION

Extractable lipids constitute only a minor proportion of sedimentary organic matter, but the structural specificity of such compounds, the biological markers, makes them particularly sensitive and useful indicators of the sources of organic matter. The environmental signatures provided by these organic geochemical markers and the preservation of such information in the geological record form the foci of this paper.

Individual lipid molecules or the distributions of homologous series of lipids are characteristic of specific biological sources, so that their recognition in sediments permits an assessment of the biological inputs and hence certain aspects of the depositional environment. Some lipids can survive early diagenetic alteration intact; others are modified and retain their diagnostic characteristics. More labile compounds, however, can be degraded or altered to the extent that evidence of their biological origin is lost. The understanding of the effect of diagenetic processes on lipid signatures and the recognition of product/precursor relationships form a major objective of current organic geochemical research aimed towards the improved assessment of past depositional environments.

In this paper, examples of the lipids useful as markers of contributions from terrestrial higher plants, from marine algae and from bacteria, and the preservation of such signatures during diagenesis, are illustrated and investigated.

EXPERIMENTAL METHODOLOGY

The lipid constituents of sediments are typically extracted with organic solvents and separated by chromatographic methods according to their functionalities into fractions composed principally of aliphatic or aromatic hydrocarbons, carboxylic acids, ketones or alcohols. Each of the fractions is analyzed by gas chromatography (GC) after derivatization where appropriate (e.g., methylation of carboxylic acids, silylation of alcohols), and individual components are identified and quantitated by gas chromatography-mass spectrometry (GC-MS). Fuller details of typical experimental procedures and analytical conditions have been published elsewhere (Barnes et al., 1979; Brassell et al., 1980a; Brassell, Gowar and Eglinton, 1980b). In GC-MS analyses, extensive use is made of mass fragmentography (MF) to aid compound recognition and quantitation (Brassell et al., 1980b). The quantitation of components discussed herein is often relative (e.g., based on MF responses) rather than absolute due to the problems of assessing contributions from coeluting components.

Sediments in Areas of Upwelling

In areas of upwelling, the high pelagic productivity induced and supported by nutrient-rich waters can be expected to be reflected in a dominance of autochthonous lipid inputs, principally derived from marine algae and bacteria, in the underlying sediments. This supposition can best be tested by comparison of the lipid distributions of contemporary sediments from areas of upwelling with those of other marine sediments.

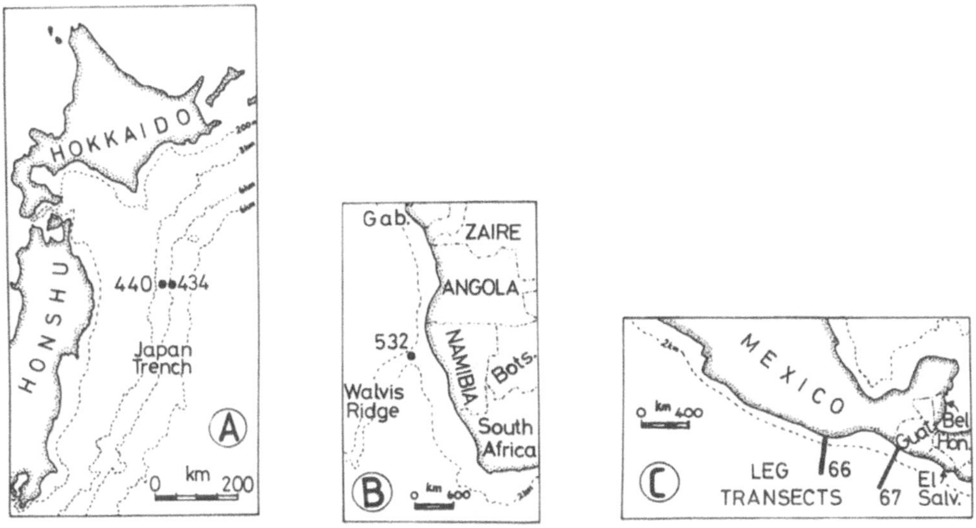

Fig. 1. Locations of DSDP sampling sites. A, Japan Trench; B, Walvis Ridge; C, Middle America Trench. Cores recovered by DSDP are sectioned and designated according to their Site (Hole)-Core-Section. Hence, 440B-3-5 refers to Section 5 of Core 3 from Hole B drilled at 440.

In this paper the lipid distributions of sediments from three environments, the Japan and Middle America trenches, and the Walvis Ridge, are considered. Details of the samples and some characteristics of the environments are given in Table 1. The locations of the sampling sites are shown in Fig. 1. In brief, the two suites of trench sediments differ from those from the Walvis Ridge principally in terms of the size of their terrestrial components and the scale of their pelagic productivity.

RESULTS AND DISCUSSION

Biological Markers

An organic geochemical marker is any compound indicative of inputs from a specific family, class or even genus of organism. Such compounds can be of direct biological origin or may have undergone a degree of diagenetic modification that has not obscured their diagnostic features. The basis of this approach is the high structural specificity of lipid molecules, but it also requires good chemotaxonomic control, which for marine organisms is sparsely documented.

Table 1. Features of environments

	Japan Trench	Walvis Ridge	Middle America Trench
DSDP Leg(s)	56/57	75	66/67
Site(s)	434, 440	532	487-500
Present Water Depths(m)	4500 - 6000	1300	2000 - 5500
Sediments	Mainly hemipelagic diatomaceous oozes	Diatomaceous and nanno-fossil oozes	Mainly hemipelagic diatomaceous muds
Age Range	Miocene - Pleistocene	Pliocene - Pleistocene	Quaternary
Environment	Hemipelagic trench sediments include turbidites and bio-genic components, notably diatoms, derived from the pro-ductive water column. Much of the trench sedimentation occurs below the CCD.	Site lies within area of high productivity asso-ciated with Benguela Upwell-ing system. Little terrigenous input.	Trench sediments in-clude terrigenous materials from slump-ing and biogenic components associated with upwelling near the Central American coast.
References	Langseth et al., 1980	Hay et al., in press; Meyers et al., this volume	Von Huene et al., 1980; Watkins et al., 1981
Published source of lipid data for samples considered herein	Brassell, 1980; Brassell et al., 1980c		Brassell et al., 1981b

Table 2. Proposed lipid indicators of direct marine inputs to oceanic sediments*

	Carbon No. Range	Structure
Straight-Chain		
Alkanes	15,17,19	Ia
Alkenes	37:3, 38:3	IIa,b
Alkadienones	37 - 39	IIIa-d
Alkatrienones	37 - 39	IIc-f
Alkanols	14 - 22	Ib
Alkenols	14 - 24	-
Alkenoic Acids	20:5, 22:6	e.g. IV
Methyl Ester	37:2	IIIe
Ethyl ester	38:2	IIIf
Acyclic Isoprenoid		
Phytol	20	V
Steroids		
5α(H)-Stanols	21-26,30,31	e.g. VIIa-f
23,24-Dimethylsterols	29,30	e.g. VIs, XIIq
$\Delta^{24(28)}$-Sterols	28 - 30	e.g. VIIw,x
22,23-Methylenesterols	29 - 31	e.g. VIIad, XIIz
4α-Methylstanols	27 - 31	e.g. XIIq,ad
Carotenoids		
Diatoxanthin	40:12	XVa
Fucoxanthin	40:9	XVb

* These proposed lipid indicators are compiled from various authors (Volkman, Eglinton, Corner and Sargent, 1980a; Brassell et al., 1980c and references therein; Simoneit, 1978; Tibbetts, 1980; Mackenzie, Brassell, Eglinton and Maxwell, 1982). They are lipids of algal origin and exclude components characteristic of marine bacteria (cf. Table 4); for structures see Appendix.

Table 3. Proposed lipid indicators of direct terrestrial inputs to oceanic sediments[a]

	Carbon No.	Structure
Straight-Chain		
Alkanes[b])		Ic
Alkanones[b])		Id
Alkanols[c])	23-33	Ie
Alkanoic acids[c])		If
Polyhydroxycarboxylic acids	e.g. 16	XVI
Iso- and Anteiso-Branched Chain		
Alkanes	27 - 33	XVIIa, XVIIIa
Diterpenoids		
Alkanes	e.g. 19	XIX
Aromatic hydrocarbons	e.g. 18,19	XX, XXI
Aromatic acids	e.g. 20	XXII
Steroids		
24R-ethylsterols	29	e.g. VIae, VIIae
Triterpenoids		
Alkanones)		e.g. XXVII, XXVIII
Alkenones)		e.g. XXIIIa-XXVI
Alkanols)	29 and 30	
Alkenols)		XXIIIb-XXVIb

[a] These proposed lipid indicators, which are principally components of higher plants, are compiled from various authors (Eglinton and Hamilton, 1967; Brassell et al., 1980c; Simoneit, 1977 and 1978; Devon and Scott, 1972; Cardoso, Eglinton and Holloway, 1977; Huang and Meinschein, 1976).

[b] With odd/even preference

[c] With even/odd preference

Table 4. Proposed lipid indicators of direct bacterial inputs to oceanic sediments[a]

	Carbon No. Range	Structures
<u>Straight-Chain</u>		
Alkanes	14-28[b]	Ig
Alkenoic acids	e.g. 18	XXIX
<u>Iso- and Anteiso-Branched Chain</u>		
Alkanes)		XVIIb, XVIIIb
Alkanols)		XVIIc, XVIIIc
Alkanoic acids)	10 - 22	XVIId, XVIIId
Alkenoic acids)		
<u>Acyclic Isoprenoids</u>		
Alkanes	25,30,40[c]	XXX - XXXIII
<u>Triterpenoids</u>		
Extended hopanoids	31 - 35	XXXIV
Hopenes[d]	30	e.g. XXXV-XXXVII
Fernenes[e]	30	e.g. XXXVIIIa-c
<u>Cyclopropanoid</u>		
Alkanes	17,19	e.g. XXXIXa
Alkanoic acids	17,19	e.g. XXXIXb

[a] These indicators, with the exception of the instances cited below, are found in sediments, but have only been recognized as constituents of bacteria (compiled from Youngblood, Blumer, Guillard and Fiore, 1971; Cranwell, 1973; Boon, 1978; Holzer, Oró and Tornabene, 1979; Ourisson, Albrecht and Rohmer, 1979; Brassell et al., 1980c and 1981a).

[b] With no odd/even preference.

[c] Lycopane has recently been suggested as a sediment input from methanogens although it has yet to be recognized in these organisms (Brassell et al., 1981).

[d] Hopenes are not uniquely synthesized by bacteria; they are also constituents of other classes of organism (e.g., Gelpi, Schneider, Mann and Oró, 1970), and are affected by diagenetic isomerisation (Ensminger, 1977). Many hopenes appear, however, to derive from bacterial inputs to sediment (Brassell et al., 1981a).

[e] The supposition that fernenes are terrestrial markers (Wardroper, 1979; Brassell et al., 1980c) has been superceded by their recognition in a bacterium (Howard, 1980) suggesting that fernenes reflect bacterial inputs to sediments (Brassell et al., 1981a).

Proposed lipid indicators of direct contributions to sediments from marine algae, terrestrial higher plant and bacterial sources are given in Tables 2 to 4 as listings, based on their structural affinities. In addition to these biolipids, many diagenetic products retaining structural features characteristic of their origins are also present in sediments. Such compounds include sterenes and sterones that retain the diagnostic side chains of their sterol precursors. To illustrate the occurrence of marker lipids in sediments, a selection of compounds indicative of inputs from terrestrial, marine and bacterial sources is discussed below.

Lipids of Terrestrial Origin

Three compound classes or lipid distributions, characteristic of terrestrial contributions to sediments are selected here to illustrate the occurrence of such components in the sediments.

(i) n-Alkanes. The distributions of *n*-alkanes in sediments from the Japan and Middle America Trenches and from Walvis Ridge (Fig. 2) all show an odd/even predominance in the C_{25}-C_{35} region with a maximum at *n*-C_{29} or *n*-C_{31}, characteristic features of higher plant waxes (Eglinton and Hamilton, 1967). Differences in the dominant *n*-alkane homology may reflect variations in the specific source of these com-

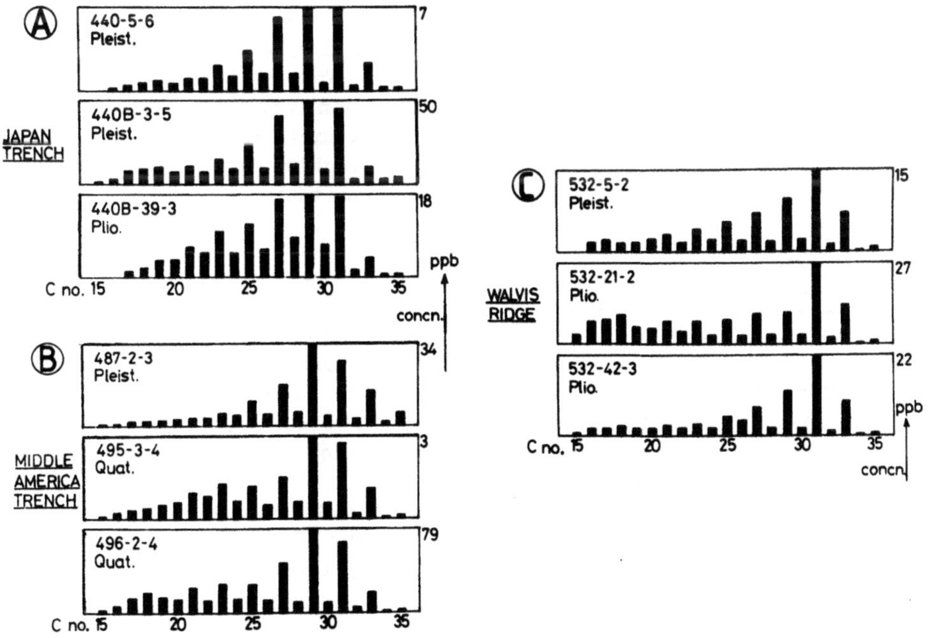

Fig. 2. Distributions of *n*-alkanes in selected sediments of various ages from the Japan Trench (A), the Middle America Trench (B), and the Walvis Ridge (C).

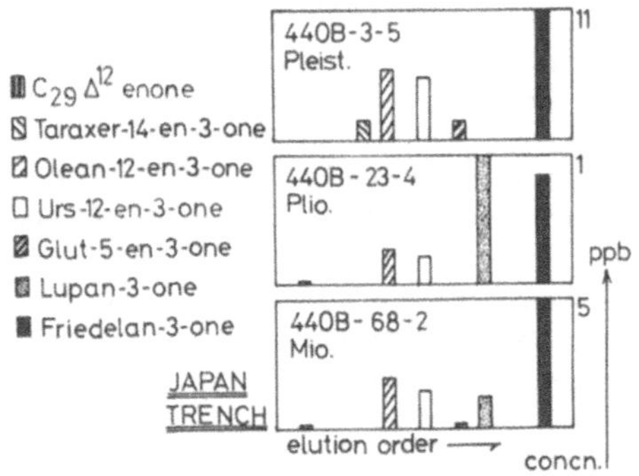

Fig. 3. Histograms of the abundance of triterpenones, arranged in GC elution order, in three sediment sections from the Japan Trench. Compound assignments were made by reference to standard spectra. Structures (see Appendix): taraxer-14-en-3-one, XXIIIa; olean-12-en-3-one, XXIVa; urs-12-en-3-one, XXVa; glut-5-en-3-one, XXVI; lupan-3-one, XXVII; friedelan-3-one, XXVIII.

pounds, although each of the three environments has its own characteristic distribution of *n*-alkanes which in all cases is principally of terrestrial origin. In contrast, algal or bacterial inputs to the sediments are not reflected in the *n*-alkane distributions, for example by a predominance of n-C_{17}. Such lack of evidence may, however, be due to the preferential biodegradation of shorter chain *n*-alkanes (e.g., Johnson and Calder, 1973; Brassell et al., 1978; Giger, Schaffner and Wakeham, 1980) in both the water column and sediment rather than low levels of algal productivity. In addition, the greater concentrations of *n*-alkanes in higher plant surface waxes than in algae makes these compounds more sensitive indicators of terrestrial than phytoplankton inputs.

(ii) Triterpenones. A number of triterpenoid structures, mainly 3-oxytriterpenoids, have been recognized only in higher plants and therefore serve as markers of terrestrial contributions to sediments (Table 3; Brassell et al., 1980c). Fig. 3 shows the distributions of non-hopanoid triterpenones in three sediments from the Japan Trench. A similar range of components was identified in the Middle America Trench sediments, although only friedelan-3-one (XXVIII) occurred in more than trace concentrations (>0.2 ng/g). Both suites of trench sediments also contained several non-hopanoid triterpenols, in particular taraxerol (XXIIIb), and β- and α-amyrins (oleanenol, XXIVb, and ursenol, XXVb). Higher plant triterpenoids were only minor components of the Walvis Ridge sediments. The Roman numerals refer to structures as listed in the Appendix.

(iii) A-Ring Degraded Triterpenoids. A variety of C_{24} tetracyclic alkenes and alkanes occur in sediments and are thought to de-

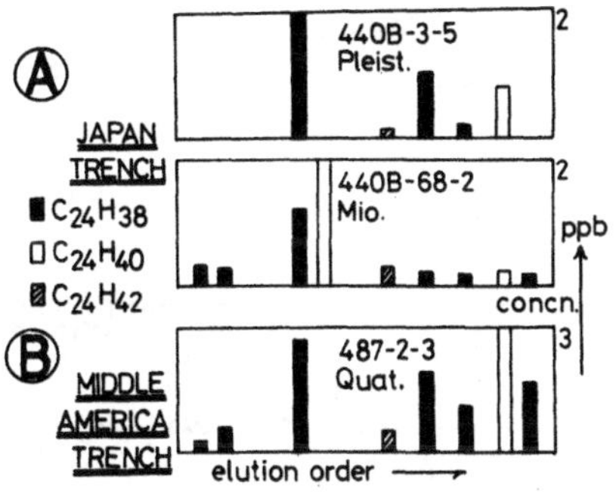

Fig. 4. Concentrations of components assigned as A-ring degraded triterpenoids, arranged in GC elution order, in sediment sections from the Japan Trench (A), and Middle America Trench (B). The only confirmed structure is that of the $C_{24}H_{42}$ alkane (XL) which appears to be derived from lupan-3-one (XXVII); see Appendix.

rive from A-ring degradation of 3-oxytriterpenoids (Corbet, Albrecht and Ourisson, 1980). Few of the structures of these compounds have been confirmed, but as degraded 3-oxytriterpenoids they, like their precursors, are markers for terrestrial contributions to sediments. Examples of the distributions of A-ring degraded triterpenoids in sediments from the Japan and Middle America trenches are given in Fig. 4. These compounds were not detected in the Walvis Ridge sediments.

Lipids of Marine Origin

Among the lipids derived from marine sources (cf. Table 2), sterols and alkadienones and alkatrienones are selected to illustrate the presence of such compounds in sediments.

(i) Sterols. The value of sterols as marine marker lipids stems from a combination of their abundance, the extensive literature concerned with the sterol compositions of marine organisms and the variety of specificity of their side chains (e.g. Schmitz, 1978; Djerassi et al., 1979; Djerassi, 1981). Several sterol side chains appear to occur only in a limited range of organisms and can therefore serve as markers for inputs from such biota to sediments (Fig. 5).

The sterol distributions of marine sediments are often highly complex (e.g., Lee, Gagosian and Farrington, 1977; Wardroper, Maxwell and Morris, 1978; Lee, Farrington and Gagosian, 1979; Brassell and Eglinton, 1981) containing perhaps more than 50 components and reflecting contributions from a variety of biota. Certain compounds can, however, be ascribed to inputs from specific organisms, including diatoms, dinoflagellates and sponges (cf. Fig. 5). Fig. 6 illustrates the occurrence of such diagnostic components among the sterol

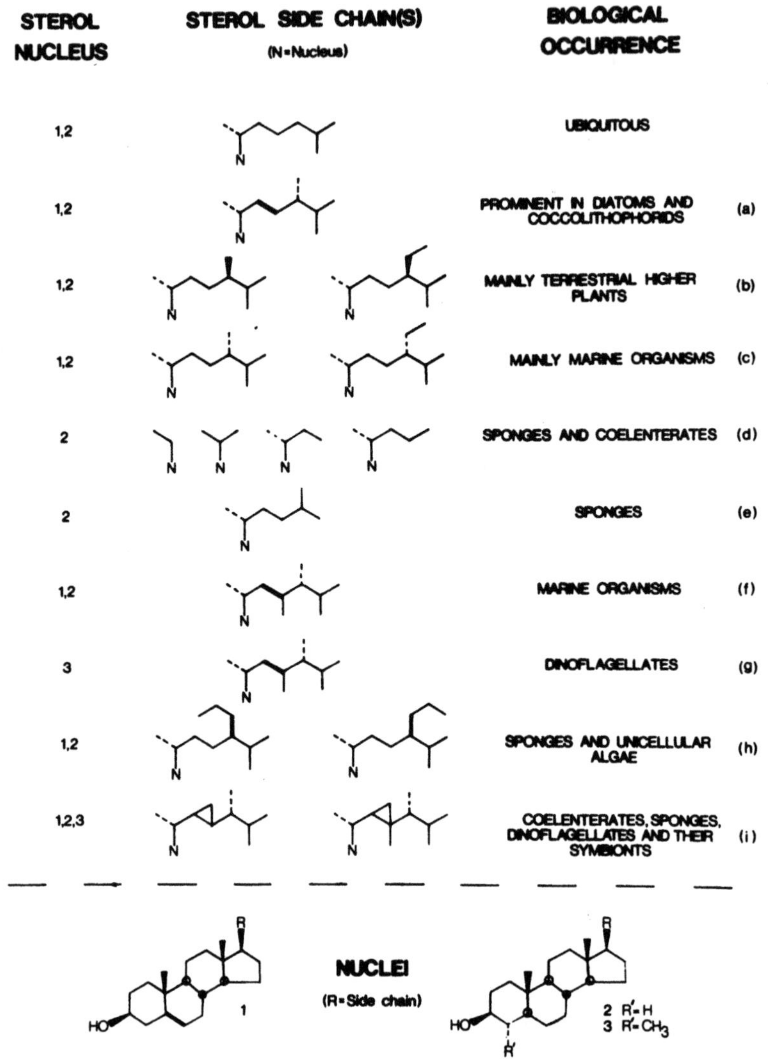

Fig. 5. Sterol types (nucleus and side chain) indicative of inputs from specific classes of organisms to sediments (from Mackenzie et al., 1982). All have been recognized in sediments (Wardroper et al., 1978; Boon et al., 1979; Wardroper, 1979; Brassell, 1980; Gagosian et al., 1980; Brassell and Eglinton, 1981). In some instances the assignment of sterols that may have originated from more than a single class of organisms may be complemented by independent information, such as the abundance of sponge spicules in the sediments (Brassell et al., 1980c) or of jellyfish (coelenterates) in the water column (Wardroper, 1979). (a) Rubinstein and Goad, 1974; Orcutt and Patterson, 1975; Volkman et al., 1981. (b) and (c) in terrestrial higher plants and in marine organisms, the stereochemistry of their sterol C-24 alkylation is dominantly β and α, respectively (Goad et al., 1974), although certain organisms do

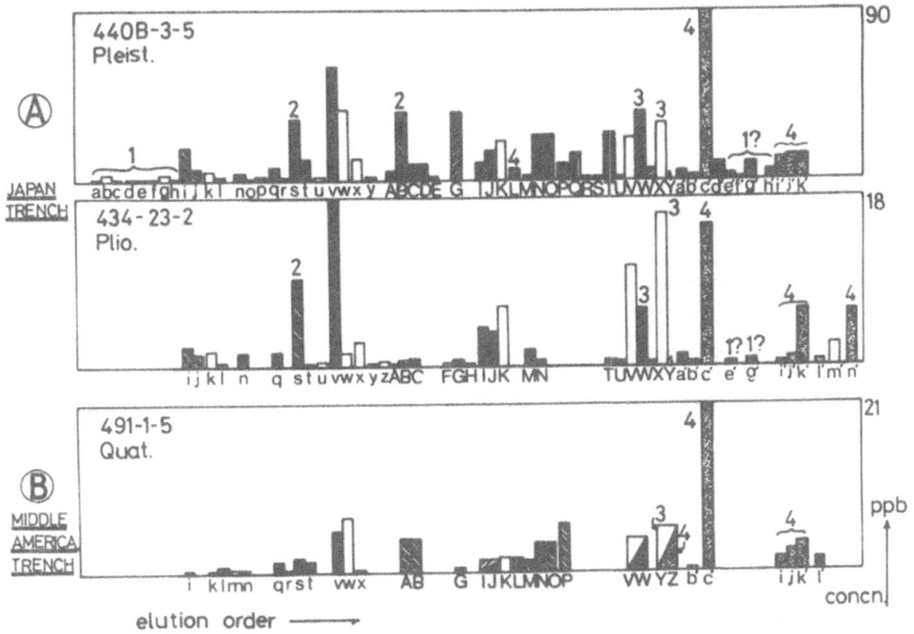

Fig. 6. Histograms of the abundances of sterols (analyzed as TMS ethers) in sediments from (A) the Japan Trench, and (B) the Middle America Trench, arranged according to their approximate GC elution order. Sterol assignments are given in Table 5. Some of the sterols thought to be markers of specific inputs to the sediments (cf., Fig. 5) are so designated. Input: 1=sponges?, 2=algae?, 3=higher plants?, 4=dinoflagellates?

←
Fig. 5 continued

not conform to this pattern. (d) of the variety of sterols with short (<C_7) side chains found in organisms (Ballantine, Williams and Burke, 1977; Carlson et al., 1978; Delseth et al., 1978) only those four shown have been identified in sediments (Brassell, 1980; Brassell and Eglinton, in press; Table 5). (e) Delseth et al., 1979. (f) 23,24-Dimethyl alkylation appears to be restricted to marine organisms (Kanazawa et al., 1974; Volkman, Eglinton and Corner, 1980b). (g) Shimizu, Alam and Kobayashi, 1976. (h) Sponges (Sheikh and Djerassi, 1974) and unicellular algae (Rohmer et al., 1980) seem a more probable source of these sterols in oceanic sediments than do scallops (e.g., Idler et al., 1976). (i) 22,23-Methylene substitution now appears to be more widespread in organisms than was originally thought (cf., Cieresko et al., 1968; Alam, Martin and Ray, 1979; Djerassi et al., 1979; Withers et al., 1979).

Table 5. Sterol identification (extended from Brassell and Eglinton, in press)

Peak (Fig. 6)	Assignment[a]	Structure
a	Unknown C_{21} sterol	
b	5α(H)-pregnan-3β-ol	VIIa
c	20-Methyl-5α(H)-pregnan-3β-ol	VIIb
d	Unknown C_{22} sterol	
e	Unknown C_{22} Δ^5-sterol	
f	24-Norchol-5-en-3β-ol ?	VIc
g	24-Nor-5α(H)-cholan-3β-ol ?	VIIc
h	5α(H)-cholan-3β-ol ?	VIId
i	24-Norcholesta-5,22-dien-3β-ol	VIe
j	24-Nor-5α(H)-cholest-22-en-3β-ol	VIIe
k	24-Nor-5α(H)-cholestan-3β-ol	VIIf
l	5β(H)-cholestan-3β-ol	IXg
m	5α(H)-cholestan-3α-ol	VIIIg
n	5β(H)-cholestan-3α-ol	Xg
o	Unknown C_{27} stera-5,22-dien-3β-ol[b]	VIh ?
p	Unknown C_{27} 5α(H)-ster-22-en-3β-ol[b]	VIIh ?
q	27-Nor-24-methylcholesta-5,22(E)-dien-3β-ol[c]	VIi
r	27-Nor-24-methyl-5α(H)-cholest-22(E)-en-3β-ol	VIIi
s	Cholesta-5,22(E)-dien-3β-ol	VIj
t	5a(H)-cholest-22(E)-en-3β-ol	VIIj
u	Unknown C_{27} 5α(H)-stan-3β-ol[b]	VIIk ?
v	Cholest-5-en-3β-ol	VIg
w	5α(H)-cholestan-3β-ol	VIIg
x	27-Nor-24-methyl-5α(H)-cholestan-3β-ol	VIIℓ
y	Cholesta-5,7-dien-3β-ol	XIVg
z	Unknown C_{28} 5α(H)-stan-3β-ol	
A	Cholesta-5,24-dien-3β-ol	VIm
B	24-Methylcholesta-5,22(E)-dien-3β-ol	VIn
C	24-Methyl-5α(H)-cholest-22(E)-en-3β-ol	VIIn
D	5α(H)-cholest-7-en-3β-ol	XIg
E	Unknown C_{28} ster-22-en-3β-ol	
F	Unknown C_{28} 5α(H)-stanol	
G	24-Methylenecholest-5-en-3β-ol	VIo
H	Unknown C_{29} sterol	
I	24-Methylene-5α(H)-cholestan-3β-ol	VIIo
J	24-Methylcholest-5-en-3β-ol	VIp
K	24-Methyl-5α(H)-cholestan-3β-ol	VIIp
L	4α,24-Dimethyl-5α(H)-cholest-22-en-3β-ol[d]	XIIn
M	23,24-Dimethylcholesta-5,22-dien-3β-ol	VIq
N	23,24-Dimethyl-5α(H)-cholest-22-en-3β-ol	VIIq
O	24-Ethylcholesta-5,22-dien-3β-ol	VIr
P	24-Ethyl-5α(H)-cholest-22-en-3β-ol	VIIr
Q	24-Methyl-5α(H)-cholest-7-en-3β-ol	XIp
R	Unknown C_{29} steradienol	
S	Unknown C_{29} 5α(H)-stanol	
T	23,24-Dimethylcholest-5-en-3β-ol	VIs
U	24(E)-ethylidenecholest-5-en-3β-ol	VIt
V	23,24-Dimethyl-5α(H)-cholestan-3β-ol	VIIs

Table 5 (cont.)

Peak	Assignment	Structure
(Fig. 6)		
W	24-Ethylcholest-5-en-3β-ol	VIu
X	24(E)-ethylidene-5α(H)-cholestan-3β-ol	VIIt
Y	24-Ethyl-5α(H)-cholestan-3β-ol	VIIu
Z	4α,24-Dimethyl-5α(H)-cholestan-3β-ol	XIIp
a'	24(Z)-ethylidenecholest-5-en-3β-ol	VIv
b'	24(Z)-ethylidene-5α(H)-cholestan-3β-ol	VIIv
c'	4α,23,24-Trimethyl-5α(H)-cholest-22(E)-en-3β-ol	XIIq
d'	24-Ethyl-5α(H)-cholest-7-en-3β-ol	XIu
e'	24(E)-propylidenecholest-5-en-3β-ol	VIw
f'	24(E)-propylidene-5α(H)-cholestan-3β-ol	VIIw
g'	24(Z)-propylidenecholest-5-en-3β-ol	VIx
h'	24(Z)-propylidene-5α(H)-cholestan-3β-ol	VIIx
i'	4α-Methyl-24-ethyl-5α(H)-cholest-8(14)-en-3β-ol	XIIIu
j'	4α,22,23-Trimethyl-5α(H)-cholestan-3β-ol[e] ?	XIIy
k'	4α,23,24-Trimethyl-5α(H)-cholestan-3β-ol[f]	XIIs
l'	22,23-Methylene-4α,24-dimethyl-5α(H)-cholestan-3β-ol	XIIz
m'	22,23-Methylene-23,24-dimethyl-5α(H)-cholestan-3β-ol	XIIad
n'	22,23-Methylene-4α,23,24-trimethyl-5α(H)-cholestan-3β-ol	XIIad

? Tentative assignments based solely on spectral interpretation and GC retention times.

[a] Assignments, unless stated otherwise, are made by comparison with reference or literature spectra of sterol TMS ethers (Brooks et al., 1968 and 1972; Gaskell and Eglinton, 1976; Idler et al., 1976; Ballantine et al., 1976 and 1977; Wardroper et al., 1978; Wardroper, 1979; Boon et al., 1979; Lee et al., 1979; Brassell, 1980; Brassell and Eglinton, 1981), and by consideraiton of GC elution orders, for example 24(E) isomers of $\Delta^{24(28)}$-sterols elute prior to their 24(Z) counterparts (Idler et al., 1976).

[b] Components may be 26,27-bisnor-23,24-dimethylcholesteroids (IIh, Ih and Ik, respectively; see Brassell and Eglinton, 1981).

[c] Assignment substantiated by coinjection with authenic standard, although coelution of cholesta-5,22(Z)-dien-3β-ol (IIaa), which possesses a similar mass spectrum, cannot be excluded.

[d] The prominence of m/z 69 in the mass spectrum of the sterol TMS ether suggests C-24(n) rather than C-23 (ab) methylation.

[e] May possess a 22,24-dimethyl (ac) rather than a 22,23-dimethyl (y) side chain; neither compound is available as a reference standard for comparison.

[f] Identification substantiated by coinjection of an authentic standard with other sedimentary sterol distribution (J.K. Volkman, personal communication).

distribution of sediments from the Japan and Middle America Trenches. In particular, dinoflagellate sterols dominate, although such compounds may derive from symbiotic rather than from free-living organisms. In contrast, substantial evidence for sterol inputs from sponges is seen only in the Japan Trench sediments in which sponge spicules are abundant.

(ii) Alkadienones and alkatrienones. At present the only organism known to contain the series of C_{37}-C_{39} methyl and ethyl alkadienones and alkatrienones found in marine sediments is the coccolithophorid *Emiliania huxleyi* (Volkman et al., 1980a). Hence, these compounds are regarded as markers for coccolithophore inputs to sediments, even where there are no skeletal remains of such organisms because the sediment was deposited below the carbonate compensation depth (Volkman et al., 1980a; Brassell et al., 1980c), as in the Japan Trench. The distribution of these long chain ketones in selected sediments from the Japan and Middle America Trenches and from Walvis Ridge are shown in Fig. 7. These ketones are usually associated with C_{37} and C_{38} alkatrienes and esters (cf. Table 2) which also occur in *E. huxleyi*. The recognition of such long chain constituents in sedi-

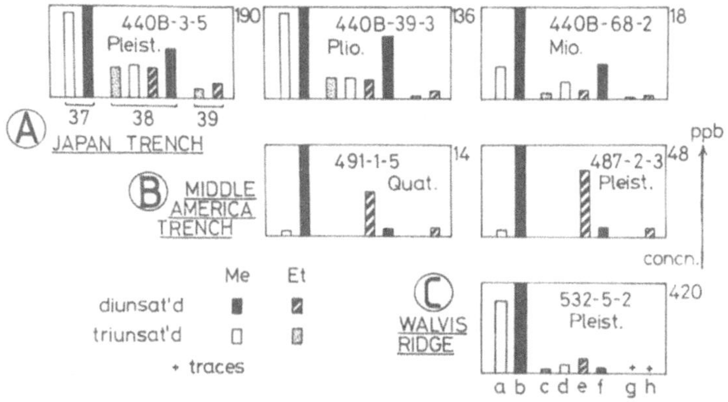

Fig. 7. Concentrations of long chain unsaturated ketones in sediments from (A) the Japan Trench, (B) the Middle America Trench, and (C) the Walvis Ridge. These alkadienones and alkatrienones, as designated for Section 532-5-2 are:

a,	heptatriaconta-8,15,22-trien-2-one	Structure: (IIIc);
b,	heptatriaconta-15,22-dien-2-one	(IIa);
c,	octatriaconta-9,16,23-trien-3-one	(IIId);
d,	octatriaconta-9,16,23-trien-2-one	(IIIe);
e,	octatriaconta-16,23-dien-3-one	(IIb);
f,	octatriaconta-16,23-dien-2-one	(IIc);
g,	nonatriaconta-10,17,24-trien-3-one	(IIIf);
h,	nonatriaconta-17,24-dien-2-one	(IId)

The configuration of the double bonds (i.e., whether E and Z) is unknown, Volkman et al. (1980a).

ments of Pliocene and Miocene age indicates that organisms no longer extant synthesized these components since the fossil record of *E. huxleyi* only extends over the past 130,000 years.

Lipids of Bacterial Origin

The evidence for bacterial activity, provided by lipid distributions is twofold. First, specific lipids that are only biosynthesized by bacteria act as markers for their contributions to sediments. Second, certain sedimentary lipids are recognizable products of bacterial degradation of other lipids. The paucity of information regarding bacterial lipid degradation processes and pathways in sediments limits the scope of the second category of compounds, hence the examples of evidence of bacterial inputs to sediments given here are all direct biological inputs to the sediments (cf. Table 4).

(i) Specific Acyclic Isoprenoids. Methanogenic bacteria and other Archaebacteria synthesize a number of acyclic isoprenoid moieties, such as compounds with head-to-head linkages, that have not been identified in other organisms. These uniquely Archaebacterial lipids include dibiphytanyl glyceryl ethers and several acyclic isoprenoid alkanes, which therefore act as markers of bacterial inputs to sediments (cf. Table 4). Three such acyclic isoprenoid alkanes, 2,6,10,15,19-pentamethyleicosane (XXX), squalane (XXXI) and lycopane (XXXII) are thought to reflect methanogenic bacterial activity in sediments (Brassell et al., 1981a). Such activity can be occurring in the sediments when sampled, but the marker compounds also preserve a sig-

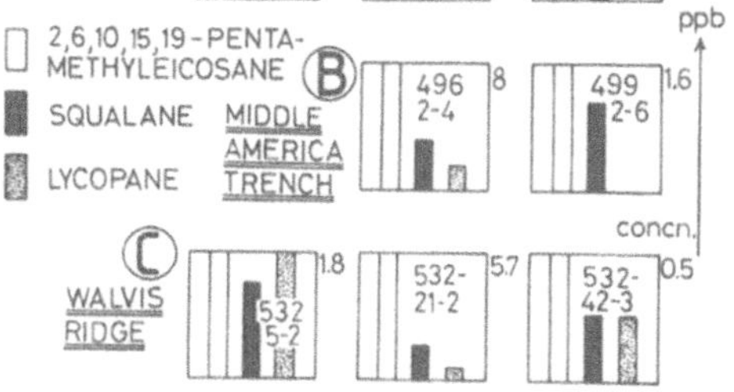

Fig. 8. Concentrations of three acyclic isoprenoid alkanes, 2,6,10,15,19-pentamethyleicosane (XXX), squalane (XXXI) and lycopane (XXXII) in sediments from the Japan Trench (A, Pleistocene to Miocene), the Middle America Trench (B, Quaternary), and Walvis Ridge (C, Pleistocene and Pliocene).

nature of methanogenic bacteria in the sediments long after their activity has ceased. Fig. 8 shows the distribution of these three components in sediments from the Japan and Middle America Trenches and the Walvis Ridge. The differences in the sedimentary distributions of these components may reflect different methanogenic bacterial populations, as methanogenic lipid compositions are species dependent (see Brassell et al., 1981a). Alternatively, the variations in sediment composition may be the function of environmental features such as the food supply for the methanogenic bacteria.

(ii) Fernenes. Until recently, the presence of fernenes in sediments was thought to reflect inputs from higher plants such as ferns (Brassell et al., 1980c), but the recognition of these compounds in a bacterium (Howard, 1980) is more in accord with their sedimentary associations. Hence, fernenes are now regarded principally as indicators of bacterial inputs to sediments. Fig. 9 illustrates the fernene distributions of sediments from the Japan and Middle America Trenches and from the Walvis Ridge. The distributions are generally similar, with only that of Section 440B-68-2, a Miocene silty claystone from the Japan Trench, differing significantly, with fern-8-ene (XXXVIIIa) rather than fern-7-ene (XXXVIIIc) dominant. The predominance of the Δ^8 fernene isomer is a feature of all Miocene and Cretaceous DSDP samples we have examined. Presumably, it either results from the greater thermal stability of this isomer or reflects the original sediment inputs suggesting an evolutionary change in bacterial fernene isomer distributions (Brassell and Eglinton, in press).

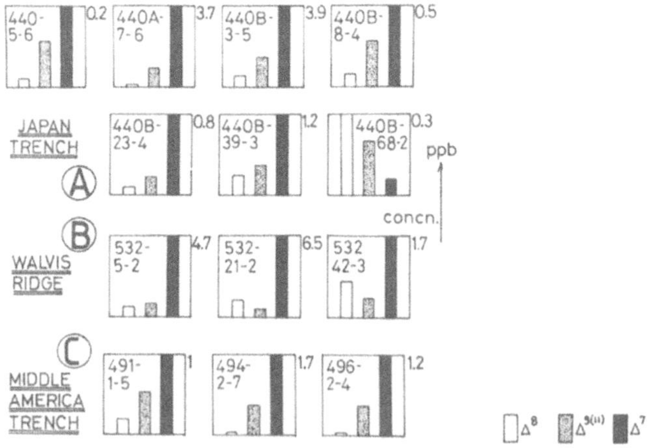

Fig. 9. Concentrations of fernenes (Δ^8,XXXVIIIa; $\Delta^{9(11)}$,XXXVIIIb; Δ^7,XXXVIIIc) in sediments from the Japan Trench (A, Pleistocene to Miocene), the Walvis Ridge (B, Pleistocene and Pliocene) and the Middle America Trench (C, Quaternary).

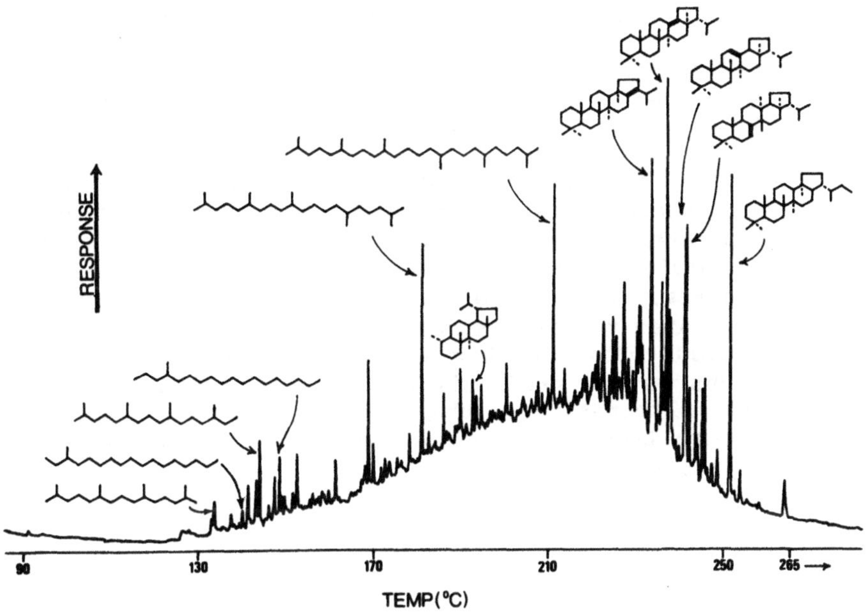

Fig. 10. Gas chromatogram of the branched/cyclic aliphatic hydrocarbons of Section 440B-8-4 from the Japan Trench (after Brassell et al., 1981a). The majority of the prominent components can be ascribed to bacterial origins. The source of the hump is uncertain, but it may represent either a natural input from a highly biodegraded thermally mature source or shipboard contamination with drilling lubricant. GC conditions: 20 m OV-1 wall-coated glass capillary, programmed from 50-265°C at 4°C min^{-1} after splitless injection at ambient temperature. By permission from Nature, Copyright (c) 1981, Macmillan Jour. Ltd.

(iii) Hopenes. The recent recognition of hop-17(21)-ene (XXXV) and neohop-13(18)-ene (XXXVI) in a bacterium (Howard, 1980) suggests that these components can be direct inputs to sediments rather than isomerization products of hop-22(29)-ene (XXXIVa) formed during diagenesis. By analogy, the presence of neohop-12-ene (XXVII) in sediments is perhaps also a reflection of bacterial inputs (Brassell et al., 1981a). The importance of such components of bacterial origin in sediment extracts is well illustrated by the distribution of branched/cyclic hydrocarbons in Japan Trench sediments (Fig. 10). Most of the prominent peaks in the chromatogram represent acyclic isoprenoid alkanes, fernenes, hopenes and an extended hopane (17β(H),21β(H)-homohopane, XXXIVb) which are derived from bacteria.

Preservation of Lipid Signatures in Japan Trench Sediments

The individual lipids that provide a signature of the biological inputs to a sediment can be preserved during diagenesis either in unaltered form or with part of their structure specifically modified.

The presence of marker compounds for marine and bacterial inputs in Upper Miocene through Pleistocene sediments from the Japan Trench (Fig. 11) illustrates their survival in an area of low geothermal gradient. Systematic downhole trends might suggest changes due to diagenetic processes, whereas the variations in the profiles observed are presumably related to the nature of the original sediment inputs. Hence, the variability in the downhole lipid concentrations may reflect fluctuations in the productivity or depositional conditions of the environment. Alternatively, the variations may be a function of lithology, although they are largely independent of the organic carbon contents of the sediment.

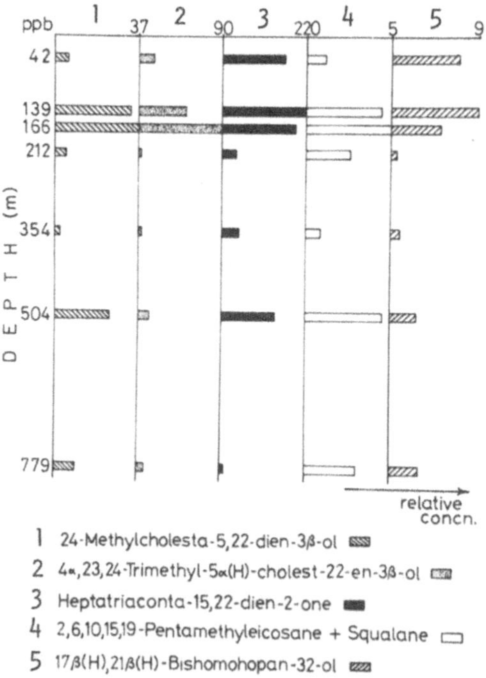

Fig. 11. Downhole concentrations of selected marker lipids at Site 440, Japan Trench, illustrating that such compounds can be preserved at depths of *ca.* 800 m under favorable conditions of diagenesis. 1 (VIn), 2 (XIIq) and 3 (IIc) are held to be markers for unicellular marine algae (Table 3), whereas 4 (XXX, XXI) and 5 (XXXIVc) are derived from bacteria (Table 4).

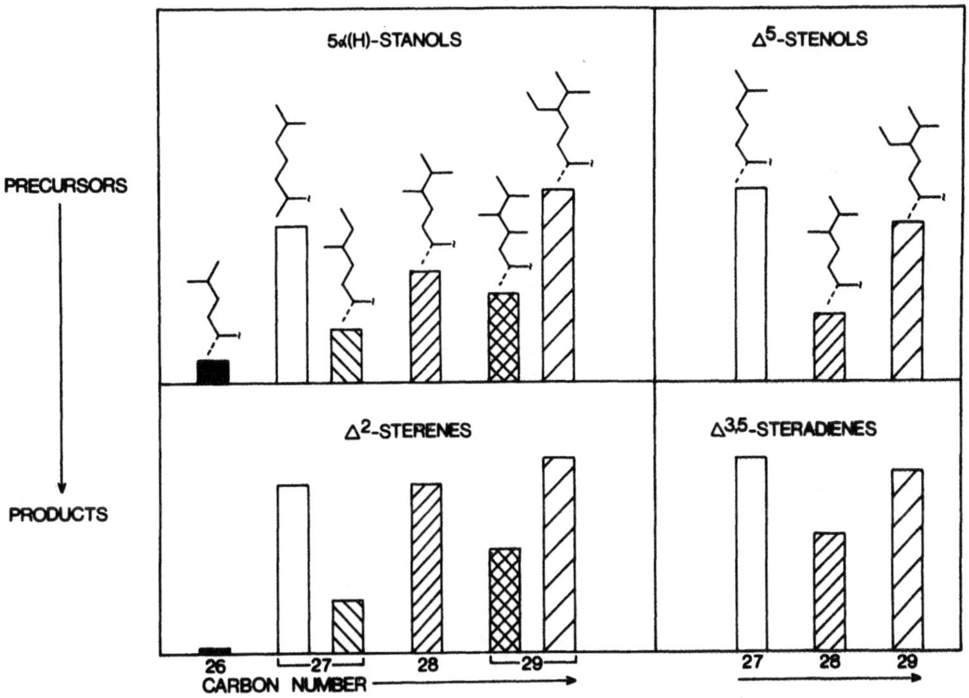

Fig. 12. Abundances of individual steroids in Section 440B-3-5 from the Japan Trench. Comparison of the side chain distributions of 5α(H)-stanols (VII) with those of Δ^2-sterenes (XLI) and of Δ^5-stenols (VI) with those of $\Delta^{3,5}$-steradienes (XLII) suggest precursor product relationships (Mackenzie et al., 1982).

The similarities in the distributions of 5α(H)-stanols (VII) and Δ^2-sterenes (XL) and of Δ^5-stenols (VI) and $\Delta^{3,5}$-steradienes (XLI), with regard to specific steroidal side chains for Section 440B-3-5, a Pleistocene diatomaceous ooze from the Japan Trench (Fig. 12), supports their proposed precursor/product relationships.

The sterene distributions observed can therefore be directly related to their precursor sterol distributions and hence the sediment inputs can be assessed. A similar understanding of the diagenetic fate of other lipid classes will extend the use of lipid markers to more ancient and mature sediments.

Environmental Comparison

The lipid compositions of sediments from the Japan and Middle America Trenches and from Walvis Ridge are dominated by components derived from marine and algal sources. The amounts of and variety of terrestrial marker lipids are greater in the suites of trench sediments than in those from the Walvis Ridge. These observations match the expected sediment inputs from environmental considerations, for example the greater downslope transport of terrigenous organic matter into trench sediment compared with that carried by wind and/or water to the Walvis Ridge.

In quantitative terms, the autochthonous lipid signal in the Walvis Ridge sediments is greater than that in the trench sediments, but the relative importance of water column productivity *versus* the preservation of organic matter in the water column and sediment is unclear. The distinction of such factors will more properly be addressed by qualitative information, namely the presences and absences and abundances of a wide range of marker lipids, than by solely quantitative data limited to a few compounds.

The variety and scope of the sediment samples discussed in this paper, however, is insufficient for detailed consideration and comparison of areas of upwelling with other productive environments. The data base for discussion of the organic geochemistry of upwelling systems is limited and needs extension. The potential of marker lipids in this respect has yet to be identified and realized. Indeed, there are no marker lipids known at present that uniquely characterize sediments from areas of upwelling and distinguish them from regions of high productivity induced by other factors.

In order to understand more fully the lipid geochemistry of upwelling systems, several features of such environments require investigation and documentation, namely (1) the lipid compositions of the organisms found within areas of upwelling and their zones of specific productivity (diatoms, dinoflagellates, etc.); (2) the qualitative and quantitative aspects of lipid changes affected by food web processes; (3) the relationship between the lipid signatures found in bottom sediments and those of the water column; and (4) the diagenetic survival of lipid signatures during sediment consolidation. Such information will help create a framework of knowledge necessary for the recognition of upwelling events in the sediment record on the basis of organic geochemical criteria.

CONCLUSIONS

Biological marker compounds have a great scope in sediment description because of their number, variety and specificity. The majority of sediment lipids, however, have yet to be characterized and exploited although those known and recognized provide a basis for further development and extension.

This paper illustrates that sedimentary lipids include marker compounds that reflect inputs of organic matter from marine, terrestrial and bacterial sources. Such lipid signatures can be preserved during diagenesis in unaltered form or by means of discrete product/precursor relationships. Comparison of the lipid signatures of sediments from the Japan and Middle America Trenches and the Walvis Ridge shows greater allochthonous inputs in the trench sediments and a stronger autochthonous lipid signal in the Walvis Ridge sediments. The effect and significance of the Benguela upwelling system in lipid terms is uncertain and the potential of organic geochemical criteria in the differentiation of high productivity induced by upwelling or by other factors has yet to be evaluated.

ACKNOWLEDGEMENT

We gratefully acknowledge financial support from the Natural Environmental Research Council (NERC GR/3/2951 and GR/3/3758) and from the National Aeronautics and Space Administration (sub-contract from NGL 05-003-003, University of California, Berkeley). We thank colleagues for helpful discussions and access to unpublished data, especially J.K. Volkman, P.A. Comet, J.R. Maxwell and V.J. Howell. The DSDP samples were provided with the assistance of the National Science Foundation.

REFERENCES

Alam, M., Martin, G.E. and Ray, S.M., 1979, Dinoflagellate sterols II: Isolation and structure of 4-methylgorgostanol from the dinoflagellate *Glenodinium foliaceum*, Journal of Organic Chemistry, 44:4466-4467.

Ballantine, J.A., Roberts, J.C. and Morris, R.J., 1976, Marine sterols. III. The sterol composition of oceanic jellyfish. The use of gas chromatographic mass spectrometric techniques to identify unresolved components, Biomedical Mass Spectrometry, 3:14-20.

Ballantine, J.A,. Williams, K. and Burke, B.A., 1977, Marine sterols. IV. C_{21} sterols from marine sources. Identification of pregnane derivatives in extracts of the sponge *Haliclona rubens*, Tetrahedron Letters, 1547-1550.

Barnes, P.J., Brassell, S.C., Comet, P.A., Eglinton, G., McEvoy, J., Maxwell, J.R, Wardroper, A.M.K., and Volkman, J.K., 1979, Preliminary lipid analyses of core sections 18, 24 and 30 from Hole 402A, in: "Initial Reports DSDP," 48, L. Montadert, D.G. Roberts et al., U.S. Government Printing Office, Washington, 965-976.

Boon, J.J., 1978, "Molecular Biogeochemistry of Lipids in Four Natural Environments," Ph.D. Thesis, University of Delft, 215 pp.

Boon, J.J., Rijpstra, W.I.C., De Lange, F., De Leeuw, J.W., Yoshioka, M. and Shimizu, Y., 1979, The Black Sea sterol - a molecular fossil for dinoflagellate blooms, Nature, 277:125-127.

Brassell, S.C., 1980, "The Lipids of Deep Sea Sediments; Their Origin and Fate in the Japan Trench," Ph.D. Thesis, University of Bristol, 265 pp.

Brassell, S.C. and Eglinton, G., 1981, Biogeochemical significance of a novel sedimentary C_{27} stanol, Nature, 290:579-582.

Brassell, S.C. and Eglinton, G., in press, Steroids and triterpenoids in deep sea sediments as environmental and diagenetic indicators, in: "Advances in Organic Geochemistry 1981", M. Bjorøy et al., eds., Wiley and Son, New York.

Brassell, S.C., Eglinton, G., Maxwell, J.R. and Philp, R.P., 1978, Natural background of alkanes in the aquatic environment, in: "Aquatic Pollutants, Transformation and Biological Effects," O. Hutzinger, I.H. Van Lelyveld and B.C.J. Zoetman, eds., Pergamon Press, Oxford, 69-86.

Brassell, S.C., Comet, P.A., Eglinton, G., McEvoy, J., Maxwell, J.R., Quirke, J.M.E., and Volkman, J.K., 1980a, Preliminary lipid analyses of cores 14, 18 and 28 from Deep Sea Drilling Project Hole 416A, in: "Initial Reports DSDP," 50, Y. Lancelot, E.L. Winterer et al., U.S. Government Printing Office, Washington, 647-664.

Brassell, S.C., Gowar, A.P. and Eglinton, G., 1980b, Computerized gas chromatography-mass spectrometry in analyses of sediments from the Deep Sea Drilling Project, in: "Advances in Organic Geochemistry 1979," A.G. Douglas and J.R. Maxwell, eds., Pergamon Press, Oxford, 421-426.

Brassell, S.C., Comet, P.A., Eglinton, G., Isaacson, P.J., McEvoy, J., Maxwell, J.R., Thomson, I.D., Tibbetts, P.J.C., and Volkman, J.K., 1980c, The origin and fate of lipids in the Japan Trench, in: "Advances in Organic Geochemistry 1979", A.G. Douglas and J.R. Maxwell, eds., Pergamon Press, Oxford, 375-391.

Brassell, S.C., Wardroper, A.M.K., Thomson, I.D., Maxwell, J.R. and Eglinton G., 1981a, Specific acyclic isoprenoids as biological markers of methanogenic bacteria in marine sediments, Nature, 290:693-696.

Brassell, S.C., Eglinton, G. and Maxwell, J.R., 1981b, Preliminary lipid analyses of two Quaternary sediments from the Middle America Trench, Southern Mexico Transect, Deep Sea Drilling Project Leg 66, in: "Initial Reports DSDP," 66, J.S. Watkins, J.C. Moore et al., U.S. Government Printing Office, Washington, 864 pp.

Brooks, C.J.W., Horning, E.C. and Young, J.S., 1968, Chracterization of sterols by gas chromatography-mass spectrometry of the trimethylsilyl ethers, Lipids, 3:391-402.

Brooks, C.J.W., Knights, B.A., Sucrow, W. and Raduchel, B., 1972, The characterization of 24-ethylidene sterols, Steroids, 20:487-497.

Cardoso, J.N., Eglinton, G. and Holoway, P.J., 1977, The use of cutin acids in the recognition of higher plant contribution to Recent sediments, in: "Advances in Organic Geochemistry 1975", R. Campos and J. Goni, eds., Enadimsa, Madrid, 273-287.

Carlson, R.M.K., Popov, S., Massey, I., Delseth, C., Ayanoglu, E., Varkony, T.H., and Djerassi, C., 1978, Minor and trace sterols in marine invertebrates. VI. Occurrence and possible origins of sterols possessing unusually short hydrocarbon side chains, Bioorganic Chemistry, 7:453-479.

Ciereskо, L.S., Johnson, M.A., Schmidt, R.W., and Koons, C.B., 1968, Chemistry of coelenterates - VI. Occurrence of gorgosterol, a C_{30} sterol, in coelenterates and their zooanthellae, Comparative Biochemistry and Physiology, 24:899-904.

Corbet, B., Albrecht, P. and Ourisson, G., 1980, Photochemical or photomimetric fossil triterpenoids in sediments and petroleum, Journal of the American Chemical Society, 102:1171-1173.

Cranwell, P.A., 1973, Branched-chain and cyclopropanoid acids in a recent sediment, Chemical Geology, 11:307-313.

Delseth, C., Carlson, R.M.K., Djerassi, C., Erdman, T.R. and Scheuer, P.J., 1978, Identification de stérols à chaines laterales courtes dans l'éponge, *Damiriaria hawaiiana*, Helvetica Chimica Acta, 61:1470-1476.

Delseth, C., Tolela, L., Scheuer, P.J., Wells, R.J. and Djerassi, C., 1979, 5α-Norcholestan-3β-ol and (24Z)-stigmasta-5,7,24(28)-trien-3β-ol, two new marine sterols from the Pacific sponges *Terpios zeteki* and *Dysidia herbacea*, Helvetica Chimica Acta, 62:101-109.

Devon, T.K. and Scott, A.I., 1972, "Handbook of Naturally-Occurring Compounds II. Terpenes," Academic Press, New York.

Djerassi, C., 1981, Recent studies in the marine sterol field, Pure and Applied Chemistry, 53:873-890.

Djerassi, C., Theobald, N., Kokke, W.C.M.C., Pak, C.S., and Carlson R.M.K., 1979, Recent progress in the marine sterol field, Pure and Applied Chemistry, 51:1815-1828.

Eglinton, G. and Hamilton, R.J., 1967, Leaf epicuticular waxes, Science, 156:1322-1335.

Ensminger, A., 1977, "Évolution de Composes Polycycliques Sédimentaries," Thèse de docteur es Sciences, Université Louis Pasteur, Strasbourg.

Gagosian, R.B., Smith, S.O., Lee, C., Farrington, J.W., and Frew, N.M., 1980, Steroid transformations in Recent marine sediments, in: "Advances in Organic Geochemistry 1979", A.G. Douglas and J.R. Maxwell, eds., Pergamon Press, Oxford, 407-419.

Gaskell, S.J. and Eglinton, G., 1976, Sterols of a contemporary lacustrine sediment, Geochimica et Cosmochimica Acta, 40:1221-1228.

Gelpi, E., Schneider, H., Mann, J. and Oró, J., 1970, Hydrocarbons of geochemical significance in microscopic algae, Phytochemistry, 8:603-612.

Giger, W., Schaffner, C. and Wakeham, S.G., 1980, Aliphatic and olefinic hydrocarbons in Recent sediments of Greifensee, Switzerland, Geochimica et Cosmochimica Acta, 44:119-129.

Goad, L.J., Lenton, J.R., Knapp, F.F. and Goodwin, T.W., 1974, Phytosterol side chain biosynthesis, Lipids, 9:582-595.

Hay, W.W., Sibuet, J.C., Baron, E.J., Boyce, R.E. Brassell, S.C., Dean, W.E., Huc, A.Y., Keating, B.H., McNulty, C.L., Meyers, P.A., Nohara, M., Schallreuter, R.E., Steinmetz, J.C., Stow, D.A.V., and Stradner, H.J., in press, Sedimentation and accumulation of organic carbon in the Angola Basin and on Walvis Ridge: Preliminary results of Deep Sea Drilling Project Leg 75, Geological Society of America Bulletin.

Holzer, G., Oró, J. and Tornabene, T.G., 1979, Gas chromatographic/mass spectrometric analyses of neutral lipids from methanogenic and thermoacidophilic bacteria, Journal of Chromatography, 186: 795-809.

Howard, D.L., 1980, "Polycyclic Triterpenes of the Anaerobic Photosynthetic Bacterium *Rhodomicrobium vannielli*," Ph.D. Thesis, University of California, Los Angeles, 272 pp.

Huang, W-Y. and Meinschein, W.G., 1976, Sterols as source indicators of organic material in sediments, Geochimica et Cosmochimica Acta, 40:323-330.

Idler, D.R., Khalil, M.W., Gilbert, J.D. and Brooks, C.J.W., 1976, Sterols of scallop Part II. Structure of unknown sterols by combination gas-liquid chromatography and mass spectrometry, Steroids, 27:155-166.

Johnson, R.W. and Calder, J.A., 1973, Early diagenesis of fatty acids and hydrocarbons in a salt marsh environment, Geochimica et Cosmochimica Acta, 37:1943-1955.

Kanazawa, A., Teshima, S., Ando, T. and Tomita, S., 1974, Occurrence of 23,24-dimethylcholesta-5,22-dien-3β-ol in a soft coral *Sarcophyta elegans*, Bulletin Japanese Society of Scientific Fisheries, 40:729.

Langseth, M., et al., eds., 1980, "Initial Reports DSDP," 56/57, Part 1, U.S. Government Printing Office, Washington, 629 pp.

Lee, C., Gagosian, R.B. and Farrington, J.W., 1977, Sterol diagenesis in recent sediments from Buzzards Bay, Massachusetts, Geochimica et Cosmochimica Acta, 41:985-992.

Lee, C., Farrington, J.W. and Gagosian, R.B., 1979, Sterol geochemistry of sediments from the Western North Atlantic Ocean and adjacent coastal area, Geochimica et Cosmochimica Acta, 43: 35-46.

MacKenzie, A.S., Brassell, S.C., Eglinton, G. and Maxwell, J.R., 1982, Chemical fossils - the geological fate of steroids, Science, 217:491-504.

Orcutt, D.M. and Patterson, G.W., 1975, Sterol, fatty acid and elemental composition of diatoms grown in chemically defined media, Comparative Biochemistry and Physiology, 50B:579-583.

Ourisson, G., Albrecht, P. and Rohmer, M., 1979, The hopanoids. Palaeochemistry and biochemistry of a group of natural products, Pure and Applied Chemistry, 51:709-729.

Rohmer, M., Kokke, W.C.M.C., Fenical, W. and Djerassi, C., 1980, Isolation of two new C_{30} sterols, (24E)-24-n-propylidenecholesterol and 24-n-propylidene cholesterol from a cultured marine chrysophyte, Steroids, 35:219-231.

Rubinstein, I. and Goad, L.J., 1974, Occurrence of (24S)-24-methylocholesta-5,22E-dien-3β-ol in the diatom *Phaeodactylum tricornutum*, Phytochemistry, 13:485-487.

Schmitz, F.J., 1978, Uncommon marine steroids, in: "Marine Natural Products, Chemical and Biological Perspectives, Vol. 1", P.J. Scheuer, ed., Academic Press, London, 241-297.

Sheikh, Y.M. and Djerassi, C., 1974, Steroids from sponges, Tetrahedron Letters, 30:4095-4111.

Shimizu, Y., Alam, M. and Kobayashi, A., 1976, Dinosterol, the major sterol with a unique side chain in the toxic dinoflagellate, *Gonyaulax tamarensis*, Journal of the American Chemical Society, 98:1059-1060.

Simoneit, B.R.T., 1977, Diterpenoid components and other lipids in deep-sea sediments and their geochemical significance, Geochimica et Cosmochimica Acta, 41:463-476.

Simoneit, B.R.T., 1978, The organic chemistry of marine sediments, in: "Chemical Oceanography", Vol. 7, J.P. Riley and R. Chester, eds., Academic Press, London, 233-311.

Tibbets, P.J.C., 1980, "The Origin of the Carotenoids of Some Quaternary and Pliocene Sediments," Ph.D. Thesis, University of Bristol.

Von Huene, R., Aubouin, J., Azema, J., Blackinton, G., Carter, J.A. Coulbourn, W.T., Cowan, D.S., Curiale, J.A., Dengo, C.A., Faas, R.W., Harrison, W., Hesse, R., Hussong, D.M., Laad, J.W., Muzylov, N., Shiki, T., Thompson, P.R., and Westberg, J., 1980, Leg 67: The Deep Sea Drilling Project Mid-America Trench transect off Guatemala, Geological Society of America Bulletin, 91: 421-432.

Volkman, J.K., Eglinton, G., Corner, E.D.S., and Sargent, J.R., 1980a, Novel unsaturated straight-chain C_{37}-C_{39} methyl and ethyl ketones in marine sediments and a coccolithophore *Emiliania huxleyi*, in: "Advances in Organic Geochemistry 1979", A.G. Douglas and J.R. Maxwell, eds., Pergamon, Oxford, 219-227.

Volkman, J.K., Eglinton, G. and Corner, E.D.S., 1980b, Sterols and fatty acids of the marine diatom *Biddulphia sinensis*, Phytochemistry, 19:1809-1813.

Volkman, J.K., Smith, D.J., Eglinton, G., Forsberg, T.E.V. and Corner, E.D.S., 1981, Sterol and fatty acid composition of four marine Haptophycean algae, Journal of the Marine Biological Association of the United Kingdom, 61:509-527.

Wardroper, A.M.K., 1979, "Aspects of the Geochemistry of Polycyclic Isoprenoids," Ph.D. Thesis, University of Bristol.

Wardroper, A.M.K., Maxwell, J.R. and Morris, R.J., 1978, Sterols of a diatomaceous ooze from Walvis Bay, Steroids, 32:203-221.

Watkins, J.S. et al., eds., 1981, "Initial Reports DSDP," 66, U.S. Government Printing Office, Washington, 864 pp.

Withers, N.W., Kokke, W.C.M.C., Rohmer, M., Fenical, W.H. and Djerassi, C., 1979, Isolation of sterols with cyclopropyl-containing side chains from the cultured marine alga *Peridinium foliaceum*, Tetrahedron Letters, 3605-3608.

Youngblood, W.W., Blumer, M., Guillard, R.R.L. and Fiore, F., 1971, Saturated and unsaturated hydrocarbons in marine benthic algae, Marine Biology, 8:190-201.

APPENDIX

STRUCTURES

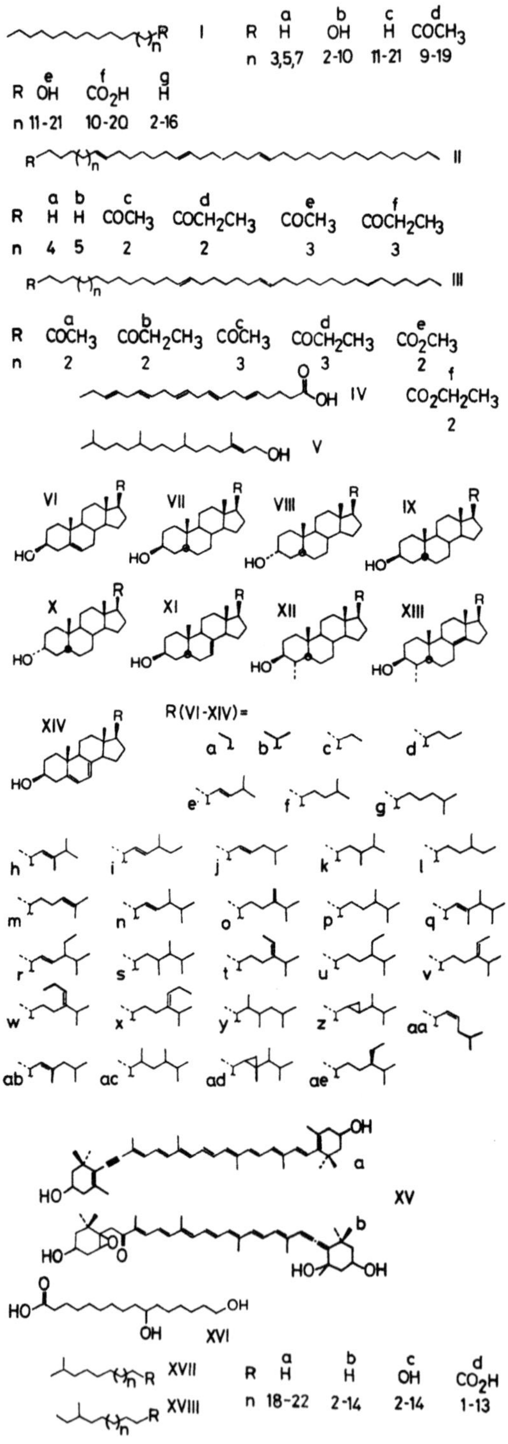

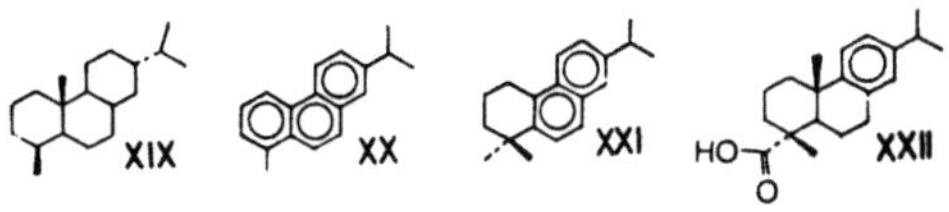

XIX XX XXI XXII

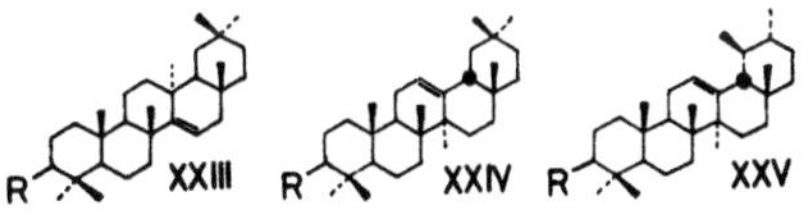

XXIII XXIV XXV

	R
a	O
b	OH

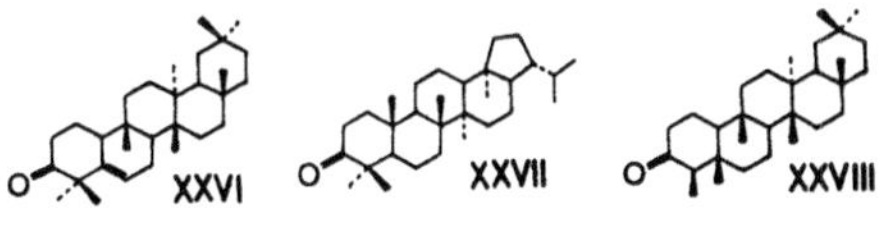

XXVI XXVII XXVIII

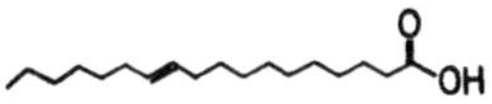

XXIX

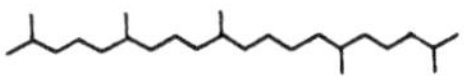

XXX

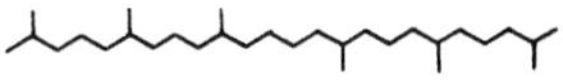

XXXI

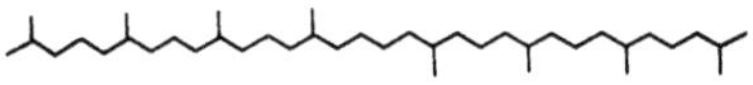

XXXII

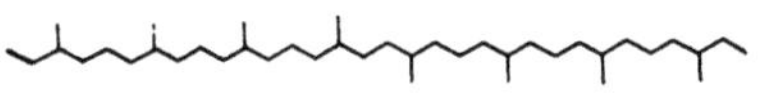

XXXIII

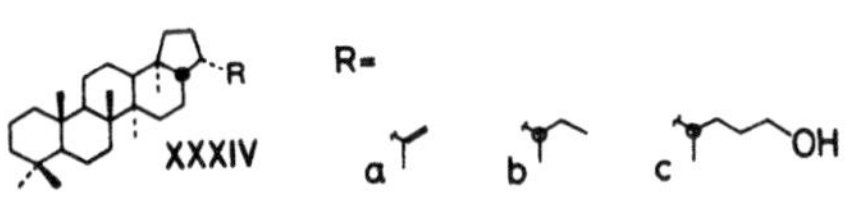

XXXIV R= a b c

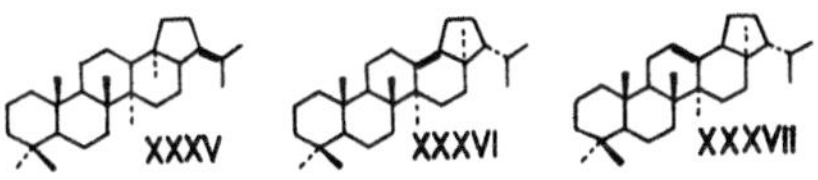

XXXV XXXVI XXXVII

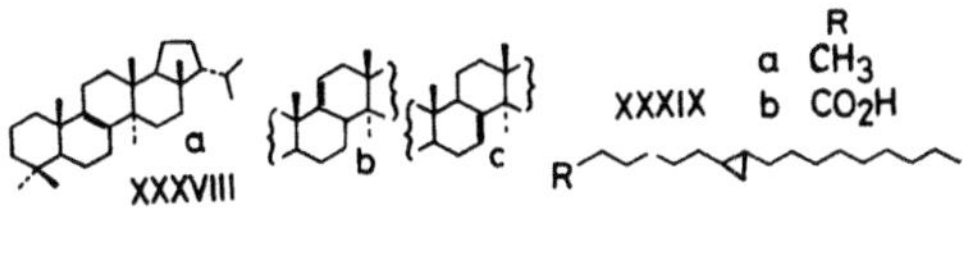

XXXVIII a b c XXXIX

	R
a	CH_3
b	CO_2H

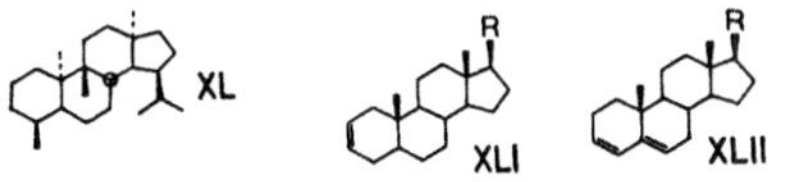

XL XLI XLII R shown in Fig. 12

METAL-STAINING OF SEDIMENTARY ORGANIC MATTER BY NATURAL PROCESSES

Venugopalan Ittekkot and Egon T. Degens

Geologisch-Paläontologisches Institut der
Universität Hamburg
Bundesstrasse 55
D-2000 Hamburg 13, Federal Republic of Germany

ABSTRACT

Electron microscopic investigations of sediments deposited in anoxic environments revealed the presence of naturally metal-stained organic matter. It appears that metal-staining is one of the mechanisms by which certain heavy metals are enriched in bituminous sediments. Based on the presence of organic matter in various stages of mineralization, it is suggested that metal-staining is the starting point for the oriented growth of various mineral phases. In addition, metal-staining also leads to the preservation of organic matter and organic structures, especially the soft tissues of organisms through geological time.

INTRODUCTION

Anoxic sediments, rich in organic matter and metals, develop in oxygen minimum zones in regions of high productivity such as upwelling areas along continental margins, and in stagnant basins due to restricted water circulation. We have studied sediments deposited in modern and ancient anoxic environments by means of transmission electron microscopy. Well-preserved organic tissues in various stages of metal fixation or mineralization were observed, revealing mechanisms of metal fixation at the molecular level. Although the majority of samples studied were formed in stagnant basins such as the Black Sea and Lake Tanganyika, the mechanisms we observed should operate in sediments formed under reducing conditions in general. The principal aim of this work is to focus attention on the metal-staining phenomena and on the way metals interact with organic matter in biogeochemical systems, to discuss briefly the structural aspects of metal fixation, and to emphasize the potential diagnostic features of metal-stained tissue structures, such as are found in upwelling sediments.

Structural Aspects of Metal Fixation

In the earth's crust, metals are principally arranged in four-fold and six-fold coordination; the main ligand is oxygen. This arrangement leads to the formation of two basic structures: the tetrahedron and the octahedron. Individual polyhedra can share ions and yield more complex structures. The elementary structure of a silicate mineral, by far the most abundant in the crust, is a tetrahedron, where Si^{4+} is surrounded by four O^{2-}. Classification of different types of silicate minerals is based on how these tetrahedra are arranged.

Metal ions associated with organic matter in biological and geochemical systems also follow a similar pattern of arrangement. Oxygen, nitrogen and sulfur donor groups necessary for the formation of coordination polyhedra are provided by functional groups such as carboxyl-, hydroxyl-, phosphate-, amino- and sulfhydril groups present in biochemical molecules. Formation of coordination polyhedra imparts a higher structural order to the participating molecules, whereby the complexes formed during this type of interaction are either bilinear, square planar, tetrahedral or octahedral with coordination numbers 2, 4 or 6 (Jernelöv and Martin, 1975). A number of biologically important metallo-structures such as enzymes, blood pigments, cytochrome and some vitamins are a result of interaction of metal ions with organic matter. For example, the enzymes carbonic anhydrase and alkaline phosphatase contain zinc in their structures. The activity of these metallo-structures is dependent on configuration and coordination introduced by metal ions. Substitution by other metals leads to the inhibition of activity due to generation of new configuration and coordinations (Brown, 1976).

Formation of minerals like carbonates, phosphates and silica in cellular systems (biomineralization) is essentially based on and initiated by organo-metal interactions and formation of coordination polyhedra. Biominerals are composed of an inorganic phase (mineral phase) and an organic phase. The organic phase acts as a template on which the epitaxial growth of the mineral phase proceeds. The concept of epitaxis assumes that the growth of one substance can take place on the surface of another provided the two substances have some similarity in structural order; the type of mineral formed is dependent on the substrate available. In biomineralization the role of substrate is taken by the organic phase (organic matrix), which is composed of substances like protein, polysaccharides and phosphoproteins. Because of their structure and functionality, these substances can act as templates for the generation of minerals by selectively forming coordination complexes with metal ions and thus providing nucleation sites for the formation and oriented growth of minerals. For example, in biological systems, acidic amino acids in the matrix protein are found to play a significant role in the formation of carbonates by acting as ligands for the coordination of calcium. In contrast, glycine and the hydroxyl-containing amino acids serine and threonine have been found to be essential for the coordination of Si^{4+} in the deposition of silica ($SiO_2 \cdot nH_2O$) (Hecky et al., 1973). Structural control in the deposition of phosphates is exerted by collagen. For a detailed discussion on the molecular aspects of biomineralization we refer to Istin (1975), Degens (1976), and Krampitz and Witt (1979).

Accumulation of heavy metals in certain organisms also appears to be related to biomineralization processes. The major part of heavy metals, for instance cadmium, cobalt, chromium, iron or manganese, in bivalves is found to be concentrated in the "less crystalline" type of intracellular minerals--granules--in the digestive glands and kidneys of these organisms (Bryan, 1973). This sort of granule formation is apparently the cellular route of detoxification of heavy metals by the organisms (Krampitz and Witt, 1979). Metal deposition mechanisms similar to the processes of biomineralization involving organic-inorganic interactions occur in certain types of bacteria (Rogers and Anderson, 1976a; 1976b; Beveridge and Murray, 1980). The precipitation of iron and manganese appears to act as a detoxification mechanism in environments enriched in these metals.

In the natural environment, mineral growth mediated by organic-inorganic interactions has also been observed, as in the formation of ooids. The process of ooid precipitation is a slow one and may proceed over a period of a few hundred years. Distribution of amino acids in consecutive HCl etchings of Bahamian ooids showed a significant stepwise increase from the outer to the inner core in aspartic acid content (Mopper and Degens, 1972). The high aspartic acid content was suggested to result from its role for Ca^{+2} fixation, and this stepwise increase in concentration to be a result of intermittent growth of the ooids. The organic matter is extracted from dissolved organic matter and the adsorbed amino acids form a thin coat around the ooid and act as a template for the nucleation and oriented growth of aragonite crystals. Recent studies (Carter, 1978) indicate that aspartic-acid-rich organic matter is specifically associated with calcium carbonate both as a part of an organic matrix of biogenic carbonates and as adsorbed organic matter on biogenic and nonbiogenic carbonates. These interactions are found to be a result of the geometrical similarity between the carboxyl group of the organic matter and the carbonate ions.

The foregoing discussion shows the similarity of processes involved in mineral formation both in cellular and geochemical systems. The structural and functional resemblance of minerals in sediments and macromolecules in cells is remarkable. Both, because of their crystalline structure and similarity, can act as templates for a wide variety of minerals and organic matter in sediments and as catalysts in cellular systems. Geochemically, these processes are important because such organic-metal and organic-inorganic interactions enhance the preservation of organic matter and other organic structures through geological time.

Organic Metal Interactions in Geochemical Systems

We have been able to discern some aspects of the above mentioned processes by doing high resolution electron microscopic investigations on sediments from anoxic environments. The method is based on the occurrence of electron-dense areas in organic matter present in

Fig. 1. X-ray photograph of *Asteropyge sp.* Broili from the lower Devonian of the Hunsrück (from Lehmann, 1934). By permission from N.J. Min., Geol. und Paläont., Copyright (c) 1934, Schweizerbart'sche Verlagsbuchhandl.

sediments caused by metal-organic interactions. Nearly 50 years ago Lehmann (1934) developed a remarkable technique which, by means of x-rays, made it possible to see, through thin fragments of rock samples, fossilized remains of marine organisms embedded in clay (Hunsrück shale; Fig. 1). The reason for this contrast (x-ray opaque regions), obtained by Lehmann, was tiny sulfide crystals, which covered the surface of organisms after deposition allowing them to be preserved through geological time. Our work is an extension of what Lehmann studied 50 years ago, except that we have used a different means of investigation, namely high resolution electron microscopy for deciphering well-preserved organic matter and minerals.

Studies were made of sediments rich in organic matter and deposited in anoxic environments which included samples from the Black Sea and Lake Tanganyika and those from the Lower Cretaceous black shales from the Wealden near Bentheim, West Germany. They were examined with an electron microscope (Philips EM 300) according to a method described by Degens, Watson and Remsen (1970). The samples were examined both directly, that is without prior heavy metal staining, and after staining with a heavy metal. The results of direct observations made on unstained samples are presented in this article. The study revealed the occurrence in sediments of different types of organic structures--membranes, bacterial cell walls, intact structures resembling native shapes of bacteria, spherical structures surrounded by sulfide crystals.

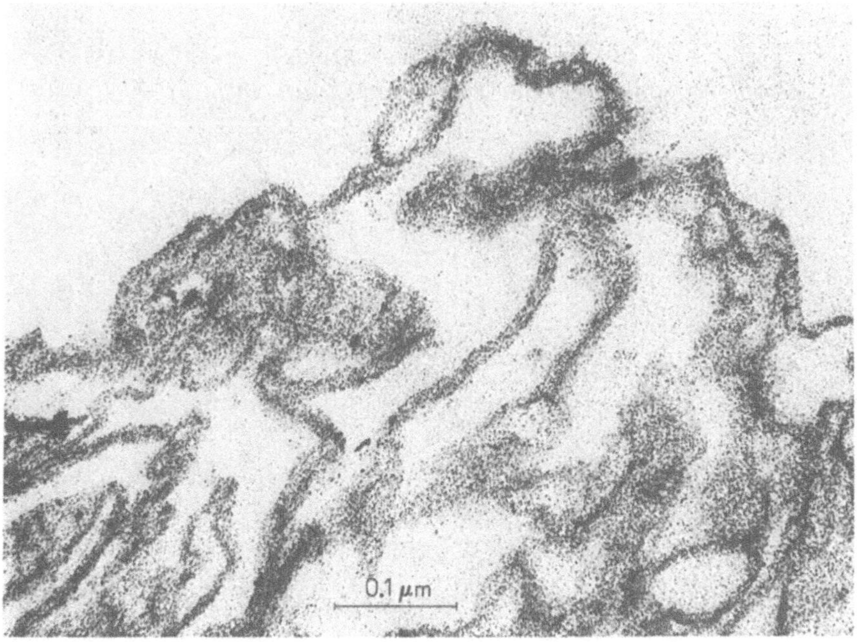

Fig. 2. Branched tubular membranes in Black Sea sediments.

Stained membrane structures. Membranes are prominent structures in all sediments investigated. Especially interesting is the presence of tubular membranes (Fig. 2) which are common in sediments but rare in present day organisms. Two kinds of tubular membranes were recognized, one 150 to 250 Å in diameter (Fig. 2) and the other approximately 700 to 800 Å in diameter (Fig. 3). The presence of such membranes have been reported in the cristae of some mitochondria, in a photosynthetic bacterium, and in a marine denitrifying bacterium (Blondin, Vail and Green, 1969). However, the tubular membranes found in the sediments were far too extensive to have been derived from any of the aforementioned sources. The presence of such membranes can best be explained by studies of the mechanism of mitochondrial swelling wherein conformational changes from lamellar to tubular membranes are observed. Principally two kinds of coordination rearrangements may proceed: one involving electron transfer or hydrolysis of ATP, and the other participation of a salt gradient (Donnan potential) (Blondin et al., 1969). The Donnan effect in

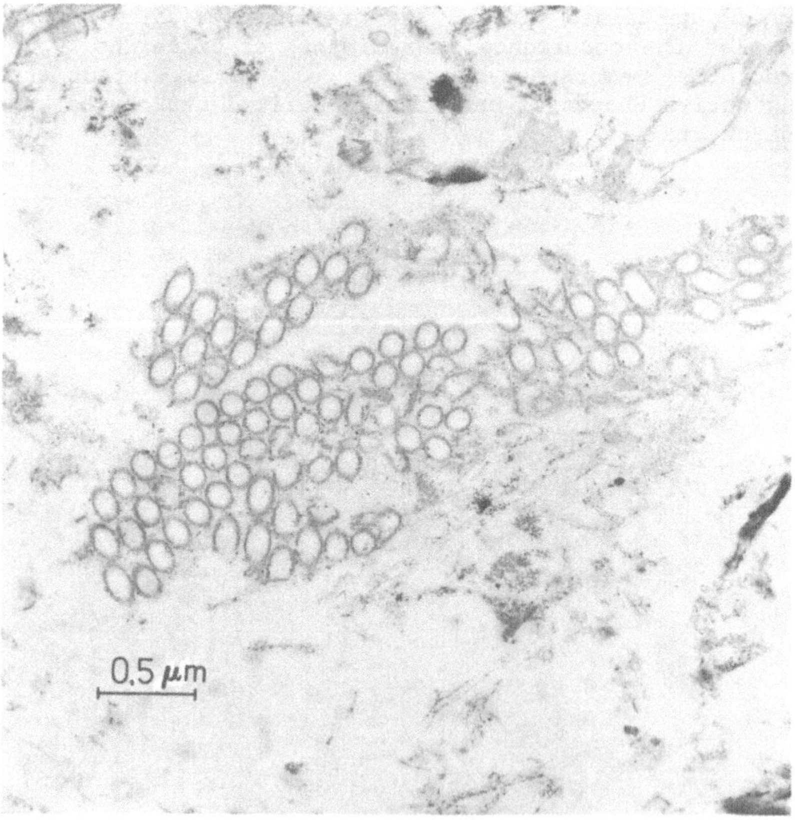

Fig. 3. Large tubular membranes with subunits in Black Sea sediments.

mitochondrial membranes is caused by the presence of polyelectrolytes and the resulting interactions between them and the metal ions. Additionally, the presence of metal ions will influence decisively the lipid-protein interactions in mitochondria. It is possible that fluctuations in the ionic environment and in the concentration of metal ions may have occurred in sediments which are similar to those fluctuations which induce configurational changes in the living cell. These environmental changes could have led to the transformation of lamellar membranes into tubular ones (Degens et al., 1970).

In an earlier study of the Black Sea sediments (Degens et al., 1970), we were not able to find any intact bacteria. However, structures resembling bacterial cell walls were abundant. Most outstanding among these structures was the occurrence of protein crystals ranging in width from 0.25 to 1 μm and in length from 1.0 to 15 μm.

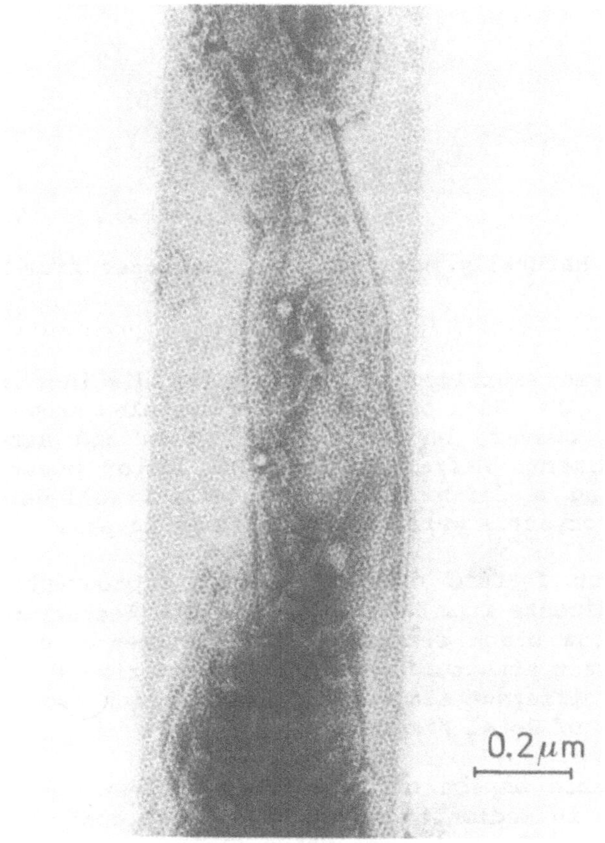

Fig. 4. Organic matter resembling a bacterial cell wall and exhibiting a crystalline arrangement of subunits in Black Sea sediments.

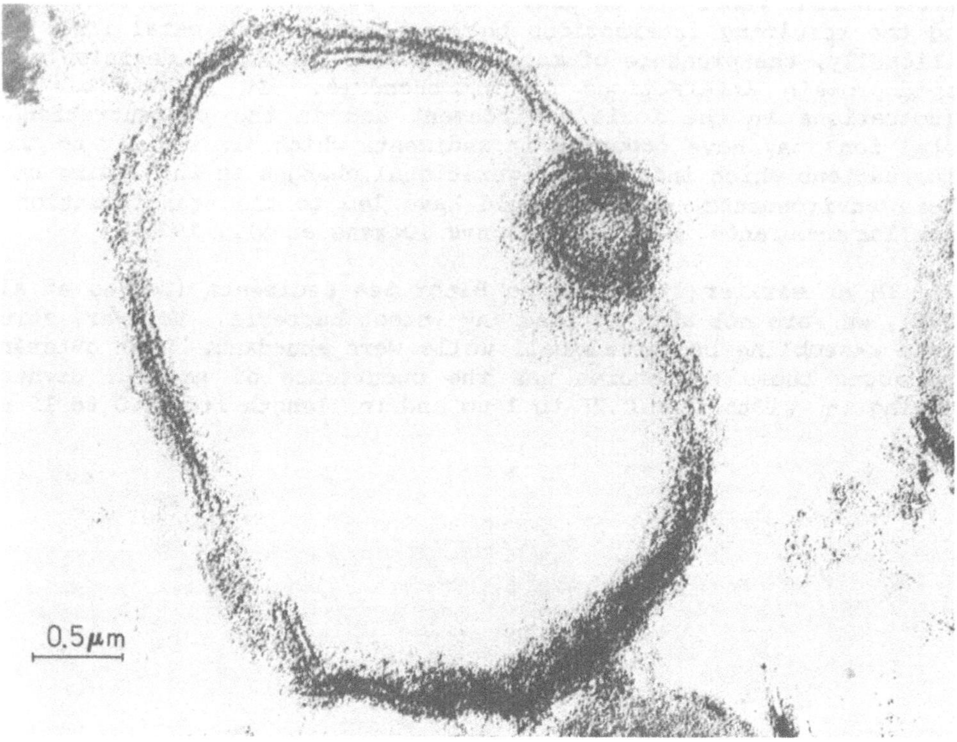

Fig. 5. Naturally metal-stained membranes from Lake Tanganyika sediments.

These structures exhibited at least three distinct arrangements of subunits (Fig. 4). Such ordered arrays are also known from bacterial cell walls. However, bacterial cell walls and structures in the Black Sea sediments differ in that the latter occur as individual crystals having a definite size range and well-developed crystal faces. These crystals were concentrated in layers.

A distinct feature of the electron micrographs obtained from studies of sediments from Lake Kivu and Lake Tanganyika and from the Lower Cretaceous black shales is the occurrence of membranes (Fig. 5), and of intact structures resembling bacteria (Fig. 6). The membranes are in different stages of disaggregation probably due to different degrees of metal fixation.

A remarkable aspect of this study is that most of the organic matter present in sediments is of high image contrast as if stained by heavy metals. In high resolution electron microscopic studies of biological specimens it has been found convenient to stain specimens with a heavy metal salt in order to obtain high contrast images.

Usually salts of uranium, lead or osmium are used. It is emphasized that our samples were studied without prior heavy metal staining and that the high contrast image observed is solely due to natural metal staining of the organic matter and its subsequent mineralization. We observed no difference in electron density between organic material stained with lead and that which was not stained. Certain heavy metals present in interstitial waters can replace alkali, earth alkali and hydrogen ions present at the membrane surface by ion exchange. In addition, metal ions in the interstitial solutions may coordinate to oxygen functions of the organic molecules with the result that polyhedra in which metal ions are coordinated to oxygen atoms will form (Matheja and Degens, 1971). Such organo-metallic interactions

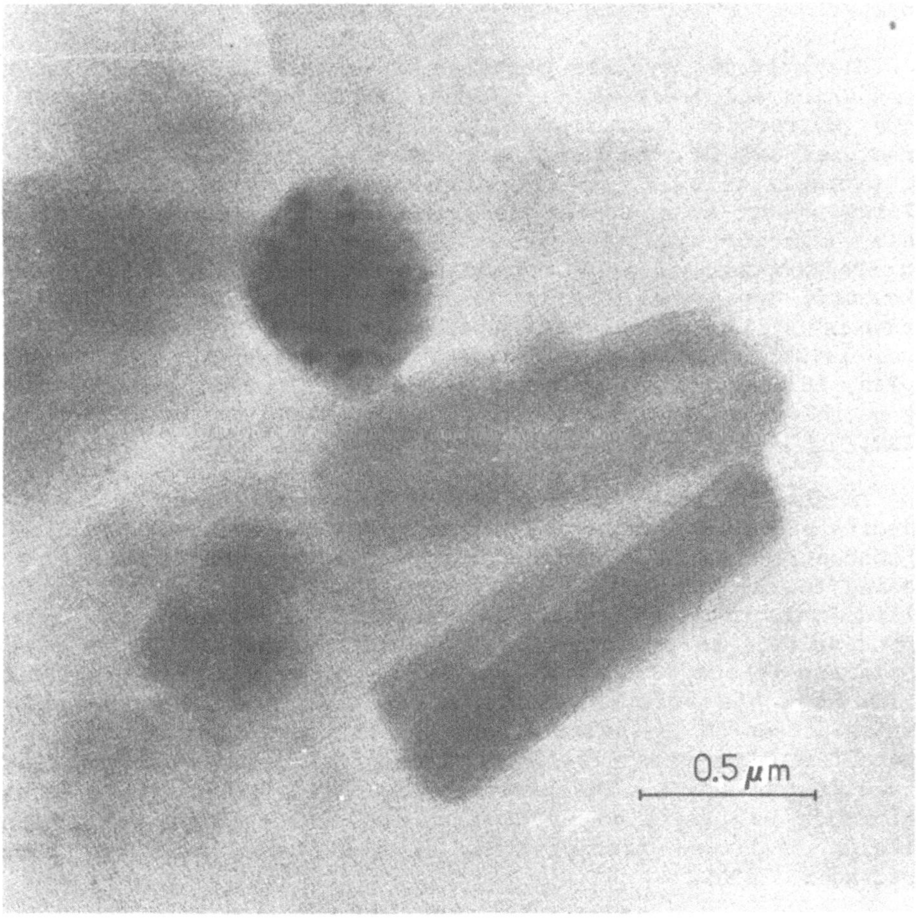

Fig. 6. Spherical structure resembling bacteria and embedded with sulfide particles from Wealden black shale, Bentheim, West Germany.

will introduce coordination changes and restructure membrane subunits leading to the pattern observed. The fact that we could observe membrane bilayers (Fig. 5; Degens and Ittekkot, 1982) suggests that metal-staining by natural processes, in addition to the mineralization observed by Lehmann (see above), is an important process taking place in the depositional environment.

Metal sulfide crystals and spheres. Finally, the occurrence of tiny sulfide crystals embedded within the organic matter could also be recognized (Figs. 7, 8 and 9). The crystals occupy specific positions in the organic matter leading to high contrast images. Spherical structures resembling bacteria are surrounded by clusters of sulfide crystals. In others, the spherical structures are full of minute particles, apparently sulfides, giving areas of high electron density.

The sulfide crystals observed by us are indeed reminiscent of those which allowed Lehmann to see fossilized remains through thin slabs of rocks. In our case, sulfide crystals surround spherical structures and in some cases are embedded in them. These findings are probably relevant to the processes leading to the formation of sulfide sphere--the so-called framboids--in bituminous sediments. These framboids are spherules ranging in diameter from 1 to 100 µm and are composed of many euhedral pyrite crystals which impart a strawberry appearance (Trudinger, 1976). The origin of these framboids is still unresolved. Both bacterial (Schneiderhöhn, 1923; Love, 1962) and inorganic (Berner, 1969; Farrand, 1970; Sweeny and Kaplan, 1973) origins have been suggested. Some of our observations may be relevant for the mechanisms involved and are briefly discussed below.

Studies relating to the effect of copper toxicity on sulfate reducing bacteria (Temple and LeRoux, 1964) have shown that high metal concentrations in the surrounding environment prevent their action leading to the precipitation of copper sulfide. In the process of sulfide precipitation, bacteria surround themselves with zones of copper sulfide as a protection. In his discussion of the paper by Temple and LeRoux (1964), Ridge (1976) points out that the isolation of bacteria via sulfide precipitation will probably prevent their gaining access to the sulfate ions and the oxygen necessary for their metabolism. The result will be "frozen" structures of the type observed in our study. The presence of sulfide granules within bacterial cells was reported by Issatchenko (1929) probably via intracellular sulfide precipitation by sulfate reducing bacteria (Trudinger, 1976).

Evidence for the involvement of organo-metal interactions in the formation of sulfide spherules was obtained from studies carried out in Lake Kivu (Degens et al., 1972). A strong particulate Zn-anomaly in the waters of this lake was found to be associated with micron-

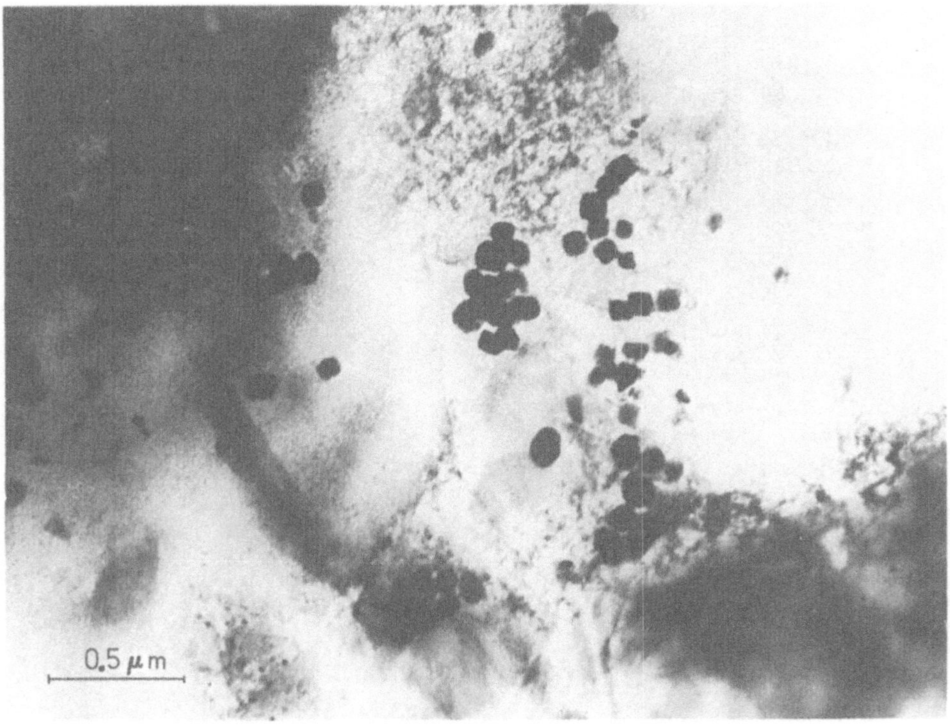

Fig. 7. Organic matter embedded with particles, apparently sulfide crystals from Black Sea sediments.

sized spheres. These spheres were hollow and had a wall thickness of 500 Å. Organic geochemical investigations of these spheres (Degens et al., 1972) showed them to consist of a complex resinous material which had little functionality except for hydroxyl groups. The spheres rise in the process of degassing of water samples at depth. Tiny gas bubbles about 1 µm in diameter act as scavengers of organic matter. The newly created resinous membranes covering the surface of gas bubbles promote selective coordination of Zn dissolved in the water column. In the prevailing H_2S regime, formation of zinc sulfide (sphalerite) crystals is induced. The size range of the crystals is 5 to 50 Å corresponding to 1 to 10 unit cells of sphalerite and suggests that hydroxylated membrane surfaces also act as templates in sphalerite formation. Similar processes were also found to occur in the Red Sea where, in addition to Zn, Cu also was associated with such spherules (Degens and Stoffers, 1977).

Formation of spherules in sediments during early diagenetic processes is also conceivable. Methane production takes place in sulfate-depleted environments (Fenchel and Blackburn, 1979), and the

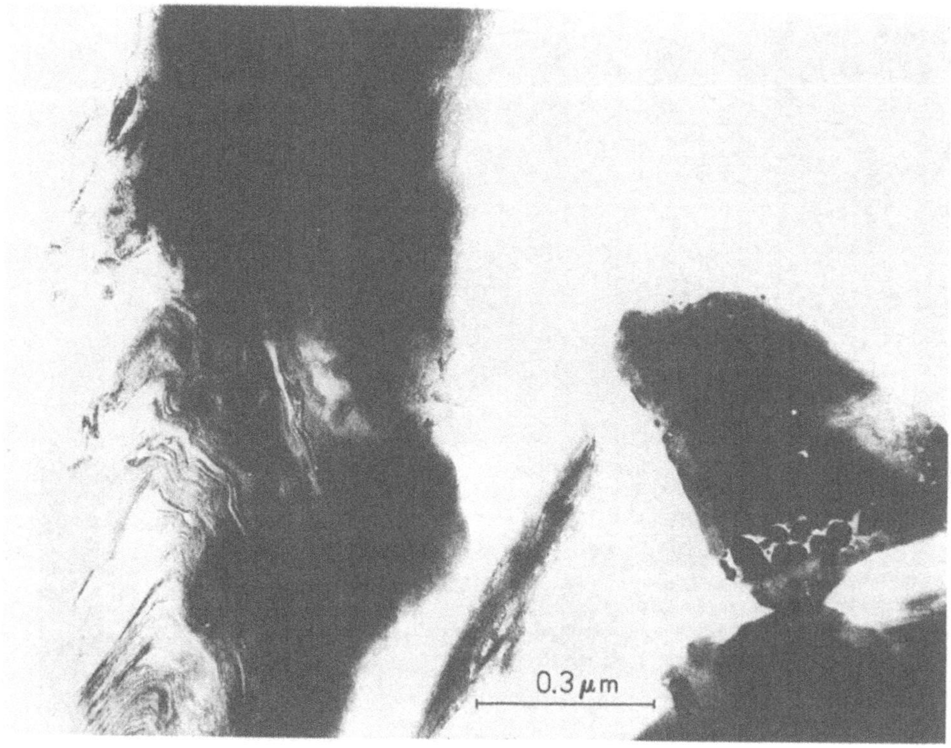

Fig. 8. Naturally metal-stained organic matter with sulfide particles (right) and without sulfide particles (left) in Lake Tanganyika sediments.

escape of methane through the overlying sulfate-rich sediments occurs via bubble transport (Martens, 1976). These bubbles may act as scavengers of organic matter present in interstitial waters. Organic coated bubbles can in turn provide ligands to coordinate metals present in these waters. Depending on the prevailing environmental conditions encountered on its passage into the overlying waters, the organic coated bubble can induce the precipitation of sulfides, carbonates or oxides, the bubble surface acting as a template for mineral deposition.

Yet another possibility is the formation of these sulfide structures by secondary processes involving the replacement of original minerals such as oxides. Certain types of bacteria are found to be capable of forming deposits of manganese and iron. The deposited metals encrust a sheath around the cells. Recent studies indicate that the process of mineral formation is probably controlled by a protein-polysaccharide-lipid composing the sheath material (Rogers and Anderson, 1976a; 1976b). It is conceivable that mineral deposition is similar to the biomineralization processes discussed earlier. The sheath material provides the necessary coordination sites for the formation of metal ion polyhedra and the nucleation of oxide minerals. Once this is accomplished, further deposition of minerals will proceed via epitaxial growth. This seems indeed to be the case because one of the findings of Rogers and Anderson is that metal deposition begins only after bacterial sheath formation has started. Once this material is synthesized by the bacterium, it is found to act in an autonomous manner in the deposition of iron and manganese.

Studies relating to metal deposition in the cell walls of bacteria show that metal binding takes place in two stages: first is an interaction of the metal with the reactive chemical groups in the wall, then deposition of increased amounts of metals (Beveridge and Murray, 1980). It was shown that carboxyl groups of glutamic acid present in cell wall materials provide a major site for mineral deposition. Here the role of glutamic acid is probably that of providing oxygen for the coordination of metal oxygen polyhedra which

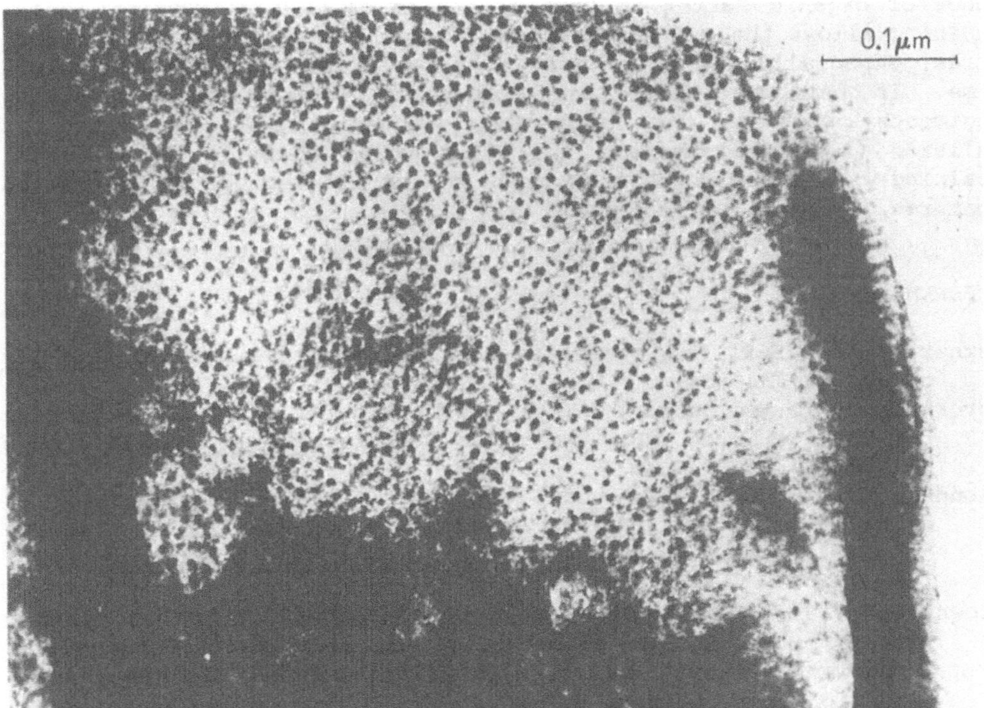

Fig. 9. Metal-stained organic matter covered with tiny sulfide particles in the Black Sea sediment.

starts the nucleation in mineral forming processes. Alteration of the carboxyl groups was found not only to limit metal deposition, but also to change the form and structure of the mineral deposit formed.

DISCUSSION: METAL FIXATION MECHANISM

From the above results and interpretation it appears that the essential mechanism of mineral formation in geochemical and cellular systems is the template mechanism. In both cases the formation of metal ion coordination polyhedra appears to be instrumental in the initiation of mineral deposition. In geochemical systems the prevailing environmental regimes and the type of microbial populations in the sedimentation area determine whether oxides, carbonates, phosphates or sulfides are the final minerals. Electron microscopic investigations of anoxic sediments have revealed the presence of naturally metal-stained organic matter. It is suggested that enrichment of certain metals like Mo and V in bituminous sediments (Vine and Tourtelot, 1970; Brumsack, 1980) is a result of such processes. In addition, metal-staining of organic matter appears to be the starting point for the oriented growth of minerals, as seen from the occurrence of organic matter in various stages of mineralization. Metal staining allows the preservation of organic matter and organic structures, especially the soft tissues of organisms, through geological time. If characteristic organic structures occur in recent upwelling sediments, such as the filamentous bacterial mats described by Gallardo (1977) and Reimers and Suess (this volume), natural metal staining will allow these structures to survive and become diagnostic features of upwelling sediments.

REFERENCES

Berner, R.A., 1969, The synthesis of framboidal pyrite, Economic Geology, 64:383-384.

Beveridge, T.J. and Murray, R.G.E., 1980, Site of metal deposition in the cell wall of *Bacillus subtilis*, Journal of Bacteriology, 141:876-887.

Blondin, G.A., Vail, W.J. and Green, D.E., 1969, The mechanism of mitrochondrial swelling II. Pseudoenergised swelling in presence of alkali metal salts, Archives Biochemistry Biophysics, 129: 158-172.

Brown, G.W., 1976, Biochemical aspects of detoxification in the marine environment, in: "Biochemical and Biophysical Perspectives in Marine Biology," Vol. 3, D.C. Malins and J.R. Sargent, eds., Academic Press, London, 302-406.

Brumsack, H.-J., 1980, Geochemistry of Cretaceous black shales from the Atlantic Ocean (DSDP Legs 11, 14, 36 and 41), Chemical Geology, 31:1-25.

Bryan, G.W., 1973, The occurrence and seasonal variation of trace metals in the scallops *Pecten maximus* and *Chlamys opercularis* (L), Journal of Marine Biological Association, (U.K.) 53: 145-151.

Carter, P.W., 1978, Adsorption of amino-acid containing organic matter by calcite and quartz, Geochimica et Cosmochimica Acta, 42: 1239-1242.

Degens, E.T., 1976, Molecular mechanisms on carbonate, phosphate and silica deposition in the living cell, Topics in Current Chemistry, 64:1-112.

Degens, E.T. and Ittekkot, V., 1982, *In situ* metal staining of biological membranes in sediments, Nature, 298:262-264.

Degens, E.T. and Stoffers, P., 1977, Phase boundaries as an instrument for metal concentration in geological systems, in: "Time- and Strata-bound Ore Deposits," D.D. Klemm and H.-J. Schneider, eds., Springer-Verlag, Berlin, 25-42.

Degens, E.T., Watson, S.W. and Remsen, C.C., 1970, Fossil membranes and cell wall fragments from a 7000-year old Black Sea sediment, Science, 168:1207-1208.

Degens, E.T., Okada, H., Honjo, S., and Hathaway, J.C., 1972, Microcrystalline sphalerite in resin globules suspended in Lake Kivu, East Africa, Mineralium Deposita, 7:1-12.

Farrand, M., 1970, Framboid pyrite precipitated synthetically, Mineralium Deposita, 5:237-247.

Fenchel, T. and Blackburn, T., 1979, "Bacteria and Mineral Cycling," Academic Press, London, 225 pp.

Gallardo, V.A., 1977, Large benthic microbial communities in sulphide biota under Peru-Chile subsurface counter-current, Nature, 268: 331-332.

Hecky, R.E., Mopper, K., Kilham, P., and Degens, E.T., 1973, The amino acid and sugar composition of diatom cell walls, Marine Biology, 19:323-331.

Issatschenko, B.L., 1929, Zur Frage der biochemischen Bildung des Pyrites, Internationale Revue für Hydrobiologie und Hydrographie, 22: 99-101.

Istin, M., 1975, The structure and formation of calcified tissue, in: "Biochemical and Biophysical Perspectives in Marine Biology," Vol. 2, D.C. Malins and J.R. Sargent, eds., Academic Press, London, 1-68.

Jernelöv, A. and Martin, A.-L., 1975, Ecological implications of metal metabolism by microorganisms, Annual Review of Microbiology, 29:61-77.

Krampitz, G. and Witt, W., 1979, Biochemical aspects of biomineralisation, Topics in Current Chemistry, 78:57-144.

Lehmann, W.M., 1934, Röntgenuntersuchung von *Asteropyge sp.* Broili aus dem rheinischen Unterdevon, Neues Jahrbuch für Mineralogie, Geologie und Paläontologie, Beilage, (B) 72:1-14.

Love, L.G., 1962, Further studies on microorganisms and the presence of syngenetic pyrite, Paleontology, 5, 444-459.

Martens, C.S., 1976, Control of methane sediment water bubble transport by microinfaunal irrigation in Cape Lookout Bight, North Carolina, Science, 192:998-1000.

Matheja, J. and Degens, E.T., 1971, "Structural Molecular Biology of Phosphates," Gustav Fischer Verlag, Stuttgart, 180 pp.

Mopper, K. and Degens, E.T., 1972, Aspects of the biogeochemistry of carbohydrates and proteins in aquatic environments, Technical Report, Woods Hole Oceanographic Institution, 72-68, Woods Hole, 118 pp.

Ridge, J.D., 1976, Origin, development, and changes in concepts of syngenetic ore deposits as seen by North American geologists, in: "Handbook of Strata-bound and Stratiform Ore Deposits," Vol. 1, K.H. Wolf, ed., Elsevier, Amsterdam, 183-297.

Rogers, S.R. and Anderson, J.J., 1976a, Measurement of growth and iron deposition in *Sphaerotilus discophorus*, Journal of Bacteriology, 126:257-263.

Rogers, S.R. and Anderson, J.J., 1976b, Role of iron deposition in *Sphaerotilus discophorus*, Journal of Bacteriology, 126:264-271.

Schneiderhöhn, H., 1923, Chalcographische Untersuchungen des Mansfelder Kupferschiefers, Neues Jahrbuch Mineralogie, Geologie, Paläontologie, 47:1-38.

Sweeney, R.E. and Kaplan, I.R., 1973, Pyrite framboid formation: laboratory synthesis and marine sediments, Economic Geology, 68:618-634.

Temple, K.L. and LeRoux, N.W., 1964, Synthesis of sulfide ores: sulfate reducing bacteria and copper toxicity, Economic Geology, 59:271-278.

Trudinger, P.A., 1976, Microbiological processes in relation to ore genesis, in: "Handbook of Strata-bound and Stratiform Ore Deposits," Vol. 1, K.H. Wolf, ed., Elsevier, Amsterdam, 135-190.

Vine, J.D. and Tourtelot, E.B., 1970, Geochemistry of black shale deposits--a summary report, Economic Geology, 65:253-272.

PARTICIPANTS

PARTICIPANTS

ABRANTES, Fatima Filomena, Servicos Geologicos de Portugal, Rua Academia das Ciencias, 19-2°, 1294 Lisboa Codex, Portugal

ALVEIRINHO DIAS, João Manuel, Servicos Geologicos de Portugal, Rua Academia das Ciencias, 19-2°, 1294 Lisboa Codex, Portugal

ARMENTROUT, John, Mobil Exploration and Production, P.O. Box 900, Dallas, Texas 75221, U.S.A.

BAIE, Lyle, Cities Service Company, P.O. Box 50408, Tulsa, Oklahoma 74110, U.S.A.

BARBER, Richard T., Marine Laboratory, Duke University, Beaufort, North Carolina 28516, U.S.A.

BRASSELL, Simon C., Organic Geochemistry Unit, School of Chemistry, University of Bristol, Bristol BS8 1TS, United Kingdom

BREMNER, J. Michael, Marine Geoscience Unit of the Geological Survey, University of Cape Town, Rondebosch 7700, South Africa

BRÖCKEL, Klaus von, Institute of Oceanography, University of Kiel, Düsternbrooker Weg 20, D-2300 Kiel 1, Federal Republic of Germany

BRONGERSMA-SANDERS, Margaretha, Houtlaan 3, 2334 CJ Leiden, The Netherlands

BRUMSACK, Hans J., Geochemisches Institut, Goldschmidtstr. 1, D-3400 Göttingen, Federal Republic of Germany

BURNETT, William C., Department of Oceanography, The Florida State University, Tallahassee, Florida 32306, U.S.A.

CALVERT, Stephen E., Department of Oceanography, The University of British Columbia, Vancouver, British Columbia, V6T 1W5, Canada

CODISPOTI, Louis A., Bigelow Laboratory for Ocean Sciences, McKown Point, W. Boothbay Harbor, Maine 04575, U.S.A.

COPELIN, Edward C., Union Oil Company of California, Science and Technology Division, 376 S. Valencia Avenue, P.O. Box 76, Brea, California 92621, U.S.A.

DENIS, Jérôme, Laboratoire de Géochimie des Eaux, Université de Paris 7, 2 Place Jussieu, F-75221 Paris Cedex 05, France

DEROO, Gérard, Institut Francais du Pétrole, 1 et 4, Avenue de Bois-Preau, B.P. 311, F-92506 Rueil-Malmaison Cedex, France

DIESTER-HAASS, Liselotte, Fachrichtung Geographie, Universität des Saarlandes, D-6600 Saarbrücken, Federal Republic of Germany

DOUGLAS, Robert G., Department of Geological Sciences, University of Southern California, Los Angeles, California 90007, U.S.A.

DUGDALE, Richard C., Department of Biological Sciences, University of Southern California, Los Angeles, California 90007, U.S.A.

DUNBAR, Robert B., Department of Geology, Rice University, Houston, Texas 77001, U.S.A.

DYER, Robin R., Robertson Research International, Ltd., 'Ty'N-Y-Coed' Llanrhos, Llandudno, Gwynedd, North Wales LL30 15A, United Kingdom

EGLINTON, Geoffrey, Organic Geochmistry Unit, School of Chemistry, University of Bristol, Contock's Close, Bristol BS8 1TS, United Kingdom

EINSELE, Gerhard, Institut und Museum für Geologie und Paläontologie, Universität Tübingen, Sigwartstrasse 10, D-7400 Tübingen, Federal Republic of Germany

FISCHER, Kathy, School of Oceanography, Oregon State University, Corvallis, Oregon 97331, U.S.A.

FIUZA, Armando, Oceanography Group, Department of Physics, University of Lisbon, Rua da Escola Politecnica 58, P-1200 Lisboa, Portugal

FLEET, Andrew, Geochemistry Branch, Exploration and Production Division, British Petroleum Research Center, Sudbury-on-Thames, Middlesex TW16 7LN, United Kingdom

FÜTTERER, Dieter, Alfred-Wegener-Institute for Polar Research, Columbus Center, D-2850 Bremerhaven, Federal Republic of Germany

GAGOSIAN, Robert B., Department of Chemistry, Woods Hole Oceanographic Institution, Woods Hole, Massachusetts 02543, U.S.A.

GANSSEN, Gerald, Geologisch-Paläontologisches Institut und Museum, Christian-Albrechts-Universität, Olshausenstrasse 40/60, D-2300 Kiel, Federal Republic of Germany

GARFIELD, Paula, Bigelow Laboratory for Ocean Sciences, McKown Point, W. Boothbay Harbor, Maine 04575, U.S.A.

GASPAR, Luis Caralho, Servicos Geologicos de Portugal, Rua Academia das Ciencias, 19-2°, 1294 Lisboa Codex, Portugal

GORSLINE, Donn S., Department of Geological Sciences, University of Southern California, Los Angeles, California 90007, U.S.A.

INGLE, James L., Department of Geology, Stanford University, Stanford, California 94305, U.S.A.

ITTEKKOT, Venu, Geologisch-Paläontologisches Institut und Museum, Universität Hamburg, Geomatikum, Bundesstrasse 55, D-2000 Hamburg, Federal Republic of Germany

JACOBS, Lucinda, Department of Oceanography, University of Washington, Seattle, Washington 98195, U.S.A.

JUILLET, Anne, Centre des Faibles Radioactivités, C.N.R.S., Place de l'église, F-91190 Gif-sur-Yvette, France

KELLER, George H., Oregon State University, School of Oceanography, Corvallis, Oregon 97331, U.S.A.

KEMPER, Edwin, Bundesanstalt für Geowissenschaften und Rohstoffe, Postfach 51 01 53, D-3000 Hannover 51, Federal Republic of Germany

KRISSEK, Lawrence A., Department of Geology, Ohio State University, Columbus, Ohio 43210, U.S.A.

LABRACHERIE, Monique, Département de Géologie et Oceanographie, Institut de Geologie du Bassin d'Aquitaine, Université de Bordeaux I, Avenue des Facultés, F-33405 Talence Cedex, France

MANGINI, Augusto, Institut für Umweltphysik, Universität Heidelberg Im Neuenheimer Feld 366, D-6900 Heidelberg, Federal Republic of Germany

MARTIN, John H., Moss Landing Marine Laboratories, P.O. Box 223, Moss Landing, California 95039, U.S.A.

MEYERS, Philip A., Department of Atmospheric and Oceanic Sciences, The University of Michigan, 2455 Haywood Avenue, Ann Arbor, Michigan 48109, U.S.A.

MITTELSTAEDT, Ekkehard, Deutsches Hydrographisches Institut, Bernhard-Nochtstrasse 78, Postfach 220, D-2000 Hamburg 3, Federal Republic of Germany

MOLINA-CRUZ, Adolfo, Instituto de Ciencias del Mar y Limnologia, Apartado Postal 70-305, 04510 Mexico, D.F., Mexico

MONTEIRO, Jose Hipolito, Servicos Geologicos de Portugal, Rua Academia das Ciencias, 19-2°, 1294 Lisboa Codex, Portugal

MOOERS, Christopher N.K., Naval Postgraduate School, Department of Oceanography, Monterey, California 93940, U.S.A.

MORRIS, Robert J., Institute of Oceanographic Sciences, Wormley, Godalming, Surrey GU8 5UB, United Kingdom

MÜLLER, German, Institut für Sedimentforschung, Universität, Heidelberg, Im Neuenheimer Feld 236, Postfach 10 30 20, D-6900 Heidelberg 1, Federal Republic of Germany

PARRISH, Judith Totman, U.S. Geological Survey, M.S. 940, Denver Federal Center, P.O. Box 25046, Denver, Colorado 80225, U.S.A.

PRELL, Warren L., Department of Geological Sciences, Brown University, Providence, Rhode Island 02912, U.S.A.

REIMERS, Clare E., Marine Biology Research Division, Scripps Institution of Oceanography, La Jolla, California 92093, U.S.A.

ROBINSON, Stephen W., U.S. Geological Survey, 345 Middlefield Road, Menlo Park, California 94025, U.S.A.

RULLKÖTTER, Jürgen, Institute of Petroleum and Organic Geochemistry, KFA Jülich G.m.b.H., P.O. Box 1913, D-5170 Jülich 1, Federal Republic of Germany

SARNTHEIN, Michael, Geologisch-Paläontologisches Institut und Museum, Christian-Albrechts-Universität, Olshausenstrasse 40/60, D-2300 Kiel, Federal Republic of Germany

SCHEIDEGGER, Kenneth F., School of Oceanography, Oregon State University, Corvallis, Oregon 97331, U.S.A.

SCHOPF, Thomas J.M., Geophysical Sciences Department, University of Chicago, 5734 S. Ellis Avenue, Chicago, Illinois 60637, U.S.A.

SCHRÖTER, Thomas, Brauerstr. 31, 1000 Berlin 45, Germany

SELLNER, Kevin G., Academy of Natural Sciences, Benedict Estuarine Research Laboratory, Benedict, Maryland 20612, U.S.A.

SHELDON, Richard P., U.S. Geologial Survey, Reston, Virginia 22092, U.S.A.

SHILLER, Alan M., Department of Earth and Planetary Sciences, Massachusetts Institute of Technology, Cambridge, Massachusetts 02139, U.S.A.

SIMONEIT, Bernd R.T., School of Oceanography, Oregon State University, Corvallis, Oregon 97331, U.S.A.

SMITH, Robert L., School of Oceanography, Oregon State University, Corvallis, Oregon 97331, U.S.A.

STARESINIC, Nick, Woods Hole Oceanographic Institution, Woods Hole, Massachusetts 02543, U.S.A.

STOFFERS, Peter, Institut für Sedimentforschung, Universität Heidelberg, Im Neuenheimer Feld 236, Postfach 10 30 20, D-6900 Heidelberg 1, Federal Republic of Germany

SUESS, Erwin, School of Oceanography, Oregon State University, Corvallis, Oregon 97331, U.S.A.

SUMMERHAYES, Colin P., BP Research Centre, Chertsey Road, Sunbury-on-Thames, Middlesex TW16 7LN, England, United Kingdom

THIEDE, Jörn, Institutt for geologi, Universitetet i Oslo, Postboks 1047, Blindern, Oslo 3, Norway

TRAGANZA, Eugene D., Naval Postgraduate School, Department of Oceanography, Monterey, California 92093, U.S.A.

VEEH, H. Herbert, School of Earth Sciences, The Flinders University of South Australia, Bedford Park, South Australia 5042, Australia

VORTISCH, Walter, Institut für Geologie und Paläontologie, Fachbereich Geowissenschaften, Phillips-Universität-Lahnberge, D-3550 Marburg/Lahn, Federal Republic of Germany

WASSMANN, Paul, Institutt for marin biologi, Universitetet i Bergen, N-5065 Blomsterdalen, Norway

WEFER, Gerold, Geologisch-Paläontologisches Institut und Museum, Christian-Albrechts-Universität, Olshausenstrasse 40/60, D-2300 Kiel, Federal Republic of Germany

WETZEL, Andreas, Institut und Museum für Geologie und Paläontologie, Universität Tübigen, Sigwartstrasse 10, D-7400 Tübingen, Federal Republic of Germany

WIEDMANN, Jost, Institut und Museum für Geologie und Paläontologie, Universität Tübigen, Sigwartstrasse 10, D-7400 Tübingen, Federal Republic of Germany

ZIMMERLE, Winfried, Deutsche Texaco AG, D-3109 Wietze/Celle, Federal Republic of Germany

INDEX

GPSR Compliance

The European Union's (EU) General Product Safety Regulation (GPSR) is a set of rules that requires consumer products to be safe and our obligations to ensure this.

If you have any concerns about our products, you can contact us on ProductSafety@springernature.com

In case Publisher is established outside the EU, the EU authorized representative is:

Springer Nature Customer Service Center GmbH
Europaplatz 3
69115 Heidelberg, Germany

Batch number: 10615115

Printed by Printforce, the Netherlands